Lecture Notes in Mathematics

Edited by A. Dold and B. Eckmann

1099

Claus Michael Ringel

T0226259

Tame Algebras and Integral Quadratic Forms

Springer-Verlag
Berlin Heidelberg New York Tokyo 1984

Author

Claus Michael Ringel
Fakultät für Mathematik, Universität Bielefeld
Postfach 8640, 4800 Bielefeld, Federal Republic of Germany

AMS Subject Classification (1980): 05C20, 06A10, 10B05, 15A30, 16A46, 16A48, 16A62, 16A64, 16-02, 18E10

ISBN 3-540-13905-2 Springer-Verlag Berlin Heidelberg New York Tokyo
ISBN 0-387-13905-2 Springer-Verlag New York Heidelberg Berlin Tokyo

Printing and binding: Beltz Offsetdruck, Hemsbach/Bergstr.
2146/3140-543210

TABLE OF CONTENTS

Introduction

The aim of these notes is twofold. On the one hand, we want to give an introduction to some parts of the new representation theory of finite dimensional algebras as it has been developed by the joint effort of several mathematicians through the last 15 years. We will present several of the basic methods in a unified way. We try to give a full account for those results which are available not too easily, and we will review rather carefully also the remaining results which are needed. On the other hand, we also want to exhibit the structure of the module categories of an exceptional class of algebras which we call tubular. These are very special algebras having 6, 8, 9, or 10 simple modules, but their representation theory may turn out to be of wider interest. The topics chosen in the introductory parts are those needed for an understanding of the module category of a tubular algebra, however they will be presented in greater generality, so to be useful also for other problems. Before we describe the content of these notes in more detail, let us give a short exposition of the general direction of investigation presented here.

We fix some (commutative) field k, and for the sake of simplicity, we usually will assume that k is algebraically closed. Given a finite-dimensional k-algebra A (associative, with 1), we consider the category A-mod of all (left) A-modules. All A-modules will be assumed to be finite-dimensional. This has the following implications: Any A-module can be decomposed as a finite direct sum of indecomposable A-modules, and the classical Krull-Schmidt theorem asserts that such a decomposition is unique up to isomorphism. Thus, we are mainly concerned with the problem of classifying the indecomposable A-modules. Of course, there are only finitely many isomorphism classes of simple (= irreducible) A-modules; however, there usually will be additional indecomposable modules. In case the number of isomorphism classes of indecomposable A-modules is finite, A is said to be <u>representation finite</u>, otherwise <u>representation infinite</u>. Actually, all the indecomposable A-modules are simple only in case A is semi-simple (= its radical vanishes). Our interest lies in the non semi-simple algebras, and in the way the indecomposable A-modules are built from the simple A-modules. Note that any A-module has a finite composition series, and the classical Jordan-Hölder theorem asserts that the multiplicity of a fixed simple A-module in a composition series of the A-module M, is independent of the chosen composition series; the various multiplicities may be collected to give an n-tupel of non-negative numbers, denoted by $\underline{\dim}\, M$ and called the dimension vector of M, with n being the number of isomorphism classes of simple A-modules. Actually, we may consider $\underline{\dim}\, M$ as the element corresponding to M in the Grothendieck group $K_0(A)$ of all A-modules modulo all short exact sequences [in fact $K_0(A)$ can be identified with $\mathbb{Z}^n$ since the isomorphism classes of the simple A-modules form a $\mathbb{Z}$-basis of $K_0(A)$]. One of the main problems of the representation theory is to determine the set of dimension vectors of the indecomposable A-modules in $K_0(A) = \mathbb{Z}^n$, and to describe the class of indecomposable A-modules having a fixed dimension vector. The algebras we will consider, will usually have finite global dimension (thus Ext^n vanishes for large n). This implies that $K_0(A)$ is endowed with a (usually non symmetric) bilinear form given by

$$\langle \underline{\dim} M, \underline{\dim} M' \rangle = \sum_{i \geq 0} (-1)^i \dim \operatorname{Ext}^i(M,M') ,$$

in particular, there is the quadratic form χ_A , with

$$\chi_A(x) = \langle x,x \rangle ,$$

defined on $K_0(A)$, called the Euler characteristic of A . One of the main objectives of these notes will be a study of the indecomposable A-modules in terms of χ_A . Let us call an element $x \in K_0(A)$ a root of χ_A provided $\chi_A(x) = 1$. Also, in case χ_A is semidefinite, an element $x \in K_0(A)$ is called a radical vector provided $\chi_A(x) = 0$. The study of indecomposable A-modules in terms of χ_A was initiated by Gabriel who showed that a hereditary algebra A is representation finite if and only if χ_A is positive definite, and that in this case, $\underline{\dim}$ furnishes a bijection between the isomorphism classes of the indecomposable A-modules and the positive roots of χ_A . For well-behaved algebras A , one may expect that a positive root of χ_A is the dimension vector of a unique indecomposable module, and a positive radical vector of χ_A should turn out to be the dimension vector of a one-parameter family of indecomposable modules. In such a situation we will say that A-mod is controlled by χ_A (for a precise definition, see section 2.4), and we have exhibited in these notes several classes of algebras A , with A-mod being controlled by χ_A . In particular, for A a tubular algebra, A-mod is controlled by χ_A . In order to be able to use these results, one needs a good insight into the set of roots and radical vectors of the corresponding quadratic forms. The first chapter of these notes is devoted to a detailed investigation of classes of quadratic forms occurring in this way in representation theory. The quadratic forms which have to be considered for representation finite algebras are usually related to the forms $\mathbb{A}_n$, $\mathbb{D}_n$, $\mathbb{E}_6$, $\mathbb{E}_7$, $\mathbb{E}_8$ (= the quadratic forms appearing in Lie theory), and the corresponding sets of roots are just the irreducible root systems in the sense of Bourbaki with all roots being of equal length.

The study of A-mod in terms of $K_0(A)$ and χ_A may be considered as a combinatorial approach to representation theory: we investigate the A-modules by dealing with the dimension vectors, thus elements of the discrete group $\mathbb{Z}^n$. There is a completely different, but also combinatorial approach, namely that of constructing the Auslander-Reiten quiver $\Gamma(A)$ of A . Recall that we are studying the set of isomorphism classes of indecomposable A-modules, and we may consider this set as the set of vertices of a more complicated structure, called a translation quiver $\Gamma(A)$, thus also as the set of vertices of a 2-dimensional simplicial complex. For a connected, and representation infinite algebra A , all components of $\Gamma(A)$ are countable, thus in this case, any indecomposable A-module M determines a countable set of indecomposable modules, those belonging to the same component as M . Actually it turns out that sometimes it is easier to describe a full component of indecomposable modules, than an individual one. There will be two types of components which we will concentrate on: directed ones (components without oriented cycles), and tubes (there are oriented cycles in the component, and the underlying topological space is homeomorphic to $S^1 \times \mathbb{R}_0^+$). As the name should suggest, most of the indecomposable modules of a tubular algebra belong to tubes, and tubes play a fundamental role for the whole categorical structure of the module category of a tubular algebra.

Let us outline now in more detail the content of these notes. The first four chapters may be considered as an introduction to representation theory. Here, two results will be presented with complete proofs. The first one are the theorems of Klejner characterizing the representation finite partially ordered sets (see 2.6), with representations in the sense of Nazarova and Rojter being considered. The original proof of Klejner used the socalled differentiation process for partially ordered sets, and, in addition, a detailed combinatorial analysis. The proof given here originates in investigations of Drozd and Ovsienko. Drozd has shown that a finite partially ordered set is representation finite if and only if a related quadratic form is weakly positive. In 2.6, we provide a new proof of this result using Auslander-Reiten quivers. Since in this way, the problem of characterizing the representation finite partially ordered sets is reduced to the study of integral quadratic forms, we have devoted the first chapter of the notes to a selfcontained study of such forms. Here, we present the two theorems of Ovsienko (1.0): one of them asserts that a critical integral quadratic form is positive definite of radical rank 1, the other asserts that the components of a positive root of a weakly positive integral quadratic form are bounded by 6, in this way giving a direct reason for the otherwise mysterious bound 6 occurring in the representation theory of representation finite algebras and partially ordered sets. We supplement these investigations of Ovsienko by providing a full classification of the graphical forms which are either critical or sincere weakly positive (1.3); the classifications immediately yield the corresponding lists of Klejner.

The second result presented with full proof is the classification of the indecomposable modules of a tame hereditary algebra, and the structure of the corresponding Auslander-Reiten quiver. An algebra is tame hereditary, if and only if it is Morita equivalent to the path algebra of a quiver whose underlying graph is a Euclidean diagram, called its type. The construction of preprojective and preinjective components is outlined in 2.3 and 2.4. The remaining components of a tame hereditary algebra form a separating tubular family. Chapter 3 is devoted to a general existence theorem for separating tubular families (3.4), starting from a socalled wing module which is dominated by some module. In this way, we are able to show directly that the tubular type of a tame hereditary algebra of Euclidean type $\tilde{\Delta}$ is given by the corresponding Dynkin diagram Δ . There is a class of algebras derived from the tame hereditary algebras, and having module categories which are similar to those of tame hereditary algebras: the tame concealed algebras. There is a complete list of all tame concealed algebras due to Happel and Vossieck which we give in an appendix. The tame concealed algebras may be characterized as the minimal representation infinite algebras having a preprojective component: any connected representation infinite algebra with a preprojective component has a factor algebra which is tame concealed (4.3), whereas any proper factor algebra of a tame concealed algebra is representation finite.

The existence theorem 3.4 for separating tubular families can be applied in particular to the algebra $A(p,q,r)$ given by the quiver $\Delta(p,q,r)$

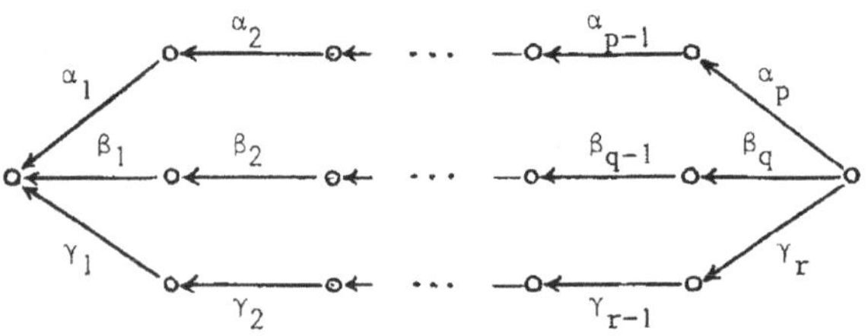

with relation

$$\alpha_p\alpha_{p-1}\cdots\alpha_1 + \beta_q\beta_{q-1}\cdots\beta_1 + \gamma_r\gamma_{r-1}\cdots\gamma_1 = 0 ,$$

we call it the canonical algebra of type (p,q,r). Let $\mathcal{P}$ be the full subcategory of $A(p,q,r)$-mod given by the indecomposable representations with all arrows of $\Delta(p,q,r)$ being represented by monomorphisms, not all of which are isomorphisms. Dually, let $\mathcal{Q}$ be the full subcategory given by the indecomposable representations with all arrows of $\Delta(p,q,r)$ being represented by epimorphisms, not all of which are isomorphisms. Let $\mathcal{T}$ be given by the remaining indecomposable $A(p,q,r)$-modules. It is a direct consequence of theorem 3.4 that $\mathcal{T}$ is a tubular family separating $\mathcal{P}$ from $\mathcal{Q}$; this means that there are no non-zero maps from modules belonging to $\mathcal{T}$ to modules in $\mathcal{P}$, no non-zero maps from modules belonging to $\mathcal{Q}$ to modules in $\mathcal{P}$ or $\mathcal{T}$, and that any map from a module in $\mathcal{P}$ to a module in $\mathcal{Q}$ factors through a direct sum of modules belonging to any fixed tube in $\mathcal{T}$. We illustrate this situation by the following picture

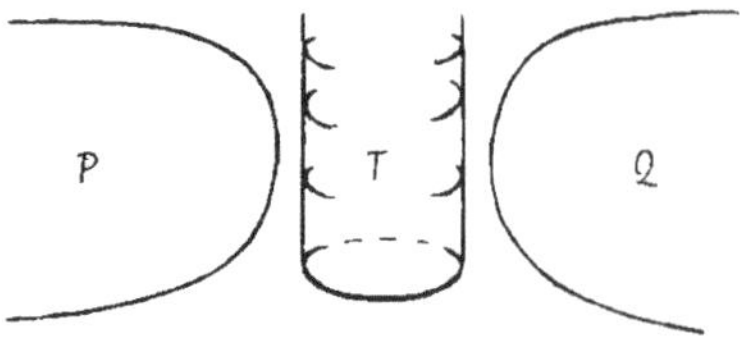

with non-zero maps being possible only from left to right. In case $\frac{1}{p}+\frac{1}{q}+\frac{1}{r}>1$, the algebra $A(p,q,r)$ is a tame concealed algebra, $\mathcal{P}$ is a preprojective component, and $\mathcal{Q}$ is a preinjective component. In case $\frac{1}{p}+\frac{1}{q}+\frac{1}{r}<1$, both subcategories $\mathcal{P}$ and $\mathcal{Q}$ are said to be wild (for example, any finite dimensional algebra can be realized as the endomorphism ring of a module in $\mathcal{P}$, or of a module in $\mathcal{Q}$), whereas the endomorphism rings of the indecomposable modules in $\mathcal{T}$ are uniserial. Actually, $\mathcal{T}$ is always controlled by the restriction of $\chi_{A(p,q,r)}$ to a hyperplane in $K_0(A(p,q,r))$. The algebras $A(p,q,r)$ with $\frac{1}{p}+\frac{1}{q}+\frac{1}{r}=1$ will be of particular interest to us: they are tubular algebras, and we will report on their module categories in describing chapter 5.

In chapter 4, we outline the use of tilting modules, and consider the properties of tubular extensions. The concealed algebras are defined in terms of tilting modules: they are the endomorphism rings of preprojective tilting modules over a representation infinite hereditary algebra A. In case A is tame hereditary, we just obtain a tame concealed algebra. For A tame hereditary, we also characterize the endomorphism rings of the tilting A-modules without an indecomposable preinjective direct summand: they are just the domestic tubular extensions of tame concealed algebras (4.9). Of course, there is the dual result for tilting A-modules without an indecomposable preprojective direct summand, whereas the endomorphism ring of a tilting A-module containing both indecomposable preinjective and indecomposable preprojective direct summands is representation finite. The study of tubular extensions is preceded by the more general consideration of components being obtained by ray insertions (4.5).

The fifth chapter is devoted to a detailed study of the module categories of the tubular algebras. By definition, tubular algebras are the tubular extensions of a tame concealed algebra of extension type (2,2,2,2), (3,3,3), (4,4,2) or (6,3,2), the extension type being called the type of the algebra. Typical examples of tubular algebras of type (3,3,3), (4,4,2) and (6,3,2) are the canonical algebras $A(3,3,3)$, $A(4,4,2)$, and $A(6,3,2)$, respectively. Examples of tubular algebras of type (2,2,2,2) are given by

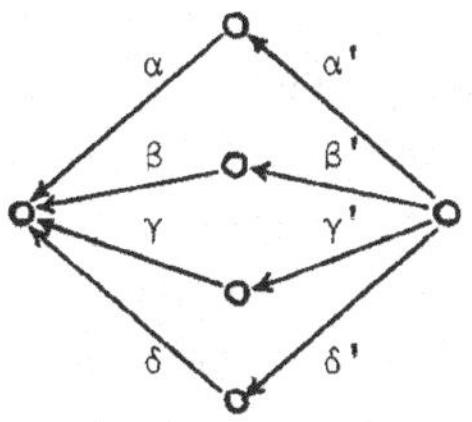

with

$$\alpha'\alpha+\beta'\beta+\gamma'\gamma=0 \quad \text{and} \quad \alpha'\alpha+\lambda\beta'\beta+\delta'\delta=0,$$

where $\lambda \in k\setminus\{0,1\}$. The structure theorem for the module category of a tubular algebra A of type $\mathbb{T}$ (5.2.4) asserts that A-mod has the following components:

first, a preprojective component $\mathcal{P}_0$, then, for any $\gamma \in \mathbb{Q}^+ \cup \{0,\infty\}$ a separating tubular family $\mathcal{T}_\gamma$, all but $\mathcal{T}_0$ and $\mathcal{T}_\infty$ being stable of type $\mathbb{T}$, and finally, a preinjective component $\mathcal{Q}_\infty$. Always $\mathcal{T}_\gamma$ separates $\mathcal{P}_\gamma$ from $\mathcal{Q}_\gamma$, where $\mathcal{P}_\gamma$ is given by the indecomposable modules which are either in $\mathcal{P}_0$ or in $\mathcal{T}_\beta$ with $\beta < \gamma$, and $\mathcal{Q}_\gamma$ is given by the indecomposable modules either in $\mathcal{T}_\delta$ with $\gamma < \delta$, or in $\mathcal{Q}_\infty$. We may visualize the structure of A-mod as follows:

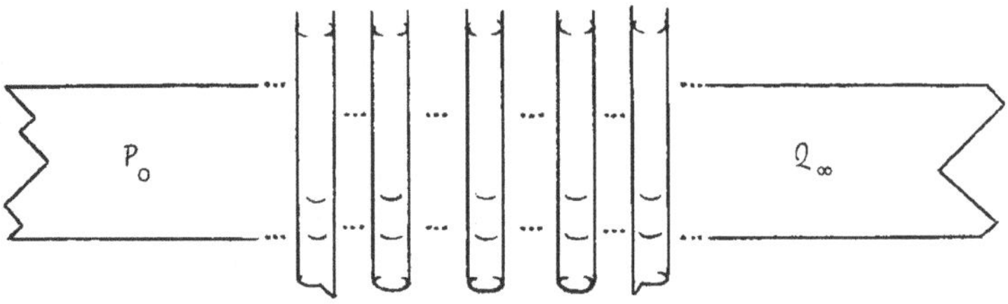

The proof rests on the main theorem 3.4 which establishes the existence of one separating tubular family, say $\mathcal{T}_1$, and the use of special tilting functors, the socalled shrinking functors, which shift $\mathcal{T}_1$ back and forth in order to obtain all the tubular families $\mathcal{T}_\gamma$ with $\gamma \in \mathbb{Q}^+$ (here, $\mathbb{Q}^+$ denotes the set of positive rational numbers). It follows that for a tubular algebra A, the category A-mod is controlled by the quadratic form χ_A. Thus, the root system for χ_A leads to a natural index set for the indecomposable A-modules. Conversely, we may use categorical properties of A-mod in order to give an interpretation of properties of the root system for χ_A. For example, the orientation of the quiver of A gives rise to a Coxeter element in the corresponding Weyl group, and this Coxeter element has order $d = 2,3,4$, or 6, in case the tubular type of A is (2,2,2,2), (3,3,3), (4,4,2), or (6,3,2), respectively. However, this just corresponds to the fact that the A-modules in the tubular families $\mathcal{T}_\gamma$, with $\gamma \in \mathbb{Q}^+$, are τ-periodic with period d, where τ is the Auslander-Reiten translation.

The final chapter deals with directed algebras. Without reference to the Auslander-Reiten theory, we may define an algebra A to be <u>directed</u> provided the category of all A-modules does not contain cycles; a cycle is a sequence $M_0, M_1, \ldots, M_n = M_0$ of indecomposable modules with $n \geq 1$, such that there exists a non-zero and non-invertible map $M_{i-1} \to M_i$, for all $1 \leq i \leq n$. A directed algebra is necessarily representation finite (2.4.9'), and a representation finite algebra A is directed if and only if its Auslander-Reiten quiver $\Gamma(A)$ does not contain cyclic paths. We call an algebra A <u>sincere</u> provided there exists an indecomposable A-module M such that any simple A-module occurs as a composition factor of M. (For a directed algebra, an indecomposable A-module is sincere if and only if it is faithful.) The main result presented in chapter 6 is the classification of all large sincere directed algebras due to Bongartz. Given an algebra A , denote by n(A) the number of isomorphism classes of simple A-modules. If A is a sincere, directed algebra, and $n(A) > 13$, then A belongs to one of 24 infinite families. As a consequence, the length of an indecomposable A-module over a directed algebra A is bounded by $2 \cdot n(A) + 48$.

One of the main tools in the representation theory of algebras is the representation theory of partially ordered sets as developed by Nazarova-Rojter and Klejner, and, more generally, the representation theory of vectorspace categories and their subspace categories. In chapter 2, we outline the basic properties of the subspace category $\check{U}(K,|\cdot|)$ of a vectorspace category $(K,|\cdot|)$; in particular, we derive for K being finite, the structure of the Auslander-Reiten sequences in $\check{U}(K,|\cdot|)$. It is well-known that subspace categories of vectorspace categories can be used very effectively for the study of one-point extensions, and we will use this device in the proof of the existence theorem for separating tubular families. On the other hand, we show that for K the path category of a finite quiver without relations, any category $\check{U}(K,|\cdot|)$ can be identified with the category of all modules generated by some tilting module $_AT$, see 4.1.8. As a consequence, we are able to calculate some subspace categories using the knowledge of corresponding module categories. In particular, this is done for what we call the tubular vectorspace categories, in 4.8. The tubular partially ordered sets have been considered by Zavadskij, who showed that they are the minimal non-domestic partially ordered sets of finite growth, and he also has determined their indecomposable representations. It is of interest that the tubular vectorspace categories in a natural way form patterns, there is a one-parameter family of patterns related to the tubular type (2,2,2,2) and there are 2, 4, and 8 patterns related to the tubular types (3,3,3) , (4,4,2) , and (6,3,2) respectively. These pattern were exhibited in my Ottawa lectures 1979, but at that time the patterns of tubular type (2,2,2,2) were not yet known to be tame [Ri4].

The chapters 2 and 4 present some results without proofs. This should not make any difficulties even for a beginner, since in all cases clear and complete proofs are easily available. First of all, this applies to two classical results, the theorems of Krull-Schmidt and of Jordan-Hölder. Similarly, we will assume the basic results concerning projective and injective resolutions, and Ext^n . We only state the existence theorem for Auslander-Reiten sequences in module categories over finite dimensional algebras, but note that in our view this is the most fundamental result of the new representation theory. Also, we state without proof Gabriel's theorem realizing any finite-dimensional basic algebra as a factor algebra of the path algebra of a quiver. Actually, this result is not used in these notes in an essential way, however, it explains why we may restrict to quivers with relations. The basic results concerning tilting modules, tilting functors, and tilted algebras are collected together in 4.1 and 4.2, most of them again without proof. Always, we try to present all the formulae necessary for applications, and we hope that the collection of formulae given in 2.4, 2.5 and 4.1 will turn out to be very useful.

There are some deviations from the by now common usage; let us mention the most essential ones. First of all, we call a minimal left almost split map for a module M just a source map for M, a minimal right almost split map a sink map, and we use the term Auslander-Reiten sequence instead of almost split sequence. By a module class $\mathcal{M}$, we understand a class of modules closed under direct sums, direct summands and isomorphisms, we will consider it as an exact category: as an additive category together with the class of all short exact sequences involving only modules from $\mathcal{M}$. Given a class $\mathcal{M}$ of modules, the smallest module class containing $\mathcal{M}$ will be denoted by $\langle\mathcal{M}\rangle$ (a more common notation would be add $\mathcal{M}$), and for module classes $\mathcal{M}_1$, $\mathcal{M}_2$, we write $\mathcal{M}_1 \vee \mathcal{M}_2$ for $\langle\mathcal{M}_1 \cup \mathcal{M}_2\rangle$, and $\mathcal{M}_1 \int \mathcal{M}_2$ denotes the module class generated by all modules M having a submodule $M_2 \in \mathcal{M}_2$ such that $M/M_2 \in \mathcal{M}_1$. Also note that we usually consider left modules and that therefore the composition of two maps $f : X \to Y$, $g : Y \to Z$ is denoted by fg (writing maps on the opposite side of the scalars), and similarly, a path of arrows $\alpha_i : x_{i-1} \to x_i$, $1 \leq i \leq \ell$, is denoted by $(x_0 \mid \alpha_1, \ldots, \alpha_\ell \mid x_\ell)$. There is the following consequence: the category of representations of a quiver Δ turns out to be equivalent to the category of modules over the path algebra $k\Delta^*$ not of Δ, but rather of the opposite quiver Δ^*. Thus, being interested in module categories, we say that the algebra A is given by the quiver Δ and the relations ρ_i, provided A is isomorphic to $k\Delta^*/\langle\rho_i^* \mid i\rangle =: A(\Delta, \{\rho_i \mid i\})$, see 2.1. Finally, following a suggestion by Gabriel, in the definition of a vectorspace category $(\mathcal{K}, |\cdot|)$, we do not assume that the functor $|\cdot|$ is faithful, see 2.5.

We stress again that our choice of topics is very personal. As we have pointed out above, it is mainly motivated by our aim of providing the prerequisites for the study of the module categories of tubular algebras, and we have included only a few additional results. There are several topics which are by now well-established in representation theory but which we will not touch at all: we should mention the classification of the representation finite self-injective (= QF) algebras, the general covering technique, the discussion of the Brauer-Thrall conjectures, the multiplicative basis theorem for representation finite and minimal representation infinite algebras, and the classification of indecomposable modules for tame algebras of infinite growth. In fact, there are several other topics which are of importance, and which should be covered in a general survey on representation theory. But this is not what we are able to give, here.

We have tried to give as complete references as possible at the end of each chapter, and we apologize for any omission. Of course, it will be easy to trace any omission of a reference to a paper which has appeared in print; however we should point out that some general ideas which have influenced the results and the methods presented here, are not available in official publications, or not even written up. First of all, there is the long collaboration with Dlab, D'Este, and Happel, from which we have drawn much advantage. In particular, we learned the importance of wing modules from investigations of Happel. The possible use of tilting functors for getting a complete classification of the indecomposable modules over a tubular algebra was suggested by S. Brenner, and it seems that such a use was one of the stimuli for her joint paper with Butler [BB] on tilting modules. Last not least, I should mention the influence of Gabriel's 1972 Ottawa lectures. In these lectures he demonstrated, in particular, the tubular structure of the regular modules of the four subspace quiver, and made us aware of the importance of tubes in representation theory. Note that the main results of the present notes are centered around the tubular structure of tame module categories.

The structure theorem for the module category of a tubular algebra was presented, with a different proof, in lectures at the N.-Copernikus-University of Torun in December 1981. The present proof was outlined in spring 1982 in lectures both at the Seminar Malliavin in Paris and at the Ukrainian Academy of Science in Kiev; however, at that time, we were not aware of the existence of the self reproductive algebras of section 5.6; instead, we had to use sets of different algebras (see the survey [Ri5]). Also, the main theorem of chapter 3 was not yet formulated in general. This formulation was presented in spring 1983 at U.N.A.M., Mexico. The chapters 1 - 3 were written in fall 1983, chapters 4 and 5 were finished during a stay at the I.H.E.S. in Bures-sur-Yvette, spring 1984. The author is grateful to all institutions for their hospitality, and for discussions which resulted in substantial improvements of the presentation. Also, we are indebted to K. Bongartz, M. C. R. Butler, D. Happel, H. Lenzing, Chr. Riedtmann and D. Vossieck for reading parts of the manuscript and spotting misprints and inaccuracies.

1. Integral quadratic forms

1.0 Two theorems of Ovsienko

A polynomial $\chi = \chi(X_1,\ldots,X_n)$ in n variables with integral coefficients will be said to be an integral quadratic form provided it is of the form

$$\chi(X_1,\ldots,X_n) = \sum_i X_i^2 + \sum_{i<j} \chi_{ij} X_i X_j$$

with $\chi_{ij} \in \mathbf{Z}$. Evaluating χ at n-tuples of integral numbers, we obtain a function $\mathbf{Z}^n \longrightarrow \mathbf{Z}$ which also will be denoted by χ. We endow $\mathbf{Z}^n$ with a partial ordering defined componentwise: $z = (z_1,\ldots,z_n) \in \mathbf{Z}^n$ is said to be positive, written $z > o$, provided $z \neq o$ and $z_i \geq o$ for all i. The integral quadratic form χ in n variables is said to be weakly positive provided $\chi(z) > o$ for all positive $z \in \mathbf{Z}^n$. Finally, an element $z \in \mathbf{Z}^n$ satisfying $\chi(z) = 1$ is called a root of χ.

Theorem 1 (Ovsienko). If $z = (z_1,\ldots,z_n)$ is a positive root of a weakly positive integral quadratic form, then $z_i \leq 6$ for all i.

Let us note that the value 6 actually does occur. Before we write down an example, we want to introduce a convenient way of describing an integral quadratic form. Note that such a form χ is given by the coefficients χ_{ij}, we therefore use the set $\{1,\ldots,n\}$ as a set of vertices, and, for $i < j$, we connect the vertex i with the vertex j by $-\chi_{ij}$ solid edges provided $\chi_{ij} < o$, and by χ_{ij} dotted edges provided $\chi_{ij} > o$ [in this way, we obtain a bigraph on $\{1,\ldots,n\}$ with possibly multiple edges but without loops]. For example,

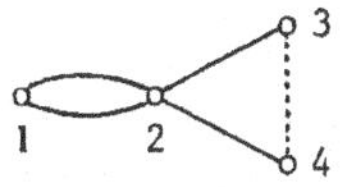

describes the quadratic form

$$\sum_{i=1}^{4} X_i^2 - 2X_1X_2 - X_2X_3 - X_2X_4 + X_3X_4 \; .$$

An interesting quadratic form is given by the so called Dynkin graph $\mathbf{E}_8$ (see 1.2)

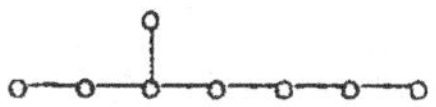

this form is positive definite (thus, weakly positive), and there is a unique maximal root, namely

Note that we use the bigraph also for writing down elements $z = (z_1,\ldots,z_n) \in \mathbb{Z}^n$, namely we replace the vertex i by the corresponding integer z_i.

We recall that a quadratic form χ in n variables with integral coefficients is said to be <u>positive semi-definite</u> provided $\chi(z) \geq 0$ for all $z \in \mathbb{Z}^n$. For a positive semi-definite quadratic form χ, the elements $z \in \mathbb{Z}^n$ satisfying $\chi(z) = 0$ are called <u>radical vectors</u>, they form a subgroup of $\mathbb{Z}^n$, the <u>radical</u> $\operatorname{rad}\chi$ of χ, and the rank of $\operatorname{rad}\chi$ is called the <u>radical rank</u> of χ. [Namely, if x,y both are radical vectors of χ, then $\chi(x+y) + \chi(x-y) = 0$, thus, both summands have to be $= 0$, since, by assumption, they are non-negative.] In case 0 is the only radical element of χ, or, equivalently, the radical rank of χ is 0, the form χ is said to be <u>positive definite</u>.

Given a quadratic form χ in n variables, and $1 \leq t \leq n$, we denote by χ^t the restriction of χ to the hyper-plane defined by $X_t = 0$, thus $\chi^t = \chi(X_1,\ldots,X_{t-1},0,X_{t+1},\ldots,X_n)$. An integral quadratic form χ in $n \geq 3$ variables is said to be <u>critical</u> provided χ is not weakly positive, however all the forms χ^t, $1 \leq t \leq n$, are weakly positive. In addition, we also call the form $\chi(X_1,X_2) = X_1^2 + X_2^2 - 2X_1X_2$ critical; this form will be denoted also by $C(1)$.

<u>Theorem 2</u> (Ovsienko). <u>A critical quadratic form is positive semi-definite with radical rank</u> 1, <u>and with a sincere positive radical vector</u>.

Here, a vector $(z_1,\ldots,z_n)$ is said to be <u>sincere</u>, provided $z_i \neq 0$ for all $1 \leq i \leq n$. The proof of the two theorems will be given in this chapter. The proof of Theorem 1 will be completed in 1.6, the proof of Theorem 2 is in 1.8. Both proofs are due to Ovsienko.

1.1 <u>Roots of an integral quadratic form</u> χ <u>and the partial derivatives</u> $D_i\chi$ <u>of</u> χ.

We denote by $e(i)$, $1 \leq i \leq n$, the canonical base vectors of $\mathbb{Z}^n$, thus $z = (z_1,\ldots,z_n) = \Sigma\, z_i e(i)$ for any $z \in \mathbb{Z}^n$.

Let $\chi = \sum_i X_i^2 + \sum_{i<j} \chi_{ij} X_i X_j$ be an integral quadratic form in n variables. Note that the $e(i)$ are positive roots, called <u>simple</u> roots. We denote by $(-,-)$ the corresponding symmetric bilinear form; it is given by a symmetric matrix with diagonal entries equal to 1, and off-diagonal entries from $\frac{1}{2}\mathbb{Z}$, namely by the matrix $\frac{1}{2}(\chi_{ij})_{ij}$, where $\chi_{ij} = \chi_{ji}$ for $i > j$, and $\chi_{ii} = 2$ for all i. Note that $\chi(z) = (z,z)$, for any z, and $\chi_{ij} = 2(e(i),e(j))$. We also note that

$$\chi(z+z') = \chi(z) + \chi(z') + 2(z,z').$$

We define

$$D_i\chi(z) := 2(e(i),z) = 2z_i + \sum_{i\neq j} \chi_{ij}\, z_j.$$

Thus $D_i\chi$ is a linear form, given by $2X_i + \sum_{j\neq i} \chi_{ij} X_j$, and we see that this is just the i-th partial derivative of χ. Note that $D_i\chi(z) \in \mathbb{Z}$ for $z \in \mathbb{Z}^n$.

(1) If χ is a positive semi-definite integral quadratic form, then $\chi(z) = 0$, if and only if $D_i\chi(z) = 0$ for all i.

Proof. If $D_i\chi(z) = 0$ for all i, then $(e(i),z) = 0$ for all i, thus $(y,z) = 0$ for all $y \in \mathbb{Z}^n$, therefore $\chi(z) = (z,z) = 0$. Conversely, assume $\chi(z) = 0$, where χ is a positive semi-definite integral quadratic form in n variables. Evaluating χ at n-tuples of real numbers, we obtain a function $\chi : \mathbb{R}^n \longrightarrow \mathbb{R}$. Now, since χ is a quadratic form, $\chi(\alpha x) = \alpha^2\chi(x)$ for all $\alpha \in \mathbb{R}$, $x \in \mathbb{R}^n$, and therefore $\chi(x) \geq 0$ for all $x \in \mathbb{Z}^n$ implies $\chi(x) \geq 0$ for all $x \in \mathbb{Q}^n$. Using the continuity of χ, we see that this implies $\chi(x) \geq 0$ for all $x \in \mathbb{R}^n$. Thus $\chi(z) = 0$ means that χ has a global minimum in z, thus

$$0 = \operatorname{grad}\chi(z) = (D_1\chi(z),\ldots,D_n\chi(z)),$$

and therefore $D_i\chi(z) = 0$ for all i.

In case χ is not necessarily semi-definite, one calls z a radical vector for χ provided $D_i\chi(z) = 0$ for all i, and again $\operatorname{rad}\chi$ denotes the set of all radical vectors.

(2) (Drozd) A weakly positive integral quadratic form χ has only finitely many positive roots.

Proof. Again, consider χ as a function $\mathbb{R}^n \longrightarrow \mathbb{R}$, and, as above, we see that $\chi(x) > 0$ for all positive $x \in \mathbb{Q}^n$ and $\chi(x) \geq 0$ for all positive $x \in \mathbb{R}^n$ (an element $x = (x_1,\ldots,x_n) \in \mathbb{R}$ being called positive provided $x \neq 0$ and $x_i \geq 0$ for all i). By an induction on n, we claim that actually $\chi(x) > 0$ for all positive $x \in \mathbb{R}^n$. This is trivial for $n = 1$, since for $0 \neq r \in \mathbb{R}$, $\chi(r) = r^2\chi(1) > 0$. Now assume there exists χ weakly positive, in $n \geq 2$ variables, and some positive $z = (z_1,\ldots,z_n) \in \mathbb{R}$ with $\chi(z) = 0$. By induction, all components $z_i > 0$, thus z lies in the interior of the positive cone of $\mathbb{R}^n$, and χ has in z a local minimum thus

$$0 = \operatorname{grad}\chi(z) = (D_1\chi(z),\ldots,D_n\chi(z)).$$

The linear forms $D_i\chi$ have integral, thus rational coefficients, and the intersection of their kernels is non-zero (z is an element of this intersection), thus there also exists a positive vector z' with rational coefficients belonging to this intersection. But this contradicts the fact that $\chi(x) > o$ for all positive $x \in \mathbb{Q}^n$. Thus, $\chi(x) > 0$ for all positive $x \in \mathbb{R}^n$. Let U be the set of positive $x \in \mathbb{R}^n$ with $\|x\| = 1$, where $\|-\|$ is the usual Euclidean norm. Then U is compact, thus the restriction of χ to U takes its minimum γ in some point of U, and therefore $\gamma > o$. Thus, for all positive $x \in \mathbb{R}^n$,

$$\gamma \leq \chi\left(\frac{x}{\|x\|}\right) = \frac{1}{\|x\|^2}\,\chi(x) ,$$

and therefore $\|y\| \leq \sqrt{\frac{1}{\gamma}}$ for any positive root y of χ. It follows that there are only finitely many positive roots for χ.

As a consequence, we see that a weakly positive integral quadratic form always has maximal positive roots. Let us consider now positive roots and their properties in more detail . Given a vector $z = (z_1,\dots,z_n) \in \mathbb{Z}^n$, we call the set $\{i \mid z_i \neq o\}$ the <u>support</u> of z. The support of z is all of $\{1,2,\dots,n\}$ if and only if z is sincere. A weakly positive integral quadratic form will be said to be <u>sincere</u> provided there exists a sincere positive root.

(3) If z is a root of the quadratic form χ, then $\Sigma z_i \cdot D_i\chi(z) = 2$.

<u>Proof.</u>

$$\Sigma z_iD_i\chi(z) = 2\Sigma z_i(e(i),z) = 2(\Sigma z_ie(i),z) = 2(z,z) = 2$$

(of course, this means that we invoke the Euler formula

$$\sum_i X_iD_i\chi = 2\chi$$

which is valid for any quadratic form χ . It also proves one implication of (1)).

(4) Let z be a positive root of a weakly positive integral quadratic form χ. Let i belong to the support of z, and assume $z \neq e(i)$. Then

$$|D_i\chi(z)| \leq 1 .$$

<u>Proof.</u> With z always $z + e(i)$ is positive; since $z_i \geq 1$, and $z \neq e(i)$, also $z - e(i)$ is positive. Thus

$$o < \chi(z \pm e(i)) = \chi(z) + \chi(e(i)) \pm 2(e(i),z)$$

$$= 1+1 \pm D_i\chi(z) ,$$

therefore

$$-2 < D_i\chi(z) < 2.$$

(5) Let z be a positive root of a weakly positive integral quadratic form χ, and assume z is not simple. Then there exists i in the support of z with $D_i\chi(z) = 1$.

<u>Proof.</u> This is an immediate consequence of (3) and (4). Namely, according to (3), there is i with $z_i D_i\chi(z) \geq 1$, then, since z is positive, $z_i \geq 1$, $D_i\chi(z) \geq 1$. Therefore, i is in the support of z, and now we obtain from (4) that $D_i\chi(z) = 1$.

Let us introduce linear transformations $\sigma_i : \mathbf{Z}^n \longrightarrow \mathbf{Z}^n$ as follows:

$$\sigma_i(z) = z - D_i\chi(z)e(i).$$

Note that $\sigma_i(z)$ and z are always comparable. The maps σ_i preserve the bilinear form $(-,-)$:

$$\begin{aligned}(\sigma_i x,\sigma_i y) &= (x - D_i\chi(x)\cdot e(i),\ y - D_i\chi(y)\cdot e(i)) \\ &= (x,y) - (x, D_i\chi(y)\cdot e(i)) - (D_i\chi(x)\cdot e(i), y) \\ &\quad + (D_i\chi(x)\cdot e(i),\ D_i\chi(y)\cdot e(i)) \\ &= (x,y) - \tfrac{1}{2}D_i\chi(x)\ D_i\chi(y) - \tfrac{1}{2}D_i\chi(x)D_i\chi(y) + D_i\chi(x)D_i\chi(y) \\ &= (x,y)\ .\end{aligned}$$

As a consequence, σ_i maps roots to roots. Also, we have $\sigma_i^2 = 1$, namely

$$\begin{aligned}\sigma_i^2(z) &= \sigma_i(z-D_i\chi(z)e(i)) \\ &= \sigma_i(z) - D_i\chi(z)\sigma_i(e(i)) \\ &= z - D_i\chi(z)e(i) - D_i\chi(z)(-e(i)) = z,\end{aligned}$$

using that $D_i\chi(e(i)) = 2$. We can reformulate (5) as follows:

(5') Let z be a positive root of a weakly positive integral quadratic form χ, and assume z is not simple. Then there exists i in the support of z with $\sigma_i(z) = z - e(i)$ (and this is again a positive root).

Using induction, we obtain:

(5") Let z be a positive root of a weakly positive integral quadratic form. Then there is a sequence $i_1,\ldots,i_m$ such that

$$z > \sigma_{i_1} z > \sigma_{i_2}\sigma_{i_1} z > \ldots > \sigma_{i_m} \ldots \sigma_{i_1} z$$

with $\sigma_{i_m} \ldots \sigma_{i_1} z$ a simple root.

Of course, (5) implies that $z = \sigma_{i_1} \dots \sigma_{i_m}(e(i))$ for some i (roots which are obtained from simple roots by applying a sequence of reflections σ_i are usually called Weyl roots; thus we see that the positive roots of a weakly positive integral quadratic form are Weyl roots).

(6) Let z be a sincere positive root of a weakly positive integral quadratic form χ. Then the following assertions are equivalent:

(i) z is a maximal root.
(ii) $\sigma_i(z) \leq z$ for all i.
(iii) $D_i\chi(z) \geq 0$ for all i.

<u>Proof.</u> By definition of σ_i, the assertions (ii) and (iii) are equivalent. (i) $\Rightarrow$ (ii): For any i, either $\sigma_i(z) \leq z$ or else $\sigma_i(z) > z$. Since with z also $\sigma_i(z)$ is a root, the maximality of z implies $\sigma_i(z) \leq z$.

(iii) $\Rightarrow$ (i): Let z' be a root with $z \leq z'$. Now

$$\begin{aligned} 1 = \chi(z') &= \chi(z'-z) + \chi(z) + 2(z'-z,z) \\ &= \chi(z'-z) + 1 + \sum_i (z'-z)_i\, 2(e(i),z) \\ &= \chi(z'-z) + 1 + \sum_i (z'-z)_i\, D_i\chi(z). \end{aligned}$$

Since $\chi(z'-z) \geq 0$, $(z'-z)_i \geq 0$ and $D_i\chi(z) \geq 0$ for all i, it follows that $\chi(z'-z) = 0$, thus $z' = z$.

Given a maximal positive root z of a weakly positive integral quadratic form χ, an index i in the support of z satisfying $D_i\chi(z) \neq 0$ will be called an <u>exceptional index</u> for z.

(7) Given a maximal positive root z of a weakly positive integral quadratic form χ, and suppose z is not a simple root. Then either there is a unique exceptional index i, and $z_i = 2$, or else there are two different exceptional indices i,j and $z_i = 1 = z_j$.

<u>Proof.</u> We may suppose that z is sincere. We have $\sum_i z_i D_i\chi(z) = 2$ by (3). Since $D_i\chi(z) \geq 0$ for all i, all summands are non-negative. Since z is not a simple root, $D_i\chi(z) = 0$ or 1. Thus, either there is a unique exceptional vertex i, thus $z_i D_i\chi(z) = 2$, and $D_i\chi(z) = 1$ implies $z_i = 2$, or else there are $i \neq j$ with $z_i D_i\chi(z) = 1 = z_j D_j\chi(z)$.

1.2 Dynkin graphs and Euclidean graphs

We consider now the case of a quadratic form χ with $\chi_{ij} \leq o$ for all $i \neq j$. In this case, χ is given by a graph J with only solid edges, having no loops but possibly multiple edges, and we call $\chi = \chi(J)$ the quadratic form of the graph J. Such quadratic forms arise very naturally in different branches of mathematics, and many of their properties are well-known. We want to recall some relevant results with an indication of the proofs.

Theorem. The quadratic form $\chi(J)$ of a connected graph J is either positive definite, or critical, or indefinite. In case $\chi(J)$ is positive definite, there exists a unique maximal root, in case $\chi(J)$ is indefinite, there exists $z > o$ with $\chi(J)(z) < o$.

A connected graph J with $\chi(J)$ positive definite, is called a Dynkin graph, a connected graph J with $\chi(J)$ critical is called a Euclidean graph.

In the following table, we exhibit the Dynkin graphs and the Euclidean graphs. For a Dynkin graph J, we replace the vertex $a \in J$ by the number z_a, where z is the unique maximal root of $\chi(J)$, and we have encircled the exceptional vertices. For a Euclidean graph J, we replace $a \in J$ by the number z_a, where z is the unique minimal positive radical vector of $\chi(J)$. For a Dynkin graph $\mathbb{A}_n, \mathbb{D}_n, \mathbb{E}_n$, the index n refers to the number of vertices. By convention, the Euclidean graphs $\tilde{\mathbb{A}}_n, \tilde{\mathbb{D}}_n, \tilde{\mathbb{E}}_n$ have $n+1$ vertices. Note that $\mathbb{E}_n, \tilde{\mathbb{E}}_n$ are only defined for $n = 6,7,8$, and $\mathbb{D}_n, \tilde{\mathbb{D}}_n$ only for $n \geq 4$.

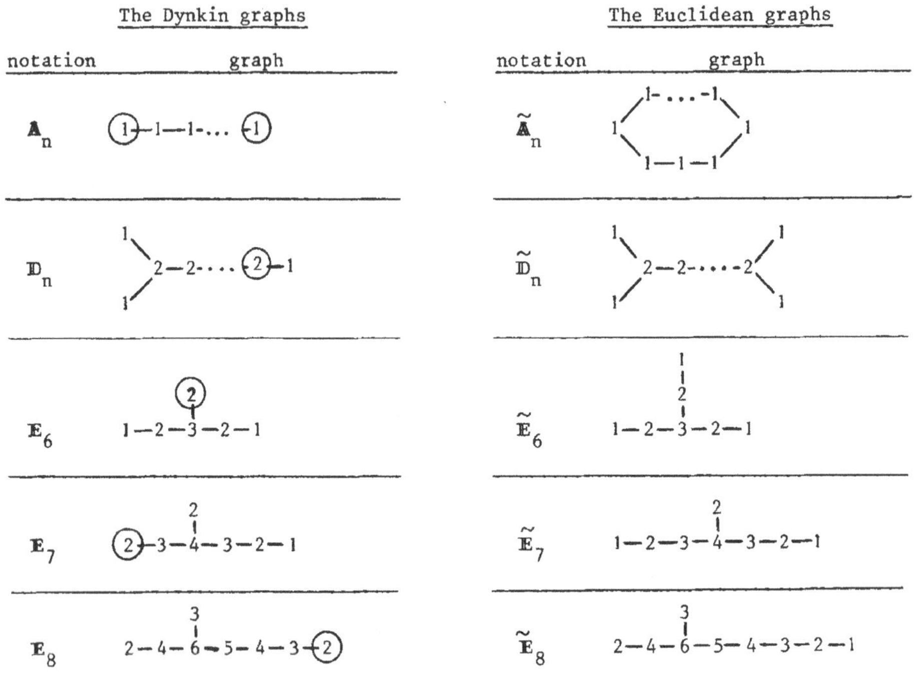
The Dynkin graphs
The Euclidean graphs
notation
graph
notation
graph
$\mathbb{A}_n$
$\tilde{\mathbb{A}}_n$
$\mathbb{D}_n$
$\tilde{\mathbb{D}}_n$
$\mathbb{E}_6$
$\tilde{\mathbb{E}}_6$
$\mathbb{E}_7$
$\tilde{\mathbb{E}}_7$
$\mathbb{E}_8$
$\tilde{\mathbb{E}}_8$

Given a root z for the quadratic form of a Dynkin or a Euclidean graph, either z or $-z$ is positive. The roots z with $z_a = 0$ or 1 for all vertices a, are always uniquely determined by their supports; in case the graph J is a tree, any connected subgraph of J occurs as support of such a root. In this way, we obtain all the positive roots for $\mathbb{A}_n$. For $\mathbb{D}_n$, there are $\frac{1}{2}(n-3)(n-2)$ roots z with $z_a > 1$, for some a, they are of the form

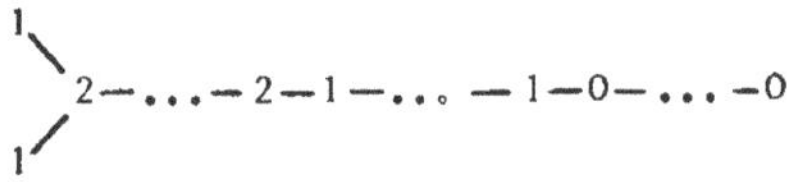

with at least one 2 and at least one 1 on the long branch. For $\mathbb{E}_6, \mathbb{E}_7, \mathbb{E}_8$, the roots z with $z_a > 1$ for some a, are easily obtained from the maximal root by applying reflections σ_i. (Note that these roots are listed in the tables of [Bou]).

All the Dynkin graphs, and the Euclidean graphs $\tilde{\mathbb{D}}_4, \tilde{\mathbb{E}}_6, \tilde{\mathbb{E}}_7, \tilde{\mathbb{E}}_8$ are stars, in the following sense: Given a function $r : \Lambda \to \mathbb{N}_1$, let us introduce the <u>star</u> $\mathbb{T}_r$ with branches indexed over Λ, the branch with index λ being of length $r(\lambda)$. It is obtained from the disjoint union of copies $\mathbb{A}_{r(\lambda)}$, $\lambda \in \Lambda$, by choosing one endpoint in any $\mathbb{A}_{r(\lambda)}$, and identifying all these endpoints to a single vertex, the <u>center</u>

of the star (of course, if $r(\lambda) = 1$, for some λ, then the corresponding branch is not visible). In case $\Lambda = \{1,\dots,s\}$, we also write $\mathbb{T}_{r(1),\dots,r(s)}$ instead of $\mathbb{T}_r$. For $n, n_1, n_2 \in \mathbb{N}_1$, we have the following equalities:

<u>Dynkin graphs:</u>	<u>Euclidean graphs</u>
$\mathbb{T}_n = \mathbb{A}_n$	
$\mathbb{T}_{n_1,n_2} = \mathbb{A}_{n_1+n_2-1}$	
$\mathbb{T}_{n,2,2} = \mathbb{D}_{n+2}$	$\mathbb{T}_{2,2,2,2} = \tilde{\mathbb{D}}_4$
$\mathbb{T}_{3,3,2} = \mathbb{E}_6$	$\mathbb{T}_{3,3,3} = \tilde{\mathbb{E}}_6$
$\mathbb{T}_{4,3,2} = \mathbb{E}_7$	$\mathbb{T}_{4,4,2} = \tilde{\mathbb{E}}_7$
$\mathbb{T}_{5,3,2} = \mathbb{E}_8$	$\mathbb{T}_{6,3,2} = \tilde{\mathbb{E}}_8$

For a <u>proof</u> of the theorem as well as the classification, one first checks directly that the quadratic forms $\mathbb{A}_n$ and $\mathbb{D}_m$ are positive definite. Of course, it is a trivial verification to check that the vector z listed for an Euclidean graph J satisfies $D_a\chi(J)(z) = 0$ for all a, thus z is a radical vector. Since for any Euclidean graph J, there exists a vertex t with $\chi(J)^t$ being the quadratic form of a disjoint union of graphs of the form $\mathbb{A}_n$ and $\mathbb{D}_m$, thus positive definite, it follows that $\chi(J)$ is positive semi-definite, with radical generated by a sincere vector. Since the graphs $\mathbb{E}_n$ $(n=6,7,8)$ are obtained from $\tilde{\mathbb{E}}_n$ by deleting one vertex, their forms have to be positive definite. Finally, let J be a connected graph which is neither Dynkin nor Euclidean. It is easy to see that J contains a Euclidean graph J', let z' be the minimal positive radical vector of J'. If there are additional edges in J connecting vertices from J', then $\chi(J)(z') < 0$. Otherwise, choose a vertex $a \in J$, not in J', which is connected by an edge to a vertex of J'. Then $\chi(J)(2z'+e(a)) < 0$. Thus, always there is some $\mathbf{z} > 0$ with $\chi(J)(z) < 0$. In particular, $\chi(J)$ is indefinite. This finishes the proof.

The positive definite integral quadratic forms always can be transformed by a base change to the quadratic form of a graph. We recall that two integral qudratic forms in n variables are said to be <u>$\mathbb{Z}$-equivalent</u> provided one is obtained from the other by a base change in $\mathbb{Z}^n$.

<u>Theorem.</u> <u>Every positive definite integral quadratic form is $\mathbb{Z}$-equivalent to a form $\chi(J)$, with J a graph; J is uniquely determined up to isomorphism of graphs and is the disjoint union of Dynkin graphs.</u>

Proof. Let χ be a positive definite integral quadratic form in n variables, and R the set of all roots of χ in $\mathbb{Z}^n$. We claim that R considered as subset of $\mathbb{R}^n$, is a (reduced) root system in the sense of Bourbaki [Bou]. First of all, R is a finite set [namely, as in 1.1, consider χ as a function $\mathbb{R}^n \longrightarrow \mathbb{R}$, and let γ be the minimum value taken by χ on the unit sphere with respect to the usual Euclidean norm $||-||$. Then $||y|| < \sqrt{\frac{1}{\gamma}}$ for any root y of χ, as in 1.1.(2).]. Of course, R generates $\mathbb{R}^n$ as an $\mathbb{R}$-vectorspace, and for $z \in R$, we have $\chi(2z) = 4$, thus $2z \notin R$, let $z^\vee = 2(z,-) : \mathbb{R}^n \longrightarrow \mathbb{R}$, thus $z^\vee(R) \subseteq \mathbb{Z}$. Also if we define $\sigma_z(x) = x - z^\vee(x)\cdot z$, then σ_z maps R into R [namely, $\chi(x-z^\vee(x)z) = \chi(x) + z^\vee(x)^2\chi(z) - 2(x,z^\vee(x)\cdot z) = 1$]. This shows that R is a reduced root system. Choosing a base B of the root system R, the $\mathbb{Z}$-span of B coincides with the $\mathbb{Z}$-span of R, thus with $\mathbb{Z}^n$. If we express χ in terms of the basis B, we obtain the quadratic form of a graph. [Namely, given a pair x,y of elements in B, the quadratic form χ restricted to the $\mathbb{Z}$-span of x,y is integral and positive definite, thus either of the form $\mathbb{A}_1 \coprod \mathbb{A}_1$ or of the form $\mathbb{A}_2$.]

1.3 Graphical forms

We consider a finite graph $I = (I_o, I_1)$ without loops or multiple edges, thus I_o is a finite (non empty) set, its elements being called vertices, and I_1 is a set of edges, an edge being a subset $\{a,b\}$ of I_o with $a \neq b$. If $\{a,b\}$ is an edge in I, then we will say that this is an edge from a to b, and that a and b are neighbors. Always, we will assume that ω is not a vertex of I.

Given a finite graph I, the extended quadratic form $\chi = \chi_I$ given by I is defined as follows: We take variables X_a, for all $a \in I_o$, and one additional variable X_ω, and define χ to be

$$\chi = X_\omega^2 + \sum_{a\in I_o} X_a^2 - \sum_{a\in I_o} X_\omega X_a + \sum_{\{a,b\}\in I_1} X_a X_b \quad ;$$

note that this is an integral quadratic form. The quadratic forms obtained in this way will be called graphical forms. Of course, a quadratic form

$$\sum_{i=1}^{n} X_i^2 + \frac{1}{2} \sum_{i,j=1}^{n} \chi_{ij} X_i X_j$$

with $\chi_{ij} = \chi_{ji}$ is graphical if and only if there is some i with $\chi_{ij} = -1$ for all j with $j \neq i$, and $\chi_{ab} = 0$ or 1 otherwise, the graph then is given by $I_o = \{1,\ldots,n\} \setminus \{i\}$, with edges $\{a,b\}$ provided $\chi_{ab} = 1$. Note that for a graphical form the index i is uniquely determined except in case $n = 2$. For $n \geq 3$, the uniquely determined index i with $\chi_{ij} = -1$ for all $j \neq i$ always will be denoted by ω. In dealing with the extended quadratic form of a graph I, we always will represent the edges of I by dotted lines, since this corresponds to the general convention introduced in 1.1.

The proof of Theorem 1 will be done by reduction to graphical forms. Also, graphical forms will be used in the representation theory of partially ordered sets (see 2.6). It will be convenient to have a complete classification both of the sincere weakly positive graphical forms, as well as the critical graphical forms.

Table 1: The sincere weakly positive graphical forms and their sincere positive roots.

For all graphs I in the following table, we exhibit one maximal sincere positive root z of χ_I by replacing the vertex $a \in I$ by the number z_a, the number z_ω being listed separately. The other maximal sincere positive roots of χ_I are obtained from z by an automorphism of I. Always, the exceptional indices (or more precisely, the corresponding numbers z_i) are encircled. For the convenience of the reader, the number of sincere positive roots of χ_I is stated in column s. Also, we note that all these forms χ_I are positive semi-definite, and the radical rank is listed in column r.r..

notation	z_ω	I	s	r.r.	notation	z_ω	I	s	r.r.
F(1)	①		1	0	F(1,1)	1		1	0
F(2)	②		2	0	F(2,2)	2	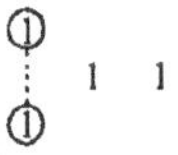	1	0
F(3)	3	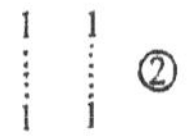	3	0	F(3,3)	3	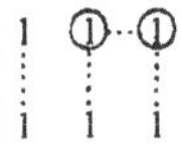	1	0
F(4')	4	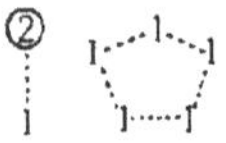	4	0	F(4,4)	4	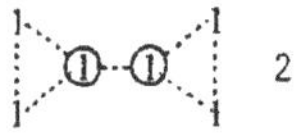	1	0
F(4)	4	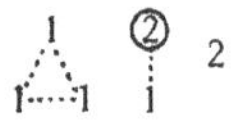	5	0	F(4',5)	4	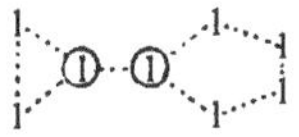	1	1
F(5')	5		5	2	F(5,5)	4	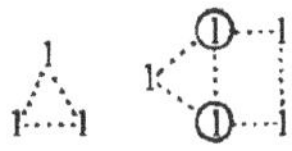	1	1
F(5)	5	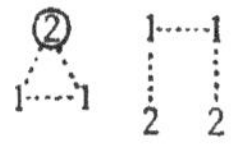	9	0	F(5,6)	5	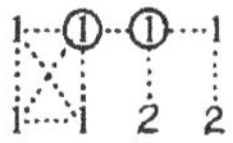	1	1
F(6)	6	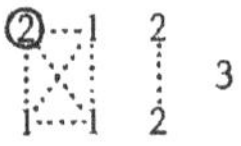	14	0					

For the weakly positive graphical forms with at least two sincere positive roots, we tabulate all the sincere positive roots. As before, the roots are exhibited in the shape of I, with z_ω added to the left separated by the symbol [.

F(2)	2[1 1 1	1[1 1 1			
F(3)	3[$\begin{smallmatrix}1&1\\1&1\end{smallmatrix}$ 2	3[$\begin{smallmatrix}1&1\\1&1\end{smallmatrix}$ 1	2[$\begin{smallmatrix}1&1\\1&1\end{smallmatrix}$ 1		
F(4')	4[$\begin{smallmatrix}2\\1\end{smallmatrix}$ 1 $\begin{smallmatrix}1\\11\end{smallmatrix}$ 1	4[$\begin{smallmatrix}1\\2\end{smallmatrix}$ 1 $\begin{smallmatrix}1\\11\end{smallmatrix}$ 1	4[$\begin{smallmatrix}1\\1\end{smallmatrix}$ 1 $\begin{smallmatrix}1\\11\end{smallmatrix}$ 1	3[$\begin{smallmatrix}1\\1\end{smallmatrix}$ 1 $\begin{smallmatrix}1\\11\end{smallmatrix}$ 1	
F(4)	4[$\begin{smallmatrix}1&2\\11&1\end{smallmatrix}$ 2	4[$\begin{smallmatrix}1&1\\11&2\end{smallmatrix}$ 2	4[$\begin{smallmatrix}1&1\\11&1\end{smallmatrix}$ 2	3[$\begin{smallmatrix}1&1\\11&1\end{smallmatrix}$ 2	3[$\begin{smallmatrix}1&1\\11&1\end{smallmatrix}$ 1
F(5')	5[$\begin{smallmatrix}2\\11\end{smallmatrix}$ 1 $\begin{smallmatrix}11\\11\end{smallmatrix}$ 1	5[$\begin{smallmatrix}1\\21\end{smallmatrix}$ 1 $\begin{smallmatrix}11\\11\end{smallmatrix}$ 1	5[$\begin{smallmatrix}1\\12\end{smallmatrix}$ 1 $\begin{smallmatrix}11\\11\end{smallmatrix}$ 1	5[$\begin{smallmatrix}1\\11\end{smallmatrix}$ 1 $\begin{smallmatrix}11\\11\end{smallmatrix}$ 1	4[$\begin{smallmatrix}1\\11\end{smallmatrix}$ 1 $\begin{smallmatrix}11\\11\end{smallmatrix}$ 1
F(5)	5[$\begin{smallmatrix}2&11\\11&22\end{smallmatrix}$	5[$\begin{smallmatrix}1&11\\21&22\end{smallmatrix}$	5[$\begin{smallmatrix}1&11\\12&22\end{smallmatrix}$	5[$\begin{smallmatrix}1&11\\11&22\end{smallmatrix}$	4[$\begin{smallmatrix}1&11\\11&22\end{smallmatrix}$
	4[$\begin{smallmatrix}1&11\\11&21\end{smallmatrix}$	4[$\begin{smallmatrix}1&11\\11&12\end{smallmatrix}$	4[$\begin{smallmatrix}1&11\\11&11\end{smallmatrix}$	3[$\begin{smallmatrix}1&11\\11&11\end{smallmatrix}$	
F(6)	6[$\begin{smallmatrix}21&2\\11&2\end{smallmatrix}$ 3	6[$\begin{smallmatrix}11&2\\21&2\end{smallmatrix}$ 3	6[$\begin{smallmatrix}11&2\\12&2\end{smallmatrix}$ 3	6[$\begin{smallmatrix}12&2\\11&2\end{smallmatrix}$ 3	6[$\begin{smallmatrix}11&2\\11&2\end{smallmatrix}$ 3
	5[$\begin{smallmatrix}11&2\\11&2\end{smallmatrix}$ 3	5[$\begin{smallmatrix}11&2\\11&1\end{smallmatrix}$ 3	5[$\begin{smallmatrix}11&1\\11&2\end{smallmatrix}$ 3	5[$\begin{smallmatrix}11&2\\11&2\end{smallmatrix}$ 2	5[$\begin{smallmatrix}11&2\\11&1\end{smallmatrix}$ 2
	5[$\begin{smallmatrix}11&1\\11&2\end{smallmatrix}$ 2	4[$\begin{smallmatrix}11&2\\11&1\end{smallmatrix}$ 2	4[$\begin{smallmatrix}11&1\\11&2\end{smallmatrix}$ 2	4[$\begin{smallmatrix}11&1\\11&1\end{smallmatrix}$ 2	

<u>Table 2</u> <u>The critical graphical forms and their minimal positive radical vectors.</u>

For the graphs I in the table below, we exhibit the minimal positive radical vectors z of χ_I by noting z_ω and replacing the vertex $a \in I$ by the number z_a.

notation	z_ω	I
C(2)	2	1 1 1 1
C(3)	3	
C(4')	4	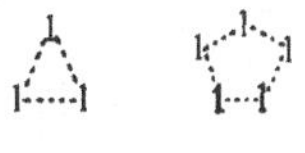
C(4)	4	 2
C(5)	5	
C(6)	6	 3

In dealing with the extended quadratic form χ_I of one of the graphs I occurring in tables 1 and 2, we will use the notation $F(i)$, $F(i,j)$, $C(i)$ both when referring to I as well as when referring to χ_I, we hope that this does not lead to confusion.

The proof of the material presented in tables 1 and 2 will be given in this chapter. In 1.5, we will show that the quadratic forms of table 1 are weakly positive, those of table 2 critical, that all these forms are positive semi-definite with radical rank 1 for the forms in table 2, and with radical rank as indicated in table 1. In 1.7, we will see that the table 1 is complete, in 1.9 that table 2 is complete.

1.4 Reduction to graphical forms

Let ℓ be a natural number. Let Φ_ℓ be the set of weakly positive integral forms χ such that $y_i \leq \ell$ for all components y_i of positive roots y of χ, and with at least one sincere positive root z, having a component $z_i = \ell$.

Lemma. *Let $\ell \geq 2$. If χ is a form in Φ_ℓ with minimal number of positive roots, then χ is graphical, and there exists a maximal sincere positive root z of χ with a unique exceptional index, and such that $z_\omega = \ell$.*

Before we are going to prove the lemma, we start with the following simple observation; always, we will use the notation of 1.1.

(1) If χ is a weakly positive form, then $\chi_{ij} \geq -1$, for all $i \neq j$.

Proof. $1 \leq \chi(e(i)+e(j)) = \chi(e(i)) + \chi(e(j)) + \chi_{ij} = 1 + 1 + \chi_{ij}$.

Now, let χ be a form in Φ_ℓ with minimal number of positive roots. Let n be the number of variables occurring in χ, and let $z \in \mathbf{Z}^n$ be a positive root of χ with a component equal to ℓ. In addition, we may assume that z is maximal. Note that z is sincere, since otherwise we omit the variables outside the support of z, and obtain a form in Φ_ℓ with a smaller number of positive roots (we have deleted some simple roots). Let ω be an index with $z_\omega = \ell$, and define I_o to be the set of remaining indices of variables occurring in χ.

(2) For all $a \neq b$ in I_o, we have $\chi_{ab} \geq 0$.

Proof. Assume $\chi_{ab} = -1$ for some $a \neq b$ in I_o. By definition, $\chi_{ab} = \chi_{ba}$, thus we may assume $z_a \leq z_b$. The following base change seems to have been considered first by Gabrielov: choose as new basis of $\mathbb{Z}^n$ the vectors $e(c)$ with $c \in I_o \cup \{\omega\}$, and $c \neq a$, and, in addition, the vector $e(a) + e(b)$. Since $e(a) + e(b)$ is a root of χ (due to the fact that $\chi_{ab} = -1$), it follows that we obtain from χ (using this base change) again an integral quadratic form, say χ'. The positive vectors with respect to the new basis are also positive vectors with respect to the old basis, namely any $y \in \mathbb{Z}^n$ can be written as

$$(*) \qquad y = y'_a(e(a) + e(b)) + \sum_{c \neq a} y'_c e(c) = (y'_a + y'_b)e(b) + \sum_{c \neq b} y'_c e(c),$$

and if all $y'_c \geq o$, then also $y'_a + y'_b \geq o$. Thus, with χ also χ' is weakly positive, and χ' has at most as many positive roots as χ. However, χ' actually has less positive roots, since $e(a)$ is a positive root for χ, but not for χ'. Also, we claim that with χ also χ' belongs to Φ_ℓ. Namely, if y is a positive root for χ', and y is written as a linear combination as in $(*)$, then $y'_a + y'_b \leq \ell$, and $y'_c \leq \ell$ for all $c \neq b$, thus also $y'_b \leq \ell$ (since $y'_a \geq o$). On the other hand, z itself remains a positive root for χ', namely

$$z = z_a(e(a) + e(b)) + (z_b - z_a)e(b) + \sum_{c \neq a,b} z_c\, e(c),$$

using that $z_a \leq z_b$, and since $a,b \in I_o$, the coefficient $z_\omega = \ell$ of z at the index ω also remains its coefficient after the base change. This shows that $\chi' \in \Phi_\ell$. However, this contradicts the minimality condition for χ.

(3) For all $a \in I_o$, we have $\chi_{\omega a} = -1$,

Proof. Since z is sincere, and $z_\omega \geq 2$, we have $z - e(a) > o$, thus

$$\begin{aligned} 1 &\leq \chi(z-e(a)) = \chi(z) + \chi(e(a)) - 2(z,e(a)) \\ &= 2 - \sum_i z_i 2(e(i),e(a)) \\ &= 2 - 2z_a - \sum_{i \neq a} z_i \chi_{ia} \end{aligned}$$

and, since $2 - 2z_a \leq o$, and all $z_i > o$, it follows that $\chi_{ia} < o$ for at least one i. However, according to (2), this is possible only for $i = \omega$, and then $\chi_{\omega a} = -1$, according to (1).

(4) For all $a,b \in I_o$, we have $\chi_{ab} \leq 1$.

Proof. Assume $\chi_{ab} \geq 2$. Without loss of generality, we may assume $D_a\chi(z) \leq D_b\chi(z)$. Note that $z + z_b e(a) - z_b e(b)$ is positive, thus

$$\begin{aligned} 1 &\le \chi(z+z_b e(a) - z_b e(b)) \\ &= 1 + z_b^2 + z_b^2 + z_b D_a\chi(z) - z_b D_b\chi(z) - z_b^2\chi_{ab} \\ &\le 1 + (2-\chi_{ab})z_b^2 . \end{aligned}$$

Using the assumption $\chi_{ab} \ge 2$, we can continue with

$$1 + (2-\chi_{ab})z_b^2 \le 1 ,$$

thus we see that $z' := z + z_b e(a) - z_b e(b)$ actually is a (positive) root. Of course, z' is a root for the restriction χ^b of χ to the hyperplane defined by $X_b = 0$, and since $z'_\omega = \ell$, we see that χ^b belongs to Φ_ℓ. However, this contradicts the minimality assumption for χ, since χ^b has less positive roots than χ.

Taking together the assertions (2), (3), (4), we see that χ is graphical. By assumption, z is a maximal sincere positive root for χ, with $z_\omega = \ell$. Now assume, there are two exceptional indices for z, thus there is $a \in I_o$ with $z_a = 1 = D_a\chi(z)$. However, in this case $\sigma_a z = z - e(a)$ is a positive root for the restriction χ^a of χ, having a component equal to ℓ, thus $\chi^a \in \Phi_\ell$, contrary to the minimality assumption for χ. This finishes the proof of the lemma.

Let us add that we can use the Gabrielov transformation encountered in step (2) of the previous proof in order to show the equivalence of certain quadratic forms.

Given a natural number r, we denote by $I(r)$ the complete graph on $\{1,2,\ldots r\}$, (with dotted edges!). The disjoint union of graphs will be denoted by $\sqcup$.

Lemma 2. *Let* $r(1),\ldots,r(s)$ *be natural numbers* ≥ 2. *The quadratic form given by* $T_{r(1),\ldots,r(s)}$ *is* $\mathbb{Z}$*-equivalent to the extended quadratic form of* $\bigsqcup_{\lambda=1}^{s} I(r(\lambda)-1)$.

For the proof, we may use inductively Gabrielov transformations removing step by step the solid edges which do not involve the center of the star. Alternatively, the base change can be described as follows: Denote the center of the star by ω, and let the branch with index λ be given by

.

Let $I_o = \{(\lambda,i) \mid 1 \leq i \leq r(\lambda)-1,\ 1 \leq \lambda \leq s\}$. The quadratic form χ of $T_{r(1),\ldots,r(s)}$ is defined on the free abelian group F with base vectors $e(\omega)$, and $e(\lambda,i)$ where $(\lambda,i) \in I$. Now, let $f(\omega) = e(\omega)$, and

$$f(\lambda,i) = \sum_{j=1}^{i} e(\lambda,j), \quad \text{with} \quad (\lambda,i) \in I_o.$$

Obviously, $f(\omega)$, and the $f(\lambda,i)$ again form a basis of F, all these elements being roots of χ, and we have

$$2(f(\omega),\ f(\lambda,i)) = -1,$$
$$2(f(\lambda,i),\ f(\lambda',i')) = \begin{cases} 1 & \text{iff} \quad \lambda = \lambda' \\ 0 & \phantom{\text{iff}} \quad \lambda \neq \lambda' \end{cases}$$

for all $(\lambda,i) \neq (\lambda',i') \in I_o$. Thus, after the base change, we obtain the extended quadratic form for a graph I with set I_o of vertices, and $I = \bigsqcup_{\lambda=1}^{s} I(r(\lambda)-1)$.

1.5 The quadratic forms occurring in tables 1 and 2.

The subgroups of $\mathbb{Z}^n$ generated by some of the base vectors $e(1),\ldots,e(n)$ will be called coordinate subspaces of $\mathbb{Z}^n$. Similarly, given a quadratic form $\chi = \chi(X_1,\ldots,X_n)$, then a form obtained from χ by setting some of the variables equal to 0 will be called a restriction of χ to a coordinate subspace. Typical such examples are the restrictions to coordinate hyperplanes, namely, the forms

$$\chi^t = \chi(X_1,\ldots,X_{t-1},0,X_{t+1},\ldots,X_n), \quad \text{for} \quad 1 \le t \le n .$$

The quadratic form $\mathbf{E}_8$ is positive definite, thus the same is true for the quadratic form $F(6)$ (since we have seen in the last section that these forms are $\mathbb{Z}$-equivalent). As restrictions to coordinate subspaces we obtain from $F(6)$ the forms $F(4)$, $F(3)$, $F(2,2)$, $F(2)$, $F(1,1)$ and $F(1)$, thus all these forms are positive definite.

It is a routine calculation to check that the forms χ_I and the vectors z listed in table 2 satisfy $D_a\chi(z) = o$ for all $a \in I \cup \{\omega\}$, thus any such z is a radical vector. [Actually, for $C(2)$, $C(3)$, $C(4')$, there is the following general argument: A graph I is called r-regular graph, provided any vertex has precisely r neighbors. Now, if I is an r-regular graph on $2(r+2)$ vertices, then the vector z with $z_a = 1$ for all $a \in I$, and $z_\omega = r+2$ is a radical vector for χ_I] Considering the form χ_I of type $C(i)$ with $i = 2,3,4,6$, there is a coordinate subspace H complementary to $\langle z\rangle$ such that the restriction of χ_I to H is of the form $F(i)$, similarly, for χ_I of type $C(5)$, there is a coordinate subspace H complementary to $\langle z\rangle$ such that the restriction of χ_I to H is of the form $F(6)$. As a consequence, the forms χ_I of type $C(2)$, $C(3)$, $C(4)$, $C(5)$, $C(6)$ are positive semi-definite with $\mathrm{rad}\chi_I = \langle z\rangle$. Also, since always z is sincere, it follows that these forms χ_I are critical.

Since the form $C(5)$ is positive semi-definite with radical rank 1, and a radical generator is sincere, it follows that all restrictions to proper coordinate subspaces are positive definite, thus $F(5)$ and $F(3,3)$ are positive definite quadratic forms. For the form χ_I of type $C(4')$ with radical vector z from table 2, there is a coordinate subspace H complementary to $\langle z\rangle$ such that the restriction of χ_I to H is of type $F(5)$, thus $C(4')$ is positive semi-definite with $\operatorname{rad}\chi_I = \langle z\rangle$. Since z is sincere, $C(4')$ is critical, and all restrictions to proper coordinate subspaces are positive definite. Thus $F(4')$ is positive definite.

Next, let us consider the extended quadratic form χ of the graph I

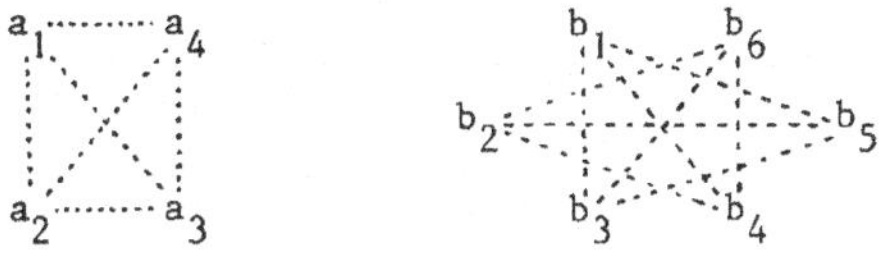

It is easy to check that

$$z = (5[\,{}^{11}_{11}\;\; 0{}^{12}_{01}2)\;,\quad z' = (5[\,{}^{11}_{11}\;\; 1{}^{22}_{00}1)\;,\quad z'' = (5[\,{}^{11}_{11}\;\; 2{}^{21}_{10}0)$$

are radical vectors (verifying that $D_a\chi(z) = o$ for all $a \in I \cup \{\omega\}$), and they are linearly independent, as the coefficients at the vertices b_1, b_2, b_3 show. The subgroup $\langle z, z', z''\rangle$ has the coordinate subspace H generated by $e(\omega)$; $e(a_i)$, $1 \le i \le 4$, and $e(b_j)$, $4 \le j \le 6$, as complement, and the restriction of χ to H is of type $F(6)$, thus $\operatorname{rad}\chi = \langle z, z', z''\rangle$ and χ is positive semi-definite of radical rank 3. The restriction χ^{a_4} of χ is of type $F(5')$, and since the corresponding coordinate subspace defined by $X_{a_4} = 0$ does not contain z, it follows that $F(5')$ is positive semi-definite of radical rank 2. Also, we claim that $F(5')$ is weakly positive. Namely, any radical vector of $F(5')$ is of the form $\alpha z + \alpha' z' + \alpha'' z''$ with $\alpha, \alpha', \alpha'' \in \mathbb{Z}$, and $\alpha + \alpha' + \alpha'' = 0$ (since the coeffient at a_4 is zero), and such a vector cannot be positive (for, if $\alpha z + \alpha' z' + \alpha'' z''$ is positive, then the coefficient at b_4 shows $\alpha \ge o$, that at b_2 shows $\alpha' \ge o$, that at b_3 shows $\alpha'' \ge o$, thus $\alpha = \alpha' = \alpha'' = o$). The further restriction $\chi^{a_4 b_6}$ of χ^{a_4} leads to the quadratic form $F(5,5)$. Thus, $F(5,5)$ is weakly positive, and positive semi-definite of radical rank ≥ 1. Since the restriction $\chi^{a_4 b_6 b_5}$ is of type $F(5)$, thus positive definite, we conclude that the radical rank of $F(5,5)$ is equal to 1.

It remains to consider the forms $F(4,4)$, $F(4',5)$ and $F(5,6)$. We need the notion of a completion of a graph (or of the corresponding graphical form). Given a graph $I = (I_0, I_1)$, a <u>completion</u> of I is a graph of the form $I' = (I_0, I_1')$ with $I_1 \subseteq I_1'$, thus I' is obtained from I by adding edges. In case $I_1 \subset I_1'$, we call I' a <u>proper</u> completion of I.

<u>Lemma.</u> Let I' be a proper completion of the graph I. If χ_I is weakly positive or critical, then $\chi_{I'}$ is weakly positive.

<u>Proof.</u> It is sufficient to consider the case of I' being obtained from I by adding one edge, say the edge a‧‧‧‧‧‧b . Then

$$\chi_{I'}(X_1,\ldots,X_n) = \chi_I(X_1,\ldots,X_n) + X_aX_b .$$

Let $z > 0$. If either χ_I is weakly positive, or else $z \notin \mathrm{rad}\chi_I$ and χ_I is critical, then

$$\chi_{I'}(z) = \chi_I(z) + z_az_b \geq \chi_I(z) > 0 .$$

On the other hand, if χ_I is critical, and $z \in \mathrm{rad}\chi_I$, then z is sincere, thus $z_az_b > 0$, and therefore

$$\chi_{I'}(z) = \chi_I(z) + z_az_b = z_az_b > 0 .$$

Now, $F(4,4)$ is a completion of $C(4)$. Similarly, $F(4',5)$ is a completion of $C(4')$, and $F(5,6)$ is a completion of $C(5)$, thus the forms $F(4,4)$, $F(4',5)$, and $F(5,6)$ are all weakly positive. [In fact, we note the following: Let χ_I be a critical graphical form, and z a positive radical vector. Assume a,b are (different) vertices of I, not connected by an edge, and with $z_a = z_b = 1$. If I' is the completion of I obtained by adding the edge $\{a,b\}$, then, on the one hand, $\chi_{I'}$ is weakly positive, on the other hand, z obviously is a positive root for $\chi_{I'}$, and also the only sincere positive root for $\chi_{I'}$. It follows that $\chi_{I'}$ is a sincere weakly positive graphical form with a unique sincere positive root. In this way, starting from a critical graphical form, we obtain the forms $F(2,2)$, $F(3,3)$, $F(4,4)$, $F(4',5)$, $F(5,5)$ and $F(5,6)$.]

In order to show that a form of type $F(4,4)$ actually is positive definite, we consider it as a restriction χ^t of a positive semi-definite integral form χ to a coordinate hyperplane. Namely, let χ be given by the bigraph

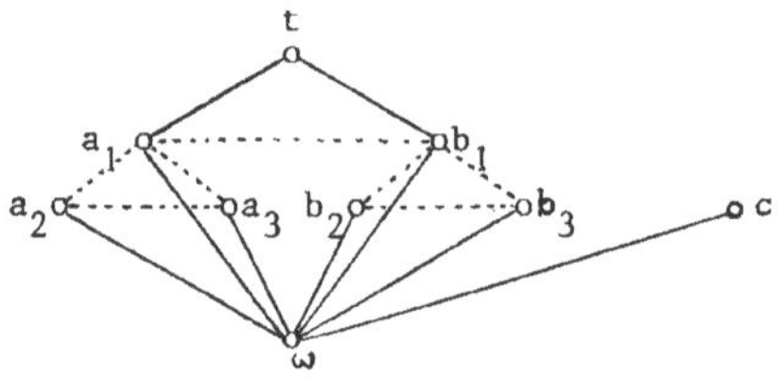

There is the following radical element r of χ

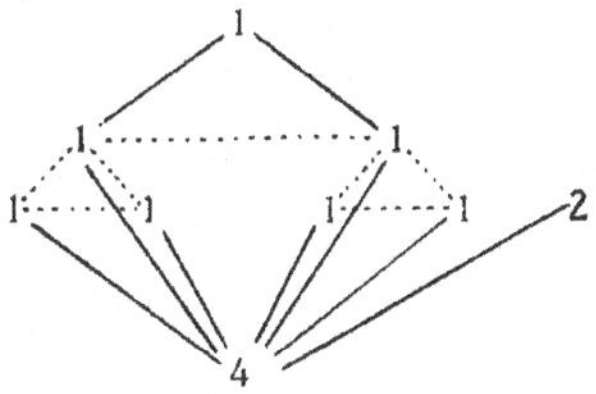

and the coordinate hyperplane defined by $X_{a_1} = 0$ is complementary to $\langle r\rangle$. If we show that χ^{a_1} is positive definite, then $\mathrm{rad}\chi = \langle r\rangle$, χ is positive semi-definite of radical rank 1, and therefore χ^t is positive definite. However, replacing the base vector $e(t)$ by $e(t) + e(b_1)$ and keeping the base vectors $e(\omega)$, $e(a_1)$, $e(a_2)$; $e(b_i)$, $1 \le i \le 3$, and $e(c)$, (this is a Gabrielov-transformation) we see that χ^{a_1} is $\mathbb{Z}$-equivalent to $F(6)$, thus positive definite.

On the other hand, we claim that the forms $F(4',5)$ and $F(5,6)$ are positive semi-definite of radical rank 1. Namely, there are the following radical vectors z

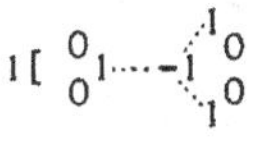 , and 

and $\langle z\rangle$ is complementary to a coordinate hyperplane of type $F(5)$, and $F(6)$, respectively.

Note that we have established that all forms in table 1 are weakly positive, those in table 2 are critical, that all these forms are positive semi-definite, with radical rank 1 for the forms of table 2, and with radical rank as indicated in table 1.

For later reference, let us collect three additional positive semi-definite graphical forms $\chi = \chi_I$ of radical rank 1; the first one being a restriction to coordinate subspaces both of $F(4',5)$ and $F(5,6)$. With any of the three graphs I, we list a radical vector z of χ_I (as in the tables above, the vertices of I are replaced by the corresponding coefficients of z)

notation	z_ω	I
$N(2)$	1	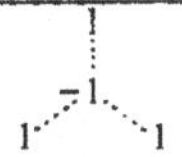
$N(3)$	2	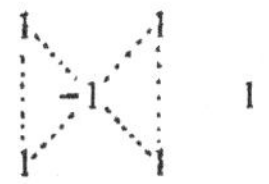
$N(4)$	3	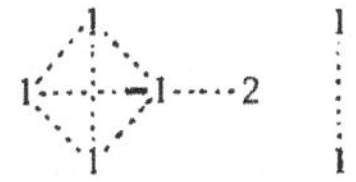

Of course, one easily checks that $D_a\chi(z) = o$ for all $a \in I \cup \{\omega\}$, thus z is a radical vector. If we denote by c the unique vertex of I with $z_c = -1$, then, for the form χ of type $\mathcal{N}(i)$, the form χ^c is of type $\mathcal{F}(i)$. Since χ^c is positive definite, it follows that χ is positive semi-definite of radical rank 1.

1.6 Maximal sincere positive roots of graphical forms with a unique exceptional index.

Lemma. *Let I be a graph with $\chi = \chi_I$ weakly positive. Let z be a maximal sincere positive root of χ with a unique exceptional index a, and let $\ell = z_\omega \geq 3$. Then either*

(i) $I = I' \sqcup I''$ with I' the complete graph on $\ell-2$ vertices, $a \in I'$, and I'' contains a full subgraph with three vertices and only one edge, moreover, $z_{a'} = 1$ for all $a' \in I'$, $a' \neq a$; or else

(ii) a has two neighbors b and c such that there is no edge from b to c, and there exists a sincere positive root z' with $z'_\omega > z_\omega \geq 5$.

Actually, we later will see that the case (ii) is impossible.

Proof. Since $z_\omega \geq 3$, and $z_a = 2$, the index a belongs to I.

First, assume there are two different neighbors b,c of a which are not joined by an edge. Then

$$1 = D_a\chi(z) \geq 2z_a - z_\omega + z_b + z_c \geq 4 - z_\omega + 1 + 1 ,$$

since z is sincere, thus $z_\omega \geq 5$. Also, one easily checks that $z' = z - e(a) + e(b) + e(c) + e(\omega)$ is a positive root, (namely $z' = \sigma_\omega\sigma_c\sigma_b\sigma_a(z)$), and also sincere, since $z_a = 2$, and $z'_\omega = z_\omega+1$. Thus, all conditions stated in (ii) are satisfied.

Now, consider the case where any two different neighbors of a are joint by an edge. Let I' be the full subgraph of I on the set $\{a\} \cup N(a)$ where $N(a)$ denotes the set of neighbors of a. By assumption I' is a complete graph. Let I'' be the full subgraph given by the vertices in $I \setminus I'$.

(1) If b is a vertex in $N(a)$, then $z_b = 1$, and all the neighbors of b lie in I'.

Proof. Let $J = I' \setminus \{a,b\}$ and J'' the set of neighbors of b in I''.

$$1 = D_a\chi(z) = 2z_a - z_\omega + z_b + \sum_{c\in J} z_c ,$$

$$0 = D_b\chi(z) = 2z_b - z_\omega + z_a + \sum_{c\in J} z_c + \sum_{d\in J''} z_d ,$$

thus

$$-1 = -D_a\chi(z) + D_b\chi(z) = -z_a + z_b + \sum_{d\in J''} z_d .$$

Taking into account that $z_a = 2$, we have

$$0 = (z_b-1) + \sum_{d\in J''} z_d .$$

Since $z_b \geq 1$, and all $z_d \geq 1$, this equality is only possible in case $z_b = 1$ and $J'' = \emptyset$.

(2) $|I'| = \ell-2$.

Proof. $$1 = D_a\chi(z) = 2z_a - z_\omega + \sum_{\substack{b\in I' \\ b\neq a}} z_b = 4 - \ell + |I'| - 1 .$$

Altogether, we see in (1) and (2) that I' is the complete graph on $\ell-2$ vertices; (1) contains the assertion that I is the disjoint union of I' and I''. It remains to consider I'' .

(3) Choose $b \in I''$ with z_b maximal. Then there is an edge $c_1 \cdots\cdots c_2$ in I'' with neither c_1 nor c_2 being a neighbor of b.

Proof. First, we note that $\sum_{c\in I''} z_c = \ell+1$. Namely.

$$0 = D_\omega\chi(z) = 2z_\omega - \sum_{c\in I'} z_c - \sum_{c\in I'} z_c ,$$

however, $\sum_{c\in I'} z_c = \ell-1$, since $z_a = 2$, $z_c = 1$ for $c \in I' \setminus \{a\}$, and $|I'| = \ell-2$. Now, denote by J the set of neighbors of b, and let J' be the set of elements in I'' different from b and not being a neighbor of b. Then:

$$\begin{aligned} 0 = D_b\chi(z) &= 2z_b - z_\omega + \sum_{c\in J} z_c \\ &= 2z_b - z_\omega + \sum_{c\in I''} z_c - z_b - \sum_{c\in J'} z_c \\ &= z_b + 1 - \sum_{c\in J'} z_c . \end{aligned}$$

Thus

$$\sum_{c\in J'} z_c = z_b+1 > z_b .$$

Since we have chosen $b \in I''$ with z_b maximal, it follows that J' consists of at least two elements, say c_1 and c_2. We claim that c_1 and c_2 are neighbors. Namely, otherwise a,b,c_1,c_2 define a full subgraph of typ $C(2)$, thus χ cannot be weakly positive. This finishes the proof.

Proof of Theorem 1. Assume there exists a weakly positive integral quadratic form χ with a positive root $z = (z_1,\ldots,z_n)$ satisfying $z_i \geq 7$ for some i. Now χ has only finitely many positive roots, according to 1.2 (2), thus there exists the supremum ℓ of all coefficients y_j, $1 \leq j \leq n$, of all positive roots y of χ. Choose some positive root z' with $z'_j = \ell$ for some j, and let χ' be the restriction of χ to the support of z'. Then $\chi' \in \Phi_\ell$, and $\ell \geq 7$. In particular, Φ_ℓ is not empty, thus choose a form $\chi'' \in \Phi_\ell$ with minimal number of positive roots. According to 1.4, χ'' is the extended quadratic form of some graph I, and there exists a maximal sincere root z'' of χ'' with unique exceptional index, and with $z''_\omega = \ell$. We use the lemma above. Since $\chi'' \in \Phi_\ell$, and $z_\omega = \ell$, the case (i) has to occur, thus $I = I' \sqcup I''$, where I' contains the complete graph $I(5)$ on five vertices, and I'' contains as a full subgraph $I(2) \sqcup I(1)$. Thus I has a full subgraph of type $C(6)$, and therefore χ'' cannot be weakly positive. This is a contradiction.

1.7 Completeness of table 1.

Let χ be a sincere weakly positive graphical form in n variables, say the extended quadratic form of the graph I. Let z be a maximal sincere positive root. By induction on n, we want to show that I and z occur in table 1. In case $n = 2$, we deal with the form $F(1)$, and $z = (1 [1)$, thus, we can assume $n \geq 3$.

(a) Assume that z has a unique exceptional index a and that we are not in case (ii) of lemma 1.6. Note that $z_\omega \geq z_i$ for all i (namely $1 \geq D_i\chi(z) \geq 2z_i - z_\omega$ shows that $z_\omega \geq 2z_i - 1 \geq z_i$), thus $z_a = 2$ implies $z_\omega \geq 2$. According to the first theorem of Ovsienko, we also have $z_\omega \leq 6$, and we consider the various cases for $\ell := z_\omega$ separately.

Case $\ell = 2$. In this case, $a = \omega$. (Namely, if a belongs to I, then

$$1 = D_a\chi(z) \geq 2z_a - z_\omega = 4 - z_\omega ,$$

thus $z_\omega \geq 3$.) Now, let $b \in I$, then

$$0 = D_b\chi(z) = 2z_b - z_\omega + \sum_{c\in N(b)} z_c = 2z_b - 2 + \sum_{c\in N(b)} z_c .$$

Since z is sincere, it follows that $z_b = 1$, and $N(b) = \emptyset$. Thus, there are no edges in I. Since

$$1 = D_\omega\chi(z) = 2z_\omega - \sum_{b\in I} z_b = 4 - |I| ,$$

we see that I consists of precisely 3 vertices, thus $I = F(2)$ and $z = (2[\,111)$.

In the remaining cases $\ell = z_\omega \geq 3$, thus $a \in I$ (since $z_a = 2$). We apply 1.6, and, by assumption only the case (i) has to be considered. Thus $I = I' \sqcup J$, with I' being a complete graph on $\ell - 2$ vertices including a, say with vertices $a = a_1, a_2, \dots, a_{\ell-2}$, and J is a graph containing a full subgraph with three vertices and one edge. Also, $z_{a_1} = 2$, and, for $2 \leq i \leq \ell-2$, we have $z_{a_i} = 1$. It remains to determine J and the restriction of z to J. Thus, we consider a graph J and positive integers z_b, $b \in J$, satisfying

(i) $\quad \sum_{b\in J} z_b = \ell + 1 ,\quad$ and

(ii) $\quad$ for any $b \in J$, $\quad 2z_b + \sum_{c\in N(b)} z_c = \ell ,$

where $N(b)$ denotes the set of neighbors of b in J. [Note that (i) follows from $D_\omega\chi(z) = 0$ taking into account the values z_{a_i}, whereas (ii) is equivalent to $D_b\chi(z) = 0$.]

<u>Case</u> $\ell = 3$. The equality (ii) shows that $z_b = 1$ for all $b \in J$, and that any b has precisely one neighbor. Since $z_b = 1$ for all b, the equality (i) states that $|J| = 4$, thus J is the 1-regular graph on 4 vertices, thus $I = F(3)$, and $z = (3[\,{}^{11}_{11}2)$.

<u>Case</u> $\ell = 4$. We have $z_b \leq 2$ for all $b \in J$, and any b with $z_b = 2$ has no neighbor. In case $z_b = 1$ for some b, then $\sum_{c\in N(b)} z_c = 2$, and for $c \in N(b)$, we must have $z_c = 1$ (since, as we have seen, the elements c with $z_c = 2$ are not neighbors), thus any b with $z_b = 1$ has precisely two neighbors. Since $\sum_{b\in J} z_b = 5$, it follows that there can be at most one b with $z_b = 2$, and in case there exists a b with $z_b = 2$, the remaining vertices form a triangle, thus $I = F(4)$ and $z = (4[\,{}^{2}_{11}\,{}^{2}_{1}\,2)$. In case $z_b = 1$ for all $b \in J$, we conclude that J is a 2-regular graph on 5 vertices, thus a pentagon, thus $I = F(4')$ and $z = (4[\,{}^{2}_{1}\;\;1{}^{1}_{11}1)$.

Case $\ell = 5$. The equality (ii) shows again that $z_b \leq 2$ for all b, and that a vertex b with $z_b = 2$ has precisely one neighbor, say c and $z_c = 1$. On the other hand, a vertex b with $z_b = 1$ has either 3 neighbors and $z_c = 1$ for $c \in N(b)$, or else 2 neighbors, say c, c' with $z_c = 1$, $z_{c'} = 2$. Assume there exists a vertex b with $z_b = 2$, and let b' be its neighbor, thus $z_{b'} = 1$. Now b' has besides b one additional neighbor, say c', and $z_{c'} = 1$. Consider the neighbors of c'. If c' has besides b' two additional neighbors c_1, c_2, then $z_{c_1} = 1 = z_{c_2}$, and b, b', c', c_1, c_2 are all the vertices of J, according to (i). However, condition (ii) is not satisfied for c_1 and c_2. Thus c' has precisely one neighbor c, and $z_c = 2$. In this way, we obtain $I = F(5)$, and $z = (5[\begin{smallmatrix} 2 & 21 \\ 11 & 21 \end{smallmatrix})$.

Now consider the case of $z_b = 1$ for all $b \in J$, thus J is a 3-regular graph on 6 vertices. The complementary graph $\bar{J}$ (with the same vertices as in J, and with an edge $\{b,c\}$ in $\bar{J}$ for $b \neq c$, if and only if b,c are not neighbors in J) is a 2-regular graph on 6 vertices. There are precisely two possibilites: Either $\bar{J}$ is the disjoint union of two triangles, or else a hexagon. However, the first case cannot occur. Namely, a triangle in $\bar{J}$ is given by three vertices b_1, b_2, b_3 without any edge in J joining them, thus a_1, b_1, b_2, b_3 is a full subgraph of I of type $C(2)$, thus χ cannot be weakly positive. There only remains the case of $\bar{J}$ being a hexagon, thus $I = F(5')$, and $z = (5[\begin{smallmatrix} 2 \\ 11 \end{smallmatrix}\ 1\begin{smallmatrix} 11 \\ 11 \end{smallmatrix}1)$.

Case $\ell = 6$. In this case, $z_b \leq 3$ for all $b \in J$. Suppose there exists some b with $z_b = 3$, then b has no neighbors. The sum of the coefficients of z for the remaining vertices of J is 4, thus there cannot exist a c with $z_c = 1$ (since $\sum_{d \in N(c)} z_d = 4$ for $z_c = 1$). As a consequence, $J \setminus \{b\}$ consists of two vertices c_1, c_2 which are neighbors, and $z_{c_1} = z_{c_2} = 2$. Thus $I = F(6)$, and $z = (6[\begin{smallmatrix} 21 & 2 \\ 11 & 2 \end{smallmatrix}\ 3)$.

Now suppose $z_b \leq 2$ for all $b \in J$. Recall that there is a full subgraph with 3 vertices b_1, b_2, c containing as only edge $\{b_1, b_2\}$. If c' is a neighbor of c, then c' is also a neighbor both of b_1 and of b_2. (Namely, if c' is neither neighbor of b_1 nor of b_2, then $a_1, a_2, b_1, b_2, c, c'$ defines a full subgraph of type $C(3)$, whereas if c' is a neighbor of b_1 but not of b_2, then $a_1, a_2, a_3, a_4, b_1, b_2, c, c'$ defines a full subgraph of type $C(5)$, thus χ is not weakly positive). Thus $N(c) \subseteq N(b_1)$, and actually $N(c) \subset N(b_1)$, since b_2 is a neighbor of b_1, but not of c. The equality (ii) applied to c and to b_1 shows that $z_c > z_{b_1}$, thus $z_c = 2$, $z_{b_1} = 1$, and similarly also $z_{b_2} = 1$. If $c' \in N(c)$, then $z_{c'} = 1$ (namely, $z_{c'} = 2$ would imply that c is the only neighbor of c', due to (ii) applied to c'), thus c has precisely two neighbors, say c_1, c_2 (apply (ii) to c). Also, c, b_1, b_2 are the only neighbors of c_1, and also of c_2 (apply (ii) to c_1).

According to (i), there has to be precisely one additional vertex d, with $z_d = 1$, and d can be neighbor only of b_1, b_2 in contrast to (ii). This contradiction shows that the case $z_b \leq 2$ for all $b \in J$ is impossible.

This finishes the consideration of case (a).

(b) <u>Assume that z has two exceptional indices.</u>

At least one of the exceptional vertices is given by a vertex in I, thus there is $t \in I$ with $z_t = D_t\chi(z) = 1$. Let I^t be obtained from I by deleting t and the edges involving t, the extended quadratic form of I^t is denoted by χ^t. Note that $y := \sigma_t(z) = z - e(t)$ is a sincere positive root for χ^t, and we want to recover I and z from I^t and y. First we note

(1) $$D_a\chi^t(y) = D_a\chi(z) - \chi_{at} \qquad \text{for all } a \in I^t.$$

Namely, $D_a\chi^t(y) = D_a\chi(y) = D_a\chi(z) - D_a\chi(e(t))$, and we have $D_a\chi(e(t)) = 2(e(a),e(t)) = \chi_{at}$. Consequently, we have

(2) $D_a\chi^t(y) \leq 0$ for all but at most one $a \in I^t$, and if $D_c\chi^t(y) = 1$ for some $c \in I^t$, then $y_c = 1$.

<u>Proof.</u> Since $\chi_{at} \geq 0$ for all $a \in I^t$, we have $D_a\chi^t(y) \leq D_a\chi(z)$ for all $a \in I^t$. Now, if $D_c\chi^t(y) = 1$, then $D_c\chi(z) = 1$, and therefore c is the second exceptional index besides t, thus c is uniquely determined in I^t, and $y_c = z_c = 1$.

(3) $$D_\omega\chi^t(y) > 0.$$

<u>Proof.</u> This is a direct consequence of 1.2 (3), and the previous assertion.

We are going to list all sincere positive roots y of weakly positive graphical forms χ^t satisfying $D_\omega\chi^t(a) > 0$. This will be easy as soon as we have shown:

(4) χ^t is one of the forms $F(1)$, $F(2)$, $F(3)$, $F(4')$, $F(4)$, $F(5')$, $F(5)$, $F(6)$.

<u>Proof.</u> First, assume y is a maximal sincere positive root of χ^t. According to (3), we know that ω is an exceptional index for y. If y has two exceptional indices, then $y_\omega = 1$, and then obviously $I^t = F(1)$. If ω is the only exceptional index for y, then $y_\omega = 2$, and $D_a\chi^t(y) = 0$ for all $a \in I^t$ implies that $I^t = F(2)$. Thus, assume y is not maximal. In particular, y is not the only sincere positive root for χ^t. Now $\chi^t \in \Phi_{\ell'}$ for some ℓ', and let y' be a maximal sincere positive root of χ^t with $y'_\omega = \ell'$. If we assume that y' has two exceptional indices, then by induction χ^t occurs in table 1. However, the forms in

table 1 with a maximal sincere positive root with two exceptional indices all have only one sincere positive root. This shows that y' has a unique exceptional index, and by induction this is possible only in case χ^t is of the $F(2)$, $F(3)$, $F(4')$, $F(4)$, $F(5')$, $F(5)$ or $F(6)$.

The following table contains all sincere positive roots y (up to automorphisms) of a weakly positive graphical form χ^t such that $D_\omega\chi^t(z) > 0$, and (again up to automorphisms) all vertices c satisfying $D_c\chi^t(y) \geq 0$. We also list the values of the function $D_\cdot\chi^t(y)$. Namely, we can recover I from I^t by using the following rules: we add to I^t a new vertex t with an edge from c to t in case $D_c\chi^t(y) = 0$, and with an edge from a to t, for $a \neq c$ in I, provided $D_a\chi^t(y) = -1$. The last column indicates the graph I obtained from I^t and c by applying these rules. Note that the vector $z = y+e(t)$ obtained in this way is the one listed in table 1.

I^t	y	$D_\cdot\chi^t(y)$	c	I
$F(1)$	$1[\ 1$	$1[\ 1$	$*$	$F(1,1)$
$F(2)$	$2[\ 1\ 1\ 1$	$1[\ o\ o\ o$	$*\ o\ o$	$F(2,2)$
$F(3)$	$3[\ \begin{smallmatrix}11\\11\end{smallmatrix}\ 1$	$1[\ \begin{smallmatrix}oo\\oo\end{smallmatrix}\ o$	$\begin{smallmatrix}*\ o\\o\ o\end{smallmatrix}\ o$	$F(3,3)$
$F(4')$	$4[\ \begin{smallmatrix}1\\1\end{smallmatrix}\ 1\begin{smallmatrix}1\\11\end{smallmatrix}1$	$1[\ \begin{smallmatrix}-1\\-1\end{smallmatrix}\ o\begin{smallmatrix}o\\oo\end{smallmatrix}o$	$\begin{smallmatrix}o\\o\end{smallmatrix}\ \ \begin{smallmatrix}*\\o\end{smallmatrix}\begin{smallmatrix}\\oo\end{smallmatrix}o$	$F(4',5)$
$F(4)$	$4[\ \begin{smallmatrix}1\\11\end{smallmatrix}\ \begin{smallmatrix}1\\1\end{smallmatrix}\ 2$	$1[\ \begin{smallmatrix}o\\oo\end{smallmatrix}\ \begin{smallmatrix}-1\\-1\end{smallmatrix}\ o$	$\begin{smallmatrix}*\\oo\end{smallmatrix}\ \begin{smallmatrix}o\\o\end{smallmatrix}\ o$	$F(4,4)$
$F(5')$	$5[\ \begin{smallmatrix}1\\11\end{smallmatrix}\ 1\begin{smallmatrix}11\\11\end{smallmatrix}1$	$1[\ \begin{smallmatrix}-1\\-1-1\end{smallmatrix}\ o\begin{smallmatrix}oo\\oo\end{smallmatrix}o$	$\begin{smallmatrix}o\\oo\end{smallmatrix}\ \begin{smallmatrix}*o\\oo\end{smallmatrix}o$	$-$
$F(5)$	$5[\ \begin{smallmatrix}1\\11\end{smallmatrix}\ \begin{smallmatrix}21\\21\end{smallmatrix}$	$1[\ \begin{smallmatrix}-1\\-1-1\end{smallmatrix}\ \begin{smallmatrix}oo\\oo\end{smallmatrix}$	$\begin{smallmatrix}o\\oo\end{smallmatrix}\ \begin{smallmatrix}o*\\oo\end{smallmatrix}$	$F(5,6)$
	$4[\ \begin{smallmatrix}1\\11\end{smallmatrix}\ \begin{smallmatrix}11\\11\end{smallmatrix}$	$1[\ \begin{smallmatrix}o\\oo\end{smallmatrix}\ \begin{smallmatrix}-1o\\-1o\end{smallmatrix}$	$\left\{\begin{matrix}\begin{smallmatrix}*\\oo\end{smallmatrix}\ \begin{smallmatrix}oo\\oo\end{smallmatrix}\\[4pt]\begin{smallmatrix}o\\oo\end{smallmatrix}\ \begin{smallmatrix}o*\\oo\end{smallmatrix}\end{matrix}\right.$	$F(4',5)$ $F(5,5)$
$F(6)$	$6[\ \begin{smallmatrix}11\\11\end{smallmatrix}\ \begin{smallmatrix}2\\2\end{smallmatrix}\ 3$	$1[\ \begin{smallmatrix}-1-1\\-1-1\end{smallmatrix}\ \begin{smallmatrix}o\\o\end{smallmatrix}\ o$	$-$	
	$5[\ \begin{smallmatrix}11\\11\end{smallmatrix}\ \begin{smallmatrix}2\\1\end{smallmatrix}\ 2$	$1[\ \begin{smallmatrix}oo\\oo\end{smallmatrix}\ \begin{smallmatrix}o\\-1\end{smallmatrix}\ -1$	$\begin{smallmatrix}*o\\oo\end{smallmatrix}\ \begin{smallmatrix}o\\o\end{smallmatrix}\ o$	$F(5,6)$

Note that for the root $4[\ \begin{smallmatrix}1\ 11\\11\ 11\end{smallmatrix}$ of $F(5)$, there are two essentially different possible choices of c, whereas for $6[\ \begin{smallmatrix}11\ 2\\11\ 2\end{smallmatrix}\ 3$, there is none. One case has to be discussed separately, namely the case $F(5')$. In this case, we obtain from I^t by adding the vertex t and edges from t to those c which satisfy $D_c\chi^t(y) = -1$,

the following graph I

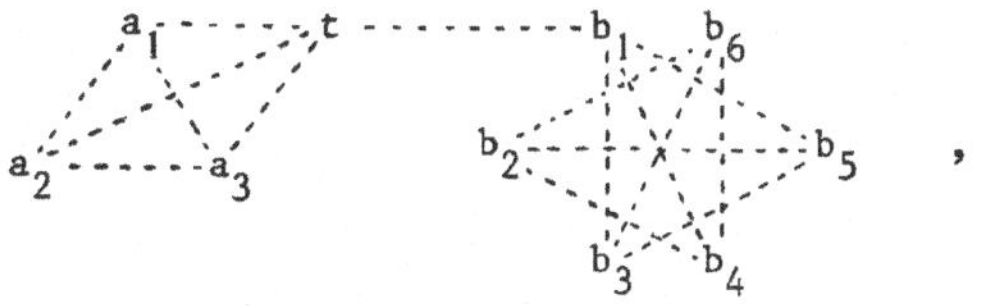

but the full subgraph of I given by $a_1, a_2, a_3, t;$ b_2, b_3, b_4, b_5 is a graph of type $F(5)$, thus the quadratic form given by I cannot be weakly positive.

Note that in all cases, we obtain as $z = y+e(t)$ the root listed in table 1. This finishes the consideration of the case (b).

(c) Assume that z has a unique exceptional index a, that a has two neighbors b and c such that there is no edge from b to c and that there exists a sincere positive root z' with $z'_\omega > z_\omega \geq 5$.

Of course, we may assume that z' is maximal. By Ovsienko's theorem, $z'_\omega = 6$ and there is no sincere positive root z'' with $z''_\omega > z'_\omega$, thus z' is one of the maximal roots classified in the previous parts of the proof. But $z'_\omega = 6$ is only possible in case $F(6)$. However, $F(6)$ has the property that any connected full subgraph is a complete graph, thus it is impossible to find a vertex a having two neighbors b,c which are not joined by an edge. This shows that case (c) cannot occur.

This finishes the proof that all sincere weakly positive graphical forms, as well as all their maximal positive roots (up to automorphism) are listed in table 1.

It is easy to obtain the remaining sincere positive roots of these forms. Namely, given a positive sincere root y, there always exists a maximal sincere positive root z and a sequence $i_1, \ldots, i_m$ such that

$$z > \sigma_{i_1} z > \sigma_{i_2}\sigma_{i_1} z > \ldots > \sigma_{i_m} \ldots \sigma_{i_1} z = y.$$

Actually, in case there is a maximal sincere positive root z with two exceptional indices, z is the only sincere positive root. Namely, the classification of all maximal sincere positive roots shows that z is the only maximal sincere positive root, and applying a reflection σ_a to z, we either keep z, or else we obtain a non sincere root. Thus, the only forms in table 1 with at least two sincere positive roots are the forms $F(2)$, $F(3)$, $F(4')$, $F(4)$, $F(5')$, $F(5)$ and $F(6)$.

1.8 Proof of Theorem 2

Let χ be a critical integral quadratic form in n variables, with $n \geq 3$. Since χ is not weakly positive, there exists some $y = (y_1,\ldots,y_n) > 0$ with $\chi(y) \leq 0$. Since all χ^t are weakly positive, $1 \leq t \leq n$, it follows that $y_t > 0$ for all $1 \leq t \leq n$. Choose such an y with $\sum_{i=1}^{n} y_i$ being minimal (in this case, we will say that y is minimal). We claim that y belongs to $\ker\chi$. Since y is minimal, $\chi(y-e(i)) > 0$ for all $1 \leq i \leq n$, hence

$$0 < \chi(y-e(i)) = \chi(y) + \chi(e(i)) - 2(y,e(i)),$$

thus, using that $\chi(e(i)) = 1$, we see that

$$2(y,e(i)) \leq \chi(y),$$

and therefore

$$2\chi(y) = 2(y,y) = \sum_{i=1}^{n} 2(y,e(i))y_i \leq \sum_{i=1}^{n} \chi(y)y_i = \chi(y) \sum_{i=1}^{n} y_i .$$

If we assume $\chi(y) < 0$, we can divide this inequality by $\chi(y)$ and obtain

$$2 \geq \sum_{i=1}^{n} y_i \geq 3 ,$$

the last inequality being due to the assumption $n \geq 3$ (and using that y is sincere). This shows that $\chi(y) < 0$ is impossible, thus $\chi(y) = 0$. Combining this with $2(y,e(i)) \leq \chi(y)$ shown above, we conclude that $(y,e(i)) \leq 0$, for all i. However,

$$0 = \chi(y) = (y,y) = \sum_{i=1}^{n} (y,e(i))y_i$$

with $y_i > 0$ and $(y,e(i)) \leq 0$ for all i, shows that actually $(y,e(i)) = 0$ for all i. Thus y is a radical vector for χ.

Now assume there is given some $z \in \mathbb{Z}^n$ with $\chi(z) \leq 0$. Consider the ratios $\frac{z_i}{y_i}$, $1 \leq i \leq n$, and choose some a with $\frac{z_a}{y_a} \leq \frac{z_i}{y_i}$ for all i. Thus $z_a y_i \leq z_i y_a$ for all i. Let $x = y_a z - z_a y$. Obviously $x_a = 0$, and $x_i \geq 0$ for all i. Now

$$\chi(x) = \chi(y_a z - z_a y) = \chi(y_a z) = y_a^2 \chi(z) \leq 0,$$

using that y is a radical vector. But $\chi(x) = \chi^a(x) \geq 0$, since χ^a is weakly positive, thus $\chi^a(x) = 0$, hence $x = 0$. This shows that z and y are linearly dependent over $\mathbb{Q}$, and consequently $\chi(z) = 0$. This shows that χ is positive semidefinite, and also that $\operatorname{rad}\chi$ is of rank 1. This finishes the proof.

1.9 Completeness of table 2

Let χ be a critical graphical form, say the extended quadratic form of the graph I. We first show:

(1) There exists $t \in I$ with χ^t being of type $F(2)$, $F(3)$, $F(4')$, $F(4)$, $F(5)$ or $F(6)$ and a positive root y of χ^t such that $\sigma_t(y)$ is a sincere positive root of χ.

Proof. First, we show that there are sincere positive roots for χ. Since χ is not weakly positive, there exists a positive vector x with $\chi(x) \leq 0$, and since χ is critical, x has to be sincere. By the second theorem of Ovsienko, χ is positive semi-definite, thus $\chi(x) = 0$ and x is a radical vector. If $a \in I$ is any vertex, then $e(a) + x$ is a positive root of χ which is sincere.

Let z be a minimal sincere positive root for χ. Note that $D_i\chi(z) \leq 2$ for any $i \in I \cup \{\omega\}$, since

$$0 \leq \chi(z - e(i)) = \chi(z) + \chi(e(i)) - D_i\chi(z) = 2 - D_i\chi(z),$$

and $D_i\chi(z) = 2$ if and only if $z - e(i)$ is a radical vector.

First, assume $D_t\chi(z) = 2$ for some $t \in I \cup \{\omega\}$, thus $z - e(t)$ is a radical vector, and non-zero, since z is sincere. Since any non-zero radical vector of a critical form is sincere, it follows that $z_t \geq 2$. Let $y := \sigma_t(z) = z - 2e(t)$. Note that $z_t = 2$, since otherwise y would be a sincere positive root with $z' < z$. In particular, $y_t = 0$, thus y is a sincere positive root of χ^t. It also follows that $t \in I$, since the only graphical form with a sincere positive root z satisfying $z_\omega = D_\omega\chi(z)$ is $F(1)$, and then $z_\omega = 1$. Since $z - e(t) = y + e(t)$ belongs to $\mathrm{rad}\chi$, we have $D_a\chi(y + e(t)) = 0$, for all $a \in I \cup \{\omega\}$, thus

$$D_a\chi^t(y) = D_a\chi(y) = D_a\chi(y + e(t)) - D_a\chi(e(t)) = -\chi_{at} \leq 0,$$

for all $a \in I \setminus \{t\} = I^t$.

Note that χ^t is a positive definite quadratic form having y as a sincere positive root, thus χ^t occurs in table 1, and has radical rank equal to 0. Also, we claim that χ^t has at least two sincere positive roots. Namely, otherwise y would be a maximal sincere positive root, and there is only one possible case of a maximal sincere positive root y with $D_a\chi^t(y) \leq 0$ for all $a \in I^t$, namely the case of I^t being of type $F(2)$. However, also in this case, χ^t has at least two sincere positive roots. Thus χ^t is of type $F(2)$, $F(3)$, $F(4')$, $F(4)$, $F(5)$ or $F(6)$.

Now assume $D_a\chi(z) \leq 1$ for all $a \in I \cup \{\omega\}$. Choose some $t \in I \cup \{\omega\}$ with

$D_t\chi(z) = 1$ (this is possible due to 1.2 (3)). Let $y := \sigma_t(z) = z - e(t)$. This is a positive root, with $y < z$. The minimality of z therefore shows that y is not sincere, thus $z_t = 1$, and y is a sincere positive root of the positive definite quadratic form χ^t. Note that we may assume $t \in I$. [Namely, if $t = \omega$, then we conclude as above that $\chi = F(1)$, thus we can consider instead the unique vertex in I.]. Now assume y is the only sincere positive root for χ^t, thus χ^t is given by one of the following graphs I^t:

$F(1)$	$F(1,1)$	$F(2,2)$	$F(3,3)$	$F(4,4)$

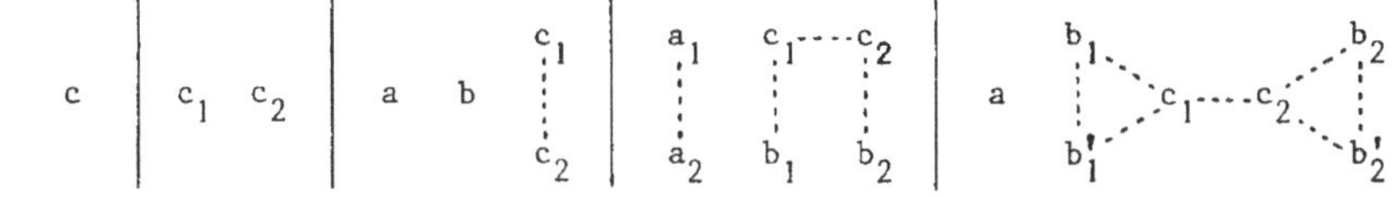

always, we have denoted exceptional vertices of y by c_1, c_2 (or c). In order to try to recover I from I^t, we note the following:

(i) If c is an exceptional vertex of y in I^t, then there is no edge from t to c.

(ii) $\sum_{a \in N(t)} y_a = y_\omega - 1$.

Namely, if c is an exceptional vertex of y in I^t, then $y_c = D_c\chi^t(y) = 1$, thus, if there exists an edge from t to c, then

$$D_c\chi(z) = D_c\chi(y + e(t)) = D_c\chi^t(y) + \chi_{ct} = 1 + 1 = 2,$$

contrary to our assumption. Also

$$1 = D_t\chi(z) = 2z_t - z_\omega + \sum_{a \in N(t)} z_a = 2 - z_\omega + \sum_{a \in N(t)} z_a$$

gives the equality (ii).

Thus, let us consider for the various cases $I^t = F(1), F(1,1), \ldots, F(4,4)$ what we obtain by adding one vertex t and suitable edges satisfying the conditions (i) and (ii). For the coefficients of y, we refer to table 1. For $F(1)$ and $F(1,1)$, we obtain $F(1,1)$, and $F(2)$, respectively. For $F(2,2)$, we have to connect by an edge one of a or b with t, thus we obtain $F(3)$.

<u>Case</u> $I^t = F(3,3)$. We see that t has to be neighbor of precisely two vertices chosen from a_1, a_2, b_1, b_2. If t is neighbor of a_1 and a_2, then we obtain $F(5)$. If t is neighbor of b_1, and b_2, then we obtain $F(4')$. Finally, if t is neighbor say of a_1 and b_1, then a_2, b_2, c_1, t is a full subgraph of type $C(2)$.

Case I^t = F(4,4). We have to add a vertex t with edges from t either to three of the vertices b_1, b_1', b_2, b_2' (the corresponding coefficients of y being 1) or else an edge from t to a (note that y_a = 2) and one from t to one of the vertices b_1, b_1', b_2, b_2'. If t is neighbor of b_1, b_1', and b_2, we see that a, b_2', c_1, t is a full subgraph of type $C(2)$, in case t is neighbor of a and say b_1, then $a \ldots t, b_1' \ldots c_1, b_2 \ldots b_2'$ is a full subgraph of type $C(3)$.

In either case, we see that any extension of I^t by a vertex and edges satisfying (i) and (ii), is either positive definite, or else has a proper restriction to a coordinate subspace which is positive semi-definite of radical rank 1. Thus, we cannot obtain as such an extension a critical form. This finishes the proof of (1).

Thus, we have to consider for I^t the following cases

F(2)	F(3)	F(4')
a b c	a_1 b_1 ⋮ ⋮ c a_2 b_2	a_1 b_1 b_2 b_5 ⋮ b_3-b_4 a_2
F(4)	**F(5)**	**F(6)**
a_1 b_1 a_2-----a_3 b_2 c	a_1 c_1----c_2 a_2----a_3 b_1 b_2	a_1---a_4 b_1 a_2----a_3 b_2 c

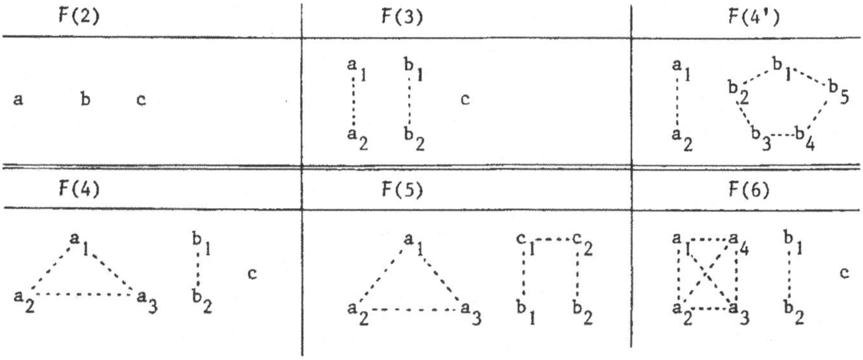

and I is obtained from I^t by adding one vertex t and suitable edges joining t with some vertices of I^t. According to (1), there exists a sincere positive root, y of χ^t such that $\sigma_t(y)$ is a sincere positive root of χ. We will see that for all possible choices of edges, satisfying this last condition we obtain either a completion of one of the graphs in table 1 or 2, or else a proper full subgraph J with χ_J having a non-zero radical vector. Note that a proper completion of a graph from table 1 or 2 always is weakly positive, according to the lemma in 1.5. Thus the only critical graphical forms are those in table 2. We consider now the various cases separately.

Case I^t = F(2). Any graph obtained from F(2) by adding one vertex and a number of edges is a completion of $C(2)$.

Case I^t = F(3). If t is a neighbor of c, then I is a completion of $C(3)$. Thus assume t is not neighbor of c. If t is not neighbor of some a_i, and also not neighbor of some b_j, then a_i, b_j, c, t is a full subgraph of type $C(2)$. If t is neighbor of both a_1, a_2 (or of both b_1, b_2), then I is a completion of F(4).

<u>Case</u> $I^t = F(4')$. If t is neighbor of both a_1, a_2, then I is a completion of $C(4')$. Thus , assume there is no edge from t to a_1, If there is a triple b_i, b_j, t without any edge, then a_1, b_i, b_j, t is a full subgraph of typ $C(2)$. Thus, we may assume that t is a neighbor of b_1, b_2, b_3. If there is an edge from t to a_2, then the full subgraph given by a_2, b_1, b_3, t is of type $N(2)$ (introduced in 1.5). Thus, assume there is no edge from t to a_2. If there is neither an edge from t to b_4 nor to b_5, then $a_1 \cdots\cdots a_2$, $b_4 \cdots\cdots b_5$, $b_2 \cdots\cdots t$ is of type $C(3)$. If there is an edge from t to b_4, then consider a positive root y of χ^t such that $z = \sigma_t y$ is a sincere positive root of χ. Then $z_t = D_t\chi(z)$, thus

$$0 = z_t - z_\omega + \sum_{c \in N(t)} z_c \geq z_t - z_\omega + \sum_{i=1}^{4} z_{b_i}.$$

Since $z_\omega \leq 4$, and $z_t \geq 1$, $z_{b_i} \geq 1$ for all i, we obtain a contradiction.

<u>Case</u> $I^t = F(4)$. If t is a neighbor both of b_1, b_2, then I is a completion of $C(4)$. If t is a neighbor of a_1, a_2, a_3, then I is a completion of $F(6)$. Thus, assume that there is no edge from t to b_1, and no edge from t to a_1. If there is also no edge from t to c, then a_1, b_1, c, t give a full subgraph of type $C(2)$. Thus, assume t is a neighbor of c. If there is an edge from t to b_2, then I is a completion of $F(5)$. Thus, assume t is not a neighbor of b_2. If there is no edge from t to one of a_2, a_3, say t is not a neighbor of a_2, then $a_1 \cdots a_2$, $b_1 \cdots b_2$, $c \cdots t$ determine a full subgraph of type $C(3)$. Thus there remains the possibility of t being neighbor precisely of a_2, a_3, c. Let y be a positive root of χ_t with $z = \sigma_t(y)$ being a sincere positive root of χ. Since $z_t = D_t\chi(z)$, we have

$$0 = z_t - z_\omega + z_{a_2} + z_{a_3} + z_c .$$

However, for any positive root y of $F(4)$, we have $y_\omega - y_c \leq 2$ (since $y_\omega - y_c$ is the coefficient of the root $\sigma_c(y)$ at the vertex c). Using that $z_\omega = y_\omega$, $z_c = y_c$, we obtain

$$z_t - z_\omega + z_{a_2} + z_{a_3} + z_c \geq z_t + z_{a_2} + z_{a_3} - 2 \geq 1 ,$$

due to the fact that z is sincere. This contradiction shows that the last possibility cannot happen.

<u>Case</u> $I^t = F(5)$. If t is a neighbor of all a_i, $1 \leq i < 3$, then I is a completion of $C(5)$. Thus, assume there is no edge from t to a_1. If t is a neighbor both of b_1, b_2, then I is a completion of $C(4')$, thus we may assume there is no edge from t to b_1. If there is no edge from t to b_2 or c_2, then a_1, b_1, b_2, t, or a_1, b_1, c_2, t respectively, are full subgraphs of type $C(2)$.

Thus, t is a neighbor both of b_2 and of c_2. If there is neither an edge from t to a_2 nor to a_3, then a_1,a_2,a_3; b_2,c_2,t; b_1 give a full subgraph of type $C(4)$. Then, we may assume that t is a neighbor of a_2. If there is an edge from t to c_1, then the full subgraph given by a_2,c_1,b_2,t is of type $N(2)$. Thus, assume there is no edge from t to c_1. If there also is no edge from t to a_3, then $a_1 \cdots a_3$, $b_1 \cdots c_1$, $t \cdots b_2$ is a full subgraph of type $C(3)$. If there is an edge from t to a_3, then the full subgraph given by a_2,a_3,t,b_2,c_2,b_1 is of type $N(3)$.

Case $I^t = F(6)$. If t is a neighbor of all a_i, $1 \leq i \leq 4$, then I is a completion of $C(6)$. Thus, assume there is no edge from t to a_1. Consider first the case of t and c not being neighbors. Then there are edges from t both to b_1, and b_2, since otherwise we obtain a full subgraph a_1,b_i,c,t of type $C(2)$. If t is a neighbor of at least two of the a_i, say of a_2 and a_3, then a_2,a_3,t,b_2,b_3,c is a full subgraph of type $N(3)$. If t is a neighbor of at most one a_i, say there are no edges from t to a_1,a_2,a_3, then a_1,a_2,a_3; b_1,b_2,t; c is a full subgraph of type $C(4)$. Second, consider the case of t and c being neighbors. If t is a neighbor of at least one b_i, then I is a completion of $F(5)$. Thus, assume that there is no edge from t neither to b_1 nor to b_2. If t is a neighbor of a_2,a_3,a_4, then a_2,a_3,a_4,t,c; b_1,b_2 is of type $N(4)$. Thus, we can assume that there is no edge from t to a_2. Then $a_1 \cdots a_2$, $b_1 \cdots b_2$, $t \cdots c$ is a full subgraph of type $C(3)$.

This finishes the proof that any critical graphical form occurs in table 2.

1.10 The extended quadratic form of a finite partially ordered set

The graphical forms we will encounter all will be derived from finite partially ordered sets. It will be convenient to have available the specialization of the tables 1 and 2 to this case.

Let $S = (S,\leq)$ be a partially ordered set. From S, we construct a graph $I(S) = (I_0,I_1)$ where $I_0 = S$, and I_1 is the set of subsets $\{a,b\} \subseteq S$ with $a < b$. For example, starting with S being a chain of r elements, we obtain as $I(S)$ the complete graph on r vertices. Given a partially ordered set $S = (S,\leq)$ we denote by $S^* = (S,\leq)$ the opposite partially ordered set having the same underlying set as $(S,\leq)$, and with $a \leq b$ in S^* if and only if $b \leq a$ in S. Of course, we have $I(S^*) = I(S)$ for any partially ordered set S. In drawing (the Hasse diagram of) a partially ordered set, we will deviate from the usual convention: In case a,b are elements of S with $a < b$, and such that $a \leq c \leq b$ implies a = c or c = b, we connect a and b by a line <u>going from left to right</u> (and not from below upwards, as usual; the reasons for this deviation will become clear later: on the one hand, most partially ordered sets considered will be derived from Auslander-Reiten quivers, the relation $\leq$ being derived from the arrows $\longrightarrow$ which usually are drawn from left to right. Also, we will associate with any partially ordered set a vectorspace category, and the relation $\leq$ there will mean nothing else than the existence of a non-trivial map $\longrightarrow$.).

Given a finite partially ordered set S, the quadratic form $\chi_{I(S)}$ will be denoted just by χ_S and called the <u>extended quadratic form</u> of S. Let us tabulate the partially ordered sets S with χ_S being either sincere weakly positive, or critical. These partially ordered sets S will usually be denoted by the type of I(S), with the following exception: there are two partially ordered sets, denoted by $\mathcal{F}(5,6)$ and $\mathcal{F}(6,5)$, and being opposite to each other, with corresponding graph $\mathcal{F}(5,6)$. Again, the number of sincere roots is listed in column s, the radical rank of χ_S in column r.r.. In the last column $\mathbb{T}$, we have added a star with quadratic form being $\mathbb{Z}$-equivalent to χ_S.

<u>Table 1'</u> <u>The partially ordered sets S with</u> χ_S <u>sincere, weakly positive.</u>

notation	S	s	r.r.	$\mathbb{T}$	notation	S	s	r.r.	$\mathbb{T}$
$\mathcal{F}(1)$		1	0	$\mathbb{A}_2$	$\mathcal{F}(1,1)$		1	0	$\mathbb{A}_3$
$\mathcal{F}(2)$		2	0	$\mathbb{D}_4$	$\mathcal{F}(2,2)$		1	0	$\mathbb{D}_5$
$\mathcal{F}(3)$		3	0	$\mathbb{E}_6$	$\mathcal{F}(3,3)$		1	0	$\mathbb{E}_7$
$\mathcal{F}(4)$		5	0	$\mathbb{E}_7$	$\mathcal{F}(4,4)$		1	0	$\mathbb{E}_8$
$\mathcal{F}(5)$	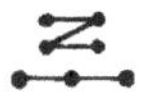	9	0	$\mathbb{E}_8$	$\mathcal{F}(5,5)$		1	1	$\tilde{\mathbb{E}}_8$
$\mathcal{F}(6)$		14	0	$\mathbb{E}_8$	$\mathcal{F}(5,6)$		1	1	$\tilde{\mathbb{E}}_8$
					$\mathcal{F}(6,5)$		1	1	$\tilde{\mathbb{E}}_8$

Table 2' The partially ordered sets S with χ_S critical.

notation	S	T
C(2)		$\tilde{\mathbb{D}}_4$
C(3)		$\tilde{\mathbb{E}}_6$
C(4)		$\tilde{\mathbb{E}}_7$
C(5)		$\tilde{\mathbb{E}}_8$
C(6)		$\tilde{\mathbb{E}}_8$

Proof. Given a partially ordered set S with I(S) occuring in table 1 or 2, we have to show that S is as listed above. Of course, I(S) is the disjoint union of complete graphs, if and only if S is the disjoint union of chains with corresponding lengths. Let us consider in detail one other example: assume I(S) contains as a full subgraph

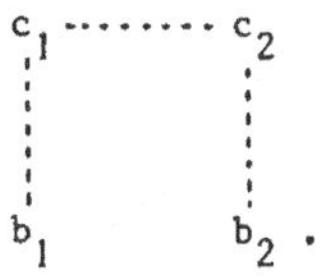

Then b_1 and c_1 are comparable in S. Assume first $b_1 < c_1$. Since c_1 and c_2 are comparable in S, whereas b_1 and c_2 are not comparable, we must have $c_2 < c_1$. Again, b_2 and c_2 are comparable, whereas c_1 and b_2 are not comparable, thus $c_2 < b_2$. Thus, in S we have the following subset

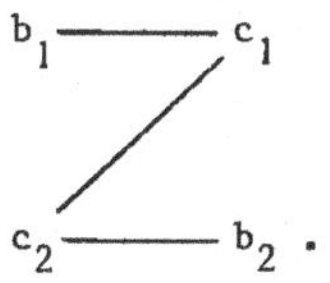

On the other hand, the assumption $c_1 < b_1$ implies that the subset in S given by b_1,b_2,c_1,c_2 is

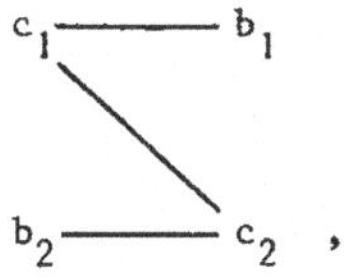

thus of the same shape as the previous partially ordered set. This finishes the consideration of the cases $F(3,3)$, $F(5)$ and $C(5)$.
Now assume, there is given an additional vertex d in $I(S)$ which is a neighbor both of b_1 and b_2, but neither of c_1 nor of c_2, thus

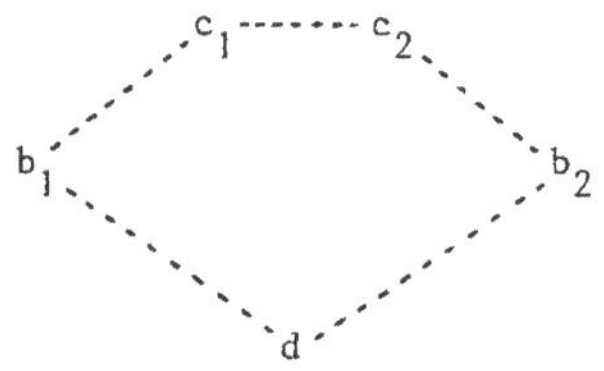

Without loss of generality, we may assume $b_1 < c_1$. Now b_1 and d are neighbors and c_1 and d are not, thus we must have $b_1 < d$. On the other hand, $c_2 < b_2$, and d is a neighbor of b_2 but not of c_2, thus $d < b_2$. From $b_1 < d < b_2$ we conclude that b_1, b_2 have to be neighbors in $I(S)$, a contradiction. This shows that $I(S)$ cannot be of the form $F(4')$, $F(4',5)$ or $C(4')$.

Similarly, we claim that

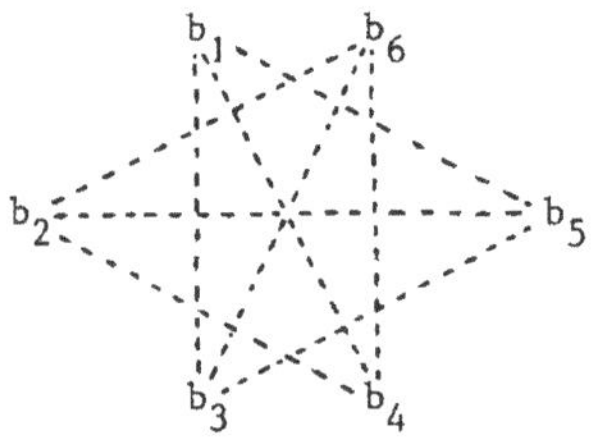

cannot be of the form $I(S)$ for a partially ordered set S. Namely, any b_i has precisely 3 neighbors; for example, b_1 has the neighbors b_3, b_4, b_5, and we claim that either $b_1 < b_3$, $b_1 < b_4$, $b_1 < b_5$, or else $b_3 < b_1$, $b_4 < b_1$, $b_5 < b_1$. Namely, if $b_1 < b_4$, then also $b_1 < b_3$, since b_3, b_4 are incomparable, and also $b_1 < b_5$, since b_4, b_5 are incomparable. Now assume $b_1 < b_3$. Then, considering the set of neighbors of b_3, we have $b_5 < b_3$. Finally, considering the set of neighbors of b_5, we conclude $b_5 < b_1$, a contradiction to $b_1 < b_5$.

It remains to check that the graphs $F(4,4)$, $F(5,5)$ and $F(5,6)$ can be realized in the form $I(S)$ precisely for the partially ordered sets S denoted by $F(4,4)$, $F(5,5)$, $F(5,6)$ and $F(6,5)$. This finishes the proof that the tables are complete.

In order to verify that χ_S is $\mathbb{Z}$-equivalent to the quadratic form defined by the respective star, we can refer to lemma 2 in 1.3, provided S is the disjoint union of chains, or, equivalently, $I(S)$ is the disjoint union of complete graphs. Since the quadratic form $C(5)$ has coordinate subspaces of the form $F(5)$ and $F(6)$, both being direct complements to the radical of the form, it follows that $F(5)$ is $\mathbb{Z}$-equivalent to $F(6)$, and $C(5)$ to $C(6)$. The radical of the quadratic form $F(5,5)$ is generated by

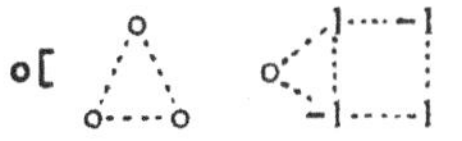

and it is complemented by a coordinate subspace of type $F(5)$, thus $F(5,5)$ is $\mathbb{Z}$-equivalent to $\tilde{\mathbf{E}}_8$. Similarly, the radical of the quadratic form $F(5,6)$ is generated by

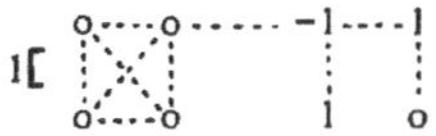

and it is complemented by a coordinate subspace of type $F(6)$, thus also $F(5,6)$ is $\mathbb{Z}$-equivalent to $\tilde{\mathbf{E}}_8$. Finally, we have seen in 1.5 that $F(4,4)$ is the coordinate subspace of a positive semidefinite form χ which is a direct complement to $\operatorname{rad}\chi$ such that some other direct complement of $\operatorname{rad}\chi$ is $\mathbb{Z}$-equivalent to $\mathbf{E}_8$. As a consequence, $F(4,4)$ itself is $\mathbb{Z}$-equivalent to $\mathbf{E}_8$. This finishes the proof of the tables.

Given a partially ordered set S, a subset S' of S is said to be <u>convex</u> provided for $a_1, a_2 \in S'$, and $a_1 < x < a_2$, also $x \in S'$. [Recall that a subset I of S is called an ideal of S provided $a \in I$ and $x < a$ imply $x \in I$; a subset F of S is called a filter of S provided $a \in F$ and $a < x$ imply $x \in F$. Note that a subset S' of S is convex if and only if there exists an ideal I and a filter F such that $S' = S \setminus (I \cup F)$.]

<u>Corallary</u>. Let S be a finite partially ordered set with χ_S not weakly positive. There exists a convex subset S' of S such that $\chi_{S'}$ is critical.

<u>Proof</u>. We can assume that S does not contain four pairwise incomparable elements, since otherwise we obtain $C(2)$ as a convex subset. By definition of a critical quadratic form, there always will exist a subset S'' (with the induced partial ordering) such that $\chi_{S''}$ is critical. Let S'' be of type $C(i)$. Let T be the convex subset of S generated by S''. Assume we have chosen S'' such that T is minimal. Consider first the cases $C(i)$, with $i = 3,4,6$. We claim that T is obtained from S'' by replacing any neighboring pair $a < b$ by a chain $a = a_1 < a_2 < \ldots < a_n = b$. Namely, let $a < b$ in S'' and assume there are incomparable elements c_1, c_2 in S with $a < c_i < b$ for $i = 1,2$. There are incomparable elements d,e in S'' which are incomparable both with a and b. But then c_1, c_2, d, e form a subset of type $C(2)$, a contradiction. As a consequence, T contains a subset S' again of type $C(i)$, which is convex in T and therefore in S.

Next, consider the case $C(5)$, say S'' being given by the vertices

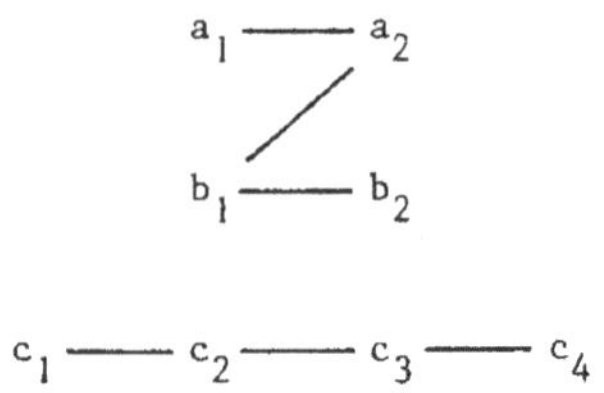

If $b_1 < d < a_2$ for some $c \in S$, then d must be comparable with a_1 or b_2 (since otherwise a_1, d, b_2, c_1 form a subset $C(2)$), thus $a_1 < d$, or $d < b_2$, and, of course, not both. In case $a_1 < d$, we consider instead of S'' the subset $\{a_1, d, b_1, b_2, c_1, c_2, c_3, c_4\}$. It generates a strictly smaller convex subset in S, contrary to the choice of S''. We argue similarly in case $d < b_2$. Thus, b_1 and a_2 are neighbors in S. As above, we see that T is obtained from S'' by replacing the neighboring pairs $a_1 < a_2$, $b_1 < b_2$, and $c_i < c_{i+1}$, $i = 1,2,3$, by chains, and therefore T contains a subset S' again of type $C(5)$ which is convex in T, and therefore in S. This finishes the proof.

<u>References</u>

The two theorems of Ovsienko (both the statements and the proofs) are taken from [Ov], also using oral communications by Ovsienko. Note that Ovsienko considers not only integral quadratic forms, but, more generally, quadratic forms with integer coefficients, thus dealing also with the quadratic forms which arise in representation theory when the base field is not algebraically closed. Drozd's lemma 1.1.2 is in [D1]. - For a treatment of the root system of a positive definite integral quadratic form, we refer to Bourbaki [Bou]. - The classification of the graphical forms which are either sincere and weakly positive, or critical, seems to be new.

2. Quivers, module categories, subspace categories (Notation, results, some proofs).

The aim of this chapter is to formulate some of the fundamental notions and results of the representation theory of finite-dimensional algebras. In as far as these results are well-presented in the literature, we only will state them and derive those consequences which are needed for our further investigations. The remaining topics will be treated in more detail.

We usually will work over an algebraically closed field k of arbitrary characteristic. Actually, all results presented here can be extended to the case of a non-algebraically-closed base field, and it should be easy to outline the necessary changes both in the formulations and the proofs for doing so. We apologize that this means for the interested reader to go again through the proofs, rewriting all the arguments for the general situation. However, the consideration of the general situation from the beginning would have made the presentation rather technical and clumsy. We hope that the restriction to the case of an algebraically closed base field makes these notes easier to read and to understand, and that the extension to the general case will be an easy exercise.

By $X \approx Y$ we denote that X,Y are isomorphic as abelian groups, or modules, or rings, whatever X,Y will be. By D, we denote the duality with respect to the base field k. Note that for a k-algebra A, and ${}_AX$ a left A-module, $D({}_AX)$ will be a right A-module, and similarly, for Y_A a right A-module, $D(Y_A)$ will be a left A-module.

We denote by $\mathbb{Z}$ the set of rational integers, by $\mathbb{N}_0$ and $\mathbb{N}_1$ the subsets of all non-negative, or positive integers, respectively. Let $\mathbb{Q}$ be the set of all rational numbers, and $\mathbb{Q}_0^+$ and $\mathbb{Q}^+$ the subsets of all non-negative, or positive rational numbers, respectively. Also, let $\mathbb{Q}_0^\infty = \mathbb{Q}_0^+ \cup \{\infty\}$. Given an algebraically closed field k, let $\mathbb{P}_1k = k \cup \{\infty\}$ denote the projective line over k.

2.1 Quivers and translation quivers

A quiver $\Delta = (\Delta_o,\Delta_1)$, or more precisely, $\Delta = (\Delta_o,\Delta_1,s,e)$ is given by two sets Δ_o,Δ_1 and two maps $s,e : \Delta_1 \longrightarrow \Delta_o$; the set Δ_o is called the set of vertices or points, the set Δ_1 is called the set of arrows, and given an arrow $\alpha \in \Delta_1$, then $s(\alpha)$ is called its starting point, and $e(\alpha)$ its end point; usually, we will write $\alpha : a \longrightarrow b$ or $a \xrightarrow{\alpha} b$, where $a = s(\alpha)$, $b = e(\alpha)$, and say that α points from a to b. The quiver Δ is said to be finite provided both Δ_o and Δ_1 are finite sets. Given a quiver $\Delta = (\Delta_o,\Delta_1,s,e)$, its underlying graph $\bar{\Delta}$ is obtained from Δ by forgetting the orientation of the arrows, thus $\bar{\Delta} = (\Delta_o,\Delta_1,\{s,e\})$, where $\{s,e\}$ is the function on Δ_1 mapping any $\alpha \in \Delta_1$ to the set $\{s(\alpha),e(\alpha)\}$. Note that $\bar{\Delta}$ may have loops and multiple edges. Given a quiver, $\Delta = (\Delta_o,\Delta_1)$, and $\Delta'_o \subseteq \Delta_o$, the full subquiver Δ' of Δ defined by Δ'_o is given by $\Delta' = (\Delta'_o,\Delta'_1)$, where Δ'_1 is the set of all arrows of Δ having both starting point and endpoint in Δ'_o. We say that Δ has no multiple arrows in case for any $x,y \in \Delta_o$, there is at most one arrow from x to y.

A path of length $\ell \geq 1$ from x to y in a quiver Δ is of the form $(x \mid \alpha_1,\ldots,\alpha_\ell \mid y)$ with arrows α_i satisfying $e(\alpha_i) = s(\alpha_{i+1})$ for all $1 \leq i < \ell$, such that x is the starting point of α_1, and y the endpoint of α_ℓ. In addition, we also define for any vertex x of Δ a path of length 0 (from x to itself), denoted by $(x|x)$. A path of length ≥ 1 from x to x is called a cyclic path. Of course, we can identify the paths of length 1 with the corresponding arrows of Δ. In case there exists a path from x to y, we say that x is a predecessor of y, and y a successor of x. In case there is an arrow $x \longrightarrow y$, we say that x is a direct predecessor of y, and y a direct successor of x. For $z \in \Delta_o$, we denote by z^- the set of direct predecessors of z, and by z^+ the set of direct sucessors of z. The direct predecessors and the direct successors of a vertex z are called the neighbors of z. The quivers we will consider always will be assumed to be locally finite, thus, by definition, any $z \in \Delta_o$ has only finitely many neighbors. A subquiver Δ' of Δ is said to be a convex subquiver provided any path $(x|\alpha_1,\ldots,\alpha_\ell|y)$ with $x,y \in \Delta'_o$ is completely contained in Δ'.

Given a quiver Δ, we define inductively full subquivers ${}_d\Delta$, with $d \geq -1$. Let ${}_{-1}\Delta$ be the empty quiver, and, for $d \geq 0$, let ${}_d\Delta$ be the full subquiver of Δ given by the vertices x with x^- contained in ${}_{d-1}\Delta$. Finally, let ${}_\infty\Delta$ be the union of all ${}_d\Delta$.

(1) *Let Δ be a (locally finite) quiver. If x is a vertex in ${}_\infty\Delta$, then x has only finitely many predecessors.*

<u>Proof</u>, by induction on d, where $x \in {}_d\Delta$. If $d = 0$, then x is the only predecessor of itself. If $d > 0$, let $y_1,\dots,y_r$ be the direct predecessors of x. Since $y_i \in {}_{d-1}\Delta$, we know by induction that any y_i has only finitely many predecessors. Since any predecessor of x different from x is also a predecessor of some y_i, it follows that x has only finitely many predecessors.

Given a quiver Δ, we can define its <u>path category</u>: it is an additive category, with objects being direct sums of indecomposable objects. The indecomposable objects in the path category of Δ are given by the vertices of Δ, and given $a,b \in \Delta_o$, the set of maps from a to b is given by the k-vectorspace with basis the set of all paths from a to b. The composition of maps is induced from the usual composition of paths:

$$(a|\alpha_1,\dots,\alpha_\ell|b)(b|\beta_1,\dots,\beta_s|c) = (a|\alpha_1,\dots,\alpha_\ell,\beta_1,\dots,\beta_s|c),$$

where $(a|\alpha_1,\dots,\alpha_\ell|b)$ is a path from a to b, and $(b|\beta_1,\dots,\beta_s|c)$ is a path from b to c. Similarly, we define the <u>path algebra</u> of Δ as being given by the k-vectorspace with basis the set of all paths in Δ. As above the product of two composable paths is defined to be the corresponding composition, the product of two non-composable paths is, by definition, zero. In this way, we obtain an associative algebra which has a unit element if and only if Δ_o is finite (the unit element being given by $\sum_{x \in \Delta_o} (x|x)$.) Note that the path algebra of Δ is finite dimensional if and only if, first of all, Δ is finite and, in addition, there is no cyclic path in Δ. We do not introduce a symbol for the path category of the quiver Δ, and $k\Delta$ will denote the path <u>algebra.</u> In $k\Delta$, we denote by $k\Delta^+$ the ideal generated by all arrows. Note that $(k\Delta^+)^n$ is the ideal generated by all paths of length $\geq n$.

We recall that a finite dimensional k-algebra A (k algebraically closed!) is called <u>basic</u> provided $A/\mathrm{rad}\, A$ is a product of copies of k. Any finite dimensional k-algebra A is Morita equivalent to a (uniquely determined) basic algebra A' (two algebras A,A' being said to be Morita equivalent provided the category of all A-modules is equivalent to the category of all A'-modules). There is the following structure theorem for basic algebras:

(2) (Gabriel) <u>Any basic finite dimensional</u> k-<u>algebra is of the form</u> $k\Delta/I$ <u>for some uniquely determined finite quiver</u> Δ <u>and some ideal</u> I <u>with</u> $(k\Delta^+)^n \subseteq I \subseteq (k\Delta^+)^2$, <u>for some</u> $n \geq 2$.

Given vertices $a,b \in \Delta_o$, a finite linear combination $\sum_w c_w w$ with $c_w \in k$, where w are paths of lengths ≥ 2 from a to b, is called a <u>relation</u> on Δ. Any ideal $I \subseteq (k\Delta^+)^2$ can be generated, as an ideal, by relations; if it is generated as an ideal by the set $\{\rho_i|i\}$ of relations, we write $I = \langle\rho_i|i\rangle$.

Example: Given a quiver Δ, a commutativity relation is a relation of the form $w-w'$, where w,w' are paths having the same starting point, and the same end point. A relation of the form w (given by the single path w) is called a zero relation.

Given a quiver $\Delta = (\Delta_0,\Delta_1)$, let Δ^* be the opposite quiver, with same set of vertices $(\Delta^*)_0 = \Delta_0$, and with a bijection $*$ between the arrows of Δ and those of Δ^* such that $\alpha^* : i \longrightarrow j$ if and only if $\alpha : j \longrightarrow i$. Given a path $w = (a\,|\,\alpha_1,\ldots,\alpha_\ell\,|\,b)$, let $w^* = (b\,|\,\alpha_\ell^*,\ldots,\alpha_1^*\,|\,a)$, given a relation $\rho = \sum_w c_w w$, let $\rho^* = \sum_{w^*} c_w w^*$. Note that the path algebra $k\Delta^*$ is just the opposite algebra of the path algebra $k\Delta$. More generally, if $\{\rho_i \mid i\}$ is a set of relations for Δ generating the ideal I in $k\Delta$, and $\{\rho_i^* \mid i\}$ generates the ideal I^* in $k\Delta^*$, then $k\Delta^*/I^*$ is the opposite algebra of $k\Delta/I$.

Given a quiver $\Delta = (\Delta_0,\Delta_1)$, a representation $V = (V_x,V_\alpha)$ of Δ over k is given by vector spaces V_x, for all $x \in \Delta_0$, and linear maps $V_\alpha : V_x \longrightarrow V_y$, for any arrow $\alpha : x \longrightarrow y$. Instead of writing V_α, we usually will write just α. If V,V' are two representations of Δ over k, a map $f = (f_x) : V \longrightarrow V'$ is given by maps $f_x : V_x \longrightarrow V'_x$ (where $x \in \Delta_0$), such that $f_x V_\alpha = V'_\alpha f_y$ for any $\alpha : x \longrightarrow y$. In this way, we obtain the category of representations of Δ. Now assume there is given a representation V of Δ. If $w = (a\,|\,\alpha_1,\ldots,\alpha_\ell\,|\,b)$ is a path in Δ, we may denote by V_w the composition $V_w = V_{\alpha_1} \cdots V_{\alpha_\ell} : V_a \longrightarrow V_b$. We say that V satisfies the relation $\rho = \sum_w c_w w$, provided $\sum_w c_w V_w = 0$ (note that, by definition, all paths w occurring in the relation have a fixed starting point, say a, and a fixed endpoint, say b, thus all V_w are linear maps from V_a to V_b, and we form the linear combination $\sum c_w V_w$ in $\operatorname{Hom}_k(V_a,V_b)$). Given any quiver Δ and a set of relations $\{\rho_i \mid i\}$, the category of all representations of Δ satisfying the relations ρ_i is abelian.

Now assume Δ is a quiver with only finitely many vertices (so that $k\Delta^*$ has a unit element: we want to consider $k\Delta^*$-modules, and, always modules will be assumed to be unital). In this case, the category of representations of Δ can be identified with the category $k\Delta^*$-mod of $k\Delta^*$-modules. Namely, given a representation $V = (V_x,V_\alpha)$ of Δ, the corresponding $k\Delta^*$-module is given by $\bigoplus_x V_x$, the direct sum being formed over all $x \in \Delta_0$, its elements may be written in the form $\sum_x v_x$, with $v_x \in V_x$; and the $k\Delta^*$-module structure on $\bigoplus_x V_x$ is given as follows: if $t \in \Delta_0^*$ ($= \Delta_0$), then $(t|t)\sum v_x = v_t$, and if $\alpha^* \in \Delta_1^*$, say with $\alpha : j \longrightarrow i$, then $\alpha^* \sum v_x = v_j V_\alpha$. Conversely, given a $k\Delta^*$-module M, let $M_x = (x|x)M$, for all $x \in \Delta_0$ ($= \Delta_0^*$), and for $\alpha : j \longrightarrow i$, let $M_\alpha : M_j \longrightarrow M_i$ be the multiplication by $\alpha^*(\in k\Delta^*)$. It is easily verified that a representation $V = (V_x,V_\alpha)$ of Δ satisfies a relation ρ if and only if the corresponding $k\Delta^*$-module is annihilated by ρ^*. Given a set of relations $\{\rho_i \mid i\}$ for Δ, let

$$A(\Delta,\{\rho_i \mid i\}) := k\Delta^*/\langle\rho_i^* \mid i\rangle,$$

and we call this algebra $A(\Delta,\{\rho_i \mid i\})$ <u>the algebra given by the quiver</u> Δ <u>and the relations</u> ρ_i. In view of the previous considerations, the $A(\Delta,\{\rho_i \mid i\})$-modules can be identified with the representations of Δ satisfying the relations ρ_i. Note that in dealing with the representations of a quiver Δ, it seems convenient to draw the arrows of Δ from right to left.

Given a finite-dimensional algebra A, then A is Morita equivalent to an algebra $A(\Delta,\{\rho_i \mid i\})$, with $(k\Delta^+)^n \subseteq \langle\rho_i \mid i\rangle$, for some $n \geq 2$. We note that Δ is uniquely determined by A, and will be called the <u>quiver</u> $\Delta(A)$ <u>of</u> A. Let us give a recipe for obtaining $\Delta(A)$ from A. Starting with a quiver Δ and a set of relations $\{\rho_i \mid i\}$ satisfying $(k\Delta^+)^n \subseteq \langle\rho_i \mid i\rangle$, it is easy to see that the only simple representations of Δ satisfying the relations ρ_i are the one-dimensional representations $E(a)$ defined for any vertex a, with $E(a)_x = k$ for $x = a$ and $= 0$ otherwise, and with all maps $E(a)_\alpha = 0$ [we recall that in an abelian category, an object is said to be simple provided it is non-zero, and zero is its only proper subobject]. Of course, $E(a)$ and $E(b)$ are isomorphic only for $a = b$. As a consequence, the vertices of Δ correspond bijectively to the (isomorphism classes of) simple A-modules. Also, given vertices a,b, any arrow $\alpha : a \longrightarrow b$ gives rise to a two-dimensional representation $M(\alpha)$ defined as follows: in case $a \neq b$, we set $M(\alpha)_x = k$ for $x = a$ or $x = b$, and $= 0$ otherwise, and $M(\alpha)_\beta = 1$ for $\beta = \alpha$, and $= 0$ otherwise; whereas in case $a = b$, let $M(\alpha)_x = k^2$ for $x = a$, and $= 0$ otherwise, and $M(\alpha)_\beta = \begin{bmatrix} 0 & 1 \\ 0 & 0 \end{bmatrix}$ for $\beta = \alpha$, and $= 0$ otherwise. It follows that there is an exact sequence

$$0 \longrightarrow E(b) \longrightarrow M(\alpha) \longrightarrow E(a) \longrightarrow 0,$$

and the exact sequences defined by $M(\alpha_1),\dots,M(\alpha_s)$, where $\alpha_1,\dots,\alpha_s$ are the arrows from a to b, form a k-basis of $\mathrm{Ext}^1(E(a),E(b))$. Thus, the number of arrows from a to b is given by $\dim_k \mathrm{Ext}^1(E(a),E(b))$.

Also, given a vertex a of Δ, there is the idempotent $(a \mid a)$ in $A(\Delta,\{\rho_i \mid i\})$, and we may consider the $A(\Delta,\{\rho_i \mid i\})$-module $P(a) := A(\Delta,\{\rho_i \mid i\})(a \mid a)$; (note, it is an indecomposable projective module, see 2.4). Any arrow $\alpha : a \longrightarrow b$ gives rise to a map

$$P(\alpha^*) : P(b) \longrightarrow P(a) ,$$

given by right multiplication by α^*. Given a representation V of Δ satisfying the relations ρ_i, we can identify $\mathrm{Hom}(P(a),V)$ and V_a, identifying the map $\varphi = (\varphi_x) : P(a) \longrightarrow V$ with its image $(a \mid a)\varphi_a$. Also, given $\varphi : P(a) \longrightarrow V$, then

$$(a \mid a)\varphi_a V_\alpha = (a \mid a)P(a)_\alpha \varphi_b = (b \mid \alpha^* \mid a)\varphi_b = (b \mid b)P(\alpha^*)_b\varphi_b ,$$

thus, the image of $v = (a \mid a)\varphi_a$ under V_α is the element $vV_\alpha = (b \mid b)P(\alpha^*)_b\varphi_b$. It follows that V_α <u>is a monomorphism if and only if</u> $\mathrm{Hom}(\mathrm{Cok}\, P(\alpha^*),V) = 0$. [Namely, if $0 \neq v \in V_a$ belongs to the kernel of V_α, and $v = (a \mid a)\varphi_a$, then the equality above gives $(b \mid b)P(\alpha^*)_b\varphi_b = 0$, thus $P(\alpha^*)\varphi = 0$, thus φ factors over the cokernel of $P(\alpha^*)$, and therefore $\mathrm{Hom}(\mathrm{Cok}\, P(\alpha^*),V) \neq 0$, and conversely.].

Now assume there is given a subquiver Δ' of Δ. Given a representation V of Δ, we say that V <u>has its support in</u> Δ' provided $V_x = 0$ for all $x \in \Delta_0 \setminus \Delta'_0$, and $V_\alpha = 0$ for all $\alpha \in \Delta_1 \setminus \Delta'_1$. In case there is given a set $\{\rho_i \mid i\}$ of relations for Δ, and $\{\rho'_j \mid j\}$ is a generating set of relations for the ideal $\langle \rho_i \mid i \rangle \cap k\Delta'$ of $k\Delta'$, then the algebra $A(\Delta',\{\rho'_j \mid j\})$ will be called <u>the restriction of</u> $A(\Delta,\{\rho_i \mid i\})$ <u>to</u> Δ'. Of course, the $A(\Delta',\{\rho'_j \mid j\})$-modules are just the $A(\Delta,\{\rho_i \mid i\})$-modules with support in Δ'. In particular, if Δ' is obtained from Δ by deleting the vertex a and all edges involving a, then $A(\Delta',\{\rho'_j \mid j\})$ is said to be <u>obtained from</u> $A(\Delta,\{\rho_i \mid i\})$ <u>by deleting the vertex</u> a. Also, given a subquiver Δ' of Δ, and a relation $\rho = \Sigma\, c_w w$ on Δ, we will say that ρ has its support in Δ', provided any arrow occuring in any path of w with $c_w \neq 0$ belongs to Δ'.

In general, given an algebra A with quiver Δ, a vertex of Δ will also be called a <u>vertex of the algebra</u> A. [Note that the vertices of an algebra correspond bijectively to the equivalence classes of the primitive orthogonal idempotents, and also to the isomorphism classes of the simple A-modules].

A translation quiver $\Gamma = (\Gamma_0,\Gamma_1,\tau)$ is given by a (locally finite) quiver (Γ_0,Γ_1) together with an injective map $\tau = \tau_\Gamma : \Gamma'_0 \longrightarrow \Gamma_0$ defined on a subset $\Gamma'_0 \subseteq \Gamma_0$ such that for any $z \in \Gamma'_0$, and any $y \in \Gamma_0$ the number of arrows from y to z is equal to the number of arrows from τz to y. (In particular, we must have $(\tau z)^+ = z^-$, and this is also sufficient in case there are no multiple arrows). If $z \in \Gamma'_0$, and $x = \tau z$, we also write $x] z$. The vertices in Γ_0 which do not belong to Γ'_0 are called <u>projective</u>, those not belonging to the image of τ are called <u>injective</u>. A non-empty translation quiver Γ is called <u>proper</u> provided z^- is non-empty for any non projective vertex z of Γ. Given a non-projective vertex z in a proper translation quiver, there is a path of length 2 from τz to z. If x,y both are non-projective vertices, and x is a direct predecessor of y, then also τy is a direct predecessor of y. Consequently, given a path

$$x_1 \longrightarrow x_2 \longrightarrow \dots \longrightarrow x_{n-1} \longrightarrow x_n$$

with no x_i being projective, $1 \leq i \leq n$, then there is a path

$$\tau x_1 \longrightarrow \tau x_2 \longrightarrow \dots \longrightarrow \tau x_{n-1} \longrightarrow \tau x_n .$$

A vertex x with $\tau^t x = x$ for some $t \geq 1$ is said to be <u>periodic</u>. Note that a finite τ-orbit either contains only periodic vertices, or else one projective vertex and one injective vertex (of course, they may coincide). An infinite τ-orbit may contain a projective vertex or an injective vertex, but not both. A τ-orbit without a projective or an injective vertex is called <u>stable.</u> In case all τ-orbits of Δ are stable, Γ itself is said to be stable. Note that $\Gamma = (\Gamma_0,\Gamma_1,\tau)$ is stable if and only if $\tau : \Gamma_0 \longrightarrow \Gamma_0$ is bijective. Given a translation quiver $\Gamma = (\Gamma_0,\Gamma_1,\tau_\Gamma)$, let $\Gamma^* = (\Gamma^*_0,\Gamma^*_1,\tau_{\Gamma^*})$ be the opposite translation quiver, with (Γ^*_0,Γ^*_1) the opposite quiver of (Γ_0,Γ_1), $\tau_{\Gamma^*} = \tau_\Gamma^{-1} : \tau_\Gamma(\Gamma'_0) \longrightarrow \Gamma_0$. If $\Gamma = (\Gamma_0,\Gamma_1,\tau_\Gamma)$ is a translation quiver, and $\Gamma'_0 \subseteq \Gamma_0$, the <u>full</u> translation subquiver Γ' of Γ defined by Γ'_0 is given by $\Gamma' = (\Gamma'_0,\Gamma'_1,\tau_{\Gamma'})$, where (Γ'_0,Γ'_1) is the full subquiver of (Γ_0,Γ_1) defined by Γ'_0, and where $\tau_{\Gamma'}z$ is defined, and equal to x, provided both z and x belong to Γ'_0 and $\tau_\Gamma z = x$. The full translation subquiver Γ' of Γ is said to be <u>mesh-complete</u> provided for any non-projective vertex z of Γ', all direct predecessors of z in Γ belong to Γ'.

Given a quiver Δ, let us define a translation quiver $\mathbb{Z}\Delta$. The set of vertices of $\mathbb{Z}\Delta$ is given by $\mathbb{Z} \times \Delta_0$, given an arrow $\alpha : a \longrightarrow b$ in Δ, there are the arrows $(z,\alpha) : (z,a) \longrightarrow (z,b)$ and $(z,\alpha)' : (z,b) \longrightarrow (z+1,a)$, and define $\tau(z,a) = (z-1,a)$ for any vertex a of Δ. Note that $\mathbb{Z}\Delta$ is a stable translation quiver; $\mathbb{Z}\Delta$ is proper if and only if Δ has no vertex without neighbors. Given a (finite or infinite) interval I in $\mathbb{Z}$, let $I\Delta$ be the full translation subquiver of $\mathbb{Z}\Delta$ defined by the set of vertices of the form (z,a) with $z \in I$, $a \in \Delta$. In particular, we will deal with $\mathbb{N}_0\Delta$ and $(-\mathbb{N}_0)\Delta$, where $\mathbb{N}_0 = \{0,1,2,\dots\}$ the set of natural numbers including 0, and $-\mathbb{N}_0 = \{-n \mid n \in \mathbb{N}_0\}$. For different quivers with same underlying graph, we may obtain in this way isomorphic translation quivers.

For example, Riedtmann has shown:

(3) If $\bar{\Delta}$ is a tree, then $\mathbb{Z}\Delta$ only depends on $\bar{\Delta}$.

Consequently, we will denote $\mathbb{Z}\Delta$ in this case also by $\mathbb{Z}\bar{\Delta}$. In particular, we will be interested in the translation quivers $\mathbb{Z}\tilde{\mathbb{D}}_n$, $\mathbb{Z}\tilde{\mathbb{E}}_6$, $\mathbb{Z}\tilde{\mathbb{E}}_7$, $\mathbb{Z}\tilde{\mathbb{E}}_8$ (for the definition of these graphs, see 1.2), and in $\mathbb{Z}\mathbb{A}_\infty$, where $\mathbb{A}_\infty$ is the infinite graph

$$\circ\!-\!\!-\!\circ\!-\!\!-\!\circ\!-\ \ \cdots\ \ -\!\circ\!-\!\!-\!\circ\!-\ \cdots$$

(with $\mathbb{N}_1 = \{1,2,3,\ldots\}$ being the set of vertices, and with edges $\{i,i+1\}$ for all $i \in \mathbb{N}_1$). If $\bar{\Delta} = \tilde{\mathbb{A}}_n$ (see again 1.2), then $\mathbb{Z}\Delta$ only depends on a pair (n_1,n_2) of numbers, where n_1 is the number of arrows of Δ going in one direction, n_2 the number of those going in the opposite direction, and $n_1 \geq n_2$; we say that Δ is of type $\tilde{\mathbb{T}}_{n_1n_2}$ (or also $\tilde{\mathbb{A}}_{n_1n_2}$), and we will denote $\mathbb{Z}\Delta$ also by $\mathbb{Z}\tilde{\mathbb{T}}_{n_1n_2}$ (note that $n_1+n_2 = n+1$).

A proper translation quiver without cyclic paths, with only finitely many τ-orbits and such that any τ-orbit contains a projective vertex, will be called a <u>preprojective</u> translation quiver. Dually, Γ is said to be a <u>preinjective</u> translation quiver, provided Γ^* is a preprojective translation quiver.

(4) <u>If</u> Γ <u>is a preprojective translation quiver, then</u> $\Gamma = {}_\infty\Gamma$.

<u>Proof.</u> Let Γ be a preprojective translation quiver. We claim that any vertex x of Γ has only finitely many predecessors. Assume not. Since Γ is locally finite, there has to exist an infinite path

$$\cdots \longrightarrow x_{i+1} \longrightarrow x_i \longrightarrow \cdots \longrightarrow x_1 \longrightarrow x_0$$

in Γ. Any τ-orbit contains a projective vertex, thus, for any i, there exists some $t_i \geq 0$ with $p_i := \tau^{t_i} x_i$ being projective. Since there are only finitely many τ-orbits, there is some projective vertex p with $I = \{i \in \mathbb{N}_0 \mid p_i = p\}$ being infinite. Let $j > i$ be two elements in I. We claim that $t_j < t_i$. Namely, if $t_j \geq t_i$, then $x_i = \tau^{(t_j-t_i)} x_j$, thus there is a path of length $2(t_j-t_i)$ from x_i

to x_j (using that Γ is assumed to be proper). Since we also have a path of length $j-i \geq 1$ from x_j to x_i, we obtain a cyclic path in Γ, impossible. However, on an infinite set I, we cannot have a strictly decreasing function $i \longmapsto t_i$ to $\mathbb{N}_o$. This contradiction shows that any x has only finitely many predecessors; we denote by $\varphi(x)$ the number of predecessors of x. By induction on $\varphi(x)$, we show that all vertices belong to ${}_\infty\Gamma$. If $\varphi(x) = 1$, then $x \in {}_o\Gamma$. Now, let $\varphi(x) > 1$, and let $y \in x^-$. Then all predecessors of y are also predecessors of x, whereas x is not a predecessor of y (since otherwise we obtain a cyclic path). Thus $\varphi(y) < \varphi(x)$. By induction, $y \in {}_\infty\Gamma$. Since x has only finitely many direct predecessors, there is some d with $y \in {}_d\Gamma$ for any $y \in x^-$. Thus $x \in {}_{d+1}\Gamma$. This finishes the proof.

Given a preprojective translation quiver Γ, we want to introduce its orbit quiver $\mathcal{O}(\Gamma)$; this is a labelled quiver,the arrows being labelled by elements of $\mathbb{N}_o$. The vertices of $\mathcal{O}(\Gamma)$ are the τ-orbits of Γ, or, equivalently, the projective vertices of Γ (note that any τ-orbit of Γ contains precisely one projective vertex). Given a projective vertex p of Γ, let $y_1,\dots,y_r$ be the direct predecessors of p. For any i, there exists $t_i \geq o$ and a projective vertex p_i with $\tau^{t_i}y_i = p_i$. Let $n(y_i,p)$ be the number of arrows from y_i to p. In $\mathcal{O}(\Gamma)$, there will be $n(y_i,p)$ arrows from p_i to p with label t_i. Since a preprojective translation quiver does not contain a cyclic path, it follows that $\mathcal{O}(\Gamma)$ is a quiver without cyclic paths.

(5) A preprojective translation quiver without injective vertices is uniquely determined by its orbit quiver.

Proof. Let Γ and Γ' be preprojective translation quivers with $\mathcal{O}(\Gamma) \approx \mathcal{O}(\Gamma')$ We are going to construct an isomorphism γ from Γ to Γ'. If x,y are vertices of Γ, or of Γ' let $n(x,y)$ be the number of arrows from x to y. Since the orbit quivers of Γ and Γ' are isomorphic, there is given a bijection γ between the projective vertices of Γ and Γ'. By induction on d, we want to define $\gamma : {}_d\Gamma \longrightarrow {}_d\Gamma'$. For $d = 0$, ${}_o\Gamma$ consists only of isolated vertices, all of which are projective in Γ, thus γ is defined on ${}_o\Gamma$. Now assume γ is defined on ${}_d\Gamma$. Choose a vertex z in ${}_{d+1}\Gamma$, not belonging to ${}_d\Gamma$. If $y \in z^-$, then $z \in {}_d\Gamma$ thus γz is defined.

Consider first the case of z not being projective. Then $\tau_\Gamma z \in {}_d\Gamma$, thus let $\gamma z := \tau_{\Gamma'}^{-1}\gamma\tau_\Gamma z$. If $y \in z^-$, then γ defines a bijection from the set of arrows $\tau_\Gamma z \to y$ to the set of arrows $\gamma\tau_\Gamma z \longrightarrow \gamma y$. Thus $n(y,z) = n(\tau z,y) = n(\gamma\tau_\Gamma z,\gamma y) = n(\gamma y,\tau_{\Gamma'}^{-1}\gamma\tau_\Gamma z) = n(\gamma y,\gamma z)$. Now assume z is projective, thus γz is defined. If $y \in z^-$, then $\tau^t y = p_i$ for some projective vertex p_i and some t, and $n(y,z)$ is the number of arrows from p_i to z in $\mathcal{O}(\Gamma)$, with label t. Since $\gamma : \mathcal{O}(\Gamma) \longrightarrow \mathcal{O}(\Gamma')$ is an isomorphism of labelled quivers, we conclude that $n(y,z) = n(\gamma y,\gamma z)$. In both cases, we see that $n(y,z) = n(\gamma y,\gamma z)$, and we define γ on the set of arrows $y \longrightarrow z$ by choosing some arbitrary bijection between this set and the set of arrows $\gamma y \longrightarrow \gamma z$. This extends γ to ${}_{d+1}\Gamma$. In this way, we obtain an isomorphism γ from Γ to Γ'.

In case we deal with a preprojective translation quiver Γ containing also injective vertices, we obtain a corresponding result when fixing, in addition, the orbit length of any τ-orbit in Γ (it may be a natural number, in case the orbit contains an injective vertex, or else ∞).

A proper translation quiver is said to be <u>hereditary</u> provided for any arrow $x \longrightarrow y$ with y projective also x is projective and there is no cyclic path containing projective vertices.

(6) <u>Let Γ be a hereditary translation quiver with at least one but only finitely many projective vertices. Then ${}_{\infty}\Gamma$ is a preprojective translation quiver, containing all projective vertices, and closed under neighbors in</u> Γ.

<u>Proof.</u> Since the set of projective vertices is non-empty, finite, closed under predecessors and does not contain a cyclic path, there exists a projective vertex p belonging to ${}_{0}\Gamma$. By induction on the number of predecessors, it follows that any projective vertex belongs to ${}_{\infty}\Gamma$. In particular, ${}_{\infty}\Gamma$ is non-empty. Let us show that ${}_{\infty}\Gamma$ is closed under neighbors. Thus, let $x \longrightarrow y$ be an arrow in Γ. If y is in ${}_{d}\Gamma$, for some d then $x \in {}_{d-1}\Gamma$. Now assume $x \in {}_{d}\Gamma$ for some $d \geq o$, and use induction on d. If y is projective, then we know that $y \in {}_{\infty}\Gamma$, thus assume y is not projective. In particular, $d \geq 1$. There is an arrow $\tau y \longrightarrow x$, thus $x \in {}_{d}\Gamma$ implies that $\tau y \in {}_{d-1}\Gamma$. Also for any other $x' \in y^{-}$, there is an arrow $\tau y \longrightarrow x'$. Since $\tau y \in {}_{d-1}\Gamma$, we know by induction that $x' \in {}_{d}\Gamma$. Consequently, $y \in {}_{d+1}\Gamma$. This shows that ${}_{\infty}\Gamma$ is closed under neighbors. In order to show that ${}_{\infty}\Gamma$ is a preprojective translation quiver, assume there is a cyclic path

$$x_o \longrightarrow x_1 \longrightarrow \dots \longrightarrow x_{n-1} \longrightarrow x_n = x_o$$

in ${}_{\infty}\Gamma$. If no x_i is projective, we can apply some power of τ in order to obtain a cyclic path containing a projective vertex, a contradiction. This finishes the proof.

Given a translation quiver $\Gamma = (\Gamma_0, \Gamma_1, \tau)$, a <u>polarization</u> of Γ is given by an injective map $\sigma : \Gamma_1' \longrightarrow \Gamma_1$ where Γ_1' is the set of all arrows $\alpha : a \longrightarrow b$ with b not projective, such that $\sigma(\alpha) : \tau b \longrightarrow a$ for $\alpha : a \to b$. In case Γ has no multiple edge, there is a unique polarization. Given a translation quiver Γ with a polarization σ, we are going to define its mesh category $k(\Gamma,\sigma)$. First, we define the mesh ideal in the path category of (Γ_0, Γ_1) as the ideal generated by the elements

$$m_z = \sum_{y \in z^-} \sum_{\alpha : y \to z} \sigma(\alpha)\alpha$$

with z a non-projective vertex. The <u>mesh category</u> $k(\Gamma,\sigma)$ is defined as the quotient category of the path category of (Γ_0, Γ_1) modulo the mesh ideal. In case Γ is a translation quiver without multiple edges, σ is uniquely determined by Γ, thus we denote the corresponding mesh category just by $k(\Gamma)$. Similarly, if Γ is a preprojective translation quiver, and σ is a polarization of Γ, then given any arrow $\alpha : x \longrightarrow y$, there exists a $t(\alpha) \geq o$ with $\sigma^{t(\alpha)}(\alpha)$ being defined, and being an arrow ending in a projective vertex. Thus, if σ_1, σ_2 are two polarizations of the preprojective translation quiver, then $\alpha \mapsto \sigma_2^{-t(\alpha)} \sigma_1^{t(\alpha)}(\alpha)$ defines an automorphism of Γ fixing all vertices, and this extends to an isomorphism of categories $k(\Gamma,\sigma_1) \approx k(\Gamma,\sigma_2)$, thus we may denote this category again just by $k(\Gamma)$.

Given a translation quiver $\Gamma = (\Gamma_0, \Gamma_1, \tau)$ with a polarization σ, we associate with it a two-dimensional simplicial complex, as suggested by Gabriel and Riedtmann. The 0-simplices are the elements of Γ_0; there are two kinds of 1-simplices, namely the elements $\alpha : x \longrightarrow y$ of Γ_1 (the "arrows"), with boundary being given by x,y, and the elements $x \rceil z$ of the graph of τ (the "extensions"), with boundary being given by x,z. Finally, for every $\alpha : y \longrightarrow z$ with $z \in \Gamma_0'$, there is a 2-simplex (or "triangle") with boundary the 1-simplices $\alpha, \sigma\alpha$, and $\tau z \rceil z$; we may denote it by $(\sigma\alpha, \alpha)$. The geometric realization of this complex will be called the <u>underlying topological space</u> $|\Gamma|$ of Γ (with respect to σ). Actually, we only will consider this complex in case Γ has no multiple edge, so that there is a unique polarization.

2.2 Krull-Schmidt k-categories

A category K is said to be (finite-dimensional) k-additive, provided K has finite direct sums, and all sets Hom(X,Y), with X,Y objects of K, are finite dimensional k-vectorspaces, with k acting centrally on Hom(X,Y), and such that the compositions $\mathrm{Hom}(X,Y) \times \mathrm{Hom}(Y,Z) \longrightarrow \mathrm{Hom}(X,Z)$ are bilinear. Note that in these notes, the composition of $f \in \mathrm{Hom}(X,Y)$ and $g \in \mathrm{Hom}(Y,Z)$ usually will be denoted by fg [the categories we will consider will almost always be categories of left modules, and we want to write homomorphisms on the opposite side of the scalars]. The direct sum of X_1,X_2 will be denoted by $X_1 \oplus X_2$; and a non-zero object X is said to be indecomposable provided $X \approx X_1 \oplus X_2$ implies $X_1 = 0$ or $X_2 = 0$. Of course, due to the finite-dimensionality of the endomorphism rings, any object of a k-additive category is (isomorphic to) the direct sum of finitely many indecomposable objects and satisfies both the descending and the ascending chain condition for direct summands. An indecomposable object X with End(X) = k will be said to be a brick. Two objects X,Y with Hom(X,Y) = 0 = Hom(Y,X) will be said to be orthogonal. If X is an object of K, the direct sum of d copies of X will be denoted by X^d, or also by $X \otimes V$, where V is a d-dimensional vectorspace over k. [Note that any vectorspace decomposition $V = V' \oplus V''$ gives rise to a direct decomposition $X \otimes V = (X \otimes V') \oplus (X \otimes V'')$. In particular, if $v_1,\dots,v_d$ is a k-basis of V, then $X \otimes V = \bigoplus_{i=1}^{d} X \otimes v_i$, with $X \otimes v_i := X \otimes kv_i \approx X$.]

A k-additive category K will be said to be a Krull-Schmidt category provided all idempotents split (i.e. if $e = e^2 \in \mathrm{Hom}(X,X)$, then there are maps $\mu: Y \longrightarrow X$, $\rho : X \longrightarrow Y$ with $\mu\rho = 1_Y$ and $\rho\mu = e$), or, equivalently, the endomorphism ring End(X) of any indecomposable object X of K is a local ring. In a Krull-Schmidt category, one has the following unicity result for direct decompositions:

Theorem of Krull-Schmidt: Let K be a Krull-Schmidt category, let X_i,Y_j be indecomposable objects in K with $1 \le i \le s$, $1 \le j \le t$, such that $\bigoplus_{i=1}^{s} X_i \approx \bigoplus_{j=1}^{t} Y_j$. Then $s = t$, and there is a permutation π of $\{1,\dots,s\}$ such that $X_i \approx Y_{\pi(i)}$ for all i.

In a Krull-Schmidt category, the classification problem for objects (as always, up to isomorphism) reduces to that for indecomposable objects. From now on, let K be a Krull-Schmidt category. We denote by $G(K)$ the Grothendieck group of K with respect to direct sums, thus $G(K)$ is the free abelian group with basis the set of isomorphism classes of indecomposable objects of K. We may define $G(K)$ also as follows: let F be the free abelian group with basis the set of all objects of K, and R the subgroup generated by the formal sums $X' - X + X''$, where $X \approx X' \oplus X''$; then $G(K) = F/R$. Given an object X in K, we denote by [X] its isomorphism class. Of course, $[X]$ can be considered as an element of $G(K)$. Note that $G(K)$ is a partially ordered group, the positive elements are the elements $[X]$ with X a non-zero object in K. In case there are only finitely many isomorphism classes of indecomposable objects in K (so that $G(K)$ is a finitely generated abelian group), K is said to be finite.

Given a Krull-Schmidt category K, a full subcategory L of K closed under direct sums and direct summands (and isomorphisms) will be called an object class in K. Note that an object class L is itself a Krull-Schmidt category, and is uniquely determined by the indecomposable objects belonging to L. Given a set N of objects in K, we denote by $\langle N\rangle$ the smallest object class containing N, it is given by the direct sums of direct summands of objects in N. [A rather common notation for $\langle N\rangle$ is also add N ; however, we will not use it]. Of course, given objects $N_1,\dots,N_t$, then $\langle N_1,\dots,N_t\rangle$ denotes the smallest object class having $N_1,\dots,N_t$ as elements. Assume now that L_1,L_2 are two object classes in K. We will write $\mathrm{Hom}(L_1,L_2) = 0$ in case $\mathrm{Hom}(X_1,X_2) = 0$ for all $X_1 \in L_1$, $X_2 \in L_2$, [and similarly for Ext^1, Tor_i, etc., provided these functors are defined for K]. From L_1,L_2, we obtain other object classes as follows: First of all, $L_1 \cap L_2$ is an object class, again. We denote by $L_1 \vee L_2$ the object class $\langle L_1 \cup L_2\rangle$, and by $L_1 \setminus L_2$ the set of all objects in L_1 with no non-zero direct summand in L_2.

Let us define the radical of the Krull-Schmidt category K. If X,Y are indecomposable, let $\mathrm{rad}(X,Y)$ be the set of non-invertible morphisms from X to Y. Of course, for X indecomposable, $\mathrm{rad}(X,X)$ is just the radical of the local ring $\mathrm{End}(X)$. Given direct sums $X = \bigoplus_{i=1}^{s} X_i$, $Y = \bigoplus_{j=1}^{t} Y_j$, any map $f : \bigoplus_{i=1}^{s} X_i \longrightarrow \bigoplus_{j=1}^{t} Y_j$ can be written in the form $f = (f_{ij})$ with $f_{ij} \in \mathrm{Hom}(X_i,Y_j)$; now, if all X_i,Y_j are indecomposable, then f will be said to belong to $\mathrm{rad}(X,Y)$ provided all f_{ij} belong to $\mathrm{rad}(X_i,Y_j)$. [Of course, this is equivalent to saying that for any split mono $\mu : X' \longrightarrow X$, and any split epi $\rho : Y \longrightarrow Y'$ with X',Y' indecomposable, $\mu f \rho \in \mathrm{rad}(X',Y')$.]

A path in K is a finite sequence $(X_0,X_1,\dots,X_m)$ of indecomposable objects X_i, $0 \le i \le m$, with $\mathrm{rad}(X_{i-1},X_i) \neq 0$ for all $1 \le i \le m$; the objects X_i are said to belong to the path, and m is called its length. In case there exists a path $(X_0,X_1,\dots,X_m)$, we write $X_0 \preceq_K X_m$, or just $X_0 \preceq X_m$; in case $m \ge 1$, we write $X_0 \prec X_m$. (Note that the relation $\preceq$ usually is not antisymmetric, even modulo isomorphism). A path $(X_0,X_1,\dots,X_m)$ of length $m \ge 1$ and with $X_0 \approx X_m$ is called a cycle. An indecomposable object N of K is said to be directing (in K), provided N does not belong to a cycle in K.

Lemma 1. An indecomposable object N in $\mathcal{K}$ is directing if and only if N is a brick and there are object classes $\mathcal{X}$, $\mathcal{N}^o$, $\mathcal{Y}$ with $\mathcal{K} = \mathcal{X} \vee \mathcal{N} \vee \mathcal{Y}$, where $\mathcal{N} = \mathcal{N}^o \vee \langle N \rangle$, such that

$$\mathrm{Hom}(N,\mathcal{N}^o) = \mathrm{Hom}(\mathcal{N}^o,N) = 0$$

$$\mathrm{Hom}(\mathcal{Y},\mathcal{X}) = \mathrm{Hom}(\mathcal{Y},\mathcal{N}) = \mathrm{Hom}(\mathcal{N},\mathcal{X}) = 0.$$

Proof. First, assume that N is directing. Let $\mathcal{X}$, $\mathcal{Y}$, $\mathcal{N}^o$ be the object classes generated by the sets of indecomposable objects Z in $\mathcal{K}$ satisfying $Z \prec N$, or $N \prec Z$, or $Z \not\preceq N \not\preceq Z$, respectively, and let $\mathcal{N} = \mathcal{N}^o \vee \langle N \rangle$. By definition any indecomposable object of $\mathcal{K}$ belongs to precisely one of $\mathcal{X}$, $\mathcal{Y}$, $\mathcal{N}^o$, $\langle N \rangle$. The first two zero-conditions follow directly from the definition of $\mathcal{N}^o$. For the verification of the remaining zero-conditions, one has to use that N does not belong to a cycle in $\mathcal{K}$.

Conversely, assume that N is a brick and that there are given object classes $\mathcal{X}$, $\mathcal{N}^o$, $\mathcal{Y}$ satisfying the conditions stated in the lemma. Since N is a brick, there is no cycle of length 1 containing N. Now assume there is a cycle $(X_o,X_1,\ldots,X_m)$ with length $m > 1$ and $X_o \approx N = X_m$. Since $\mathrm{Hom}(N,X_1) \neq 0$ and $X_1 \not\approx N$, it follows that X_1 belongs to $\mathcal{Y}$. Using induction, we see that all X_i, $1 \leq i \leq m$, belong to $\mathcal{Y}$, since $\mathrm{Hom}(X_{i-1},X_i) \neq 0$. But this contradicts $X_m = N$. Thus, N is directing.

Remark. Given a directing object N, the object classes $\mathcal{X}$, $\mathcal{N}^o$, $\mathcal{Y}$ of the lemma usually are not uniquely determined. [Consider for example for $\mathcal{K}$ the module category A-mod where A is given by

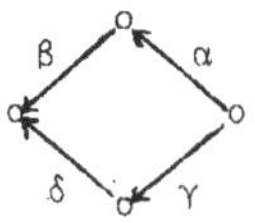

with $\alpha\beta = 0 = \gamma\delta$

and $\underline{\dim}\, N = {}_{o}\overset{1}{\underset{o}{}}{}_{o}$.] However, there always is a smallest possible class $\mathcal{X}$, namely $\mathcal{X} = \langle X \text{ indecomposable} \mid X \prec N \rangle$, and similarly, a smallest possible class $\mathcal{Y}$.

In a Krull-Schmidt category, an object N will be said to be separating provided N is directing, and any map $X \longrightarrow Y$ with $X \preceq N \preceq Y$ factors through a direct sum of copies of N.

Let $\mathcal{K}$ be a Krull-Schmidt category, and X,Y objects in $\mathcal{K}$. Then $\mathrm{rad}^2(X,Y)$ is given by the set of maps of the form fg with $f \in \mathrm{rad}(X,M)$, $g \in \mathrm{rad}(M,Y)$, for some object M. Note that $\mathrm{rad}^2(X,Y) \subseteq \mathrm{rad}(X,Y) \subseteq \mathrm{Hom}(X,Y)$ are k-subspaces, and, in fact End(X)- End(Y)-subbimodules of Hom(X,Y), and we let

$$\mathrm{Irr}(X,Y) = \mathrm{rad}(X,Y)/\mathrm{rad}^2(X,Y).$$

By construction, this End(X)-End(Y)-bimodule is annihilated from the left by rad(X,X), from the right by rad(Y,Y). In case

$$\dim_k \mathrm{Irr}(X,Y) \leq 1$$

for all indecomposable objects X,Y in K, we will say that K has <u>trivial modulation</u>.

Given X,Y objects in K, a morphism $f : X \longrightarrow Y$ is said to be <u>irreducible</u>, provided, on the one hand, f is neither split mono, nor split epi, and, on the other hand, for any factorization $f = f'f''$, the map f' is split mono, or the map f'' is split epi. Note that in case both X and Y are indecomposable, then $f : X \longrightarrow Y$ is irreducible if and only if $f \in \mathrm{rad}(X,Y) \setminus \mathrm{rad}^2(X,Y)$, thus there exists an irreducible map from X to Y if and only if $\mathrm{Irr}(X,Y) \neq 0$; thus, the bimodule Irr(X,Y) is a measure for the multiplicity of irreducible maps and it is called the <u>bimodule of irreducible maps</u>.

If K is a Krull-Schmidt category, and L is a full subcategory, then L itself is a Krull-Schmidt category, provided L is closed under direct sums and direct summands. If this is the case, and X,Y are indecomposable objects in L, then $\mathrm{rad}_K(X,Y) = \mathrm{rad}_L(X,Y)$, however, the definition of rad^2, and therefore of Irr depends on the whole category K or L; in general, we only have an obvious epimorphism

$$\mathrm{Irr}_L(X,Y) \twoheadrightarrow \mathrm{Irr}_K(X,Y).$$

The <u>quiver</u> $\Delta(K)$ of a Krull-Schmidt category K is defined as follows: its vertices are the isomorphism classes [X] of the indecomposable objects X in K, and there is an arrow $[X] \to [Y]$ in case $\mathrm{Irr}(X,Y) \neq 0$. In case $d_{XY} := \dim_k \mathrm{Irr}(X,Y) > 1$, we will mark the arrow $[X] \to [Y]$ with the multiplicity d_{XY}; or else, we may draw d_{XY} different arrows from [X] to [Y]. However, usually we will deal with Krull-Schmidt categories with trivial modulation. A <u>component</u> of K is, by definition, the object class generated by the indecomposable objects belonging to a connected component of $\Delta(K)$.

Given a Krull-Schmidt category K, and X an object in K, a <u>source map</u> for X in K is a map $f : X \longrightarrow Y$ satisfying the following properties:

(α) f is not split mono;

(β) given $f' : X \longrightarrow Y'$, not split mono, there exists $\eta : Y \to Y'$ with $f' = f\eta$; and

(γ) if $\gamma \in \mathrm{End}(Y)$ satisfies $f\gamma = f$, then γ is an automorphism.

Note that if there exists a source map $f : X \longrightarrow Y$, then X has to be indecomposable. [Namely, $X \neq 0$, since otherwise f is split mono, and if $X = X_1 \oplus X_2$, with projections $p_i : X \longrightarrow X_i$, and $X_i \neq 0$, for $i = 1,2$, then p_i is not split mono, thus $p_i = f\eta_i$ for some $\eta_i : Y \longrightarrow X_i$, thus $1_X = [p_1,p_2] = f[\eta_1,\eta_2]$ shows that f is split mono, again a contradiction.].

Dually, let Z be an (indecomposable) object in K. Then a <u>sink map</u> for Z in K is a map $g : Y \longrightarrow Z$ satisfying the following properties:

(α*) g is not split epi;

(β*) given $g' : Y' \longrightarrow Z$, not split epi, there exists $\eta : Y' \longrightarrow Y$ with $g' = \eta g$; and

(γ*) if $\gamma \in \mathrm{End}(Y)$ satisfies $\gamma g = g$, then γ is an automorphism.

It is obvious that source maps and sink maps, if they exist, are unique up to isomorphism. (They have been introduced by Auslander and Reiten under the names "minimal left almost split map" and "minimal right almost split map"). We will say that the Krull-Schmidt category K <u>has source maps</u> (or <u>has sink maps</u>) provided for any indecomposable object in K, there exists a source map (or a sink map, respectively).

<u>Lemma 2.</u> A finite Krull-Schmidt category has source maps and sink maps.

<u>Proof.</u> Let K be a finite Krull-Schmidt category, say with indecomposable objects $X_1,\ldots,X_n$, and let X be one of these indecomposable objects. Let $\bar{f}_{ij}$, $1 \leq j \leq d_i$, be a basis of the k-vectorspace $\mathrm{rad}(X,X_i)$, let $\bar{f}_i : X \longrightarrow \bar{Y}_i := X_i \otimes \mathrm{Hom}(X,X_i)$ be defined by $x\bar{f}_i = \sum_j x\bar{f}_{ij} \otimes \bar{f}_{ij}$, and $\bar{f} = (\bar{f}_i)_i : X \longrightarrow \bar{Y} = \bigoplus_{i=1}^{n} \bar{Y}_i$. Then $\bar{f}$ satisfies the conditions (α) and (β) of a source map. Let Y be a minimal direct summand of $\bar{Y}$ through which f factors [the existence of such a minimal direct summand follows from the descending chain conditions on direct summands of $\bar{Y}$]. Obviously, the induced map $f : X \longrightarrow Y$ satisfies all three conditions (α), (β), (γ) of a source map. Dually, one shows the existence of a sink map for X.

The precise relationship between source maps and the bimodules of irreducible maps is given in the following lemma. (Of course, there also is the dual statement concerning sink maps).

<u>Lemma 3.</u> Assume there exists a source map for X. Let $Y_1,\ldots,Y_e$ be pairwise non-isomorphic indecomposable objects, and assume there are given maps $f_{ij} : X \longrightarrow Y_i$, $1 \leq j \leq d_i$, $1 \leq i \leq e$, with residue class $\bar{f}_{ij}$ in $\mathrm{Irr}(X,Y_i)$. Then $f = (f_{ij})_{ij} : X \longrightarrow \bigoplus_{i=1}^{e} Y_i^{d_i}$ is a source map for X if and only if $\bar{f}_{i1},\ldots,\bar{f}_{id_i}$

is a k-basis of $Irr(X,Y_i)$, for all $1 \leq i \leq e$, and any indecomposable object Y' with $Irr(X,Y') \neq 0$ is isomorphic to some Y_i.

Proof. First, assume that f is a source map. Let Y' be indecomposable, and $f' : X \longrightarrow Y'$ a map in $rad(X,Y')$. Since f' belongs to $rad(X,Y')$, we can factor it through f, say $f' = f\eta$ for some $\eta = (\eta_{ij})_{ij} : Y = \bigoplus_{i=1}^{e} Y_i^{d_i} \longrightarrow Y'$. Now assume f' does not belong to $rad^2(X,Y')$. It follows from $f' = f\eta = \sum_{i,j} f_{ij}\eta_{ij}$, that at least one η_{ij} is outside $rad(Y_i,Y')$, thus an isomorphism. This shows that any indecomposable object Y' with $Irr(X,Y') \neq 0$ is isomorphic to some Y_i. Now, assume Y' is of the form Y_i, say $Y' = Y_1$. Then any η_{1j} is an endomorphism of Y_1, thus $\eta_{1j} = c_j 1_{Y_1} + \eta'_{1j}$ for some $c_j \in k$ and $\eta'_{ij} \in rad(Y_1,Y_1)$. Since all η_{ij}, with $i \geq 2$, are in $rad(Y_i,Y_1)$, and all $f_{ij} \in rad(X,Y_i)$, it follows that:

$$f' = \sum_{ij} f_{ij}\eta_{ij} \equiv \sum_{j} f_{1j}c_j \pmod{rad^2(X,Y_1)},$$

thus $\bar{f}_{11},\ldots,\bar{f}_{1d_1}$ generate $Irr(X,Y_1)$ as a k-space. Let $t_i = \dim Irr(X,Y_i)$. Then $t_i \leq d_i$. Let $\bar{h}_{ir}$, $1 \leq r \leq t_i$, be a basis of $Irr(X,Y_i)$, with $h_{ir} : X \longrightarrow Y_i$, thus $\bar{f}_{ij} = \sum_{r=1}^{t_i} \bar{h}_{ir} b_{irj}$ for some $b_{irj} \in k$. Let $h = (h_{ir})_{ir} : X \longrightarrow \bigoplus_{i=1}^{e} Y_i^{t_i}$, $b_i : Y^{t_i} \longrightarrow Y^{d_i}$ the map given by the matrix with r-j-entry equal to b_{irj}, and $b = \bigoplus_{i=1}^{e} b_i$. Then $f - hb \in rad^2(X,Y)$, say $f - hb = uv$ for some $u \in rad(X,Z)$, $v \in rad(Z,Y)$. Since f is a source map, we can factor both h and u through f, thus $h = fh'$, $u = fu'$, and therefore

$$f = hb + uv = f(h'b + u'v).$$

Condition (γ) of a source map asserts that $h'b + u'v$ is an automorphism, thus $h'b$ is an automorphism (since $v \in rad(Z,Y)$). As a consequence, b is split epi, thus, using that $t_i \leq d_i$, we see that $t_i = d_i$ for all i, and that b actually is an isomorphism. We therefore conclude, on the one hand, that the generating set $\bar{f}_{i1},\ldots,\bar{f}_{id_i}$ of $Irr(X,Y_i)$ actually is a k-basis, and, on the other hand, that h is a source map (Namely, $h = f(1 - u'v)b^{-1}$, and $(1 - u'v)b^{-1}$ is an isomorphism). This finishes the proof.

Given a translation quiver Δ, we have defined full translation subquivers ${}_d\Delta$ for all integers $d \geq -1$, and also for $d = \infty$. Given a Krull-Schmidt category K, define ${}_dK$ for all integers $d \geq -1$, and also for $d = \infty$, inductively as follows: First, ${}_{-1}K = \langle 0 \rangle$. Assume ${}_{d-1}K$ is defined, then an indecomposable object X of K

belongs to ${}_d\mathcal{K}$ if and only if any indecomposable object Y with $\mathrm{rad}(Y,X) \neq 0$ belongs to ${}_{d-1}\mathcal{K}$. Finally, let ${}_\infty\mathcal{K} = \bigsqcup_{d \in \mathbb{N}_0} {}_d\mathcal{K}$.

Lemma 4. Let $\mathcal{K}$ be a Krull-Schmidt category. Then any indecomposable object in ${}_\infty\mathcal{K}$ is directing.

Proof. Assume X is indecomposable and in ${}_\infty\mathcal{K}$. Choose d minimal with X in ${}_d\mathcal{K}$. If $(X_0,X_1,\ldots,X_m)$ is a path in $\mathcal{K}$ with $X_m \approx X$, then $X_i \in {}_{d-m+i}\mathcal{K}$, using inductively the lemma. In particular, $X_0 \in {}_{d-m}\mathcal{K}$. Thus, for $m \geq 1$, we see that X_0 cannot be isomorphic to X, since otherwise we would obtain a contradiction to the minimality of d.

Lemma 5. Let $\mathcal{K}$ be a Krull-Schmidt category with sink maps, and let X be an indecomposable object of $\mathcal{K}$. Then X belongs to ${}_d\mathcal{K}$ if and only if $[X]$ belongs to ${}_d\Delta(\mathcal{K})$. If X belongs to ${}_d\mathcal{K}$ and X' is indecomposable with $\mathrm{Hom}(X',X) \neq 0$, then there is a path from $[X']$ to $[X]$ in $\Delta(\mathcal{K})$.

Proof, by induction on d. Let $g : Y \longrightarrow X$ be a sink map for X, and let $0 \neq g' \in \mathrm{rad}(X',X)$. By definition of a sink map, there exists $\eta : X' \longrightarrow Y$ with $g' = \eta g$. Let $Y = \oplus Y_i$, with all Y_i indecomposable, $\eta = (\eta_i)_i$, $g = (g_i)_i$, where $\eta_i : X' \longrightarrow Y_i$, and $g_i : Y_i \longrightarrow X$. Assume $X \in {}_d\mathcal{K}$, thus all Y_i belong to ${}_{d-1}\mathcal{K}$. Since $0 \neq g' = \Sigma \eta_i g_i$, there exists some j with $\eta_j \neq 0$. In particular, $Y \neq 0$, thus $d \geq 1$. If $\eta_j \notin \mathrm{rad}(X',Y_j)$, then η_j is an isomorphism, thus with Y_j also X' belongs to ${}_{d-1}\mathcal{K}$. Otherwise, by induction $X' \in {}_{d-2}\mathcal{K} \subseteq {}_{d-1}\mathcal{K}$. This finishes the proof.

Lemma 6. Let $\mathcal{K}$ be a Krull-Schmidt category with sink maps. Assume, $\Delta(\mathcal{K})$ has a finite component $\mathcal{C}$. If X is an indecomposable object not belonging to $\mathcal{C}$, then $\mathrm{Hom}(X,\mathcal{C}) = 0$.

Proof. Assume $\mathrm{Hom}(X,C) \neq 0$ for some indecomposable object C in $\mathcal{C}$, say there is $0 \neq h : X \longrightarrow C$. Since $X \notin \mathcal{C}$, the map h is not split epi, thus, we can factor h through the sink map of C, say $g : D \longrightarrow C$ with $D = \oplus D_i$, where all D_i are indecomposable. Now $0 \neq h = h'g$ with $h' : X \longrightarrow D$ yields some $h_i' : X \longrightarrow D_i$, $g_i : D_i \longrightarrow C$ with $h_i'g_i \neq 0$. Let $C = C^{(0)}$, $D_i = C^{(1)}$, $g_i = g^{(1)}$, $h_i' = h^{(1)}$, thus $h^{(1)}g^{(1)} \neq 0$. By induction, we define $h^{(i)} : X \longrightarrow C^{(i)}$, $g^{(i)} : C^{(i)} \longrightarrow C^{(i-1)}$, with $C^{(i)}$ indecomposable, $g^{(i)}$ irreducible and $h^{(i)}g^{(i)} \ldots g^{(1)} \neq 0$. However, this is impossible, since in a finite Krull-Schmidt category $\mathcal{C}$, we have $\mathrm{rad}^d\, \mathcal{C} = 0$ for some d.

2.3 Exact categories

The Krull-Schmidt categories we will be interested in usually will be object classes in abelian categories, and it will be necessary to use additional information derived from such an embedding. In particular, given an object class L in the abelian category A, we will need to know the pairs (f,g) of maps $f : X' \longrightarrow X$, $g : X \longrightarrow X''$ in L such that

$$(*) \qquad 0 \longrightarrow X' \xrightarrow{f} X \xrightarrow{g} X'' \longrightarrow 0$$

is an exact sequence in A (we will say that (f,g) is a short exact sequence in A belonging to L). Since there may be different embeddings of L into abelian categories A_1, A_2 such that the short exact sequences of A_1 belonging to L coincide with those of A_2 belonging to L, we do not want to fix an embedding but just the set of pairs (f,g) in L which become short exact sequences in the larger category. Also note that the object classes L of an abelian category A which we deal with usually will be closed under extensions (i.e. if $(*)$ is an exact sequence in A, and X', X'' belong to L, then also X is in L). We introduce the following definitions:

The pair (K,S) will be called a Krull-Schmidt category with short exact sequences provided K is a Krull-Schmidt category, and S is a set of pairs (f,g) of maps in K such that f is a kernel of g in K, and g is a cokernel of f in K (this means that for $(f,g) \in S$, we have $fg = 0$, and if $f'g = 0$ for an f' in K, then $f' = \eta f$ for a unique η in K, and if $fg' = 0$ for a g' in K, then $g' = g\gamma$ for a unique γ in K). The pair (K,S) will be said to be an exact category provided K is a Krull-Schmidt category, such that there exists a full embedding $K \subseteq A$ with A an abelian category such that S is the set of all short exact sequences of A belonging to K, and such that, moreover, K is closed under extensions in A. Of course, an exact category is a Krull-Schmidt category with short exact sequences.

Given a Krull-Schmidt category (K,S) with short exact sequences, then $(f,g) \in S$ with $f : X' \longrightarrow X$, $g : X \longrightarrow X''$ is said to be split provided f is a split monomorphism (thus $ff' = 1$ for some f') or, equivalently, g is a split epimorphism (thus $g'g = 1$ for some g'). [For the proof of equivalence, note that it follows from $ff' = 1$ that f is the inclusion of a direct summand, thus $1-f'f = h'h$ for some h,h' and $hh' = 1$, $hf' = 0$, $fh' = 0$. However, then h' is a cokernel of f, thus there is an isomorphism ζ with $g = h'\zeta$. As a consequence, $g'g = 1$ where $g' = \zeta^{-1}h$. Of course, there also is the dual implication.]

Lemma 1. Let (K,S) be a Krull-Schmidt k-category with short exact sequences, and $(f,g) \in S$ not split, $f : X' \longrightarrow X$, $g : X \longrightarrow X''$. Then $\dim_k \operatorname{End}(X) < \dim_k \operatorname{End}(X' \oplus X'')$.

Proof. Let $(f,g) \in S$. Let $R_o = \operatorname{End}(X)$, let $R_1 = \{\zeta \in \operatorname{End}(X) \mid$ there exists ζ' with $f\zeta = \zeta' f\}$, and $R_2 = \{\zeta \in \operatorname{End}(X) \mid f\zeta = 0,\ \zeta g = 0\}$. Note that f is a (categorical) monomorphism in K, and g a (categorical) epimorphism in K, thus the k-linear map $\varphi_2 : \operatorname{Hom}(X'',X') \longrightarrow R_2$ given by $\varphi_2(\alpha) = g\alpha f$ for $\alpha \in \operatorname{Hom}(X'',X')$, is bijective. Also, define $\varphi_o : R_o \longrightarrow \operatorname{Hom}(X',X'')$ by $\varphi_o(\zeta) = f\zeta g$ for $\zeta \in R_o$, and note that its kernel is R_1. Similarly, define $\varphi_1 : R_1 \longrightarrow \operatorname{End}(X') \times \operatorname{End}(X'')$ by $\varphi_1(\zeta) = (\zeta',\zeta'')$ where $f\zeta = \zeta' f$, $\zeta g = g\zeta''$, and note that its kernel is R_2. Finally, we claim that φ_1 is surjective only in case (f,g) splits. Namely, assume there is $\zeta \in R_1$ with $\varphi_1(\zeta) = (1,0)$. Since $\zeta g = 0$, we can factor ζ through f, say $\zeta = f'f$, and then $ff' = 1$. Thus (f,g) splits. Thus, for (f,g) non-split, we have

$$\dim R_1/R_2 < \dim \operatorname{End}(X') + \dim \operatorname{End}(X''),$$

and always

$$\dim R_2 = \dim \operatorname{Hom}(X'',X'),\quad \dim R_o/R_1 \leq \dim \operatorname{Hom}(X',X'').$$

This finishes the proof.

Given an exact category (K,S), and X',X'' objects of K, there is defined $\operatorname{Ext}^1_{(K,S)}(X'',X')$ as the set of equivalence classes of all pairs (f,g), with $f : X' \longrightarrow X$, $g : X \longrightarrow X''$ for some X; here, (f_1,g_1) and (f_2,g_2) are called equivalent provided there is some map η with $f_1\eta = f_2$ and $g_1 = \eta g_2$. Also, given (f,g) in S, say $f : X' \longrightarrow X$, $g : X \longrightarrow X''$ and $\xi' : X' \longrightarrow Y'$, or $\xi'' : Z'' \longrightarrow X''$, there do exist (uniquely up to equivalence) corresponding induced exact sequences: Namely, choose an embedding of K as an extension closed full subcategory of an abelian category A and such that S is the set of all short exact sequences of A belonging to K. In A, we can construct the induced exact sequences

$$\begin{array}{ccccccccc} o & \longrightarrow & X' & \xrightarrow{f} & X & \xrightarrow{g} & X'' & \longrightarrow & o \\ & & \downarrow{\scriptstyle \xi'} & & \downarrow & & \downarrow{\scriptstyle 1} & & \\ o & \longrightarrow & Y' & \longrightarrow & Y & \longrightarrow & X'' & \longrightarrow & o \end{array}$$

and

$$\begin{array}{ccccccccc} 0 & \longrightarrow & X' & \longrightarrow & Z & \longrightarrow & Z'' & \longrightarrow & 0 \\ & & \downarrow 1 & & \downarrow & & \downarrow \xi'' & & \\ 0 & \longrightarrow & X' & \xrightarrow{f} & X & \xrightarrow{g} & X'' & \longrightarrow & 0 \end{array},$$

and since K is extension closed in A, we see that with Y', X'' also Y, and with X', Z'' also Z belongs to K. Thus, the induced exact sequences again belong to S, and they are independent of the embedding, since Y is characterized as a push out in K, and Z as a pullback in K. It follows that $\operatorname{Ext}^1_{(K,S)}(-,-)$ is functorial, covariant in the second variable and contravariant in the first, with values being k-vectorspaces (note that the addition in $\operatorname{Ext}^1_A(X'',X')$ can be defined using direct sums and induced exact sequences), and given $X' \xrightarrow{f} X \xrightarrow{g} X''$ in S, there exist connecting homomorphisms $\operatorname{Hom}(Y,X'') \longrightarrow \operatorname{Ext}^1_{(K,S)}(Y,X')$ and $\operatorname{Hom}(X',Y) \longrightarrow \operatorname{Ext}^1_{(K,S)}(X'',Y)$ for all Y in K, such that the corresponding long sequences involving Hom and $\operatorname{Ext}^1_{(K,S)}$ are exact.

Let (K,S) be a Krull-Schmidt category with short exact sequences. The pair (f,g) will be called an Auslander-Reiten sequence of (K,S) provided (f,g) belongs to S, with f being a source map in K, and g a sink map in K. Given an Auslander-Reiten sequence (f,g) of (K,S), with $f : X \longrightarrow Y$, $g : Y \longrightarrow Z$, we note that both X and Z are indecomposable, and that the isomorphism class $[X]$ of X is uniquely determined by $[Z]$, and $[Z]$ is uniquely determined by $[X]$. We denote $[X]$ by $\tau_{(K,S)}[Z]$ (or also we write $X = \tau_{(K,S)}Z$), and similarly $[Z]$ by $\tau^-_{(K,S)}[X]$, and call $\tau_{(K,S)}$ the Auslander-Reiten translation of (K,S). If we decompose $Y = \bigoplus_i Y_i^{d_i}$ with Y_i indecomposable, and pairwise non-isomorphic (and $d_i \geq 1$), then $\dim_k \operatorname{Irr}(X,Y_i) = d_i = \dim_k \operatorname{Irr}(Y_i,Z)$ for all i, according to lemma 3 in 2.2, thus there are d_i arrows $[X] \to [Y_i]$ and d_i arrows $[Y_i] \to [Z]$. Note that $\Delta(K)$ becomes a translation quiver by adding $\tau_{(K,S)}$. The translation quiver $(\Delta(K), \tau_{(K,S)})$ will be called the Auslander-Reiten quiver $\Gamma(K,S)$ of (K,S). Note that in case (K,S) is an exact category, then in order to know that $(f,g) \in S$ is an Auslander-Reiten sequence we only have to verify that f is a source map or that g is a sink map, because of the following lemma:

Lemma 2. *Let (K,S) be an exact category, and $(f,g) \in S$. Then f is a source map in K if and only if g is a sink map in K.*

Proof. We consider the case that $f : X \longrightarrow Y$ is a source map in K and show that its cokernel $g : Y \longrightarrow Z$ is a sink map. The opposite implication follows by duality.

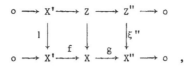

First, we show that Z is indecomposable. Let Z_1 be an indecomposable direct summand of Z in K with inclusion map $u : Z_1 \longrightarrow Z$ such that the induced exact sequence

$$\begin{array}{ccccccccc} o & \longrightarrow & X & \xrightarrow{f} & Y & \xrightarrow{g} & Z & \longrightarrow & o \\ & & {\scriptstyle 1}\big\uparrow & & \big\uparrow{\scriptstyle u'} & & \big\uparrow{\scriptstyle u} & & \\ o & \longrightarrow & X & \xrightarrow{f'} & M_1 & \xrightarrow{g'} & Z_1 & \longrightarrow & o \end{array}$$

does not split (such a summand must exist, since (f,g) does not split). Since f' is not split mono, and f is a source map in K, there exists η_1 with $f' = f\eta_1$, thus $f = f'u' = f\eta_1 u'$, and therefore $\eta_1 u'$ is an automorphism of Y. In particular, u' is split epi. Now, Z_1 is a direct summand of Z, say $Z = Z_1 \oplus Z_2$, let $p : Z \longrightarrow Z_2$ be the canonical projection. Then $u'gp = g'up = o$. Since u' is a (categorical) epimorphism, $gp = o$, since g is a cokernel, even $p = o$. This shows that $Z_1 = Z$.

Let $g' : Z' \longrightarrow Z$ be a map in K, not split epi. Construct the induced exact sequence in K

$$\begin{array}{ccccccccc} o & \longrightarrow & X & \xrightarrow{f} & Y & \xrightarrow{g} & Z & \longrightarrow & o \\ & & {\scriptstyle 1}\big\uparrow & & \big\uparrow{\scriptstyle y} & & \big\uparrow{\scriptstyle g'} & & \\ o & \longrightarrow & X & \xrightarrow{x} & M & \xrightarrow{y'} & Z' & \longrightarrow & o \end{array} .$$

We claim that the induced exact sequence splits. Assume not, then there exists $\eta : Y \longrightarrow M$ with $x = f\eta$, since f is a source map. Since $f\eta y = xy = f$, and g is a cokernel of f, we see that $\eta y - 1_Y$ factors through g, say $\eta y - 1_Y = g\zeta$, and similarly $f\eta y' = xy' = o$ implies $\eta y = g\zeta'$ for some ζ'. Thus

$$g(\zeta g + 1_Z) = (g\zeta + 1_Y)_g = \eta yg = \eta y'g' = g\zeta' g' \quad ,$$

and since g is epi, we conclude

$$1_Z = \zeta' g' - \zeta g \quad .$$

However, this is impossible, since neither g nor g' is split epi, and $\operatorname{End}(Z)$ is a local ring. Thus, the induced exact sequence splits: there exists η' with $\eta' y' = 1_{Z'}$, thus $g' = 1_{Z'} g' = \eta' y' g' = (\eta' y)g$.

Finally, assume there exists $\gamma \in \operatorname{End}(Y)$ with $g = \gamma g$ and not being an automorphism. Then there exists a non-trivial decomposition $Y = Y_1 \oplus Y_2$ with restriction of g to Y_2 being zero. Thus, we can factor the inclusion map $u_2 : Y_2 \longrightarrow Y$ through the kernel f of g, say $u_2 = \eta f$ for some $\eta : Y_2 \longrightarrow X$, and we obtain

$1 = u_2p_2 = \eta(f,p_2)$, where $p_2 : Y \longrightarrow Y_2$ is the projection map. Thus η is split mono. Since X is indecomposable and $Y_2 \neq 0$, this implies that η is an isomorphism, and therefore f is a split monomorphism, a contradiction. This finishes the proof.

A Krull-Schmidt category (K,S) with short exact sequences is said to be standard provided the categories K and $k(\Gamma(K,S))$ are equivalent.

Lemma 3. Let (K,S) be a Krull-Schmidt category with short exact sequences, and having sink maps. Let $\Gamma(K,S)$ be a preprojective translation quiver. Assume the sink map in K of any indecomposable object Z with $[Z]$ a projective vertex in $\Gamma(K,S)$, is a monomorphism in K. Then (K,S) is standard.

Proof. Let us define a functor $F : k(\Gamma(K,S)) \longrightarrow K$. Recall that the vertices of $\Gamma(K,S)$ are the isomorphism classes of indecomposable objects in K, and for any vertex x of $\Gamma(K,S)$, we choose one representative $F(x) \in x$. Also, if X is an indecomposable object in K, fix some isomorphism $\varphi_X : X \longrightarrow F([X])$. For $\alpha : y \rightarrow z$ in $\Gamma(K,S)$, we define $F(\alpha)$ by induction on the number $p(z)$ of predecessors of z. Assume $p(z) = d \geq 1$, and $F(\alpha')$ is defined for all $\alpha' : y' \longrightarrow z'$ with $p(z') < d$. Let $g : Y \longrightarrow F(z)$ be a sink map for $F(z)$ in K, let $Y = \bigoplus Y_i^{m_i}$ with all Y_i indecomposable and pairwise non-isomorphic, let $y_i = [Y_i]$ and define $\varphi = \bigoplus_i (\varphi_{Y_i})^{m_i} : \bigoplus_i Y_i^{m_i} \longrightarrow \bigoplus F(y_i)^{m_i}$. Note that there are m_i arrows from $[Y_i]$ to z, say $\alpha_{i1},\ldots,\alpha_{im_i}$. Consider first the case of z being a projective vertex in $\Gamma(K,S)$. Now $\varphi^{-1}g : \bigoplus F(y_i)^{m_i} \longrightarrow F(z)$ also is a sink map, and we define $F(\alpha_{ij})$ as the restriction of $\varphi^{-1}g$ to the j-th copy of $F(y_i)$ in $F(y_i)^{m_i}$. Note that for any i, the residue classes of $F(\alpha_{i1}),\ldots,F(\alpha_{im_i})$ modulo $\mathrm{rad}^2(F(y_i), F(z))$ give a basis of $\mathrm{Irr}(F(y_i), F(z))$. Next, let τz be defined, thus we can assume that there exists $(f,g) \in S$ with $f : X \longrightarrow Y$ being a source map of K. Let $x = [X]$. Then also $\varphi_X^{-1}f\varphi : F(x) \longrightarrow \bigoplus_i F(y_i)^{m_i}$ is a source map. Let $\beta_{i1},\ldots,\beta_{im_i}$ be the arrows from $[X]$ to $[Y_i]$, and note that $p(y_i) < d$, thus $F(\beta_{ij})$ is defined for all i,j. By induction, we assume that the maps $F(\beta_{ij})$ are defined in such a way that, for any i, the residue classes of $F(\beta_{i1}),\ldots,F(\beta_{im_i})$ modulo $\mathrm{rad}^2(F([X]),F([Y_i]))$ are a basis of $\mathrm{Irr}(F(x),F(y_i))$. As a consequence, the map $(F(\beta_{ij}))_{ij} : F(x) \longrightarrow \bigoplus_i F(y_i)^{m_i}$ is a source map, thus there exists an automorphism ψ of $\bigoplus F(y_i)^{m_i}$ with $\varphi_X^{-1}f\varphi\psi = (F(\beta_{ij}))_{ij}$. Let $F(\alpha_{ij})$ be the restriction of $\psi^{-1}\varphi^{-1}g : \bigoplus_i F(y_i)^{m_i} \longrightarrow F(z)$ to the j-th copy of $F(y_i)$ in $F(y_i)^{m_i}$. Then, for any i, $F(\alpha_{i1}),\ldots, F(\alpha_{im_i})$ give, modulo $\mathrm{rad}^2(F(y_i),F(z))$, a basis of $\mathrm{Irr}(F(y_i),F(z))$, and $\sum_{i,j} F(\beta_{ij})F(\alpha_{ij}) = 0$.

In this way, we have defined a k-linear functor F from the path category of $\Delta(K)$ to K. Actually, the last equality shows that F satisfies the mesh relations, thus $F : k(\Gamma(K,S)) \longrightarrow K$.

Note that F is dense, since for any indecomposable object X of K, there is the isomorphism $\varphi_X : X \longrightarrow F([X])$.

Let us show that F is full. Given $h : F(z') \longrightarrow F(z)$, we have to show that h is image under F. We use induction on $p(z)$. If $z' = z$, then h is an automorphism of an indecomposable object, thus a scalar multiplication, say by $c \in k$, and therefore $h = F(c\cdot 1_z)$. [Here, we use that $z \in {}_\infty\Gamma(K,S)$, thus $F(z) \in {}_\infty K$, and therefore $F(z)$ is directing]. If $z' \neq z$, then $F(z')$ and $F(z)$ are not isomorphic, thus $h \in \mathrm{rad}(F(z'),F(z))$, and therefore h can be factored through any sink map of $F(z)$. Let $\alpha_i : y_i \longrightarrow z$ be the arrows ending in z (here, the y_i are not assumed to be pairwise different), then, by construction of F, the map $(F(\alpha_i))_i : \oplus F(y_i) \longrightarrow F(z)$ is a sink map. We obtain $h = \sum_i h_i F(\alpha_i)$ for suitable $h_i : F(z') \longrightarrow F(y_i)$. Now $p(y_i) < p(z)$, thus, by induction, $h_i = F(v_i)$ for some $v_i \in k(\Gamma(K,S))$, thus $h = F(\sum_i v_i\alpha_i)$ is in the image of F.

Finally, we show that F is faithful. Consider a non-zero element $v = \sum_w c_w w$ in the path category of $\Delta(K)$, say with all w being paths from z' to z and $c_w \in k$. Assuming $F(v) = 0$, we want to show that v is in the ideal generated by the mesh relations. Of course, z' has to be a proper predecessor of z (since for $z' = z$, there is only one path from z to itself, and its image under F is the identity morphism of $F(z)$). Again, we use induction on $p(z)$. Let $\alpha_i : y_i \longrightarrow z$, $1 \leq i \leq t$, be the arrows ending in z, thus w can be written as $w = w'\alpha_w$ for some path w' and $\alpha_w \in \{\alpha_1,\dots,\alpha_t\}$. For $1 \leq i \leq t$, let $J(i) = \{w \mid \alpha_w = \alpha_i\}$, thus

$$v = \sum_w c_w w = \sum_{i=1}^{t} \Big(\sum_{w\in J(i)} c_w w' \Big)\alpha_i = \sum_{i=1}^{t} v_i\alpha_i$$

where $v_i = \sum_{w\in J(i)} c_w w'$. We have

$$(*) \qquad 0 = F(v) = \sum_{i=1}^{t} F(v_i)F(\alpha_i) ,$$

and $(F(\alpha_i))_i : \bigoplus_{i=1}^{t} F(y_i) \longrightarrow F(z)$ is a sink map. Now, if z is projective in $\Gamma(K,S)$, then by assumption this sink map is a monomorphism, thus $F(v_i) = 0$ for all i. Since $p(y_i) < p(z)$, we see by induction that all v_i belong to the ideal generated by the mesh relations, thus also $v = \sum v_i\alpha_i$. Otherwise, there exists an Auslander-Reiten sequence $(f,g) \in S$, with a commutative diagram

$$\begin{array}{ccccc} X & \xrightarrow{f} & Y & \xrightarrow{g} & Z \\ \downarrow{\scriptstyle \varphi_X} & & \downarrow & & \downarrow{\scriptstyle \varphi_Z} \\ F(x) & \xrightarrow{(F(\beta_i))_i} & \oplus F(y_i) & \xrightarrow{(F(\alpha_i))_i} & F(z) \end{array}$$

with $x = [X] = \tau z$, and $\beta_i : x \longrightarrow y_i$, $1 \leq i \leq t$, being the arrows starting at x. Note that the vertical maps all are isomorphisms. Thus, since f is a kernel of g in K, we see that $(F(\beta_i))_i$ is a kernel of $(F(\alpha))_i$ in K. As a consequence, the equality (*) gives a map $h : F(z') \longrightarrow F(x)$ with $F(v_i) = hF(\beta_i)$ for all i, and, since F is full, $h = F(v')$ for some v'. Thus $F(v_i - v'\beta_i) = o$ for all i. Since $p(y_i) < p(z)$, we know by induction that $v_i - v'\beta_i$ is in the ideal generated by the mesh relations. Also, $\Sigma\, \beta_i\alpha_i$ is a mesh relation, thus

$$v = \sum_i v_i\alpha_i = \sum_i (v_i - v'\beta_i)\alpha_i + v'(\sum_i \beta_i\alpha_i)$$

belongs to the ideal generated by the mesh relations.

<u>Remark.</u> Let us denote by $k\langle T_1,T_2\rangle$ the free associative algebra in two (non-commuting) variables T_1,T_2. Of course, the category $k\langle T_1,T_2\rangle$-mod of all (finite-dimensional left) $k\langle T_1,T_2\rangle$-modules can be identified with the category of representations of the quiver

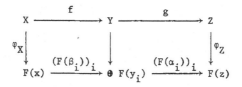

with one vertex and two loops, thus with the category of pairs of endomorphisms of k-vector-spaces. (The category $k\langle T_1,T_2\rangle$-mod is usually considered as very difficult to handle: there are families of pairwise non-isomorphic indecomposable modules depending on arbitrarily many parameters, any finite-dimensional k-algebra occurs as the endomorphism ring of some $k\langle T_1,T_2\rangle$-module, and for any n there are full exact embeddings

$$k\langle T_1,\ldots,T_n\rangle\text{-mod} \longrightarrow k\langle T_1,T_2\rangle\text{-mod},$$

with $k\langle T_1,\ldots,T_n\rangle$ the free associative algebra in the n variables $T_1,\ldots,T_n$.) A Krull-Schmidt category (K,S) with short exact sequences is said to be <u>strictly wild</u> provided there is a full exact embedding

$$k\langle T_1,T_2\rangle\text{-mod} \longrightarrow (K,S).$$

We usually will leave aside strictly wild categories.

2.4 Modules over (finite dimensional) algebras

Let A be a (finite dimensional) k-algebra. We usually will consider (finite dimensional) left modules, and call them just modules. The category of all A-modules will be denoted by A-mod; note that A-mod is a Krull-Schmidt category. Given a homomorphism $f : X \longrightarrow Y$ of A-modules, we denote by Ker f, Im f, Cok f its kernel, image, and cokernel, respectively.

An object class in A-mod will be called a module class (it is a full subcategory of A-mod closed under direct sums, direct summands and isomorphisms). We always will consider any module class $\mathcal{M}$ as a Krull-Schmidt category with short exact sequences, the short exact sequences being those in A-mod belonging to $\mathcal{M}$. In particular, the Auslander-Reiten quiver $\Gamma(\mathcal{M})$ is defined. Given an indecomposable module Z in $\mathcal{M}$, there may exist both the Auslander-Reiten translate $\tau_{\mathcal{M}}Z$ in $\mathcal{M}$ as well as the Auslander-Reiten translate $\tau_{A\text{-mod}}Z$; we always will denote $\tau_{A\text{-mod}}Z$ just by τZ or $\tau_A Z$, and will refer to $\tau_{\mathcal{M}}Z$ as the relative Auslander-Reiten translate. Similarly, the Auslander-Reiten quiver $\Gamma(A\text{-mod})$ of A-mod will just be denoted by $\Gamma(A)$, and for $\mathcal{M}$ a proper module class in A-mod, we will call $\Gamma(\mathcal{M})$ a relative Auslander-Reiten quiver.

Assume $\mathcal{M}_1, \mathcal{M}_2$ are two module classes. Let $\mathcal{M}_1 \int \mathcal{M}_2$ be the set of direct summands of modules M with a submodule M_2 in $\mathcal{M}_2$ such that $M/M_2 \in \mathcal{M}_1$. More generally

$$\mathcal{M}_1 \int \mathcal{M}_2 \int \cdots \int \mathcal{M}_m$$

will denote the set of direct summands of modules M having a chain of submodules

$$(*) \qquad M = M_1 \supseteq M_2 \supseteq \cdots \supseteq M_{m+1} = 0 \qquad \text{with } M_i/M_{i+1} \in \mathcal{M}_i.$$

Note that in case $\mathrm{Hom}(\mathcal{M}_i, \mathcal{M}_j) = 0$ for $i > j$, the chain (*) is uniquely determined by M, and the set of modules M having such a chain is closed under direct summands.

Let us consider some special module classes in A-mod. First of all, A-proj = $\langle_A A\rangle$ denotes the set of all projective modules [of course, the projective modules P are characterized by the following lifting property: given an epimorphism $\varepsilon : X \longrightarrow Y$ and $f : P \longrightarrow Y$, then there exists $f' : P \longrightarrow X$ with $f'\varepsilon = f$]. Given any A-module M, there exists a projective cover $\varphi : P \longrightarrow M$ [this is a surjective map φ, with P being projective, such that any endomorphism $\alpha \in \mathrm{End}(P)$ with $\alpha\varphi = \varphi$ is an automorphism], and it is unique up to isomorphism. By a projective presentation of M we mean an exact sequence

$$P_1 \xrightarrow{\ \delta\ } P_o \xrightarrow{\ \varepsilon\ } M \longrightarrow 0$$

with P_o, P_1 being projective; it will be said to be <u>minimal</u> provided $\varepsilon : P_o \longrightarrow M$ and $\delta : P_1 \longrightarrow \mathrm{Im}(\delta)$ are projective covers. Similarly, a <u>projective resolution</u> of M is an exact sequence

$$(*) \qquad \cdots P_{i+1} \xrightarrow{\delta_i} P_i \;\cdots\; P_1 \xrightarrow{\delta_o} P_o \xrightarrow{\varepsilon} M \longrightarrow 0$$

with all P_i being projective, and it is said to be <u>minimal</u>, provided $\varepsilon : P_o \longrightarrow M$ and all $\delta_i : P_{i+1} \longrightarrow \mathrm{Im}(\delta_i)$ are projective covers. In case M has a minimal projective resolution (*) with $P_d \neq 0$, $P_{d+1} = 0$ (and therefore $P_j = 0$ for $j > d$), we will say that the projective dimension of M is d, and write $\mathrm{proj.dim.}M = d$; in case all $P_i \neq 0$, we write $\mathrm{proj.dim.}M = \infty$. Note that $\mathrm{proj.dim.}M \leq d$ if and only if $\mathrm{Ext}^i(M,Y) = 0$ for all $i > d$ and all A-modules Y. The supremum of the projective dimension of all modules, or, equivalently, of all simple modules, is called the global dimension of A, and denoted by $\mathrm{gl.dim.}A$. We note that $\mathrm{gl.dim.}A$ can also be defined to be the smallest number d with $\mathrm{Ext}^{d+1}(X,Y) = 0$ for all X,Y.

Next, A-inj = $\langle D(A_A)\rangle$ denotes the set of all <u>injective</u> modules [the injective modules Q are characterized by the following extension property: given a mono map $\mu : X \longrightarrow Y$ and $f : X \longrightarrow Q$, there exists $f' : Y \longrightarrow Q$ with $\mu f' = f$]. Given any A-module M, there exists an <u>injective envelope</u> $\eta : M \longrightarrow Q$ [i.e. a monomorphism η with Q being injective such that any $\alpha \in \mathrm{End}(Q)$ with $\eta\alpha = \eta$ is an automorphism], and η is unique up to isomorphism. An <u>injective presentation</u> of M is an exact sequence

$$0 \longrightarrow M \xrightarrow{\mu} Q_o \xrightarrow{\delta_o} Q_1 ,$$

with Q_o, Q_1 being injective; an <u>injective resolution</u> of M is an exact sequence

$$(**) \qquad 0 \longrightarrow M \xrightarrow{\mu} Q_o \xrightarrow{\delta_o} Q_1 \;\cdots\; Q_i \xrightarrow{\delta_i} Q_{i+1} \;\cdots$$

with all Q_i injective; such an injective presentation or injective resolution is said to be <u>minimal</u> provided μ and the maps $Q_i/\mathrm{Ker}(\delta_i) \longrightarrow Q_{i+1}$ induced by δ_i are injective envelopes. In case M has a minimal injective resolution (**) with $Q_d \neq 0$, $Q_{d+1} = 0$ (and therefore $Q_j = 0$ for $j > d$), we will say that the injective dimension of M is d, and write $\mathrm{inj.dim.}M = d$; in case all $Q_i \neq 0$, we write $\mathrm{inj.dim.}M = \infty$. One has $\mathrm{inj.dim.}M \leq d$ if and only if $\mathrm{Ext}^i(X,M) = 0$ for all $i > d$ and all A-modules X. Consequently, $\mathrm{gl.dim.}\ A$ may also be defined as the supremum of the injective dimension of all modules, or also of all simple modules.

Note that the categories A-proj and A-inj are equivalent. In fact, consider the endofunctor $\nu := D\,\mathrm{Hom}_A(-,{}_AA)$ of A-mod, it will be called the <u>Nakayama functor</u>.

Its restriction to A-proj defines an equivalence from A-proj to A-inj, the inverse being given by $\nu^- := \mathrm{Hom}_A(D(A_A),-)$.

Given a finite-dimensional algebra A, we have introduced in 2.1 its quiver $\Delta(A)$, and for any Krull-Schmidt category K, there is defined its quiver $\Delta(K)$, according to 2.2. In particular, we may compare the quiver $\Delta(A)$ of A and the quiver $\Delta(A\text{-proj}) \approx \Delta(A\text{-inj})$, the last isomorphism being obtained from the Nakayama functor ν. In fact,

$$\Delta(A\text{-proj}) \approx \Delta(A)^*.$$

[Namely, without loss of generality, $A = A(Q,\{\rho_i \mid i\})$ for some finite quiver Q, and some set of relations ρ_i, with $(kQ^+)^n \subseteq \langle\rho_i \mid i\rangle$ for some $n \geq 2$. Of course, the vertices of $Q = \Delta(A)$ correspond bijectively to the (isomorphism classes of) indecomposable projective A-modules, the vertex a being associated to the module $P(a) = A(Q,\{\rho_i \mid i\})(a \mid a)$. Given vertices $a,b \in \Delta(A)$, let $\alpha_1,\ldots,\alpha_s$ be the arrows in $\Delta(A)$ from a to b. Then one easily may check that $P(\alpha_1^*),\ldots,P(\alpha_s^*)$ is a basis of $\mathrm{Irr}_{A\text{-proj}}(P(b),P(a))$; thus, in $\Delta(A\text{-proj})$, there are precisely s arrows from $[P(b)]$ to $[P(a)]$. This shows that $\Delta(A\text{-proj}) \approx Q^*$.]

Next, consider the module class of semi-simple modules, a module being <u>semi-simple</u> provided it is a direct sum of simple modules. Given any module M, the sum of all simple submodules of M is semi-simple, and is called the <u>socle</u> of M, denoted by soc M. The intersection of all maximal submodules of M is called the <u>radical</u> rad M of M. Note that M/rad M is the largest semi-simple factor module of M, called the <u>top</u> of M.

There are only finitely many isomorphism classes of simple A-modules, and we denote by $E(i)$, $1 \leq i \leq n$, a complete set of (representatives of the isomorphism classes of) simple A-modules. Note that $E(i)$, $1 \leq i \leq n$, is a set of pairwise orthogonal bricks. Let $P(i) \longrightarrow E(i)$ be a projective cover of $E(i)$ and $E(i) \longrightarrow Q(i)$ an injective envelope. Thus, $E(i) \approx \mathrm{top}\, P(i) \approx \mathrm{soc}\, Q(i)$. Then $P(i)$, $1 \leq i \leq n$, is a complete set of indecomposable projective A-modules, and $Q(i)$, $1 \leq i \leq n$, is a complete set of indecomposable injective A-modules, and $\nu P(i) \approx Q(i)$.

Any module M has a chain of submodules

$$0 = M_0 \subseteq M_1 \subseteq \cdots \subseteq M_\ell = M$$

with M_i/M_{i-1} being simple for all i; such a chain is called a <u>composition series</u>, the modules M_i/M_{i-1} the <u>composition factors</u>, and $\ell =: |M|$ the length of M. The number of times $E(j)$ occurs as a composition factor of M in a given composision series is given by $\dim_k \mathrm{Hom}(P(j),M)$, or also by $\dim_k \mathrm{Hom}(M,Q(j))$; in particular, it does not depend on the choice of the composition series ("Jordan-Hölder-theorem");

we will denote this number by $(\underline{\dim} M)_j$. In this way, we define the dimension vector $\underline{\dim} M$ of M as an element of $\mathbb{N}_o^n$, and $\ell = \sum_j (\underline{\dim} M)_j$. There is another, slightly different way for introducing $\underline{\dim}$. Let F be the free abelian group with basis the set of all A-modules, and R the subgroup generated by the formal sums $M' - M + M''$, where

$$0 \longrightarrow M' \longrightarrow M \longrightarrow M'' \longrightarrow 0$$

is an exact sequence. Then $K_o(A) := F/R$ is called the Grothendieck group of A, and, by the Jordan-Hölder-theorem, it is a free abelian group with basis the images $e(i)$ of the modules $E(i)$ under the canonical map $F \longrightarrow K_o(A)$. Using this basis, we may identify $K_o(A)$ with $\mathbb{Z}^n$. In this way, given any A-module M, the image of M (considered as a basis element of F) under the canonical map $F \longrightarrow K_o(A)$ is just $\underline{\dim} M$. Sometimes, it will be convenient to embed $K_o(A)$ into $K_o(A,\mathbb{Q}) := K_o(A) \otimes \mathbb{Q} = \mathbb{Q}^n$.

A module M is said to be sincere provided any simple module occurs as a composition factor of M, or, equivalently, $\underline{\dim} M$ is a sincere vector in $K_o(A) = \mathbb{Z}^n$. A module class $\mathcal{M}$ is said to be sincere provided there exists a (not necessarily indecomposable) sincere module M in $\mathcal{M}$. Finally, the algebra A is called sincere provided there exists a sincere indecomposable module.

We recall that a module X is generated by a module M if and only if X is isomorphic to a factor module of a direct sum of copies of M , or, equivalently, if the sum of the images of all maps $M \to X$ is equal to X . Dually, X is cogenerated by M if and only if X is isomorphic to a submodule of a direct sum of copies of M , if and only if the intersection of the kernels of all maps $X \to M$ is zero. An A-module M is said to be faithful provided the only element $a \in A$ satisfying $aM = 0$, is the element $a = 0$. The A-module M is faithful if and only if ${}_AA$ is cogenerated by M. Of course, a faithful module always is sincere.

We denote by $C = C_A$ the <u>Cartan matrix</u> of A, it is an $n \times n$-matrix, its i-j-entry is given by $\dim_k \mathrm{Hom}(P(i),P(j))$, thus the j-th column is given by $p(j)^T$, where $p(j) := \underline{\dim}\, P(j)$. Also since ν is an equivalence on A-proj, and $Q(i) = \nu P(i)$ for all i, we have

$$(\underline{\dim}\, Q(i))_j = \dim_k \mathrm{Hom}(Q(i),Q(j)) = \dim_k \mathrm{Hom}(P(i),P(j)),$$

thus the i-th row of C is given by $q(i) := \underline{\dim}\, Q(i)$. Thus

(a) $$p(i) = e(i)C^T, \qquad q(i) = e(i)C.$$

The algebras A we deal with usually will have the property that the Cartan matrix C_A is invertible (over $\mathbb{Q}$, or even over $\mathbb{Z}$). For example, if A has finite global dimension, then C_A is invertible over $\mathbb{Z}$. [Namely, if M has a finite projective resolution

$$0 \longrightarrow P_p \longrightarrow P_{p-1} \longrightarrow \dots \longrightarrow P_o \longrightarrow M \longrightarrow 0,$$

then $\underline{\dim}\, M = \sum_{j=o}^{p} (-1)^j\, \underline{\dim}\, P_j$, thus $\underline{\dim}\, M$ is an integral linear combination of the dimension vectors $p(i)$, $1 \le i \le n$. If A has finite global dimension, then every A-module has a finite projective resolution. In particular, the basis vectors $e(j) = \underline{\dim}\, E(j)$ of $K_o(A)$ are integral linear combinations of the $p(i)$, $1 \le i \le n$. In this way, we obtain a left inverse for C_A^T, thus C_A is invertible.] Also, if C_A is an upper triangular matrix, then C_A is invertible at least over $\mathbb{Q}$, and, in this case, C_A is invertible over $\mathbb{Z}$ if and only if $\mathrm{End}(P(i)) = k$, for all i. [A typical case of an algebra A with a Cartan matrix which is invertible over $\mathbb{Q}$, but not over $\mathbb{Z}$ is a local algebra A with non-zero radical].

Now assume that $C = C_A$ is invertible. The matrix C^{-T} defines a (usually non-symmetric) <u>bilinear form</u> $\langle -,- \rangle$ on $K_o(A,\mathbb{Q})$, namely

$$\langle x,y \rangle = xC^{-T}y^T,$$

the symmetrized bilinear form will be denoted by $(-,-)$, it is given by

$$(x,y) = \frac{1}{2}\, x\, (C^{-1} + C^{-T})y^T,$$

and the corresponding quadratic form will be denoted by $\chi = \chi_A$, thus $\chi(x) = (x,x) = \langle x,x \rangle$. The bilinear form $\langle -,- \rangle$ has the following homological interpretation.

Lemma. Let X,Y *be A-modules, with* proj.dim. $X < \infty$, *or* inj.dim. $Y < \infty$. *Then*

$$\langle \underline{\dim}\, X, \underline{\dim}\, Y\rangle = \sum_{t\geq o} (-1)^t \dim \mathrm{Ext}^t(X,Y) .$$

Proof. We assume proj.dim. $X = d < \infty$, the case of Y having finite injective dimension being dual. We use induction on d. For $d = o$, the module X is projective, and we assume, in addition, that X is indecomposable, say $X = P(i)$. Let $y = \underline{\dim}\, Y$, then

$$\langle \underline{\dim}\, X, \underline{\dim}\, Y\rangle = \langle p(i), y\rangle = p(i)C^{-T}y^T = e(i)y^T$$

is the i-th component of y, thus = dim Hom(P(i),Y), whereas $\mathrm{Ext}^t(P(i),Y) = 0$ for all $t > o$. Now consider the cases $d > o$. Take a projective cover $P \longrightarrow X$, with kernel X', then proj.dim. $X' = d-1$. Applying Hom(-,Y) to the exact sequence $0 \longrightarrow X' \longrightarrow P \longrightarrow X \longrightarrow 0$, we obtain a long exact Ext-sequence. Calculating dimensions, we get

$$\sum_{t\geq o} (-1)^t \dim \mathrm{Ext}^t(X,Y) = \sum_{t\geq o} (-1)^t \dim \mathrm{Ext}^t(P,Y) - \sum_{t\geq o} (-1)^t \dim \mathrm{Ext}^t(X',Y)$$

$$= \langle \underline{\dim}\, P, \underline{\dim}\, Y\rangle - \langle \underline{\dim}\, X', \underline{\dim}\, Y\rangle,$$

using induction. The latter however is equal to $\langle \underline{\dim}\, X, \underline{\dim}\, Y\rangle$, since $\underline{\dim}\, X = \underline{\dim}\, P - \underline{\dim}\, X'$. This finishes the proof.

In particular, if A is hereditary, then

$$\langle e(i), e(j)\rangle = \delta_{ij} - \dim \mathrm{Ext}^1(E(i),E(j)),$$

with δ_{ij} the Kronecker function, and it follows that χ_A coincides with the quadratic form $\chi(\overline{\Delta(A)})$ of the graph $\overline{\Delta(A)}$, as defined in 1.2.

We add some formal calculations in $K_o(A,\mathbb{Q})$ which can be done in case the Cartan matrix C_A is invertible. The meaning of these calculations will become clear at the end of this section. Let $\Phi = \Phi_A := -C^{-T}C$, and call it the *Coxeter matrix*, or, its action on $K_o(A,\mathbb{Q})$, the Coxeter transformation. We note the following properties:

(b) $$p(i)\Phi = -q(i), \quad \text{for all } 1 \leq i \leq n.$$

This is a direct consequence of (a). We may reformulate this property as follows:

(b') For $P \in A$-proj, $(\underline{\dim}\, P)\Phi = -\underline{\dim}\, \nu P$.

Namely, P is a direct sum of copies of the various P(i), and $Q(i) \approx \nu P(i)$.

(c) For all $x,y \in K_o(A,\mathbb{Q})$, $\langle x,y\rangle = -\langle y, x\Phi\rangle = \langle x\Phi, y\Phi\rangle$.

Proof: We have

$$\langle x,y\rangle = xC^{-T}y^T = (xC^{-T}y^T)^T = yC^{-1}x^T$$

$$= yC^{-T}C^TC^{-1}x^T = -\ yC^{-T}\Phi^Tx^T = -\langle y,x\Phi\rangle\ ;$$

this proves the first assertion. Applying it twice, we obtain the second assertion.

(d) For $w \in K_o(A,\mathbb{Q})$, one has $w \in \text{rad}\chi_A$ if and only if $w\Phi = w$.

Namely, we have $(w,x) = o$ for all x if and only if $w(C^{-1} + C^{-T}) = 0$, if and only if $wC^{-1} = -\ wC^{-T}$, if and only if $w\Phi = w$.

Given a non-zero Φ-invariant element $w \in K_o(A,\mathbb{Q})$, we define

$$\iota_w : K_o(A,\mathbb{Q}) \longrightarrow \mathbb{Q} \quad \text{by} \quad \iota_w(x) = \langle w,x\rangle.$$

Then ι_w is Φ-invariant.

$$\iota_w(x) = \iota_w(x\Phi) \qquad \text{for all} \quad x \in K_o(A,\mathbb{Q}).$$

[Namely, $\iota_w(x) = \langle w,x\rangle = \langle w\Phi,x\Phi\rangle = \langle w,x\Phi\rangle = \iota_w(x\Phi)$]. But this means that for all $x \in K_o(A,\mathbb{Q})$, the element $x\Phi - x$ belongs to the kernel of ι_w.

Let us consider now source maps and sink maps in module classes. First, dealing with all of A-mod, the main result is the following:

Theorem of Auslander-Reiten. Any indecomposable A-module has a source map and a sink map in A-mod.

In fact, if M is indecomposable and projective, then the inclusion map rad $M \longrightarrow M$ is a sink map, if M is indecomposable and not projective, then the sink map for M is an epimorphism. Dually, if M is indecomposable and injective, then the projection map $M \longrightarrow M/\text{soc } M$ is a source map, if M is indecomposable and not injective, then the source map for M is a monomorphism. Thus, given an indecomposable non-projective module Z, with sink map $g : Y \longrightarrow Z$, let $f : X \longrightarrow Y$ be its kernel. Since g is surjective, (f,g) is an exact sequence in A-mod. Applying lemma 2 of 2.3 for $\mathcal{K}$ = A-mod and $\mathcal{S}$ = {all short exact sequences in A-mod}, we see that X is indecomposable (and non-injective) and f is a source map. Dually, starting with an indecomposable non-injective module X, with source map $f : X \longrightarrow Y$ and cokernel $g : Y \longrightarrow Z$, then Z is indecomposable (and non-projective), and g is a sink map. In this way, we obtain a bijection between the isomorphism classes of indecomposable non-projective modules Z and the isomorphism classes of indecomposable non-injective modules X, given by the Auslander-Reiten translate $\tau = \tau_{A\text{-mod}}$.

Of importance is the actual construction of τM, where M is indecomposable (here, we allow M to be projective; however, if M is projective, we will obtain $\tau M = 0$). We start with a minimal projective presentation of M, say

$$(+) \qquad 0 \longrightarrow M'' \longrightarrow P_1 \xrightarrow{p} P_0 \longrightarrow M \longrightarrow 0$$

(we have added, to the left, $M'' = \text{Ker } p$). Then τM is given by the kernel of νp. Note that the Nakayama functor ν is right exact, thus we obtain the following exact sequence

$$(++) \qquad 0 \longrightarrow \tau M \longrightarrow \nu P_1 \xrightarrow{\nu p} \nu P_0 \longrightarrow \nu M \longrightarrow 0 .$$

In case M is indecomposable and not projective, τM is indecomposable, and we obtain in this way a minimal injective presentation of τM (with the cokernel νM added to the right). [Actually, we do not need to start with a minimal projective presentation (+); it is sufficient to have an exact sequence (+) with $p : P_1 \longrightarrow \text{Im}(p)$ a projective cover.]

There is the dual construction for τ^-. Given an indecomposable A-module N, let

$$(+++) \qquad 0 \longrightarrow N \longrightarrow Q_0 \xrightarrow{q} Q_1 \longrightarrow N'' \longrightarrow 0$$

be a minimal injective presentation, with $N'' = \text{Cok } q$. Let $\tau^- N = \text{Cok}\nu^- q$, thus the left exact functor ν^- gives the following exact sequence

$$0 \longrightarrow \nu^- N \longrightarrow \nu^- Q_0 \xrightarrow{\nu^- q} \nu^- Q_1 \longrightarrow \tau^- N \longrightarrow 0 .$$

If N is injective, then $\tau^- N = 0$. Otherwise, $\tau^- N$ is indecomposable, and we obtain in this way a minimal projective presentation of $\tau^- N$ (with the kernel $\nu^- N$ added to the left). [Again, we only need that (+++) is an exact sequence, with Q_0 injective, and $\text{Im}(q) \hookrightarrow Q_1$ an injective envelope.]

There are several consequences:

(0) *If* $\text{proj.dim.} M \le 1$, *then* $\tau M \approx D\, \text{Ext}^1(M, {}_A A)$.

(0*) *If* $\text{inj.dim.} M \le 1$, *then* $\tau^- M \approx \text{Ext}^1(DM, {}_A A)$.

Proof. Assume $\text{proj.dim.} M \le 1$, and let (+) be a minimal projective presentation, thus $M'' = 0$. Application of $\text{Hom}(-, {}_A A)$ gives the exact sequence

$$0 \longrightarrow \text{Hom}(M, {}_A A) \longrightarrow \text{Hom}(P_0, {}_A A) \longrightarrow \text{Hom}(P_1, {}_A A) \longrightarrow \text{Ext}^1(M, {}_A A) \longrightarrow 0$$

of right A-modules. If we dualize this sequence, and use that $\nu = D\,\text{Hom}(-, {}_A A)$ we obtain

$$0 \longrightarrow D\, \text{Ext}^1(M, {}_A A) \longrightarrow \nu P_1 \longrightarrow \nu P_0 \longrightarrow \nu M \longrightarrow 0.$$

This gives (0). Dually, one shows (0*).

(1) <u>For any module</u> M, <u>we have</u> proj.dim. $M \leq 1$ <u>if and only if</u> $\mathrm{Hom}(D(A_A),\tau M) = 0$.

<u>Proof.</u> We may suppose M being indecomposable. Applying the left exact functor $\nu^- = \mathrm{Hom}(D(A_A),-)$ to (++), we obtain the exact sequence

$$0 \longrightarrow \nu^-\tau M \longrightarrow \nu^-\nu P_1 \xrightarrow{\nu^-\nu p} \nu^-\nu P_o .$$

Now, $\nu^-\nu p \approx p$, thus $M'' = 0$ if and only if $\mathrm{Hom}(D(A_A),\tau M) = \nu^-\tau M = 0$. Of course, $M'' = 0$ if and only if proj.dim. $M \leq 1$, since (+) is a minimal projective presentation.

There is the dual assertion:

(1*) <u>For any module</u> N, <u>we have</u> inj.dim. $N \leq 1$ <u>if and only if</u> $\mathrm{Hom}(\tau^- N, {}_A A) = 0$.

Also, the construction of τM outlined above allows us to give a concise formula for the dimension vector $\underline{\dim}\,\tau M$. Using the notation of (+), one has

(2) $$\underline{\dim}\,\tau M = (\underline{\dim}\,M)\Phi - (\underline{\dim}\,M'')\Phi + \underline{\dim}\,\nu M.$$

<u>Proof.</u> From the projective presentation (+) of M, we obtain

$$-\underline{\dim}\,P_1 + \underline{\dim}\,P_o = \underline{\dim}\,M - \underline{\dim}\,M''.$$

Applying Φ to this sequence, and using $(\underline{\dim}\,P)\Phi = -\underline{\dim}\,\nu P$ for P projective, we obtain

$$\underline{\dim}\,\nu P_1 - \underline{\dim}\,\nu P_o = (\underline{\dim}\,M)\Phi - (\underline{\dim}\,M'')\Phi.$$

From the injective presentation of τM, we obtain

$$\begin{aligned}\underline{\dim}\,\tau M &= \underline{\dim}\,\nu P_1 - \underline{\dim}\,\nu P_o + \underline{\dim}\,\nu M \\ &= (\underline{\dim}\,M)\Phi - (\underline{\dim}\,M'')\Phi + \underline{\dim}\,\nu M .\end{aligned}$$

We note the following corollaries:

(3) <u>If</u> proj.dim. $M \leq 2$, <u>then</u> $\underline{\dim}\,\tau M \geq (\underline{\dim}\,M)\Phi$. <u>If, in addition,</u> inj.dim. $\tau M \leq 2$, <u>then the difference</u> $\underline{\dim}\,\tau M - (\underline{\dim}\,M)\Phi$ <u>is the dimension vector of an injective module</u>.

<u>Proof.</u> If proj.dim. $M \leq 2$, then M'' is projective, thus $(\underline{\dim}\,M'')\Phi = -\underline{\dim}\,\nu M''$, and therefore $\underline{\dim}\,\tau M - (\underline{\dim}\,M)\Phi$ is equal to the dimension vector of $\nu M'' \oplus \nu M$, and therefore non-negative. Note that $\nu M''$ is injective. If, in addition, inj.dim. $\tau M \leq 2$, then also νM is injective, since it is the cokernel of the map $\nu p : \nu P_1 \longrightarrow \nu P_o$, with both $\nu P_1, \nu P_o$ being injective.

Remark. Note that the proof above shows that for algebras of global dimension ≤ 2, the Nakayama functor ν maps the whole category A-mod to the full subcategory A-inj of the injective A-modules.

(4) If proj.dim. $M \leq 1$ and $\mathrm{Hom}(M,{}_AA) = 0$, then $\underline{\dim}\,\tau M = (\underline{\dim}\,M)\Phi$.

Proof. If proj.dim. $M \leq 1$, then $M'' = 0$, since we deal with a minimal presentation, and by assumption $\nu M = D\,\mathrm{Hom}(M,{}_AA) = 0$. Thus this is a direct consequence of (2).

It follows from (the proof of) (3) that for M with proj.dim. $M \leq 2$, also the converse is true: we have $\underline{\dim}\,\tau M = (\underline{\dim}\,M)\Phi$ only provided proj.dim. $M \leq 1$ and $\mathrm{Hom}(M,{}_AA) = 0$. [Namely, the equality means that $\nu M'' \oplus \nu M = 0$, and since M'' is projective, we have $\nu M'' = 0$ if and only if $M'' = 0$]. Of course, we should point out that for non-zero M, the two conditions proj.dim. $M \leq 1$ and $\mathrm{Hom}(M,{}_AA) = 0$ together immediately imply proj.dim. $M = 1$.

We want to stress that assertion (4) gives us a module theoretic interpretation of the Coxeter transformation Φ, under special assumptions on modules, and that (2) and (3) give precise formulas for the perturbation in general. Let us formulate the corresponding results for τ^- and Φ^{-1}; we use the notations of (+++):

(2*) $\underline{\dim}\,\tau^-N = (\underline{\dim}\,N)\Phi^{-1} - (\underline{\dim}\,N'')\Phi^{-1} + \underline{\dim}\,\nu^-N$.

(3*) If inj.dim. $N \leq 2$, then $\underline{\dim}\,\tau^-N \geq (\underline{\dim}\,N)\Phi^{-1}$. If, in addition, proj.dim. $\tau^-N \leq 2$, then the difference $\underline{\dim}\,\tau^-N - (\underline{\dim}\,N)\Phi^{-1}$ is the dimension vector of a projective module.

(4*) If inj.dim. $N \leq 1$ and $\mathrm{Hom}(D(A_A),N) = 0$, then $\underline{\dim}\,\tau^-N = (\underline{\dim}\,N)\Phi^{-1}$.

One of the main interests for the use of τ and τ^- lies in the fact that it allows us to reduce the computation of Ext-groups to that of Hom-groups. Given modules X,Y, let $\underline{\mathrm{Hom}}(X,Y)$ be the factor group of $\mathrm{Hom}(X,Y)$ modulo the subgroup of all maps $X \longrightarrow Y$ which factor through a projective module. Similarly, let $\overline{\mathrm{Hom}}(X,Y)$ be the factor group of $\mathrm{Hom}(X,Y)$ modulo the subgroup of all maps $X \longrightarrow Y$ which factor through an injective module. The following result is basic for the use of τ and τ^-; for a proof, we have to refer to [AR1].

(5) For all modules X,Y, one has

$$\mathrm{Ext}^1(X,Y) \approx D\,\overline{\mathrm{Hom}}(Y,\tau X) \approx D\,\underline{\mathrm{Hom}}(\tau^-Y,X),$$

Combining (1) and (1*) with (3), we obtain the following consequence: if we assume that proj.dim. $X \leq 1$, then $\mathrm{Hom}(D(A_A),\tau X) = 0$, according to (1), thus any homomorphism $Y \longrightarrow \tau X$ factoring through an injective module has to be zero, thus $\overline{\mathrm{Hom}}(Y,\tau X) = \mathrm{Hom}(Y,\tau X)$. This shows:

(6) *If* proj.dim. $X \leq 1$, *and* Y *arbitrary, then*

$$\mathrm{Ext}^1(X,Y) \approx D\,\mathrm{Hom}(Y,\tau X).$$

(6*) *If* inj.dim. $Y \leq 1$, *and* X *arbitrary, then*

$$\mathrm{Ext}^1(X,Y) \approx D\,\mathrm{Hom}(\tau^- Y,X).$$

We recall that an indecomposable object of a Krull-Schmidt category K is called directing provided it does not belong to a cycle in K. Thus, given an algebra A, an indecomposable A-module will be said to be *directing* provided it does not belong to a cycle in A-mod. [Note that a cyclic path in the Auslander-Reiten quiver $\Gamma(A)$ of A gives rise to a cycle in A-mod; however, there usually will exist additional cycles both consisting of modules belonging to a single component as well as cycles consisting of modules belonging to different components of $\Gamma(A)$.]

(7) *Let* N *be a sincere, directing* A-*module. Then* proj.dim. $N \leq 1$, inj.dim.$N \leq 1$ *and* gl.dim.$A \leq 2$. *Also, either* N *projective or* $\underline{\dim}\, \tau N = (\underline{\dim}\, N)\Phi$, *and similarly, either* N *injective or* $\underline{\dim}\, \tau^- N = (\underline{\dim}\, N)\Phi^{-1}$.

Proof. Assume proj.dim.$N \geq 2$. According to (1), we have $\mathrm{Hom}(Q,\tau N) \neq 0$ for some indecomposable injective A-module Q. Since N is sincere, $\mathrm{Hom}(N,Q) \neq 0$, thus $N \preceq Q \preceq \tau N \prec N$ shows that N belongs to a cycle. Dually, inj.dim.$N \leq 1$.

Also, assume gl.dim.$A \geq 3$, then $\mathrm{Ext}^3(U,V) \neq 0$ for some indecomposable A-modules U,V. Consider a projective cover $P \longrightarrow U$ of U, say with kernel U'. Then $0 \neq \mathrm{Ext}^3(U,V) \approx \mathrm{Ext}^2(U',V)$, thus there exists an indecomposable direct summand U'' of U' with $\mathrm{Ext}^2(U'',V) \neq 0$. In particular, proj.dim. $U'' \geq 2$, thus again $\mathrm{Hom}(Q',\tau U'') \neq 0$ for some indecomposable injective A-module Q'. Since U'' is a submodule of P, there exists an indecomposable direct summand P' of P with $\mathrm{Hom}(U'',P') \neq 0$. Also, $\mathrm{Hom}(N,Q') \neq 0$ and $\mathrm{Hom}(P',N) \neq 0$, since N is sincere. Thus, $N \preceq Q' \preceq \tau U'' \prec U'' \preceq P' \preceq N$ shows that N belongs to a cycle.

Finally, if N is not projective, then $\mathrm{Hom}(N,P) = 0$, for any indecomposable projective module P, since otherwise we obtain a cycle $P \preceq N \prec P$. Thus $\underline{\dim}\, \tau N = (\underline{\dim}\, N)\Phi$, according to (4).

(7') *Let* M *be a directing* A-*module, and* e_0, e_1 *two primitive idempotents. Let* $a = e_1 a e_0 \in A$ *be non-zero. Then the multiplication map* $a_M : e_0 M \to e_1 M$ *is mono or epi. As a consequence, any sincere and directing module is faithful.*

Proof. Let $P_i = Ae_i$, $i = 0,1$, and let $\bar{a} : P_1 \to P_0$ be given by left multiplication with a. Note that $e_i M$ $\mathrm{Hom}(P_i,M)$ $D\,\mathrm{Hom}(M,\nu P_i)$, canonically, and that the map a_M may be identified in this way with $\mathrm{Hom}(\bar{a},M)$ and with $D\,\mathrm{Hom}(M,\nu\bar{a})$. Now assume a_M is neither mono nor epi. In particular, $\bar{a}$ is not invertible, let C be its cokernel, then there is an exact sequence

$$P_1 \xrightarrow{\bar{a}} P_0 \longrightarrow C \longrightarrow 0$$

which actually is a minimal projective presentation of C, since both P_0, P_1 are indecomposable projective. Applying $\mathrm{Hom}(\ ,M)$ to this sequence, we see that the kernel of $\mathrm{Hom}(\bar{a},M)$ is given by $\mathrm{Hom}(C,M)$. According to (++), we obtain the exact sequence

$$0 \longrightarrow \tau C \longrightarrow \nu P_1 \xrightarrow{\nu\bar{a}} \nu P_0 ,$$

and applying $D\,\mathrm{Hom}(M,-)$, we see that the cokernel of $D\,\mathrm{Hom}(M,\nu\bar{a})$ is given by $D\,\mathrm{Hom}(M,\tau C)$. Since we assume that a_M is neither mono nor epi, $\mathrm{Hom}(C,M) \neq 0$ and $\mathrm{Hom}(M,\tau C) \neq 0$, thus we obtain a cycle $M \preceq \tau C \prec C \preceq M$, impossible.

Now assume M is both sincere and directing. Given a primitive idempotent e, we have $eM \neq 0$, thus the multiplication map $a_M : e_0M \to e_1M$ considered above is non-zero. Consequently, for any non-zero element b of A, the linear transformation $M \to M$ given by left multiplication by b is non-zero (write $1 = \sum_{i=1}^{n} e_i$ with primitive and orthogonal idempotents e_i, then $M = \oplus\, e_iM$).

<u>Remark</u>: Let us reformulate (7') in terms of representations of quivers. Let Δ be a finite quiver, $\{\rho_i \mid i\}$ a set of relations, and $A = A(\Delta,\{\rho_i^* \mid i\})$. Let $M = (M_x, M_\alpha)$ be an A-module, considered as a representation of Δ. Now assume M is directing, and let w be a path from a to b in Δ, then either w^* belongs to the ideal $\langle \rho_i^* \mid i\rangle$ in $k\Delta^*$, or else the induced map $M_w : M_a \to M_b$ is mono or epi. In particular, all the maps M_α with α an arrow in Δ, are mono or epi.

(8) <u>Let</u> N <u>be a directing</u> A-<u>module, Then</u> $\mathrm{End}(N) = k$, <u>and</u> $\mathrm{Ext}^i(N,N) = 0$ <u>for all</u> $i \geq 1$. (<u>Thus, if</u> $\mathrm{gl.dim.}A < \infty$, <u>then</u> $\underline{\dim}\, N$ <u>is a root for</u> χ_A). <u>Also, if</u> N' <u>is an indecomposable</u> A-<u>module with</u> $\underline{\dim}\, N = \underline{\dim}\, N'$, <u>then</u> $N \approx N'$.

<u>Proof.</u> Since there is no cycle of length 1 containing N, we see that N is a brick. In order to calculate $\mathrm{Ext}^i(N,N)$, let us denote by $\mathcal{X}'$ the module class defined by all indecomposable modules X satisfying $X \preceq N$. By induction on j, we show $\mathrm{Ext}^j(\mathcal{X}',N) = 0$. Thus, let X be an indecomposable module with $X \preceq N$. The existence of any non-zero element in $\mathrm{Ext}^1(X,N)$ would imply $N \prec X$, thus we would obtain a cycle containing N. Now assume $\mathrm{Ext}^j(\mathcal{X}',N) = 0$, for some $j \geq 1$ and let $P \longrightarrow X$ be a projective cover of X, with kernel X'. Then $\mathrm{Ext}^{j+1}(X,N) \approx \mathrm{Ext}^j(X',N) = 0$, since with X also X' belongs to $\mathcal{X}'$. This finishes the induction proof, and therefore $\mathrm{Ext}^i(N,N) = 0$ for all $i \geq 1$. Of course, if $\mathrm{gl.dim.}A < \infty$, then χ_A is defined and $\chi_A(\underline{\dim}\, N) = \sum_{i\geq o} (-1)^i \dim \mathrm{Ext}^i(N,N) = \dim \mathrm{End}(N) = 1$.

Now, let N' be an indecomposable A-module with $\underline{\dim}\, N = \underline{\dim}\, N'$. We want to show that N and N' are isomorphic. Without loss of generality, we may assume that N (and thus also N') is sincere. [Namely, replace A by A/I, where I

is the ideal of A generated by all idempotents e of A satisfying $eN = 0$.] As a consequence, $\text{proj.dim.} N \leq 1$, $\text{inj.dim.} N \leq 1$, and $\text{gl.dim.} A \leq 2$. In particular, the bilinear form $\langle -,- \rangle$ is defined. Using that $\underline{\dim}\, N = \underline{\dim}\, N'$ is a root of χ_A, and that $\text{proj.dim.} N \leq 1$, we see

$$\begin{aligned} 1 &= \langle \underline{\dim}\, N, \underline{\dim}\, N \rangle = \langle \underline{\dim}\, N, \underline{\dim}\, N' \rangle \\ &= \dim \operatorname{Hom}(N,N') - \dim \operatorname{Ext}^1(N,N'), \end{aligned}$$

thus $\operatorname{Hom}(N,N') \neq 0$. Similarly, using that $\text{inj.dim.} N \leq 1$, we conclude $\operatorname{Hom}(N',N) \neq 0$. However, since N is directing, we can have $N \preceq N'$ and $N' \preceq N$ only in case $N \approx N'$.

An algebra A will be said to be <u>directed</u> provided every indecomposable A-module is directing.

(9) <u>Let A be a finite dimensional directed algebra. If $\text{gl.dim.} A \leq 2$ (for example, if A is sincere), then χ_A is weakly positive, and $\underline{\dim}$ furnishes a bijection between the indecomposable A-modules and the positive roots of χ_A.</u>

<u>Proof.</u> Let x be positive in $K_0(A)$, thus $x = \underline{\dim}\, X$ for some non-zero A-module X, and we choose X with smallest possible $\dim \operatorname{End}(X)$. Let $X = \bigoplus_i X_i$ with all X_i indecomposable. Then $\operatorname{Ext}^1(X_i, X_j) = 0$ for $i \neq j$, according to 2.3. Since the X_i are directing, also $\operatorname{Ext}^1(X_i, X_i) = 0$ for all i, thus $\operatorname{Ext}^1(X,X) = 0$. Therefore

$$\chi_A(x) = \chi_A(\underline{\dim}\, X) = \dim \operatorname{End}(X) + \dim \operatorname{Ext}^2(X,X) > 0.$$

This shows that χ_A is weakly positive. Also, if x is, in addition, a root of χ_A, then $1 = \chi_A(x) = \dim \operatorname{End}(X) + \dim \operatorname{Ext}^2(X,X)$ shows that $\operatorname{End}(X) = k$, thus X is indecomposable. This shows that the map $\underline{\dim}$ from the set of isomorphism classes of indecomposable modules to the set of positive roots is not only injective (see (8)), but also surjective. This finishes the proof.

(9') <u>Corollary.</u> <u>A directed algebra is representation-finite.</u>

<u>Proof.</u> First, assume that A is directed, and in addition, sincere. According to (9), χ_A has only finitely many positive roots, and $\underline{\dim}$ furnishes a bijection between the indecomposable A-modules and the positive roots of χ_A. Thus A is representation-finite. In general, if A is directed, but not necessarily sincere, we consider for any subset I of $\Delta(A)_0$ the set S_I of indecomposable modules M with support of $\underline{\dim}\, M$ being equal to I. These modules are sincere over a suitable factor algebra A_I of A (modulo an ideal generated by an idempotent), and with A

also A_I is directed. If S_I is non-empty, then A_I is sincere, thus by the previous consideration, S_I consists of only finitely many isomorphism classes. Since any indecomposable module belongs to some S_I, and there are only finitely many possible subsets I, it follows that A is representation-finite.

(9") Corollary. Let A be a directed algebra, and M an indecomposable A-module. Then, the components of $\underline{\dim} M$ are bounded by 6.

Proof. We can assume that M is, in addition, sincere. Thus A is directed and sincere, and therefore χ_A is weakly positive. Since $\underline{\dim} M$ is a positive root, we can apply Theorem 2 of Ovsienko (see 1.0).

A vector x in $K_o(A)$ is said to be connected provided its support is connected with respect to the quiver of A. Of course, given an indecomposable A-module X, then $\underline{\dim} X$ always is connected. In case χ_A is weakly positive, we will say that A-mod is controlled by χ_A provided $\underline{\dim}$ furnishes a bijection between the indecomposable A-modules and the connected positive roots of χ_A. Thus, a directed algebra A with gl.dim.$A \leq 2$ is controlled by χ_A (and every positive root of χ_A is connected). Similarly, if U is a subset of $K_o(A)$ such that the restriction $\chi_A|U$ of χ_A to U is positive semi-definite, and $\mathcal{X}$ is a module class in A-mod, we will say that $\mathcal{X}$ is controlled by $\chi_A|U$ provided the following conditions are satisfied:

(α) For any indecomposable A-module X in $\mathcal{X}$, $\underline{\dim} X$ is either a (connected, positive) root or a (connected, positive) radical vector of χ_A in U.

(β) For any connected, positive root x of χ_A in U, there is precisely one isomorphism class of indecomposable A-modules X in $\mathcal{X}$ satisfying $\underline{\dim} X = x$.

(β') For any connected, positive radical vector x of χ_A in U, there is an infinite family of isomorphism classes of indecomposable A-modules X in $\mathcal{X}$ satisfying $\underline{\dim} X = x$.

By abuse of language, the Auslander-Reiten quiver Γ(A-mod) of A-mod will also be called the Auslander-Reiten quiver of A, and denoted just by $\Gamma(A)$.

(10) (Auslander) Let A be a connected, finite-dimensional algebra, and $\mathcal{C}$ a finite component of $\Gamma(A)$. Then $\mathcal{C} = \Gamma(A)$.

Proof. According to Lemma 6 in 2.2, any indecomposable module X with $\mathrm{Hom}(X,\mathcal{C}) \neq o$ or $\mathrm{Hom}(\mathcal{C},X) \neq o$ belongs to $\mathcal{C}$. Choose some indecomposable module C in $\mathcal{C}$. There is some indecomposable projective module P with $\mathrm{Hom}(P,C) \neq o$, thus $P \in \mathcal{C}$. Since A is connected, there are sequences of paths in A-mod connecting any two indecomposable projective modules, thus all belong to $\mathcal{C}$. Since for any indecomposable A-module M, there is a non-zero map from an indecomposable projective module to M, also $M \in \mathcal{C}$.

A component $\mathcal{C}$ of A-mod will be called a preprojective component, or a preinjective component, provided $\Gamma\mathcal{C}$ is a preprojective, or preinjective translation quiver, respectively.

(11) Let $\mathcal{C}$ be a preprojective component of A-mod. Then $\mathcal{C}$ is standard. Given an indecomposable module X in $\mathcal{C}$, then X is directing, and any indecomposable module Y with $Y \preceq X$ also belongs to $\mathcal{C}$.

Of course, there are the dual assertions for a preinjective component.

Proof. According to 2.1.4, a preprojective component $\mathcal{C}$ of A-mod is contained in ${}_\infty$(A-mod), thus any indecomposable module X in $\mathcal{C}$ is directing (2.2.4), and $\mathrm{Hom}(Y,X) \neq 0$, with Y indecomposable, implies $Y \in \mathcal{C}$ (2.2.5). By induction, even $Y \preceq X$ implies $Y \in \mathcal{C}$. The first assertion follows from 2.3.3, since the sink map for the indecomposable projective module P is just the inclusion map $\mathrm{rad}\, P \longrightarrow P$, and therefore a monomorphism.

Recall that a finite-dimensional k-algebra A is said to be hereditary provided submodules of projective modules are projective, again; or, equivalently, $\mathrm{gl.dim}\, A \leq 1$, or equivalently provided A is Morita equivalent to the path algebra of a finite quiver without cyclic paths. Using the first characterization, it is easy to see that A is hereditary if and only if its Auslander-Reiten quiver $\Gamma(A)$ is a hereditary translation quiver.

(12) Let A be a (finite dimensional) connected hereditary algebra. Then A has a preprojective component containing all indecomposable projective modules. If A is Morita equivalent to $k\Delta$, where Δ is some quiver, then the orbit quiver of A is given by Δ^* with all labels being zero.

Proof. Without loss of generality, we can assume $A = k\Delta$. The indecomposable projective modules are of the form $P(x) = A(x|x)$, with $x \in \Delta_0$. Note that $P(x)$ has a k-basis consisting of the set of all paths of Δ^* ending in x. Any arrow $\alpha : y \longrightarrow z$ in Δ_1 gives rise to a monomorphism $\alpha^* : P(z) \longrightarrow P(y)$, given by right multiplication by α^*. Note that if $\alpha_1,\dots,\alpha_n$ are all the arrows from y to z, then the residue classes of the corresponding maps $\alpha_1^*,\dots,\alpha_n^* : P(y) \longrightarrow P(z)$ modulo $\mathrm{rad}^2(P(y),P(z))$ form a basis of $\mathrm{Irr}(P(y),P(z))$. It follows that the full subquiver of $\Gamma(A)$ given by the projective vertices can be identified with Δ^*. Using 2.1 (4), we see that ${}_\infty\Gamma(A)$ is a preprojective translation quiver containing all projective vertices, and closed under neighbors in Δ. Since Δ is connected, also ${}_\infty\Gamma(A)$ is connected, thus it is a component of $\Gamma(A)$.

(13) Theorem of Gabriel. Let Δ be a finite quiver. Then Δ is representation finite if and only if the underlying graph $\bar{\Delta}$ of Δ is a disjoint union of Dynkin graphs. In this case, $\underline{\dim}$ yields a bijection between the indecomposable representations of Δ and the positive roots of $\chi(\bar{\Delta})$.

Proof. Let $A = k\Delta$, let $\chi = \chi_A = \chi(\bar{\Delta})$ and let $C = {}_\infty\Gamma(A)$ be the preprojective component of A-mod. Any indecomposable A-module M in C is directing, according to 2.2. First, assume C is finite. Without loss of generality, we can assume A to be connected. According to (10), C = A-mod, thus A is representation finite and any indecomposable module is directing. As a consequence, χ is weakly positive, and $\underline{\dim}$ gives a bijection between indecomposable A-modules and positive roots. However, $\chi(\bar{\Delta})$ is weakly positive if and only if $\bar{\Delta}$ is a disjoint union of Dynkin graphs. Conversely, assume C is infinite then, using (8), we obtain as $\{\underline{\dim}\, C \mid C$ indecomposable in $C\}$ an infinite set of positive roots of $\chi(\bar{\Delta})$, thus $\chi(\bar{\Delta})$ is not weakly positive. This finishes the proof.

(14) Let A be a connected, hereditary algebra which is representation infinite, with quiver $\Delta = \Delta(A)$. In A-mod, there is a preprojective component P, containing all projective modules, and no indecomposable injective module, and there is a preinjective component Q, containing all injective modules, and no indecomposable projective module. Also, $P \approx k(\mathbb{N}_0\Delta^*)$, and $Q \approx k((-\mathbb{N}_0)\Delta^*)$.

Proof. Let P be the preprojective component of A-mod. Using (11) for the opposite algebra A^{op} of A, we see that there also is a preinjective component Q, and that Q contains all injective modules. If P contains any indecomposable injective module, then it follows $P = Q$, thus any τ-orbit in P is finite. Since P has only finitely many τ-orbits, P would be a finite component, thus A would be representation finite, according to (10). Thus P contains no indecomposable injective module, and $\Gamma P = \mathbb{N}_0\Delta^*$ (note that A is Morita equivalent to $k\Delta^*$). Dually, Q contains no indecomposable projective module, and $\Gamma Q = (-\mathbb{N}_0)\Delta^*$. Since preprojective, and preinjective components are standard, it follows that $P \approx k(\mathbb{N}_0\Delta^*)$, $Q \approx k((-\mathbb{N}_0)\Delta^*)$.

Let us stress that for A being hereditary, the two constructions τ, τ^- both are functorial [namely, as we have seen in (0) and (0*), they are given by $D\mathrm{Ext}^1(-,{}_AA)$, and $\mathrm{Ext}^1(D-,{}_AA)$, respectively], and that τ^- is left adjoint to τ.

2.5 Subspace categories and one-point extensions of algebras

Let K be a Krull-Schmidt category, and $|\cdot| : K \to$ k-mod an additive functor. The pair $(K,|\cdot|)$ will be called a vectorspace category.

By $\check{U}(K,|\cdot|)$ we denote the category of triples $V = (V_0,V_\omega,\gamma_V : V_\omega \longrightarrow |V_0|)$, where V_0 is an object in K, V_ω is a (finite dimensional) k-space, and γ_V is k-linear. Given two such triples V,W, a map $V \longrightarrow W$ is given by a pair $f = (f_0,f_\omega)$ with $f_0 : V_0 \longrightarrow W_0$ a map in K, and $f_\omega : V_\omega \longrightarrow W_\omega$ being k-linear, such that $\gamma_V|f_0| = f_\omega\gamma_W$. By $U(K,|\cdot|)$ we denote the full subcategory of $\check{U}(K,|\cdot|)$ consisting of all triples $V = (V_0,V_\omega,\gamma_V)$ with γ_V being mono; it is called the subspace category of $(K,|\cdot|)$. [Note that we do not assume that the functor $|\cdot|$ is faithful.] Obviously, both $\check{U}(K,|\cdot|)$ and $U(K,|\cdot|)$ are Krull-Schmidt categories again, and the only indecomposable object of $\check{U}(K,|\cdot|)$ not belonging to $U(K,|\cdot|)$ is $E_K(\omega) = E(\omega) := (0,k,o)$. We also may characterize the objects V in $U(K,|\cdot|)$ as those objects of $\check{U}(K,|\cdot|)$ satisfying $\mathrm{Hom}(E(\omega),V) = 0$.

There is a general duality principle for subspace categories. Given a category K, denote by K^{op} the opposite category, the objects of K^{op} being the same as those in K, and with $\mathrm{Hom}_{K^{op}}(X,Y) = \mathrm{Hom}_K(Y,X)$. If $|\cdot| : K \longrightarrow$ k-mod is a functor, then $|\cdot| : K^{op} \longrightarrow$ k-mod is a contravariant functor, thus $D|\cdot| : K^{op} \longrightarrow$ k-mod is a functor again. Of course, if K is an additive category, and $|\cdot|$ is an additive functor, then also $D|\cdot| : K^{op} \longrightarrow$ k-mod is an additive functor. We have

(1) The contravariant functor $* : U(K,|\cdot|) \longrightarrow U(K^{op},D|\cdot|)$ sending $V = (V_0,V_\omega,\gamma_V)$ to $V^* = (V_0,\mathrm{Ker}\, D\gamma_V,u)$, with $u : \mathrm{Ker}\, D\gamma_V \longrightarrow D|V_0|$ the inclusion map, defines a duality between $U(K,|\cdot|)$ and $U(K^{op},D|\cdot|)$.

The proof ist straightforward.

A map $f = (f_0,f_\omega)$ in $\check{U}(K,|\cdot|)$ will be said to be K-split mono, provided f_0 is split mono in K, and f_ω is mono; and f is said to be K-split epi, provided f_0 is split epi in K, and f_ω is epi. Note that for μ, a K-split mono map in $U(K, |\cdot|)$, we can construct its cokernel in $\check{U}(K, |\cdot|)$, and similarly, for ε being K-split epi, we can construct its kernel in $\check{U}(K,|\cdot|)$. In both situations, we obtain a sequence

$$(*) \qquad 0 \longrightarrow U \xrightarrow{\mu} V \xrightarrow{\varepsilon} W \longrightarrow 0 ,$$

with

$$0 \longrightarrow U_o \xrightarrow{\mu_o} V_o \xrightarrow{\varepsilon_o} W_o \longrightarrow 0,$$

being split exact in K, and

$$0 \longrightarrow U_\omega \xrightarrow{\mu_\omega} V_\omega \xrightarrow{\varepsilon_\omega} W_\omega \longrightarrow 0$$

being exact in k-mod, such a sequence will be said to be K-<u>split exact</u>. Let $S(K)$ be the set of all K-split exact sequences.

(2) $(\check{U}(K,|\cdot|), S(K))$ <u>is an exact category</u>.

<u>Proof.</u> Let A be the category of contravariant functors from K to k-mod. Given an object X in K, the functor $D\,\mathrm{Hom}_K(X,-)$ belongs to A and actually is an injective object of A. In this way, we obtain a full embedding of K into A such that K is an object class of A containing only injective objects. Also, $D|\cdot|$ is an object of A, and the restriction of the functor $\mathrm{Hom}(D|\cdot|,-)$ to K coincides with $|\cdot|$. Thus, we may assume that K is an object class in an abelian category A that K contains only injective objects, and that there is an object R in A with $|\cdot|$ being the restriction of $\mathrm{Hom}(R,-)$ to K. Let B be the (abelian) category of all triples $V = (V_0,V_\omega,\gamma_V)$, where V_0 is an object in A, V_ω is a finite-dimensional k-vectorspace, and $\gamma_V : V_\omega \longrightarrow \mathrm{Hom}(R,V_0)$ is k-linear, with maps $(f_0,f_\omega) : V \longrightarrow W$ being given by a map $f_0 : V_0 \longrightarrow W_0$ in A and a k-linear map $f_\omega : V_\omega \longrightarrow W_\omega$ such that $\gamma_V|f_0| = f_\omega\,\gamma_W$. [In case A is a Krull-Schmidt k-category, we just have $B = \check{U}(A,\mathrm{Hom}(R,-))$. For example, this happens in case K is a finite Krull-Schmidt k-category].

Of course, the full subcategory of B given by all triples $V = (V_0,V_\omega,\gamma_V)$ with V_0 in K is just $\check{U}(K,|\cdot|)$, thus we have a full embedding of $\check{U}(K,|\cdot|)$ into the abelian category B. Note that for any extension

$$0 \longrightarrow U \xrightarrow{\mu} V \xrightarrow{\varepsilon} W \longrightarrow 0$$

in B with U,W in $\check{U}(K,|\cdot|)$, the map μ_0 is split mono in B, since the objects in K all are injective in A. Thus, V belongs to $\check{U}(K,|\cdot|)$, and the exact sequence (μ,ε) is K-split exact. This shows that $\check{U}(K,|\cdot|)$ is an extension closed full subcategory of the abelian category B, such that the K-split exact sequences of $\check{U}(K,|\cdot|)$ are just the short exact sequences of B belonging to $\check{U}(K,|\cdot|)$. This finishes the proof.

In case K is a finite Krull-Schmidt k-category, the category A is just the category of finite-dimensional A-modules, where $A = \mathrm{End}(X)$, with X the direct sum of a complete set of indecomposable objects of K, and, in this case, the embedding defined above identifies K with the module class A-inj of all injective A-modules. Let us state this explicitly:

(2') *Let* K *be a finite Krull-Schmidt k-category, and* $|\cdot| : K \to$ k-mod *an additive functor. Then there exists a finite-dimensional k-algebra* A *and an A-module* R *such that we can identify* K *with* A-inj *and* $|\cdot|$ *with the restriction of* Hom(R,-) *to* A-inj. *In this way,* $\check{U}(K,|\cdot|)$ *is identified with the object class in the abelian category* $\check{U}$(A-mod,Hom(R,-)) *formed by all triples* $V = (V_o, V_\omega, \gamma_V)$ *with* V_o *an injective* A-*module.*

In case K can be identified with A-inj, and $|\cdot|$ with Hom(R,-), we call (A-inj,Hom(R,-)) an *injective realization* of $(K,|\cdot|)$.

Let K be a Krull-Schmidt k-category, and $|\cdot| : K \longrightarrow$ k-mod an additive functor. Since $(\check{U}(K,|\cdot|),S(K))$ is an exact category, there is defined $\mathrm{Ext}^1_{\check{U}(K,|\cdot|),S(K)}(V,W)$ for any V,W in $\check{U}(K,|\cdot|)$ and we denote this Ext-group just by $\mathrm{Ext}^1_K(V,W)$. Note that $\mathrm{Ext}^1_K(V,W)$ is given by the set of all K-split exact sequences

$$(*) \qquad 0 \longrightarrow W \longrightarrow * \longrightarrow V \longrightarrow 0$$

modulo the usual equivalence. Any K-split exact sequence $(*)$ is equivalent to a sequence

$$\begin{array}{ccccccccc}
0 \to & W_\omega & \xrightarrow{[1\,0]} & W_\omega \oplus V_\omega & \xrightarrow{\left[\begin{smallmatrix}0\\1\end{smallmatrix}\right]} & V_\omega & \to 0 \\
& \downarrow{\scriptstyle \gamma_W} & & \downarrow{\scriptstyle \left[\begin{smallmatrix}\gamma_W & 0\\ \delta & \gamma_V\end{smallmatrix}\right]} & & \downarrow{\scriptstyle \gamma_V} & \\
0 \to & |W_o| & \xrightarrow{[1\,0]} & |W_o| \oplus |V_o| & \xrightarrow{\left[\begin{smallmatrix}0\\1\end{smallmatrix}\right]} & |V_o| & \to 0
\end{array}$$

with δ a k-linear map $V_\omega \to |W_o|$, and we just denote this sequence by $[\delta]$. It is easy to check that for $\delta,\delta' : V_\omega \longrightarrow |W_o|$, the sequences $[\delta]$, $[\delta']$ are equivalent if and only if $\delta - \delta' = \gamma_V|\beta_o| - \beta_\omega\gamma_W$, for some $\beta_o : V_o \longrightarrow W_o$ in K and some k-linear map $\beta_\omega : V_\omega \to W_\omega$. Thus, for V,W in $\check{U}(K,|\cdot|)$ we obtain the following exact sequence

$$(3) \qquad 0 \to \mathrm{Hom}(V,W) \xrightarrow{\nu_1} \mathrm{Hom}_K(V_o,W_o) \times \mathrm{Hom}_k(V_\omega,W_\omega) \xrightarrow{\nu_2} \mathrm{Hom}_k(V_\omega,|W_o|) \xrightarrow{\nu_3} \mathrm{Ext}^1_K(V,W) \to 0$$

where ν_1 is the inclusion map, ν_2 sends (β_o,β_ω) to $\gamma_V|\beta_o| - \beta_\omega\gamma_W$ and finally, ν_3 sends α to $[\alpha]$.

This exact sequence has several consequences. First of all, we derive some exactness properties of the functor Ext^1_K.

(3') If $f : V' \longrightarrow V$ is a map in $\check{U}(K,|\cdot|)$ with f_ω being injective, then $\mathrm{Ext}^1_K(f,W)$ is surjective for all W. If $g : W \longrightarrow W'$ is a map in $\check{U}(K,|\cdot|)$ with $|g_o|$ being surjective, then $\mathrm{Ext}^1_K(V,g)$ is surjective for all V.

Proof. We use the following two commutative diagrams

$$\begin{array}{ccc} \mathrm{Hom}_k(V_\omega,|W_o|) & \longrightarrow & \mathrm{Ext}^1_{\mathcal{K}}(V,W) \\ {\scriptstyle \mathrm{Hom}_k(f_\omega,|W_o|)}\downarrow & & \downarrow{\scriptstyle \mathrm{Ext}^1_{\mathcal{K}}(f,W)} \\ \mathrm{Hom}_k(V'_\omega,|W_o|) & \longrightarrow & \mathrm{Ext}^1_{\mathcal{K}}(V',W) \end{array}$$

and

$$\begin{array}{ccc} \mathrm{Hom}_k(V_\omega,|W_o|) & \longrightarrow & \mathrm{Ext}^1_{\mathcal{K}}(V,W) \\ {\scriptstyle \mathrm{Hom}_k(V_\omega,|g_o|)}\downarrow & & \downarrow{\scriptstyle \mathrm{Ext}^1_{\mathcal{K}}(V,g)} \\ \mathrm{Hom}_k(V_\omega,|W'_o|) & \longrightarrow & \mathrm{Ext}^1_{\mathcal{K}}(V,W') \end{array}$$

Since the horizontal maps are surjective, the surjectivity of the left vertical map implies that of the right vertical map. This proves the assertions.

Now assume $\mathcal{K}$ is a finite Krull-Schmidt k-category. Given $V = (V_o, V_\omega, \gamma_V)$ in $\check{\mathcal{U}}(\mathcal{K},|\cdot|)$, let us define its dimension vector $\underline{\dim}_{\mathcal{K}}\, V := ([V_o], \dim_k V_\omega) \in G(\mathcal{K}) \times \mathbb{Z}$. Note that $G(\mathcal{K})$ is a finite rank torsion free abelian group with basis the set of isomorphism classes, say $[X_1],\ldots,[X_n]$, of indecomposable objects in $\mathcal{K}$. In $G(\mathcal{K}) \times \mathbb{Z}$, we choose $(0,1)$ as an additional basis vector, and denote it by $e(\omega)$. We define on $G(\mathcal{K}) \times \mathbb{Z}$ a bilinear form $\langle -,- \rangle$ as follows:

$$\begin{aligned} \langle [X_i],[X_j] \rangle &= \dim_k \mathrm{Hom}(X_i,X_j) \;, \\ \langle [X_i],e(\omega) \rangle &= 0 \\ \langle e(\omega),[X_i] \rangle &= -\dim_k |X_i| \\ \langle e(\omega),e(\omega) \rangle &= 1 \;. \end{aligned}$$

The corresponding quadratic form will be denoted by $\chi_{(\mathcal{K},|\cdot|)}$ or just by $\chi_{\mathcal{K}}$, with $\chi_{\mathcal{K}}(x) = \langle x,x \rangle$. Note that $\chi_{\mathcal{K}}$ is an integral quadratic form (in the sense of 1.0) if and only if $\mathrm{End}(X) = k$ for every indecomposable object X in $\mathcal{K}$. Let $(A\text{-inj}, \mathrm{Hom}(R,-))$ be an injective realization of $(\mathcal{K},|\cdot|)$; let $C = C_A$ be the Cartan matrix of A, and $r = \underline{\dim}\, R$. Then the bilinear form $\langle -,- \rangle$ on $G(A\text{-inj}) \times \mathbb{Z}$ is given by the following matrix

$$\begin{bmatrix} C & 0 \\ -r & 1 \end{bmatrix} .$$

(since for indecomposable injective A-modules $Q(i)$, $Q(j)$, we have $\dim_k \mathrm{Hom}(Q(i),Q(j)) = e(i)Ce(j)^T$).

(3") <u>If K is a finite Krull-Schmidt k-category and</u> $|\cdot| : K \longrightarrow$ k-mod <u>an additive functor, then for</u> V,W <u>objects in</u> $U(K,|\cdot|)$, <u>we have</u>

$$\langle \underline{\dim}_K V, \underline{\dim}_K W\rangle = \dim_k \operatorname{Hom}(V,W) - \dim_k \operatorname{Ext}^1_K(V,W).$$

<u>Proof.</u> According to (3), the right hand side of the equality is equal to

$$\dim \operatorname{Hom}_K(V_o,W_o) + \dim \operatorname{Hom}_k(V_\omega,W_\omega) - \dim \operatorname{Hom}_k(V_\omega,|W_o|)$$

thus to $\langle [V_o],[W_o]\rangle + \dim_k V_\omega \cdot \dim_k W_\omega - \dim_k V_\omega \cdot \dim_k |W_o|$,

and, by definition of $\langle -,-\rangle$, this is equal to

$$\langle ([V_o],\dim V_\omega),([W_o],\dim W_\omega)\rangle.$$

(4) <u>Let K be a finite Krull-Schmidt category,</u> $|\cdot| : K \longrightarrow$ k-mod <u>an additive functor. If</u> V <u>is directing in</u> $\check{U}(K,|\cdot|)$, <u>then</u> End(V) = k, $\operatorname{Ext}^1_K(V,V) = 0$, <u>thus</u> $\chi_K(\underline{\dim}_K V) = 1$. <u>Also, if</u> V' <u>is any other indecomposable object in</u> $\check{U}(K,|\cdot|)$, <u>with</u> $\underline{\dim}_K V' = \underline{\dim}_K V$, <u>then</u> $V' \approx V$.

<u>Proof.</u> Any non-zero, non-invertible endomorphism of V gives a cycle (V,V) of length 1. Any non-split K-split exact sequence

$$0 \longrightarrow V \longrightarrow W \longrightarrow V \longrightarrow 0 ,$$

say, with $W = W' \oplus W''$ where W' is indecomposable, gives a cycle (V,W',V) of length 2. This shows that for V directing, we have

$$\operatorname{End}(V) = k,\ \operatorname{Ext}^1_K(V,V) = 0,$$

thus $\chi_K(\underline{\dim}_K V) = 1$. Now assume V is directing and V' is indecomposable, $\underline{\dim}_K V' = \underline{\dim}_K V$. Then

$$\begin{aligned}\dim \operatorname{Hom}(V,V') &= \langle \underline{\dim}_K V, \underline{\dim}_K V'\rangle + \dim \operatorname{Ext}^1_K(V,V') \\ &\geq \chi_K(\underline{\dim}_K V) = 1 ,\end{aligned}$$

thus $\operatorname{Hom}(V,V') \neq o$. Similarly, $\operatorname{Hom}(V',V) \neq o$. If V' and V would not be isomorphic, then V would belong to the cycle (V,V',V), impossible.

Now, we are going to determine source maps and sink maps in categories of the form $\check{U}(K,|\cdot|)$. First, we will show that source maps and sink maps of K can be lifted to corresponding source maps and sink maps in $\check{U}(K,|\cdot|)$. This is rather fundamental for the constructions exhibited in these notes. We consider an arbitrary (not necessarily finite) Krull-Schmidt category. Note that there are two essentially different full embeddings of K into $\check{U}(K,|\cdot|)$: First of all, given $X \in K$, we may send it to $(X,0,o)$, and usually we will denote $(X,0,o)$ just by X. Second, we may send X to $(X,|X|,1_{|X|}) =: \bar{X}$. We always will identify K with the full subcategory of $\check{U}(K,|\cdot|)$ given by the triples $V = (V_o,V_\omega,\gamma_V)$ with $V_\omega = 0$, and we will denote by $\bar{K}$ the full subcategory of $\check{U}(K,|\cdot|)$ given by all triples $V = (V_o,V_\omega,\gamma_V)$ with γ_V bijective.

(5) Let K be a Krull-Schmidt category. If $f : X \longrightarrow Y$ is a source map in K, then $(f,|1_X|) : \bar{X} \longrightarrow (Y,|X|,|f|)$ is a source map in $\check{U}(K,|\cdot|)$. If $g : Y \longrightarrow Z$ is a sink map in K, then $(g,o) : (Y,\mathrm{Ker}|g|,u) \longrightarrow Z$ (with $u : \mathrm{Ker}|g| \longrightarrow |Y|$ the inclusion map) is a sink map in $\check{U}(K,|\cdot|)$.

Proof. Consider first the case of $f : X \longrightarrow Y$ being a source map in K. Of course, $(f,1_{|X|}) : \bar{X} \longrightarrow (Y,|X|,|f|)$ is not split mono, since otherwise f is split mono. Let $v = (v_o,v_\omega) : \bar{X} \longrightarrow V$ be a map in $\check{U}(K,|\cdot|)$, which is not split mono. Note that v_o itself is not split mono. [Namely, if there exists v_o' with $v_ov_o' = 1_X$, then $vv' = 1_{\bar{X}}$ for $v' = (v_o',\gamma_V|v_o'|)$.] Since f is a source map in K, there exists $\eta_o : Y \longrightarrow V_o$ with $f\eta_o = v_o$. Let $\eta = (\eta_o,v_\omega) : (Y,|X|,|f|) \longrightarrow V$. This is a map in $\check{U}(K,|\cdot|)$, since $v_\omega\gamma_V = |v_o| = |f||\eta_o|$, and it satisfies $v = (f,1_{|X|})\eta$. Finally, let ζ be an endomorphism of $(Y,|X|,|f|)$ with $(f,1_{|X|})\zeta = (f,1_{|X|})$. Then $f\zeta_o = f$ shows that ζ_o is an automorphism of Y, and $1_{|X|}\zeta_\omega = 1_{|X|}$ shows that $\zeta_\omega = 1_{|X|}$. Thus ζ is an automorphism.

Next, consider the case of $g : Y \longrightarrow Z$ being a sink map in K. Again, $(g,0) : (Y,\mathrm{Ker}|g|,u) \longrightarrow Z$ cannot be split epi, since otherwise g is split epi. Let $w = (w_o,w_\omega) : W \longrightarrow Z$ be a map in $\check{U}(K,|\cdot|)$ which is not split epi. Of course, $w_\omega = 0$. Note that w_o itself is not split epi. Since g is a sink map in K, there exists $\eta_o : W \longrightarrow Z$ with $w_o = \eta_o g$. Now $\gamma_W|\eta_o||g| = \gamma_W|w_o| = w_\omega \cdot 0 = 0$ shows that $\gamma_W|\eta_o|$ factors through $u : \mathrm{Ker}|g| \to |Y|$, say $\gamma_W|\eta_o| = \eta_\omega u$. Then $\eta = (\eta_o,\eta_\omega) : W \longrightarrow (Y,\mathrm{Ker}|g|,u)$ satisfies $w = \eta(g,0)$. Also, if ζ is an endomorphism of $(Y,\mathrm{Ker}|g|,u)$ satisfying $\zeta(g,o) = (g,o)$, then $\zeta_o g = g$, thus ζ_o is an automorphism of Y, and $\zeta_\omega u = u|\zeta_o|$ shows that ζ_ω is the restriction of $|\zeta_o|$ to $\mathrm{Ker}|g|$, thus ζ is an automorphism.

It is interesting to see that these lifted source maps and lifted sink maps fit together rather well. Namely, assume (K,S) is a Krull-Schmidt category with short exact sequences. A sequence

$$0 \longrightarrow U \xrightarrow{\mu} V \xrightarrow{\varepsilon} W \longrightarrow 0$$

in $\check{U}(K,|\cdot|)$ will be said to be (K,S)-exact provided

$$0 \longrightarrow U_0 \xrightarrow{\mu_0} V_0 \xrightarrow{\varepsilon_0} W_0 \longrightarrow 0$$

belongs to S, and

$$0 \longrightarrow U_\omega \xrightarrow{\mu_\omega} V_\omega \xrightarrow{\varepsilon_\omega} W_\omega \longrightarrow 0$$

is an exact sequence of k-mod. Let $S(K,S)$ be the set of (K,S)-exact sequences. It is easy to check that $(\check{U}(K,|\cdot|),S(K,S))$ is a Krull-Schmidt category with short exact sequences. A functor $|\cdot| : K \longrightarrow$ k-mod will be said to be left exact on S provided it maps any sequence in S to a left exact sequence of vector spaces.

(6) Let (K,S) be a Krull-Schmidt category with short exact sequences, and $|\cdot| : K \longrightarrow$ k-mod an additive functor which is left exact on S. If

$$0 \longrightarrow X \xrightarrow{f} Y \xrightarrow{g} Z \longrightarrow 0$$

is an Auslander-Reiten-sequence in (K,S), then

$$(*) \qquad 0 \longrightarrow \bar{X} \xrightarrow{(f,1_{|X|})} (Y,|X|,|f|) \xrightarrow{(g,o)} Z \longrightarrow 0$$

is an Auslander-Reiten-sequence in $(\check{U}(K,|\cdot|),S(K,S))$.

Proof. It is clear that the sequence given by $(f,1_{|X|})$ and (g,o) is (K,S)-exact. According to (5), we know that $(f,1_{|X|})$ is a source map. Since $|\cdot|$ is left exact on S, we know that $|f| : |X| \longrightarrow |Y|$ is a kernel of $|g|$, thus (g,o) is a sink map.

We will say that the Auslander-Reiten sequence $(*)$ is lifted from (K,S).

We consider now the case of K being finite and non-zero. Let us introduce one particular object $F(\omega)$ in $U(K,|\cdot|)$. For any indecomposable object X of K, let $t(X)$ be a subspace of $|X|$ which is complementary to $\Sigma\ \mathrm{Im}|f|$, where f ranges over all maps in $\mathrm{rad}(Y,X)$, with Y arbitrary. Now, let $X_1,\ldots,X_m$ be representatives of the isomorphism classes of the indecomposable objects in K, and define

$$F(\omega) = F_K(\omega) := (\bigoplus_{i=1}^{m} X_i \underset{k}{\otimes} Dt(X_i), k, \gamma_{F(\omega)}),$$

with $\gamma_{F(\omega)} = (\gamma_i)_i$ being given by the canonical maps

$$\gamma_i : k \longrightarrow t(X_i) \underset{k}{\otimes} Dt(X_i) \subseteq |X_i| \underset{k}{\otimes} Dt(X_i).$$

[with image of 1 being of the form $\sum_j x_{ij} \otimes x'_{ij}$, where $x_{i1}, x_{i2}, \ldots$ is a k-basis of $t(X_i)$, and $x'_{i1}, x'_{i2}, \ldots$ the dual basis in $Dt(X_i)$.] It is easy to check that $F(\omega)$ is indecomposable and does not depend on the choice of the subspaces $t(X)$. Actually, this also follows directly from the next assertion. Recall that we have denoted by $*$ the duality functor from $U(K,|\cdot|)$ to $U(K^{op}, D|\cdot|)$.

(7) <u>Assume</u> K <u>to be finite and non-zero. The map</u> $(1,0) : F(\omega)_o \longrightarrow F(\omega)$ <u>is a sink map, and we obtain in</u> $\check{U}(K^{op}, D|\cdot|)$ <u>a</u> K^{op}<u>-split exact sequence</u>

$$0 \longrightarrow F(\omega)^* \xrightarrow{(1,0)^*} F(\omega)^*_o \xrightarrow{q} E_{K^{op}}(\omega) \longrightarrow 0$$

<u>with</u> $(1,0)^*$ <u>being a source map, and</u> q <u>a sink map in</u> $\check{U}(K^{op}, D|\cdot|)$.

For the proof, one observes that any map $\delta : k \longrightarrow |X_o|$ with X_o in K can be factored as $\delta = \gamma_{F(\omega)}|\delta'|$, with $\delta' : F(\omega)_o \longrightarrow X_o$ in K, thus any map $f = (f_o, f_\omega) : X \longrightarrow F(\omega)$ in $\check{U}(K,|\cdot|)$ with $f_\omega \neq 0$ is split epi.

Given a finite dimensional algebra A_o, and an A_o-module R, we denote by $A_o[R]$ the <u>one-point extension</u> of A_o by R, namely the algebra

$$A_o[R] = \begin{bmatrix} A_o & R \\ 0 & k \end{bmatrix}$$

(thus the set of all matrices of the form $\begin{bmatrix} a & r \\ 0 & b \end{bmatrix}$, with $a \in A_o$, $r \in R$, $b \in k$, subject to the usual addition and multiplication of matrices). The quiver of $A_o[R]$ contains the quiver of A_o as a full subquiver, and there is an additional vertex ω, called the <u>extension vertex</u> of $A_o[R]$; it is always a source. The category of $A_o[R]$-modules can be described as follows:

(8) $A_o[R]\text{-mod} \approx \check{\mathcal{U}}(A_o\text{-mod},\mathrm{Hom}(R,-))$.

<u>Proof.</u> The triple $V = (V_o, V_\omega, \gamma_V)$ in $\check{\mathcal{U}}(A_o\text{-mod},\mathrm{Hom}(R,-))$ corresponds to the $A_o[R]$-module $\begin{bmatrix} V_o \\ V_\omega \end{bmatrix}$, with $\begin{bmatrix} 0 & R \\ 0 & 0 \end{bmatrix}$ operating on it via the map $\bar{\gamma}_V : R \otimes_k V_\omega \longrightarrow V_o$ adjoint to γ_V.

Usually, we will not distinguish between these categories and we will call a triple $V = (V_o, V_\omega, \gamma_V)$ in $\check{\mathcal{U}}(A_o\text{-mod},\mathrm{Hom}(R,-))$ just an $A_o[R]$-module, and V_o will be called its <u>restriction</u> to A_o. Note that $E(\omega) := (0,k,o)$ is a simple injective $A_o[R]$-module, and it is the only indecomposable $A_o[R]$-module with restriction to A_o being zero.

Note that if $\mathcal{M}$ is a module class in A_o-mod, and R is an arbitrary A_o-module (not necessarily belonging to $\mathcal{M}$), then we obtain $\check{\mathcal{U}}(\mathcal{M},\mathrm{Hom}(R,-))$, and also $\mathcal{U}(\mathcal{M},\mathrm{Hom}(R,-))$ as module classes in $A_o[R]$-mod. Of particular interest is the case $\mathcal{M} = A_o$-inj, since, as we have seen in (2'), any subspace category $\check{\mathcal{U}}(\mathcal{K},|\cdot|)$ with $\mathcal{K}$ being finite, can be realized as $\check{\mathcal{U}}(A_o\text{-inj},\mathrm{Hom}(R,-))$ for a suitable algebra A_o and some A_o-module R. However, also other module classes $\mathcal{M}$ will have to be considered.

Let A_o be a finite dimensional algebra, and R an A_o-module and $A = A_o[R]$. Assume the Cartan matrix C_o of A_o is invertible. We may consider both $K_o(A)$ with its bilinear form C_A^{-T}, and $G(A_o\text{-inj}) \times \mathbb{Z}$ with its bilinear form related to the vectorspace category $(A_o\text{-inj},\mathrm{Hom}(R,-))$. There is the canonical map

$$G(A_o\text{-inj}) \times \mathbb{Z} \xrightarrow{\begin{bmatrix} C_o & 0 \\ 0 & 1 \end{bmatrix}} K_o(A)$$

It maps the pair $([Q],z)$ onto $(\underline{\dim}\, Q, z)$, where $Q \in A_o$-inj, and $z \in \mathbb{Z}$. Given

$V = (V_o, V_\omega, \gamma_V) \in \check{U}(A_o\text{-inj}, \mathrm{Hom}(R,-))$, the image of $\underline{\dim}_{A_o\text{-inj}}(V_o)$ under the canonical map is just $\underline{\dim}\begin{bmatrix} V_o \\ V_\omega \end{bmatrix}$.

(a) <u>The canonical map</u> $G(A_o\text{-inj}) \times \mathbb{Z} \to K_o(A)$ <u>is an isometry</u>.

<u>Proof</u>: We only have to check

$$\begin{bmatrix} C_o & 0 \\ -r & 1 \end{bmatrix} = \begin{bmatrix} C_o & 0 \\ 0 & 1 \end{bmatrix} \begin{bmatrix} C_o^{-T} & 0 \\ -rC_o^{-T} & 1 \end{bmatrix} \begin{bmatrix} C_o & 0 \\ 0 & 1 \end{bmatrix}^T .$$

(b) <u>Under the canonical map</u> $G(A_o\text{-inj}) \times \mathbb{Z} \to K_o(A)$, <u>the image of a positive vector is a positive vector</u> (but the converse is not true, in general).

<u>Proof</u>: All entries of $\begin{bmatrix} C_o & 0 \\ 0 & 1 \end{bmatrix}$ are non-negative, thus the image of a positive vector has only non-negative entries. Also, by assumption, C_o is invertible, thus the image of a positive vector is also non-zero. If A_o is semisimple, then C_o is the identity matrix, thus a vector in $G(A_o\text{-inj}) \times \mathbb{Z}$ is positive if and only if its canonical image in $K_o(A)$ is positive. On the other hand, assume there exists an A_o-module M with $\mathrm{inj.dim.}M = 1$, let

$$0 \to M \to Q_o \to Q_1 \to 0$$

be a minimal injective resolution of M . Then $\underline{\dim}\, M$ is positive in $K_o(A)$, and it is the imgae of $([Q_o]-[Q_1],0) \in G(A_o\text{-inj}) \times \mathbb{Z}$. Note that $[Q_o]-[Q_1]$ is positive in $G(A_o\text{-inj})$ only in the exceptional case that Q_1 is a direct summand of Q_o .

(c) Assume R is sincere. Then the canonical image in $K_o(A)$ of a positive vector $([Q],z_\omega) \in G(A_o\text{-inj}) \times \mathbb{Z}$, with $z_\omega \neq 0$, is connected.

<u>Proof</u>: Let $Q(i)$ be an indecomposable injective A_o-module, and $\overline{Q_o(i)} = (Q_o(i), \mathrm{Hom}(R,Q_o(i)), 1) \in \check{U}(A_o\text{-inj}, \mathrm{Hom}(R,-))$. It is indecomposable, and $\mathrm{Hom}(R,Q_o(i)) \neq 0$, since R is supposed to be sincere. We may consider $\overline{Q_o(i)}$ as an $A_o[R]$-module, and the support of its dimension vector $\underline{\dim}\, \overline{Q_o(i)} \in K_o(A)$ is obtained from the support of $\underline{\dim}\, Q_o(i)$ by adding the extension vertex ω . Since $\overline{Q_o(i)}$ is indecomposable, the support of $\underline{\dim}\, \overline{Q_o(i)}$ is connected. For an arbitrary module $Q \in A_o\text{-inj}$, and $0 < z_\omega \in \mathbb{Z}$, the support of the canonical image of $([Q],z_\omega)$ in $K_o(A)$ is the union of connected subsets all containing the extension vertex ω , thus connected.

(c') <u>Assume</u> R <u>is sincere. Let</u> z <u>be a positive vector in</u> $G(A_o\text{-inj}) \times \mathbb{Z}$, <u>with</u> $\chi_{A_o\text{-inj}}(z) \leq 1$. <u>Then the canonical image of</u> z <u>in</u> $K_o(A)$ <u>has connected support</u>.

<u>Proof</u>: Using (c), we only have to consider the case $z = ([Q],0)$, where Q is in $A_o\text{-inj}$. Then $\chi_{A_o\text{-inj}}(z) = \dim \mathrm{End}(Q)$, according to (3"). Since $Q \neq 0$, it follows that Q is indecomposable. The canonical image of z in $K_o(A)$ is $\underline{\dim}\, Q$, and the dimension vector of an indecomposable module is connected.

In dealing with the one-point extension $A_o[R]$, we can use (6) due to the fact that the functor Hom(R,-) is left exact (on all short exact sequences of A_o-modules). Thus any Auslander-Reiten sequence in A_o-mod gives a lifted Auslander-Reiten sequence in A-mod, all other Auslander-Reiten sequences in A-mod being split when restricted to A_o-mod.

(9) <u>Let K be a finite Krull-Schmidt category.</u>

<u>Every indecomposable object</u> U <u>in</u> $\check{U}(K,|\cdot|)$ <u>has a source map</u> $f : U \longrightarrow V$. <u>If</u> U <u>is neither in</u> $\bar{K}$, <u>nor of the form</u> $E(\omega)$, <u>then</u> f <u>is K-split mono. If, in addition,</u> $U \neq F_{K^{op}}(\omega)^*$, <u>then the cokernel of</u> f <u>is in</u> $U(K,|\cdot|)$.

<u>Every indecomposable object</u> W <u>in</u> $\check{U}(K,|\cdot|)$ <u>has a sink map</u> $g : V \longrightarrow W$. <u>If</u> W <u>is neither in</u> K, <u>nor of the form</u> $F(\omega)$, <u>then</u> g <u>is K-split epi, and the kernel of</u> g <u>is in</u> $U(K,|\cdot|)$.

<u>Proof.</u> First, let us show that $\check{U}(K,|\cdot|)$ has source maps. We use the canonical realization of $(K,|\cdot|)$ with K = A-inj, $|\cdot|$ = Hom(R,-), where A is a finite dimensional algebra, and R an A-module. Given U in $\check{U}(K,|\cdot|)$, let $f' : U \longrightarrow V' = (V'_o, V'_\omega, \gamma_{V'})$ be a source map in A[R]-mod. Let $f''_o : V'_o \longrightarrow V''_o$ be an injective envelope of V'_o in A-mod, and let $V'' = (V''_o, V'_\omega, \gamma_{V'}|f''_o|)$, $f'' = (f''_o, 1_{V'_\omega})$. Note that V'' belongs to $\check{U}(K,|\cdot|)$, and we consider the map $f'f'' : U \longrightarrow V''$. First of all, $f'f''$ is not split mono, since otherwise f' would be split mono. Second, let $h : U \longrightarrow X$ be any map in $\check{U}(K,|\cdot|)$, not split mono. Then $h = f'\eta$ for some $\eta : V' \longrightarrow X$, since f' is a source map in A[R]-mod. Now X_o is an injective A-module, and f''_o is mono, thus $\eta_o = f''_o\eta'_o$ for some $\eta'_o : V''_o \longrightarrow X_o$. Then $\eta' = (\eta'_o, \eta_\omega) : V'' \longrightarrow X$ is a map in $\check{U}(K,|\cdot|)$, and $\eta = f''\eta'$. Thus $h = (f'f'')\eta'$. Taking a minimal direct summand V of V'' containing the image of $f'f''$, we obtain a source map $f = f'f'' : U \longrightarrow V$ in $\check{U}(K,|\cdot|)$.

Now assume, U is not in $\bar{K}$. The map $(1_{U_o}, \gamma_U) : U \longrightarrow \bar{U}_o$ cannot be split mono, since otherwise $\gamma_U : U_\omega \rightarrow |U_o|$ would be an isomorphism. Thus, there exists $\zeta : V \longrightarrow \bar{U}_o$ with $(1_{U_o}, \gamma_U) = f\zeta$. In particular, $1_{U_o} = f_o\zeta_o$, thus f_o is split mono in K. Also, $\gamma_U = f_\omega\zeta_\omega$ shows that with γ_U also f_ω is mono. However, for U indecomposable, γ_U always is mono except in case $U = E(\omega)$. This shows that for U neither in $\bar{K}$, nor of the form $E(\omega)$, the source map $U \longrightarrow V$ in $\check{U}(K,|\cdot|)$ is K-split mono.

Now, let U be indecomposable, with a K-split mono source map $f : U \longrightarrow V$ in $\check{U}(K,|\cdot|)$. We may form the cokernel $g : V \longrightarrow W$ in $\check{U}(K,|\cdot|)$, and according to (2.3), we know that W is indecomposable and g a sink map. If W does not belong to $U(K,|\cdot|)$, then $W = E(\omega)$, and therefore $U = F_{K^{op}}(\omega)^*$, according to (7). This finishes the proof of the first part of (9).

It follows that any indecomposable object U in $\mathcal{U}(K,|\cdot|)$ has a source map in $\mathcal{U}(K,|\cdot|)$. Namely, if $f : U \longrightarrow V$ is a source map in $\check{\mathcal{U}}(K,|\cdot|)$, let $\pi : V_\omega \to V_\omega/\mathrm{Ker}\,\gamma_V$ be the canonical projection, and $\gamma_V = \pi\gamma'_V$. Then $(f_o, f_\omega\pi) : U \longrightarrow (V_o, V_\omega/\mathrm{Ker}\,\gamma_V, \gamma_V')$ is a source map in $\mathcal{U}(K,|\cdot|)$. Actually, if U is not in $\bar{K}$, there is no irreducible map $U \longrightarrow E(\omega)$ in $\check{\mathcal{U}}(K,|\cdot|)$, thus V does not split off a copy of $E(\omega)$, and therefore the source map in $\check{\mathcal{U}}(K,|\cdot|)$ for V lies in fact in $\mathcal{U}(K,|\cdot|)$.

Applying the duality functor $* : \mathcal{U}(K^{op},D|\cdot|) \longrightarrow \mathcal{U}(K,|\cdot|)$, and using the previous results for $\mathcal{U}(K^{op},D|\cdot|)$, we see that any indecomposable object W in $\mathcal{U}(K,|\cdot|)$ has a sink map in $\mathcal{U}(K,|\cdot|)$. Also, if W is neither in K nor of the form $F_K(\omega)$, then W^* is neither in K^{op} nor of the form $F_K(\omega)^*$, thus, we have a K^{op}-split exact sequence

$$0 \longrightarrow W^* \xrightarrow{g^*} V^* \xrightarrow{f^*} U^* \longrightarrow 0$$

in $\mathcal{U}(K^{op},D|\cdot|)$, with g^* being the source map in $\mathcal{U}(K^{op},D|\cdot|)$. Applying $*$, we obtain the K-split exact sequence

$$0 \longrightarrow U \xrightarrow{f} V \xrightarrow{g} W \longrightarrow 0$$

in $\mathcal{U}(K,|\cdot|)$, thus the sink map for W in $\mathcal{U}(K,|\cdot|)$ is K-split epi.

It remains to note that for any indecomposable W in $\mathcal{U}(K,|\cdot|)$ its sink map in $\mathcal{U}(K,|\cdot|)$ is even a sink map in $\check{\mathcal{U}}(K,|\cdot|)$, since $\mathrm{Hom}(E(\omega),W) = 0$. Of course, also $E(\omega)$ has a sink map in $\check{\mathcal{U}}(K,|\cdot|)$, as we have seen in (7). This finishes the proof.

Given a finite Krull-Schmidt category K, we denote by $\Gamma(\check{\mathcal{U}}(K,|\cdot|))$ the Auslander-Reiten quiver of $\check{\mathcal{U}}(K,|\cdot|)$ with respect to the K-split exact sequences. We obtain from (9) the following corollary:

(9') <u>The projective vertices of</u> $\Gamma(\check{\mathcal{U}}(K,|\cdot|))$ <u>are the isomorphism classes</u> [X], <u>with</u> X <u>indecomposable in</u> K, <u>and</u> $[F(\omega)]$. <u>The injective vertices of</u> $\Gamma(\check{\mathcal{U}}(K,|\cdot|))$ <u>are the isomorphism classes</u> $[\bar{X}]$, <u>with</u> X <u>indecomposable in</u> K, <u>and</u> $[E(\omega)]$.

<u>Proof.</u> If X is indecomposable in K, or $X = F(\omega)$, then the sink map for X is known, in particular, it is not a K-split mono. Thus [X] has to be projective in $\Gamma(\check{\mathcal{U}}(K,|\cdot|))$. Conversely, for the remaining indecomposable objects W in $\check{\mathcal{U}}(K,|\cdot|)$, the sink map $g : V \longrightarrow W$ is K-split epi, according to (9), thus we can form the kernel in $\check{\mathcal{U}}(K,|\cdot|)$ and obtain a K-split exact sequence

$$0 \longrightarrow U \xrightarrow{f} V \xrightarrow{g} W \longrightarrow 0$$

According to 2.3, f is a source map, thus this sequence is an Auslander-Reiten sequence.

Similarly, one proves the second assertion.

(10) Let K be a finite Krull-Schmidt category and $|\cdot|$ an additive functor which does not vanish on any component of K. If a component of $\Gamma(\check{U}(K,|\cdot|))$ is finite, then this is all of $\check{U}(K,|\cdot|)$.

Proof. Let C be a finite component of $\Gamma(\check{U}(K,|\cdot|))$. Let V be in C. If $U \in \check{U}(K,|\cdot|)$ is indecomposable and $\mathrm{Hom}(U,V) \neq o$ or $\mathrm{Hom}(V,U) \neq o$, then $U \in C$, according to lemma 6 in 2.2. Now, if $V = (V_o, V_\omega, \gamma_V)$, and V_o has a non-zero direct summand in some component K_i of K, and X is indecomposable in K_i, then also X belongs to C. In K_i, there is some indecomposable X_i with $|X_i| \neq o$. Thus we have non-zero maps $X' \longrightarrow \bar{X}' \longrightarrow E(\omega)$, and therefore $E(\omega)$ is in C. On the other hand, given any other component K_j of K, there is some indecomposable X_j with $|X_j| \neq o$, thus the non-zero maps $X_j \longrightarrow \bar{X}_j \longrightarrow E(\omega)$ imply that X_j belongs to C. Since K belongs to C, it follows that all of $\check{U}(K,|\cdot|)$ is in C.

The vectorspace category $(K,|\cdot|)$ will be said to be subspace finite provided $\check{U}(K,|\cdot|)$ is a finite category, and otherwise subspace infinite.

Let us consider the one-point extension $A = A_o[R]$ of A_o. We want to determine the Cartan matrix and the Coxeter matrix of A. We will consider $K_o(A_o,\mathbb{Q})$ as a subspace of $K_o(A,\mathbb{Q})$, with $e(\omega) = \underline{\dim}\, E(\omega)$ denoting the additional canonical base vector. Let C_o be the Cartan matrix of A_o, and $r = \underline{\dim}\, R$. Then

$$C = C_A := \begin{bmatrix} C_o & r^T \\ 0 & 1 \end{bmatrix}$$

is the Cartan matrix of A. Note that C is invertible if and only if C_o is invertible. If we assume that C_o is invertible, the bilinear form $\langle -,- \rangle$ on $K_o(A,\mathbb{Q})$ is given by

$$C^{-T} = \begin{bmatrix} C_o^{-T} & 0 \\ -rC_o^{-T} & 1 \end{bmatrix} ,$$

the symmetrized bilinear form $(-,-)$ is given by the matrix

$$\frac{1}{2}(C^{-1}+C^{-T}) = \frac{1}{2}\begin{bmatrix} C_o^{-1}+C_o^{-T} & -C_o^{-1}r^T \\ -rC_o^{-T} & 2 \end{bmatrix},$$

and the Coxeter matrix

$$\Phi_A = -C^{-T}C = \begin{bmatrix} \Phi_o & -C_o^{-T}r^T \\ -r\Phi_o & \chi(r)-1 \end{bmatrix}$$

with Φ_o being the Coxeter matrix of A_o. We will be interested in elements of $K_o(A,\mathbb{Q})$ which are invariant under Φ_A.

(11) <u>For</u> $w_o \in K_o(A_o,\mathbb{Q})$, <u>the following properties are equivalent:</u>

(i) $w_o + e(\omega)$ <u>is in</u> $\mathrm{rad}\chi_A$.

(ii) $r = w_o(I-\Phi_o^{-1})$ <u>and</u> $\chi(w_o) = 1$.

(iii) <u>The two linear forms</u> $\langle r,-\rangle$ <u>and</u> $2(w_o,-)$ <u>on</u> $K_o(A_o,\mathbb{Q})$ <u>coincide, and</u> $\chi(w_o) = 1$.

<u>Proof.</u> The equivalence of (ii) and (iii) is straightforward, since the forms $\langle r,-\rangle$ and $2(w_o,-)$ coincide if and only if $rC_o^{-T} = w_o(C_o^{-1}+C_o^{-T})$, if and only if $r = w_o(C_o^{-1}C_o^T+I) = w_o(-\Phi_o^{-1}+I)$. In order to see that these conditions are equivalent to (i), let us calculate $2(w_o+e(\omega),-)$

$$[w_o \ \ 1]\begin{bmatrix} C_o^{-1}+C_o^{-T} & -C_o^{-1}r^T \\ -rC_o^{-T} & 2 \end{bmatrix} = [w_o(C_o^{-1}+C_o^{-T}) - rC_o^{-T},\ -w_oC_o^{-1}r^T+2] .$$

The first component vanishes if and only if $2(w_o,-) = \langle r,-\rangle$, the second component vanishes if $w_oC_o^{-1}r^T = 2$. However, in case the first component vanishes, we have $2\chi(w_o) = 2(w_o,w_o) = \langle r,w_o\rangle = w_oC_o^{-1}r^T$, thus, in this case, the second component vanishes if and only if $2\chi(w_o) = 1$.

(12) <u>For</u> $w_o \in K_o(A_o,\mathbb{Q})$, <u>the following are equivalent:</u>

(i) w_o, <u>considered as element of</u> $K_o(A,\mathbb{Q})$, <u>belongs to</u> $\mathrm{rad}\chi_A$.

(ii) $w_o \in \mathrm{rad}\chi_{A_o}$, <u>and</u> $\langle w_o,r\rangle = 0$.

<u>Proof.</u> Let us calculate $2(w_o,-)$ in $K_o(A,\mathbb{Q})$

$$[w_o,0]\begin{bmatrix} C_o^{-1}+C_o^{-T} & -C_o^{-1}r^T \\ -rC_o^{-T} & 2 \end{bmatrix} = [w_o(C_o^{-1}+C_o^{-T}),-w_oC^{-1}r^T] .$$

The first component vanishes if and only if $2(w_o,-)$ vanishes on $K_o(A_o,\mathbb{Q})$. If the first component vanishes, the second is just $\langle w_o,r\rangle = w_oC^{-T}r = -w_o C_o^{-1}r^T$.

2.6 Subspace categories of directed vectorspace categories and representations of partially ordered sets

A vectorspace category $(K,|\cdot|)$ is said to be directed provided there is no cycle in K, or, equivalently, every indecomposable object in K is directing. Note that for a directed vectorspace category $(K,|\cdot|)$, the set of isomorphism classes of indecomposable objects in K forms a partially ordered set with respect to $\preceq$, and we denote it by $S(K)$.

A vectorspace category $(K,|\cdot|)$ is said to be linear provided the functor $|\cdot|$ is faithful, and $\dim_k|X| = 1$ for any indecomposable object X in K. In a linear vectorspace category $(K,|\cdot|)$, we have $\dim_k \mathrm{Hom}_K(X,Y) \leq 1$ for any two indecomposable objects X,Y of K [namely, since $|\cdot|$ is faithful, the map $|\cdot| : \mathrm{Hom}_K(X,Y) \longrightarrow \mathrm{Hom}_k(|X|,|Y|)$ is injective]; also, if X,Y,Z are indecomposable, and $f : X \longrightarrow Y$, $g : Y \longrightarrow Z$ are non-zero maps in K, then fg is non-zero, again [namely, $|fg| = |f|\cdot|g| \neq 0$, since the image of $|f|$ is all of $|Y|$.] As a consequence, given indecomposable objects X, Y, then $[X] \preceq [Y]$ if and only if $\mathrm{Hom}_K(X,Y) \neq 0$, and any indecomposable object is directing. Thus, a linear vectorspace category always is directed. Note that for a linear vectorspace category $(K,|\cdot|)$, the quadratic form χ_K on $G(K) \times \mathbb{Z}$ coincides with the extended quadratic form $\chi_{S(K)}$ of the partially ordered set $S(K)$ as introduced in 1.10.

Let $S = (S,\leq)$ be a partially ordered set. We define its incidence category $K(S)$ over k as follows: its indecomposable objects are of the form $O(a)$, with $a \in S$,

$$\mathrm{Hom}(O(a), O(b)) = \begin{cases} k & \text{if} \quad a \leq b \\ 0 & \phantom{\text{if}} \quad a \not\leq b \end{cases}$$

and the composition of homomorphisms is given by the multiplication in k. In case S is a finite partially ordered set, $K(S)$ is nothing but the category $A(S)$-proj of $A(S)$-projective modules, where $A(S)$ is the usual incidence algebra of S. [There are several different ways of defining $A(S)$: first of all $A(S)$ is the endomorphism ring of $\bigoplus_{a\in S} O(a)$; also, $A(S)$ has as k-basis the set $\{(a,b) \mid a,b \in S,\ a \leq b\}$, and $(a,b)(a',b') = \delta_{ba'}(a,b')$ with $\delta_{ba'}$ being the Kronecker function. In this way, we may consider $A(S)$ as a subalgebra of the algebra of all $n \times n$-matrices over k, where n is the number of elements of S, the rows and columns being indexed by the elements of S. The algebra $A(S)$ has a canonical module, consisting of the set of column-vectors of length n. Of course, the quiver of $A(S)$ has as vertices the elements of S, and with an arrow $a \longleftarrow b$ if and only if a is a direct predecessor of b in S (this means that $a < b$, and that $a \leq a' < b$ implies $a = a'$), and $A(S)$ is the quotient of the path algebra of this quiver modulo the ideal generated by all commutativity relations.] Let $|\cdot|$ be the canonical functor given by $|O(a)| = k$ for any $a \in S$, and $|\alpha| = \alpha$ for $\alpha \in k = \mathrm{Hom}(O(a),O(b))$, where $a \leq b$. We will call $(K(S),|\cdot|)$ the vectorspace category given by the partially ordered set S. Obviously, $(K(S),|\cdot|)$ is a linear vectorspace category, and $S(K(S)) = S$. In case S is finite, and $K(S) = A(S)$-inj, then $|\cdot|$ is given by $\mathrm{Hom}(M(S),-)$, where $M(S)$ is the canonical $A(S)$-module.

We denote $\check{U}(K(S),|\cdot|)$ just by $\check{U}(S)$, and call the objects in $\check{U}(S)$ representations of the partially ordered set S (this is the notion of a representation of S as introduced by Nazarova and Rojter [NR1]). Similarly, $U(S) := U(K(S),|\cdot|)$.

(1) Let $(K,|\cdot|)$ be a finite linear vectorspace category. Let Z_o be indecomposable in K, considered as an object in $\check{U}(K,|\cdot|)$. Let $Y \longrightarrow Z_o$ be a sink map in $\check{U}(K,|\cdot|)$. If $[Z_o]$ is minimal in $S(K)$, then $Y = 0$; otherwise Y is indecomposable.

Proof. Since K is finite, we know that K has sink maps. Given Z_o, indecomposable in K, let $g : Y_o \longrightarrow Z_o$ be the sink map for Z_o in K, decompose $Y_o = \bigoplus_{i=1}^{t} Y_o^{(i)}$, with all $Y_o^{(i)}$ indecomposable. Let $g = (g_i)_i$, with $g_i : Y_o^{(i)} \longrightarrow Z_o$. Note that all the maps g_i are non-zero. Clearly, $[Y_o^{(i)}]$, $1 \leq i \leq t$, are just all direct predecessors of $[Z_o]$ in $S(K)$. In particular, the elements $[Y_o^{(i)}]$ are pairwise incomparable in $S(K)$. Now we apply 2.5.5 in order to determine Y. Namely, $Y = (Y_o, Y_\omega, u)$, with $Y_\omega = \mathrm{Ker}|g|$, and u being the inclusion map. In particular, if $[Z_o]$ is minimal in $S(K)$, then $Y_o = 0$, thus also $Y = 0$. Now, assume $[Z_o]$ is not minimal, thus $Y_o \neq 0$. Let $Y = Y' \oplus Y''$ be a direct decomposition of Y. Since $\mathrm{Hom}_K(Y_o^{(i)}, Y_o^{(j)}) = 0$ for $i \neq j$, there is $I' \subseteq \{1,\ldots,t\}$ with $Y_o' = \bigoplus_{i\in I'} Y_o^{(i)}$, $Y_o'' = \bigoplus_{i\in I''} Y_o^{(i)}$, where $I'' = \{1,\ldots,t\} \setminus I'$. Since $Y_\omega = (Y_\omega \cap |Y_o'|) \oplus (Y_\omega \cap |Y_o''|)$, and $\dim Y_\omega = -1+\dim|Y_o|$, we have $|Y_o'| \subseteq Y_\omega$ or $|Y_o''| \subseteq Y_\omega$. Thus assume $|Y_o''| \subseteq Y_\omega$. Now, Y_ω is the kernel of $|g|$, thus g vanishes on any $Y_o^{(i)}$ with $i \in I''$, this is possible only for $I'' = \emptyset$, thus $Y = Y'$.

(2) Let $(K, |\cdot|)$ be a finite linear vectorspace category. Then $_\infty\mathcal{U}(K,|\cdot|)$ is closed under neighbors. Also, $\Gamma(_\infty\mathcal{U}(K,|\cdot|))$ is a preprojective translation quiver with orbit quiver being a tree.

Proof. We can assume that K is non-zero. We recall from 2.5 that the projective vertices of $\Gamma(\mathcal{U}(K,|\cdot|))$ are those of the form $[X]$, with X indecomposable in K, and also $[F(\omega)]$. Also note that the sink map for $F(\omega)$ is given by $F(\omega)_o \longrightarrow F(\omega)$, and $F(\omega)_o$ is the direct sum of all X, with $[X]$ being minimal in $S(K)$. We have seen in (1) that for $[X]$ being minimal in $S(K)$, X belongs to $_o\mathcal{U}(K,|\cdot|)$. Since all indecomposable objects in $_o\mathcal{U}(K,|\cdot|)$ give projective vertices in $\Gamma(\mathcal{U}(K,|\cdot|))$, we see that $_o\mathcal{U}(K,|\cdot|)$ is the additive subcategory generated by the objects X of K with $[X]$ minimal in $S(K)$. Also note that $F(\omega)$ belongs to $_1\mathcal{U}(K,|\cdot|)$. Now, let us show by induction on d that given Y,Z indecomposable in $\mathcal{U}(K,|\cdot|)$, with an irreducible map $Y \longrightarrow Z$, then $Y \in {}_d\mathcal{U}(K,|\cdot|)$ implies $Z \in {}_{d+1}\mathcal{U}(K,|\cdot|)$. This is trivially true for $Z \in K$, since in this case $Y \longrightarrow Z$ is the sink map for Z, according to (1). Also, it is true for $Z = F(\omega)$. Thus, we only have to consider the case of $[Z]$ not being a projective vertex in $\Gamma(\mathcal{U}(K,|\cdot|))$. Let $[Y^{(i)}]$, $1 \leq i \leq t$, be the vertices in $[Z]^-$, and $[X] = \tau[Z]$. By assumption, Y is one of the $Y^{(i)}$, say $Y \approx Y^{(1)}$. Since $Y \in {}_d\mathcal{U}(K,|\cdot|)$ and there is an irreducible map $X \longrightarrow Y$, we see that X belongs to $_{d-1}\mathcal{U}(K,|\cdot|)$. Using induction, we conclude from the existence of irreducible maps $X \longrightarrow Y^{(j)}$ that all $Y^{(j)}$ belong to $_d\mathcal{U}(K,|\cdot|)$. Thus Z is in $_{d+1}\mathcal{U}(K,|\cdot|)$. This shows that $\mathcal{U}(K,|\cdot|)$ is closed under neighbors. Of course, $\Gamma(_\infty\mathcal{U}(K,|\cdot|))$ is a preprojective translation quiver. Also, its orbit quiver is a tree. Namely, let m be the number of minimal elements of $S(K)$. Then the orbit quiver contains the star $\mathbb{T}_{2,\ldots,2}$ with m branches and all labels being 0, with arrows starting at the center of the star [the center is the τ-orbit of $[F(\omega)]$, the remaining vertices are the τ-orbits of the minimal elements $[X]$ of $S(K)$], and the orbit quiver is obtained from this star by adding successively vertices with precisely one neighbor constructed before.

We recall from 2.2 that the indecomposable objects in ${}_{\infty}\mathcal{U}(K,|\cdot|)$ are directing in $\mathcal{U}(K,|\cdot|)$. In 2.5.4, we have seen that for a directing object X, we have $\mathrm{End}(X) = k$, $\mathrm{Ext}^1_K(X,X) = 0$, thus $\chi_K(\underline{\dim}_K X) = 1$, and if X' is any other indecomposable object in $\mathcal{U}(K,|\cdot|)$ with $\underline{\dim}_K X = \underline{\dim}_K X'$, then $X \approx X'$.

(3) Theorem of Drozd. Let $(K,|\cdot|)$ be a linear vectorspace category. Then $(K,|\cdot|)$ is subspace-finite if and only if χ_K is weakly positive. In this case, $\underline{\dim}_K$ yields a bijection between the indecomposable objects in $\mathcal{U}(K,|\cdot|)$ and the positive roots of χ_K.

Proof. If $\mathcal{U}(K,|\cdot|)$ is finite, then ${}_{\infty}\mathcal{U}(K,|\cdot|) = \mathcal{U}(K,|\cdot|)$, according to 2.5.10, thus all indecomposable objects X in $\mathcal{U}(K,|\cdot|)$ are bricks with $\mathrm{Ext}^1_K(X,X) = 0$, and, in particular, $\underline{\dim}_K X$ is a positive root for χ_K. Let y be a positive element in $G(K) \times \mathbb{Z}$, choose Y with $\underline{\dim}_K Y = y$ and smallest possible k-dimension of $\mathrm{End}(Y)$. Let $Y = \oplus Y_i$ with all Y_i indecomposable. According to lemma 1 of 2.3, we have $\mathrm{Ext}^1_K(Y_i,Y_j) = 0$ for $i \neq j$. Since also $\mathrm{Ext}^1_K(Y_i,Y_i) = 0$ for all i, we have $\mathrm{Ext}^1_K(Y,Y) = 0$. Thus $\chi_K(y) = \chi_K(\underline{\dim}\, Y) = \dim_k \mathrm{End}(Y) > 0$, and therefore χ_K is weakly positive. Also, in case y is a positive root of χ_K, then $\dim_k \mathrm{End}(Y) = \chi_K(y) = 1$, thus Y is indecomposable. This shows that $\underline{\dim}_K$ is a surjection from the set of isomorphism classes of indecomposable objects in $\mathcal{U}(K,|\cdot|)$ onto the set of positive roots of χ_K. On the other hand, it follows from 2.5.4 that $\underline{\dim}_K$ is also injective.

Conversely, assume that χ_K is weakly positive. If ${}_{\infty}\mathcal{U}(K,|\cdot|)$ would be infinite, the dimension vectors $\underline{\dim}_K X$, with X indecomposable, would furnish infinitely many positive roots for χ_K, impossible. Thus ${}_{\infty}\mathcal{U}(K,|\cdot|)$ is finite, and therefore $\mathcal{U}(K,|\cdot|) = {}_{\infty}\mathcal{U}(K,|\cdot|)$, according to 2.5.10. Thus $\mathcal{U}(K,|\cdot|)$ is finite. This finishes the proof.

As a corollary of this theorem and 1.10, we will obtain the two fundamental theorems of Klejner on representations of finite partially ordered sets. In order to introduce the notion of a critical vectorspace category, we will have to compare the subspace categories $\mathcal{U}(K,|\cdot|_1)$, and $\mathcal{U}(K,|\cdot|_2)$, where $|\cdot|_1$ and $|\cdot|_2$ are two additive functors on a Krull-Schmidt category K.

Let $|\cdot|_1, |\cdot|_2 : K \to$ k-mod be two additive functors, and $\eta : |\cdot|_1 \longrightarrow |\cdot|_2$ a natural transformation. Then η defines a functor $\mathcal{U}(K,|\cdot|_1) \longrightarrow \mathcal{U}(K,|\cdot|_2)$ which again will be denoted by η, as follows: given $V = (V_o,V_\omega,\gamma_V)$ in $\mathcal{U}(K,|\cdot|_1)$, let $\eta(V) = (V_o,V_\omega,\gamma_V\eta_{V_o})$, given a map $u = (u_o,u_\omega) : V \longrightarrow W$ in $\mathcal{U}(K,|\cdot|_1)$, let $\eta(u_o,u_\omega) = (u_o,u_\omega)$. Also, η defines a functor $\eta^* : \mathcal{U}(K,|\cdot|_2) \longrightarrow \mathcal{U}(K,|\cdot|_1)$. Namely, let $X = (X_o,X_\omega,\gamma_X) \in \mathcal{U}(K,|\cdot|_2)$, we form the pullback

$$\begin{array}{ccc} X'_\omega & \xrightarrow{\eta'_{X_o}} & X_\omega \\ \Big\downarrow{\scriptstyle \gamma'_X} & & \Big\downarrow{\scriptstyle \gamma_X} \\ |X_o|_1 & \xrightarrow{\eta_{X_o}} & |X_o|_2 \end{array},$$

and let $\eta^{\cdot}(X) = (X_o, X'_\omega, \gamma'_X)$. Given a map $f = (f_o, f_\omega) : X \longrightarrow Y$ in $\check{\mathcal{U}}(K, |\cdot|_2)$, then

$$\eta'_{X_o} f_\omega \gamma_Y = \eta'_{X_o} \gamma_X |f_o|_2 = \gamma'_X \eta_{X_o} |f_o|_2 = \gamma'_X |f_o|_1 \eta_{Y_o},$$

thus there exists a unique $f'_\omega : X'_\omega \longrightarrow Y'_\omega$ satisfying

$$f'_\omega \eta'_{Y_o} = \eta'_{X_o} f_\omega, \quad \text{and} \quad f'_\omega \gamma'_Y = \gamma'_X |f_o|_1,$$

in particular, $(f_o, f'_\omega) : \eta^{\cdot}(X) \longrightarrow \eta^{\cdot}(Y)$ is a map. Note that always with γ_X also γ'_X is mono, thus $\eta^{\cdot}$ maps $\mathcal{U}(K, |\cdot|_2)$ into $\mathcal{U}(K, |\cdot|_1)$. Also, if the transformation η is mono, then the induced functor η maps $\mathcal{U}(K, |\cdot|_1)$ into $\mathcal{U}(K, |\cdot|_2)$.

(4) Let $\eta : |\cdot|_1 \longrightarrow |\cdot|_2$ be a natural transformation. If η is mono, then the induced functor $\eta : \check{\mathcal{U}}(K, |\cdot|_1) \longrightarrow \check{\mathcal{U}}(K, |\cdot|_2)$ is a full exact embedding. If η is epi, then the functor $\eta^{\cdot} : \mathcal{U}(K, |\cdot|_2) \longrightarrow \mathcal{U}(K, |\cdot|_1)$ is a full exact embedding.

<u>Proof</u>. Clearly, $\eta : \check{\mathcal{U}}(K, |\cdot|_1) \longrightarrow \check{\mathcal{U}}(K, |\cdot|_2)$ is an exact embedding. Assume that η is mono. We have to show that in this case the functor η is full. Thus, let $V, W \in \check{\mathcal{U}}(K, |\cdot|_1)$, and $u = (u_o, u_\omega) : \eta(V) \longrightarrow \eta(W)$ a map in $\check{\mathcal{U}}(K, |\cdot|_2)$. Then

$$u_\omega \gamma_W \eta_{W_o} = u_\omega \gamma_{\eta(W)} = \gamma_{\eta(V)} |u_o|_2 = \gamma_V \eta_{V_o} |u_o|_2 = \gamma_V |u_o|_1 \eta_{W_o}.$$

Since η_{W_o} is mono, we conclude $u_\omega \gamma_W = \gamma_V |u_o|_1$, thus $(u_o, u_\omega) : V \longrightarrow W$ is a map in $\check{\mathcal{U}}(K, |\cdot|_1)$.

Next, consider the functor $\eta^{\cdot} : \mathcal{U}(K, |\cdot|_2) \longrightarrow \mathcal{U}(K, |\cdot|_1)$, and assume that η is epi. Clearly, $\eta^{\cdot}$ is faithful. For , given a map $u = (u_o, u_\omega) : X \longrightarrow Y$ in $\mathcal{U}(K, |\cdot|_2)$ with $\eta^{\cdot}(u) = (u_o, u'_\omega)$ being zero, it follows from $\eta'_{X_o} u_\omega = u'_\omega \eta'_{Y_o} = 0$ that $u_\omega = 0$, since η'_{X_o} is epi. Next, let us show that $\eta^{\cdot}$ is full. Thus, let $X, Y \in \mathcal{U}(K, |\cdot|_2)$, and $(f_o, f'_\omega) : \eta^{\cdot}(X) \longrightarrow \eta^{\cdot}(Y)$ a map in $\mathcal{U}(K, |\cdot|_1)$. We have

$$(*) \qquad \begin{aligned} f'_\omega \eta'_{Y_o} \gamma_Y &= f'_\omega \gamma'_Y \eta_{Y_o} = \gamma'_X |f_o|_1 \eta_{Y_o} \\ &= \gamma'_X \eta_{X_o} |f_o|_2 = \eta'_{X_o} \gamma_X |f_o|_2 , \end{aligned}$$

thus $(\mathrm{Ker}\ \eta'_{X_o})f'_\omega\eta'_{Y_o}\gamma_Y = 0$, and since by assumption γ_Y is mono, we even have $(\mathrm{Ker}\ \eta'_{X_o})f'_\omega\eta'_{Y_o} = 0$. It follows that we can factor $f'_\omega\eta'_{Y_o}$ through the epimorphism η'_{X_o}, so there is a unique f_ω with

$$f'_\omega\eta'_{Y_o} = \eta'_{X_o}f_\omega .$$

Also, using again (*), we have

$$\eta'_{X_o}f_\omega\gamma_Y = f'_\omega\eta'_{Y_o}\gamma_Y = \eta'_{X_o}\gamma_X|f_o|_2 ,$$

and since η'_{X_o} is epi, we conclude that (f_o, f_ω) is a map in $U(K,|\cdot|_2)$, with image under $\eta^{\cdot}$ being (f_o, f'_ω). Thus $\eta^{\cdot}$ is full. Finally, in order to **show** that $\eta^{\cdot}$ is exact, note that the definition of $\eta^{\cdot}(X)$ gives a commutative diagram with exact rows

$$\begin{array}{ccccccccc}
0 & \longrightarrow & \mathrm{Ker}\ \eta_{X_o} & \longrightarrow & X'_\omega & \xrightarrow{\eta'_{X_o}} & X_\omega & \longrightarrow & 0 \\
 & & \| & & \downarrow{\scriptstyle\gamma'_X} & & \downarrow{\scriptstyle\gamma_X} & & \\
0 & \longrightarrow & \mathrm{Ker}\ \eta_{X_o} & \longrightarrow & |X_o|_1 & \xrightarrow{\eta_{X_o}} & |X_o|_2 & \longrightarrow & 0
\end{array}$$

Now, given a K-split exact sequence

$$0 \longrightarrow X \xrightarrow{f} Y \xrightarrow{g} Z \longrightarrow 0$$

in $U(K,|\cdot|_2)$, we obtain a commutative diagram

$$\begin{array}{ccccccccc}
 & & 0 & & 0 & & 0 & & \\
 & & \downarrow & & \downarrow & & \downarrow & & \\
0 & \longrightarrow & \mathrm{Ker}\ \eta_{X_o} & \longrightarrow & \mathrm{Ker}\ \eta_{Y_o} & \longrightarrow & \mathrm{Ker}\ \eta_{Z_o} & \longrightarrow & 0 \\
 & & \downarrow & & \downarrow & & \downarrow & & \\
0 & \longrightarrow & X'_\omega & \xrightarrow{f'_\omega} & Y'_\omega & \xrightarrow{g'_\omega} & Z'_\omega & \longrightarrow & 0 \\
 & & \downarrow & & \downarrow & & \downarrow & & \\
0 & \longrightarrow & X_\omega & \xrightarrow{f_\omega} & Y_\omega & \xrightarrow{g_\omega} & Z_\omega & \longrightarrow & 0 \\
 & & \downarrow & & \downarrow & & \downarrow & & \\
 & & 0 & & 0 & & 0 & &
\end{array}$$

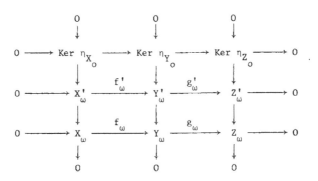

with exact columns. The third row is exact. Also, the first row is obtained from the split exact sequence with maps f_o and g_o by applying the functor $\mathrm{Ker}\ \eta_-$, thus also this row is exact. It follows that the second row also is exact. This shows that $\eta^{\cdot}$ is an exact functor, and finishes the proof.

Note that in general the induced functor $\eta^{\cdot} : \mathcal{U}(K,|\cdot|_2) \longrightarrow \mathcal{U}(K,|\cdot|_1)$ is not full, even if η is epi. [Namely, let X_o be indecomposable in K, with η_{X_o} being a proper epimorphism, and consider $\overline{X}_o = (X_o, |X_o|_2, 1)$ in $\mathcal{U}(K,|\cdot|_2)$. Note that $\eta^{\cdot}(\overline{X}_o) = (X_o, |X_o|_1, 1)$, and $\eta^{\cdot}(E(\omega)) = E(\omega)$. We have

$$\dim \operatorname{Hom}_{\mathcal{U}(K,|\cdot|_i)}((X_o, |X_o|_i, 1), E(\omega)) = \dim |X_o|_i ,$$

thus the map

$$\eta^{\cdot} : \operatorname{Hom}(\overline{X}_o, E(\omega)) \longrightarrow \operatorname{Hom}(\eta^{\cdot}(\overline{X}_o), \eta^{\cdot}(E(\omega)))$$

cannot be surjective.]

Given a vectorspace category $(K,|\cdot|)$, let $K(|\cdot|)$ be obtained from K by factoring out the ideal in K given by all maps f with $|f| = 0$. We may consider the functor $|\cdot|$ as being defined on $K(|\cdot|)$, and $|\cdot|$ is a faithful functor on $K(|\cdot|)$. Of course, the canonical functor $K \longrightarrow K(|\cdot|)$ induces a canonical functor $\mathcal{U}(K,|\cdot|) \longrightarrow \mathcal{U}(K(|\cdot|),|\cdot|)$.

(5) <u>The canonical functor</u> $\mathcal{U}(K,|\cdot|) \longrightarrow \mathcal{U}(K(|\cdot|),|\cdot|)$ <u>maps the objects in</u> $\mathcal{U}(K,|\cdot|)$ <u>which are of the form</u> $(X,0,o)$ <u>with</u> $X \in K$, <u>and</u> $|X| = 0$, <u>to zero, and it gives a bijection between the remaining indecomposable objects in</u> $\mathcal{U}(K,|\cdot|)$ <u>and the indecomposable objects in</u> $\mathcal{U}(K(|\cdot|),|\cdot|)$.

<u>Proof.</u> The canonical functor $K \longrightarrow K(|\cdot|)$ is full, therefore the same is true for the canonical functor $\mathcal{U}(K,|\cdot|) \longrightarrow \mathcal{U}(K(|\cdot|),|\cdot|)$. Given an indecomposable object $V = (V_o, V_\omega, \gamma_V)$ in $\mathcal{U}(K,|\cdot|)$, its image is either zero or indecomposable. The image can be zero only in case the identity map 1_{V_o} of V_o in K satisfies $|1_{V_o}| = 0$, thus only in case $|V_o| = 0$. However, for an indecomposable object $V = (V_o, V_\omega, \gamma_V)$ with $|V_o| = 0$, we either have $V_\omega = 0$, and V_o is indecomposable or else $V = E(\omega)$. Conversely, we can write any object of $\mathcal{U}(K(|\cdot|),|\cdot|)$ as image of an object of $\mathcal{U}(K,|\cdot|)$ of the form $(V_o, V_\omega, \gamma_V)$, with no indecomposable summand V_o' of V_o satisfying $|V_o'| = 0$. Finally, assume there are given $V = (V_o, V_\omega, \gamma_V)$, $W = (W_o, W_\omega, \gamma_W)$ in $\mathcal{U}(K,|\cdot|)$ such that neither V_o nor W_o have indecomposable summands annihilated by $|\cdot|$, and let V, W become isomorphic in $\mathcal{U}(K(|\cdot|),|\cdot|)$. Since the canonical functor is full, there is a map $f = (f_o, f_\omega) : V \longrightarrow W$ which becomes an isomorphism in $\mathcal{U}(K(|\cdot|),|\cdot|)$. However, this is possible only in case f is an isomorphism even in $\mathcal{U}(K,|\cdot|)$, since neither V_o nor W_o have indecomposable summands annihilated by $|\cdot|$. This finishes the proof.

A vectorspace category $(K,|\cdot|)$ will be called <u>critical</u> provided the following conditions are satisfied:

(α) $|\cdot|$ is a faithful functor.

(β) $(K,|\cdot|)$ is subspace infinite.

(γ) For any proper subfunctor $|\cdot|'$ of $|\cdot|$, the vectorspace category $(K,|\cdot|')$ is subspace finite.

(γ*) For any proper factor functor $|\cdot|'$ of $|\cdot|$, the vectorspace category $(K,|\cdot|')$ is subspace finite.

We are going to characterize the critical directed vectorspace categories. Also, we will classify all subspace finite sincere directed vector space categories. Here, an object $V = (V_o, V_\omega, \gamma_V)$ in $\check{U}(K,|\cdot|)$ is said to be <u>sincere</u> provided any indecomposable object of K is a direct summand of V_o, and $V_\omega \neq 0$. A sincere object $\check{U}(K,|\cdot|)$ can only exist in case K is finite, and, $V \in \check{U}(K,|\cdot|)$ is sincere if and only if $\underline{\dim}_K V$ is a sincere element of $G(K) \times \mathbb{Z}$. Also, the vectorspace category $(K,|\cdot|)$ is called <u>sincere</u> provided the functor $|\cdot|$ is faithful and there exists an indecomposable sincere object in $\check{U}(K,|\cdot|)$.

Besides vectorspace categories given by partially ordered sets, we also will have to consider one additional vectorspace category, namely $(k\text{-mod}, \mathrm{Hom}(k^2,-))$ which we call <u>the vectorspace category of type</u> $C(1)$. Note that its quadratic form is just the form $C(1)$ introduced in 1.0. The objects in $U(k\text{-mod}, \mathrm{Hom}(k^2,-))$ are called Kronecker modules; they are given by triples $(V_o, V_\omega, \gamma_V)$, where V_o, V_ω are vectorspaces, and a k-linear map $\gamma_V : V_\omega \longrightarrow \mathrm{Hom}(k^2, V_o) = V_o \oplus V_o$, or, equivalently, two vectorspaces V_o, V_ω and two k-linear maps $\gamma_1, \gamma_2 : V_\omega \longrightarrow V_o$. We will study this category in 3.2. in detail.

<u>Theorem 1 of Klejner.</u> <u>A directed vectorspace category is critical if and only if it is of type</u> $C(1)$, <u>or given by a partially ordered set of type</u> $C(2)$, $C(3)$, $C(4)$, $C(5)$, <u>or</u> $C(6)$.

<u>Theorem 2 of Klejner.</u> <u>A directed vectorspace category is subspace finite and sincere if and only if it is given by a partially ordered set of type</u> $F(1), F(2), F(3), F(4), F(5), F(6); F(1,1), F(2,2), F(3,3), F(4,4), F(5,5), F(5,6)$, <u>or</u> $F(6,5)$.

In case $(K,|\cdot|)$ is given by the partially ordered set S of type $C(2)\ldots,$ $C(6)$, $F(1),\ldots, F(6)$, $F(1,1),\ldots, F(6,5)$, we will call the type of S also the <u>type</u> of $(K,|\cdot|)$.

<u>Proof of the theorems of Klejner.</u> It is easy to see that the vectorspace categories of type $C(1),\ldots, C(6)$ all are critical [the full structure of $\check{U}(K,|\cdot|)$ will be derived in 3.2, for $(K,|\cdot|)$ of type $C(1)$, and in 4.4, for $(K,|\cdot|)$ of type $C(2),\ldots, C(6)$]. Also, the vectorspace categories given by the partially

ordered sets S with χ_S sincere and weakly positive all are sincere and subspace finite, according to the theorem of Drozd.

Conversely, let $(K,|\cdot|)$ be a directed vectorspace category, and assume there is an indecomposable object X in K with $\dim_k|X| \geq 2$. In case there is no indecomposable object Y in K with $|Y| \neq 0$ and $X \prec Y$, we define a functor $|\cdot|' : K \to k\text{-mod}$ as follows: let $|X|' = k^2$, and $|Z|' = 0$ for Z indecomposable, and $Z \not\approx X$. Obviously, $|\cdot|'$ can be considered as a subfunctor of $|\cdot|$, and $(K(|\cdot|'),|\cdot|')$ is of the form $C(1)$, thus subspace infinite. If we assume that $(K,|\cdot|)$ is critical, it follows that $(K,|\cdot|)$ itself is of the form $C(1)$. In general, define a functor $|\cdot|''$ on K as follows: given an indecomposable object Y in K, let $|Y|'' = 0$ in case $X \prec Y$, and $|Y|'' = |Y|$ otherwise. Clearly, $|\cdot|''$ is a factor functor of $|\cdot|$, and $(K,|\cdot|'')$ is subspace infinite by the previous considerations ($\dim|X|'' \geq 2$, and there is no indecomposable object Y with $|Y|'' \neq 0$ and $X \prec Y$). If we assume that $(K,|\cdot|)$ is critical, it follows that $|\cdot| = |\cdot|''$, and therefore $(K,|\cdot|)$ is of the form $C(1)$.

It remains to consider the case that $\dim_k|X| \leq 1$ for all indecomposable objects X in K. We assume that $(K,|\cdot|)$ is either critical, or subspace finite and sincere, therefore the functor $|\cdot|$ is faithful, by assumption. Thus $(K,|\cdot|)$ is a linear vectorspace category. If $(K,|\cdot|)$ is sincere, then obviously K is finite. Also, for $(K,|\cdot|)$ critical, K has to be finite, since otherwise we easily construct a subfunctor or a factor functor $|\cdot|'$ of $|\cdot|$ with $(K,|\cdot|')$ being subspace infinite. If $(K,|\cdot|)$ is subspace finite and sincere, then the quadratic form $\chi_{S(K)} = \chi_K$ is weakly positive and sincere, thus $S(K)$ is one of the listed partially ordered sets, according to 1.10. If $(K,|\cdot|)$ is critical, then $\chi_{S(K)}$ is not weakly positive, thus according to the corollary in 1.10, there is a convex subset S' of $S(K)$ with $\chi_{S'}$ being critical. Note that $S' = S(K) \setminus (S_1 \cup S_2)$, where S_1 is an ideal and S_2 a filter in $S(K)$. Let $|\cdot|' : K \to k\text{-mod}$ be defined by $|X|' = |X|$ for X indecomposable and in S', and $|X|' = 0$ for X indecomposable and not in S'. Then $|\cdot|'$ is a factor functor of a subfunctor of $|\cdot|$. Since $(K,|\cdot|')$ is still subspace-infinite, and $(K,|\cdot|)$ is critical, it follows that $S(K) = S'$. Using also in this case 1.10, we again conclude that $S(K)$ is one of the listed partially ordered sets.

Finally, we have to verify that a linear vectorspace category $(K,|\cdot|)$ with $S(K)$ of the form $C(2),\ldots, C(6)$; $F(1),\ldots, F(6)$; $F(1,1),\ldots, F(6,5)$ is actually given by the corresponding partially ordered set. Let (A-inj,Hom(R,-)) be an injective realization of $(K,|\cdot|)$. Note that in all cases A is hereditary, and R is a module with dimension vector $(1,1,\ldots,1)$. Actually, always $|\cdot|$ is faithful, thus R is a faithful A-module. Note that in all cases but $F(5,5)$, there is a single faithful A-module with dimension vector $(1,1,\ldots,1)$. It remains to consider the case $F(5,5)$. We can assume that A is the path algebra of the quiver

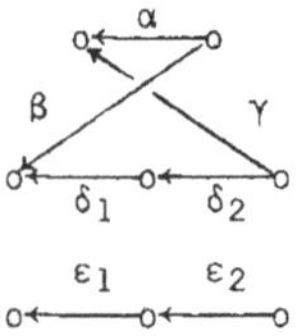

The faithful representations of this quiver with dimension vector $(1,1,\dots,1)$ are of the form

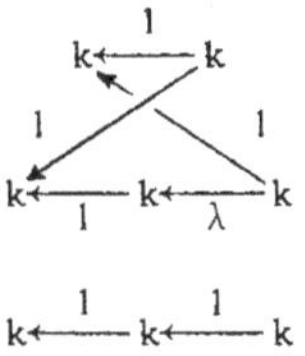

where $0 \neq \lambda \in k$. However, we may replace in the path algebra one of the generators by a non-zero multiple in order to change the form of R. If we replace the arrow δ_2 by $\lambda\delta_2$, then R has the form

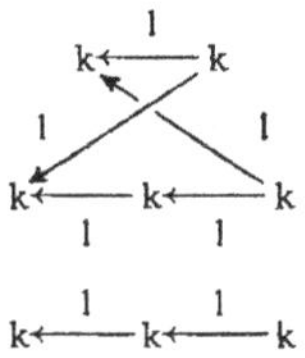

but this means that $(K,|\cdot|)$ is given by the partially ordered set $S(K)$. This finishes the proof.

The finite partially ordered sets with $A(S)$ being hereditary, have some special properties which are of interest to us. Let us first give a criterion for $A(S)$ being hereditary:

Lemma Let S be a finite partially ordered set. Then $A(S)$ is hereditary if and only if the following condition is satisfied: given a,b_1,b_2,c in S with $a < b_1 < c$, $a < b_2 < c$, then b_1, b_2 are comparable.

Proof. We have noted above that $A(S)$ is the quotient of a quiver Δ modulo all commutativity relations. Thus, $A(S)$ is hereditary, if and only if given two vertices of Δ, there is at most one path joining them. In Δ, there is a path from c to a if and only if $a \leq c$, and there is a unique path from c to a, provided the elements b with $a \leq b \leq c$ form a chain. This finishes the proof.

Note that for all the partially ordered sets S occurring in the theorem of Klejner (both the sincere representation finite ones, as well as the critical ones), A(S) is hereditary.

(6) Let S be a finite partially ordered set with A(S) hereditary. Then $_\infty\check{\mathcal{U}}(S)$ is standard.

Proof. According to lemma 3 of 2.3 and using 2.5.9', it is sufficient to show that the sink maps for the objects O(a), a ∈ S, and F(ω) in $\check{\mathcal{U}}(S)$ are monomorphisms. The sink map for F(ω) has been calculated in 2.5.7, it is given by $(1,0)\colon F(\omega)_o \to F(\omega)$, thus it is K-split mono. Now, consider O(a), with a ∈ S. We call b ∈ S a direct predecessor of a in S, provided b < a, and $b \le c < a$ implies b = c. Let $b_1,\dots,b_t$ be the direct predecessors of a. Clearly, the canonical map

$$g = (g_i) : \bigoplus_{i=1}^{t} O(b_i) \to O(a)$$

(with $g_i = 1 \in k = \mathrm{Hom}(O(b_i),O(a))$, for all i) is a sink map for O(a) in $K(S)$. According to 2.5.5, the sink map for O(a) in $\check{\mathcal{U}}(S)$ is given by $(g,0) : Y \to O(a)$, with $Y_o = \bigoplus_{i=1}^{t} O(b_i)$, $Y_\omega = \mathrm{Ker}\,|g|$, and γ_Y the inclusion map. Let $u : X \to Y$ be a map in $\check{\mathcal{U}}(S)$, with $u(g,0) = 0$. Decompose X_o into indecomposable direct summands, say $X_o = \bigoplus_{j=1}^{s} O(d_j)$, with not necessarily pairwise different $d_j \in S$. The map

$$u_o : \bigoplus_{j=1}^{s} O(d_j) \to \bigoplus_{i=1}^{t} O(b_i)$$

is of the form $u_o = (u_{ji})_{ji}$, with $u_{ji} \in \mathrm{Hom}(O(d_j),O(b_i)) \subseteq k$. Since A(S) is hereditary, and the elements $b_1,\dots,b_t$ are pairwise incomparable predecessors of a, it follows that any d_j is predecessor of at most one b_i; thus, for given j, there is at most one i with $u_{ji} \neq 0$. However, by assumption, $u_o g = 0$, and the restriction of $u_o g$ to $O(d_j)$ is just $\sum_{i=1}^{t} u_{ji}$. It follows that all $u_{ji} = 0$, therefore u_o is the zero map. Now, $u_\omega \gamma_Y = \gamma_X u_o = 0$, and γ_Y is an inclusion map, thus also $u_\omega = 0$, therefore $u = 0$. This finishes the proof.

<u>The Auslander-Reiten quivers</u> $\Gamma(\check{\mathcal{U}}(S))$ <u>for the sincere representation finite partially ordered sets</u> (with the dimension vector of any indecomposable representation V which is sincere or for which [V] is a projective vertex):

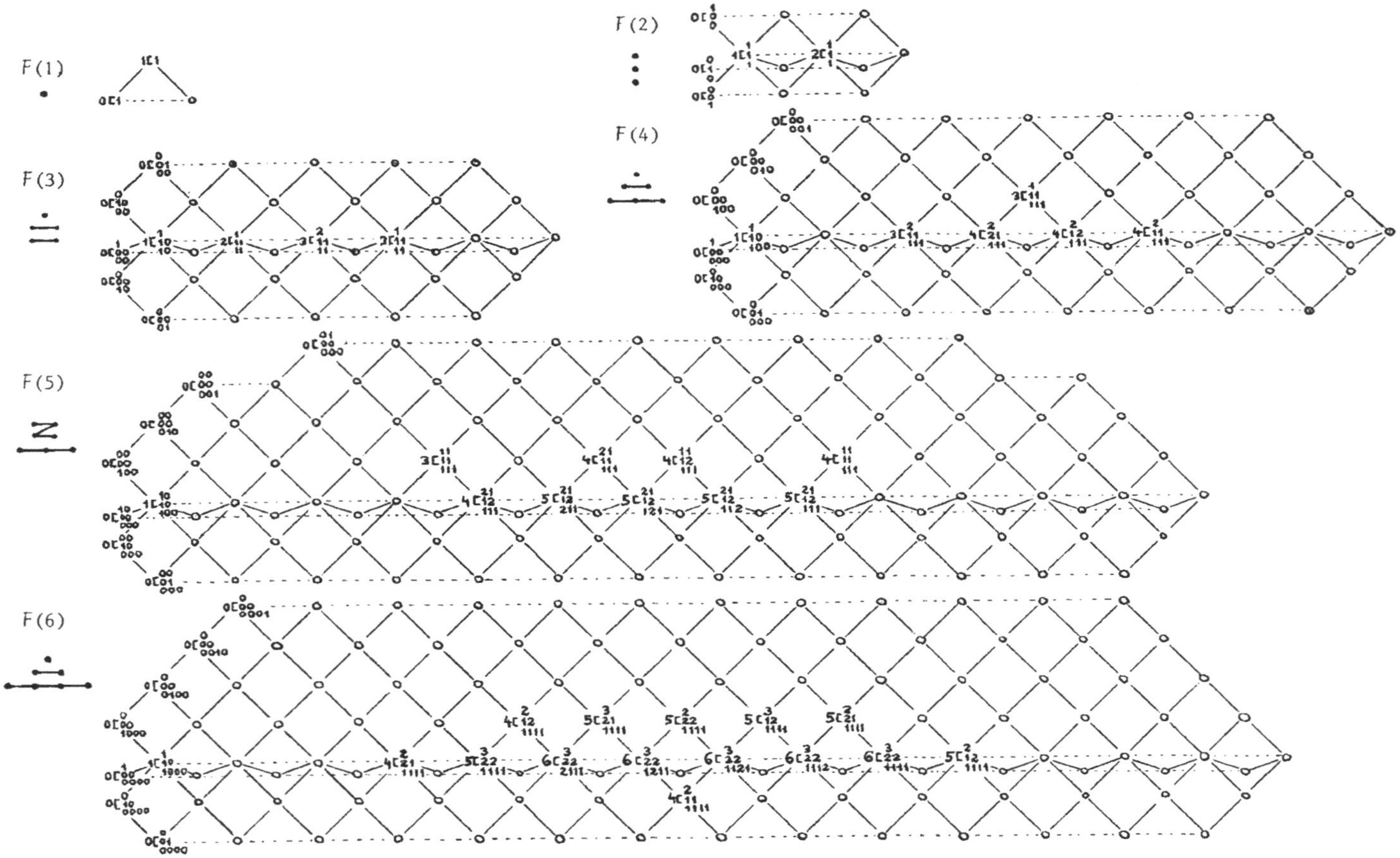

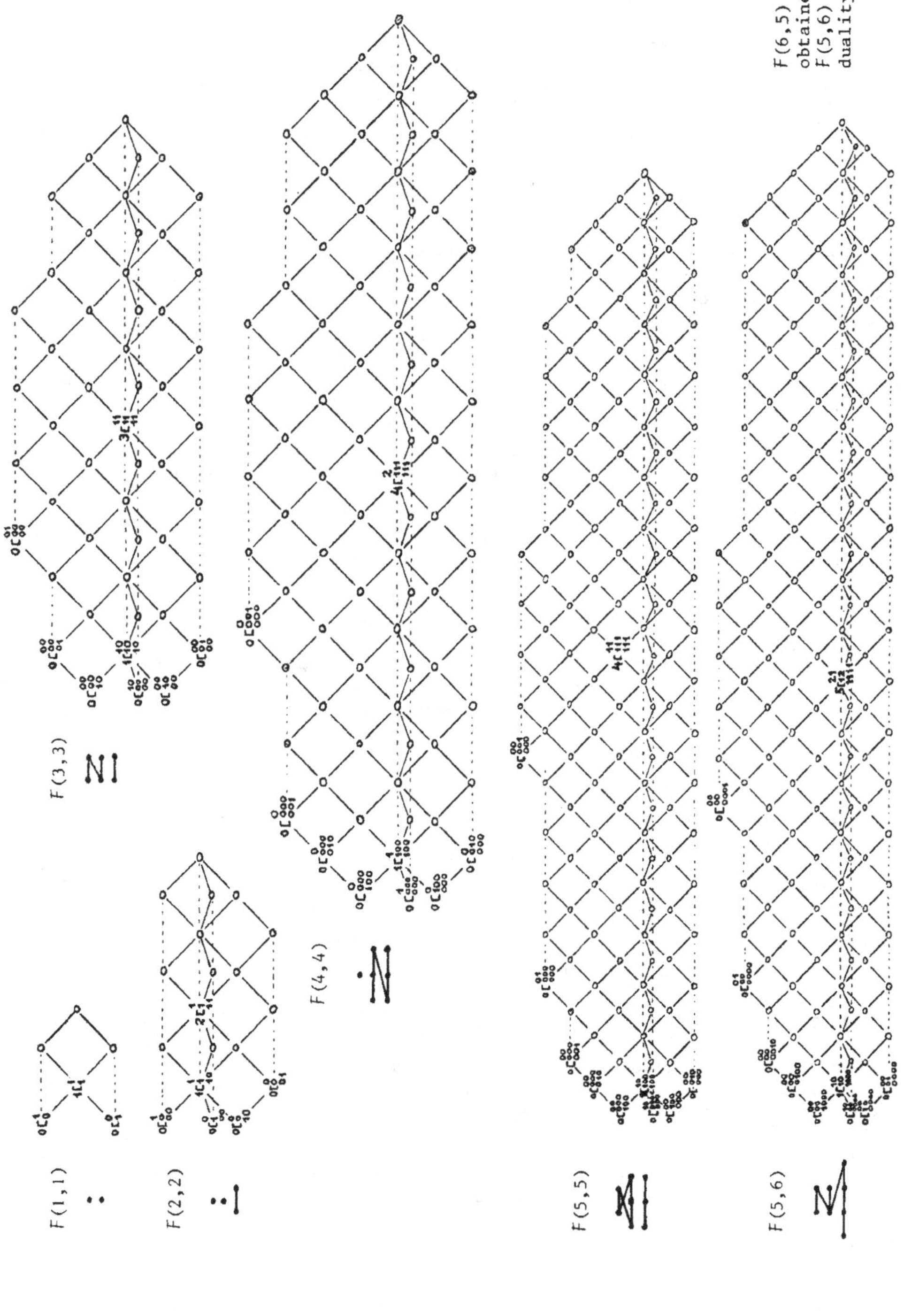

F(6,5) can be obtained from F(5,6) using duality

References and comments

The representation theory of quivers was initiated by Gabriel in order to reduce various problems of algebra and geometry to problems in linear algebra. In [G1] he gave the famous criterion 2.4.12 for a quiver to be representation finite, and he outlined in [G2] the use of quivers in the representation theory of algebras. In particular, there one finds the important structure theorem 2.1.2 for basic finite dimensional k-algebras, a concise proof of 2.1.2 may be found in his report [G4], section 4.3. (Actually, more generally, [G2] considers algebras over arbitrary base fields, and even arbitrary length categories, and then species instead of quivers.). The notions of a translation quiver (with the additional assumption that there are no multiple arrows and no loops), of the corresponding mesh category, and of the realization of a translation quiver as a two-dimensional simplicial complex are due to Riedtmann [Rm], we refer to this paper also for the proof of 2.1.3.

The basic notions concerning rings, modules and homology which we will assume can be found in nearly any text book. In particular, this applies to proofs of the theorems of Krull-Schmidt and Jordan-Hölder, and to the use of projective and injective resolutions. For example, we may refer to the books of Curtis-Reiner[CR],Anderson-Fuller[AF] and Cartan-Eilenberg [CE]. The theory of Morita equivalence was develop by Bass [B], it also may be found in [AF]. The standard references to Grothendieck groups, and $K_0(A)$, are the books of Bass [B] and Swan [Sw]. The definition of a bilinear form on $K_o(A)$ using C_A^{-T}, where C_A is the Cartan matrix of A, was suggested by Brenner-Butler [BB], the definition of the Coxeter matrix using 2.4.b was suggested (in the case of hereditary algebras) by Auslander-Platzeck [AP]. The concept of an exact category is due to Heller [H], (who used the name "abelian category"; note that for the similarly defined "exact" categories in Quillen [Q], idempotents are not supposed to split), the use of Krull-Schmidt categories is everywhere standard.

The question whether an indecomposable A-module is determined by its dimension vector, is very natural. We answer this question in the affirmative for directing modules, in 2.4.8. The results 2.4.7, 2.4.8 and 2.4.9 are improvements of earlier joint results with Happel [HR1]; lemma 2.3.1, which is used in the proof, is taken from [Ri2]. Note that there is a recent different approach due to Auslander-Reiten [AR3]. A common device for characterizing, for a well-behaved algebra A, the set of dimension vectors of indecomposable modules, is the use of the quadratic form χ_A. We should stress that our definition for a module class to be controlled by χ_A is adjusted to the needs of these notes, only: it works well both for concealed algebras and for tubular algebras. However, in general more sophisticated concepts have to be used, as is apparent, for example, from the investigations of wild hereditary algebras by Kac, we refer to his survey [Ka]. Also, for algebras of global dimension > 2, it seems appropriate to work with a different quadratic form χ_A' being obtained by truncation from χ_A, as was demonstrated by Bongartz [Bo3]; in particular, using χ_A' instead of χ_A, assertion 2.4.9 remains valid without the restriction on the global dimension of A.

Auslander-Reiten sequences were introduced (under the name "almost split sequences") by Auslander-Reiten in [AR1], the concepts of irreducible maps, source maps (called "minimal left almost split maps") and sink maps (called "minimal right almost split maps"), and their relations to Auslander-Reiten sequences are given in [AR2]. There are several existence proofs for Auslander-Reiten sequences in module categories available, the first one, starting from the formula 2.4.5, is given in [AR1]; a good reference is also [G4]. From [G4], we have copied the proof of 2.3.2, lemma 2.2.3 is taken from [Ri3]. The description of the Auslander-Reiten translation τ for a module category in terms of the exact sequence 2.4(++) is already in [AR1] (under the name "dual of transpose"). Assertion 2.4.1, characterizing modules of projective dimension 1 using τ, seems to have been noticed first by

Bongartz [Bo1], assertions similar to 2.4.3 were frequently used by Bautista, see for example [Ba1]. Theorem 2.4.10. of Auslander is taken from [A2].

Preprojective components of Auslander-Reiten quivers are always easily constructed; this is outlined, for example, in [G4], section 6.6. As we have noted after the proof of 2.1.5, a preprojective translation quiver is uniquely determined by its orbit quiver and the lengths of the τ-orbits. Actually, in case a preprojective translation quiver is a component of the Auslander-Reiten quiver of an algebra, then the lengths of the τ-orbits may be calculated from the orbit quiver, using the additive functions starting at the projective vertices. In case the underlying graph of the orbit quiver is a tree, one may consider instead of the orbit quiver its underlying graph together with a grading, as outlined in [BG]. Versions of 2.3.3 seem to have been known independently at various places, see for example [G4] and [Ba1]; we have chosen a formulation which also can be applied to representations of partially ordered sets, see 2.6.6. Note that there is a general theory of preprojective modules due to Auslander-Smalø [AS], the modules in a preprojective component always are preprojective modules, due to 2.1.4, however note that a component which only contains preprojective modules does not have to be a preprojective component.

Many of the notions defined for general finite dimensional algebras are generalizations of concepts which first were found fertile for hereditary algebras. The representation theory of hereditary algebras started with Gabriels theorem 2.4.12, and its conceptual proof due to Bernstein-Gelfand-Ponomarev [BGP]. Part of the technique was presented before by Gelfand-Ponomarev [GP] in their study of the four-subspace quiver. Given a connected quiver Δ without cyclic paths, the representations of Δ belonging to the preprojective component were introduced in [BGP] as the "(+)-irregular" representations, those belonging to the preinjective component as "(-)-irregular", and essential parts of 2.4.11 and 2.4.13 (and all of 2.4.12) have been shown in this paper. The relation between τ and the Coxeter functors of [BGP] has been determined by Gabriel in [G4], section 5.4.

A useful induction technique for algebras whose quiver has no cyclic path is the method of one-point extensions. This was the main device in Gabriel's first paper [G1] and later was used by several authors ([L1],[M1]). A detailed account may be found in [Ri4]. This technique reduced the study of module categories to that of subspace categories of vectorspace categories, thus often to the study of representations of partially ordered sets.

The representation theory of partially ordered sets as outlined in 2.6 was developed by Nazarova-Rojter [NR1] (this theory should not be confused with the representation theory of the incidence algebras of partially ordered sets, as considered, for example by Loupias, see [L2]). In addition, given a partially ordered set S, Gabriel [G3] has introduced the category $L(S)$ of S-spaces which differs only slightly from $\check{\mathcal{U}}(S)$; it is representation equivalent to $\check{\mathcal{U}}(S) \setminus K(S)$, see Drozd [D1]. Actually, according to Simson [Si], $\check{\mathcal{U}}(S)/K(S) \approx L(S)$. Vectorspace categories and their subspace categories were introduced by Nazarova and Rojter [NR2].

The theorems of Klejner were given in [Kl1,2], his proof uses the differentiation process due to Nazarova-Rojter [NR1], it involves rather long combinatorial considerations. A homological interpretation of the differentiation process was given by Gabriel in [G3], and he and Bongartz, gave an outline for the combinatorial part of the proof of theorem (2) [unpublished]. For theorem (2), there also exists a proof due to Kerner [Ke] using the Zavadskij algorithm [Z1]. The proof for both theorems of Klejner presented here originates in investigations of Drozd and Ovsienko. We use theorem 2.6.3 of Drozd, and the theory of integral quadratic forms as developed in chapter 1. Drozd's theorem has appeared, with a different proof, in [D1]. The original formulation of the theorems of Klejner deals only with partially ordered sets, and not with directed vectorspace categories, thus his critical list does not include $C(1)$. Of course, in this way, one also may shorten our proof considerably: the result follows directly from the theorem of Drozd and 1.10, without reference to 2.6.4 or 2.6.5. - Let us give a justification for the markings $F(1),..,F(6)$; $C(1),...,C(6)$; and $F(1,1),..,F(5,6)$. The index a of the critical directed vectorspace category $C(a)$ refers to the fact that the positive radical generator for $\chi_{C(a)}$ is of the form $(a[h_1,...,h_n)$, even with $\Sigma h_i = 2a$. Similarly, any $F(a)$ has a unique maximal root, and it is of the form $(a[x_1,...,x_n)$. Note that the $F(a,b)$ are just those representation finite partially ordered sets S which have a unique sincere positive root $x = (u[x_1,...,x_n)$, and such that, in addition, $\Sigma x_i = 2u$. Note that the root x has two exceptional vertices, say i,j both belonging to the partially ordered set S, and the sets $S\setminus\{i\}$ and $S\setminus\{j\}$ are of the form $F(a)$ and $F(b)$.

The phenomenon of wild behaviour of module categories seems to be noticed first by Krugliak [Kr]. In [Br], Brenner has shown that the quivers with indefinite quadratic forms are strictly wild. Similarly, Nazarova [N3] has determined the minimal partially ordered sets which are strictly wild. There is a general result due to Drozd [D2] asserting that any representation infinite algebra is either tame or wild.

3. Construction of stable separating tubular families

This chapter is devoted to a procedure for constructing algebras A having components in their Auslander-Reiten quivers which are of the form $\mathbb{Z}\mathbb{A}_\infty/n$. Here, for $n \in \mathbb{N}_1$, the translation quiver $\mathbb{Z}\mathbb{A}_\infty/n$ is obtained from $\mathbb{Z}\mathbb{A}_\infty$ by identifying any vertex $x \in \mathbb{Z}\mathbb{A}_\infty$ with $\tau^n x$, and any arrow $x \longrightarrow y$ with the arrow $\tau^n x \longrightarrow \tau^n y$. The translation quiver $\mathbb{Z}\mathbb{A}_\infty/n$ is called a stable tube. Actually, we will construct not just single stable tubes but rather one-parameter families $\mathcal{T}$ of stable tubes, always indexed over the projective line $\mathbb{P}_1 k$, and these families will be "separating": the category A-mod will be divided into three module classes $\mathcal{P},\mathcal{T},\mathcal{Q}$, with $\mathcal{T}$ being the constructed $\mathbb{P}_1 k$-family of tubes, such that there are no non-zero maps from $\mathcal{Q}$ to $\mathcal{P}$, from $\mathcal{Q}$ to $\mathcal{T}$ or from $\mathcal{T}$ to $\mathcal{P}$, whereas, on the other hand, any map from $\mathcal{P}$ to $\mathcal{Q}$ factors through $\mathcal{T}$, even through any one of the tubes. One may illustrate this situation by the following picture

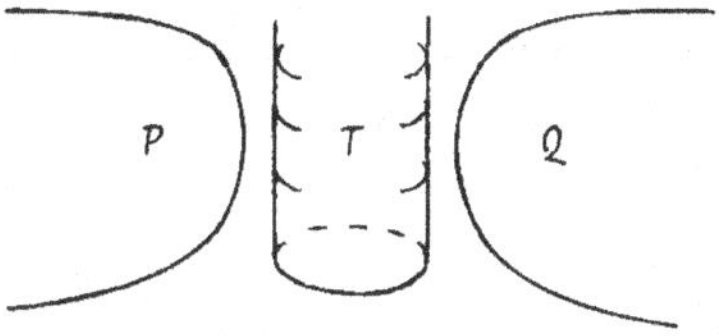

with non-zero maps being possible only from left to right. In addition, the module classes $\mathcal{P},\mathcal{T},\mathcal{Q}$ are closed under the Auslander-Reiten-translations τ_A,τ_A^{-1} and under extensions, and $\mathcal{T}$ is, in fact, standard. Let us first formulate the relevant definitions.

3.1 Separating tubular families

A translation quiver Γ without multiple arrows is said to be a tube provided $|\Gamma| = S^1 \times \mathbb{R}_0^+$ (where S^1 is the unit circle and $\mathbb{R}_0^+$ the set of non-negative real numbers), and Γ contains a cyclic path. In case Γ is a tube, the subset $S^1 \times \{o\}$ of $|\Gamma|$ is called its mouth. Examples of tubes are the translation quivers of the form $\mathbb{Z}\mathbb{A}_\infty/n$. Recall that a translation quiver is said to be stable provided it does not contain either projective or injective vertices.

(0) A translation quiver Γ is a stable tube if and only if Γ is of the form $\mathbb{Z}\mathbb{A}_\infty/n$ for some $n \in \mathbb{N}_1$.

Proof. Given a stable tube Γ, we first note that no arrow $x \to y$ of can belong to the mouth of Γ, for, it determines the triangle $(\tau y,x,y)$, since y is not projective, and the triangle $(x,y,\tau^- x)$, since x is not injective. Thus, all 1-simplices belonging to the mouth of Γ are extensions. Since the mouth of Γ (being compact) is formed by finitely many, say n simplices, it follows that the vertices of Γ lying on the mouth form a τ-orbit of length n. We fix some vertex $x = x[1]$ lying on the mouth of Γ. There is a unique arrow starting at $x[1]$, let $x[2]$ be its end point. Assume, for some $i \geq 2$ we have defined a chain of arrows

$$x[1] \longrightarrow x[2] \longrightarrow \dots \longrightarrow x[i]$$

with $\tau x[j+1] \neq x[j-1]$ for all $1 < j < i$. Then $x[i]$ does not belong to the mouth of Γ, thus there are precisely two arrows starting at $x[i]$. One of these two arrows is $x[i] \longrightarrow \tau^{-}x[i-1]$, and we denote by $x[i+1]$ the endpoint of the remaining arrow starting at $x[i]$. In this way, we obtain an infinite chain

$$x[1] \longrightarrow x[2] \longrightarrow \dots \longrightarrow x[i] \longrightarrow x[i+1] \longrightarrow \dots$$

with $\tau x[j+1] \neq x[j-1]$ for all $j \geq 2$. Also, let $\vec{\mathbb{A}}_\infty$ be the quiver with $\mathbb{N}_1$ as set of vertices, and with arrows $i \longrightarrow i+1$, for all $i \in \mathbb{N}_1$, thus $\mathbb{A}_\infty$ is the underlying graph of $\vec{\mathbb{A}}_\infty$. We define a map from $\mathbb{Z}\vec{\mathbb{A}}_\infty$ to Γ by sending the vertex (z,i) of $\mathbb{Z}\vec{\mathbb{A}}_\infty$ to $\tau^{-z}x[i]$ (where $z \in \mathbb{Z}$, $i \in (\vec{\mathbb{A}}_\infty)_0 = \mathbb{N}_1$). It is easy to see that this map induces an isomorphism from $\mathbb{Z}\vec{\mathbb{A}}_\infty/n$ onto Γ.

Given a stable tube $\Gamma = \mathbb{Z}\mathbb{A}_\infty/n$, the number n will be called the rank of Γ; a stable tube of rank 1 will also be said to be homogeneous. Note that the vertices of a stable tube Γ can be written uniquely in the form $x[\ell]$, with x a vertex lying on the mouth of Γ, and $\ell \in \mathbb{N}_1$.

Now, let A be a finite dimensional k-algebra. We recall that an A-module X is said to be a brick if $\mathrm{End}(X) = k$, and the A-modules X_1, X_2 are said to be orthogonal provided $\mathrm{Hom}(X_1,X_2) = \mathrm{Hom}(X_2,X_1) = 0$. Given pairwise orthogonal bricks $E_1,\dots,E_n$, we define $\mathcal{E}(E_1,\dots,E_n)$ to be the set of A-modules X having a chain of submodules

$$X = X_0 \supset X_1 \supset X_2 \supset \dots \supset X_m = 0$$

for some $m \in \mathbb{N}_0$, with X_{i-1}/X_i being isomorphic to one of $E_1,\dots,E_n$, for any $1 \leq i \leq m$.

(1) Given pairwise orthogonal bricks $E_1,\dots,E_n$, the full subcategory $\mathcal{E} = \mathcal{E}(E_1,\dots,E_n)$ of A-mod is an abelian category, with $E_1,\dots,E_n$ being the simple objects of $\mathcal{E}$.

Proof. Clearly, $\mathcal{E}$ is closed under direct sums. Given X,Y in $\mathcal{E}$, and some map $f : X \longrightarrow Y$, we have to show that $\mathrm{Ker}\, f$, $\mathrm{Im}\, f$, and $\mathrm{Cok}\, f$ all belong to $\mathcal{E}$. There are given chains of submodules

$$X = X_0 \supset X_1 \supset \dots \supset X_m = 0, \qquad Y = Y_0 \supset Y_1 \supset \dots \supset Y_t = 0,$$

with all X_{i-1}/X_i, Y_{i-1}/Y_i belonging to $\{E_1,\dots,E_n\}$. By induction on $m+t$, we show that Im f belongs to $\mathcal{E}$. This is trivially true for $m+t \leq 1$, thus assume $m+t \geq 2$. If $X_{m-1}f = 0$, then we replace X by X/X_{m-1}, and use induction. Thus we may assume $X_{m-1}f \neq 0$. Choose i maximal with $X_{m-1}f \subseteq Y_{i-1}$, and let $p : Y_{i-1} \longrightarrow Y_{i-1}/Y_i$ be the canonical projection. Then

$$X_{m-1} \xrightarrow{f} Y_{i-1} \xrightarrow{p} Y_{i-1}/Y_i$$

is a non-zero map, thus an isomorphism. As a consequence,

$$X_{m-1}f \oplus Y_i = Y_{i-1} .$$

Replacing the submodule chain of Y by

$$Y = Y_0 \supset Y_1 \supset \ldots \supset Y_{i-1} \supset Y_i' \supset \ldots \supset Y_{t-1}' \supset Y_t = o ,$$

with $Y_j' = X_{m-1}f \oplus Y_{j+1}$, for $i \leq j \leq t-1$, we see that we can assume $X_{m-1}f = Y_{t-1}$. Now we apply induction to the induced map $\bar{f} : X/X_{m-1} \longrightarrow Y/Y_{t-1}$, and note that the image of f is an extension of the image of $\bar{f}$ by X_{m-1}.

Also, the kernel of f belongs to $\mathcal{E}$. Namely, if $X_{m-1} \subseteq \text{Ker } f$, then Ker f is an extension of the kernel of the induced map $X/X_{m-1} \longrightarrow Y$ by X_{m-1}. Otherwise, if $X_{m-1} \nsubseteq \text{Ker } f$, then, as above, we can assume $X_{m-1}f = Y_{t-1}$, and the kernel of f is isomorphic to the kernel of the induced map $\bar{f} : X/X_{m-1} \longrightarrow Y/Y_{t-1}$.

Similarly, the cokernel of f belongs to $\mathcal{E}$. Of course, $E_1,\ldots,E_n$ are simple objects in $\mathcal{E}$, and any other object of $\mathcal{E}$ has a chain of subobjects with factors of the form $E_1,\ldots,E_n$. This finishes the proof.

An abelian category $\mathcal{A}$ is said to be <u>serial</u> provided any object in $\mathcal{A}$ has finite length and any indecomposable object in $\mathcal{A}$ has a unique composition series.

(2) <u>Let</u> $E_1,\ldots,E_n$ <u>be pairwise orthogonal bricks in some module category</u> A-mod, <u>with</u> $\tau E_i \approx E_{i-1}$ <u>for all</u> $1 \leq i \leq n$, <u>where</u> $E_0 = E_n$. <u>Assume</u> $\text{Ext}^2(E_i,E_j) = 0$ <u>for all</u> $1 \leq i, j \leq n$. <u>Then</u> $\mathcal{E} = \mathcal{E}(E_1,\ldots,E_n)$ <u>is serial, it is a standard component of</u> $\Gamma(A)$, <u>and</u> $\Gamma(\mathcal{E}) = \mathbb{Z}\mathbb{A}_\infty/n$.

<u>Proof.</u> In the following, the index i will always run through $\mathbb{Z}/n$. Let $E_i[1] := E_i$ and $E_i[o] := o$ for all i. By induction on j, we want to construct serial objects $E_i[j]$, $1 \leq i \leq n$, $j \in \mathbb{N}_1$ in $\mathcal{E}(E_1,\ldots,E_n)$, and maps

$$u_{ij} : E_i[j-1] \rightarrow E_i[j], \quad p_{ij} : E_i[j] \rightarrow E_{i+1}[j-1]$$

such that for $j \geq 2$ all the sequences

$$\text{(a)}\quad o \longrightarrow E_i[j-1] \xrightarrow{\quad u_{ij}\quad} E_i[j] \xrightarrow{\quad p'_{ij}\quad} E_{i+j-1}[1] \longrightarrow o$$

$$\text{(b)}\quad o \longrightarrow E_i[1] \xrightarrow{\quad u'_{ij}\quad} E_i[j] \xrightarrow{\quad p_{ij}\quad} E_{i+1}[j-1] \longrightarrow o$$

$$\text{(c)}\quad o \longrightarrow E_i[j-1] \xrightarrow{[p_{i,j-1},u_{i,j}]} E_{i+1}[j-2] \oplus E_i[j] \xrightarrow{\begin{bmatrix} u_{i+1,j-1} \\ p_{i,j} \end{bmatrix}} E_{i+1}[j-1] \longrightarrow o$$

are exact, where $p'_{ij} = p_{i,j} \;\cdots\; p_{i+j-2,2}$ and $u'_{ij} = u_{i2} \cdots u_{ij}$, and such that the sequences (c), in fact, are Auslander-Reiten sequences, for all $j \in \mathbb{N}_1$. Note that it follows from (a) that $E_i[j-1]$ embeds into $E_i[j]$ as a maximal subobject in $\mathcal{E}$ (its cokernel being one of the given bricks), and the seriality of $E_i[j]$ implies that this is the only maximal subobject of $E_i[j]$, or, equivalently, that $u_{ij}f = o$ for any $f : E_i[j] \to E_t$. For $j = 1$, nothing has to be done. We consider first the case $j = 2$. Now using 2.4 (5),

$$1 \leq \dim \operatorname{Ext}^1(E_{i+1}[1],E_i[1]) = \dim D\,\overline{\operatorname{Hom}}(E_i[1],E_i[1]) \leq \dim \operatorname{End}(E_i[1]) = 1,$$

we see that $\operatorname{Ext}^1(E_{i+1}[1],E_i[1]) = k$, and we take any non-split exact sequence

$$o \longrightarrow E_i[1] \xrightarrow{u_{i2}} E_i[2] \xrightarrow{p_{i2}} E_{i+1}[1] \longrightarrow o,$$

this defines $E_i[2]$, u_{i2}, p_{i2}, and gives us all three sequences (a), (b), (c). Of course, $E_i[2]$ being of length 2 in $\mathcal{E}$ is indecomposable and therefore serial in $\mathcal{E}$, since otherwise the defining sequence would split. Now assume, $E_i[j]$, u_{ij}, p_{ij} are defined for all $j < d$, where d is some fixed number, $d \geq 3$. In particular, the solid part of the following diagram is given

$$\begin{array}{ccccccccc}
 & & & & o & & o & & \\
 & & & & \downarrow & & \downarrow & & \\
o & \longrightarrow & E_i[1] & \xrightarrow{u'_{i,d-1}} & E_i[d-1] & \xrightarrow{p_{i,d-1}} & E_{i+1}[d-2] & \longrightarrow & o \\
 & & \Vert & & \downarrow{\scriptstyle u_{id}} & & \downarrow{\scriptstyle u_{i+1,d-1}} & & \\
o & \longrightarrow & E_i[1] & \dashrightarrow & E_i[d] & \xrightarrow{-p_{id}} & E_{i+1}[d-1] & \longrightarrow & o \\
 & & & & \downarrow & & \downarrow{\scriptstyle p'_{i+1,d-1}} & & \\
 & & & & E_{i+d-1}[1] & = & E_{i+d-1}[1] & & \\
 & & & & \downarrow & & \downarrow & & \\
 & & & & o & & o & &
\end{array}$$

Since $\operatorname{Ext}^2(E_{i+d-1}[1],E_i[1]) = o$, it follows that this diagram can be completed in order to be commutative and having exact rows and columns. [Namely, the vanishing of this Ext^2-term shows that the map $\mathrm{Ext}^1(u_{i+1,d-1},E_i[1])$ is surjective.] In this way, we obtain $E_i[d], u_{id}$, and p_{id}. Also, a kernel of p_{id} is given by $u'_{i,d-1}u_{id} = u'_{id}$, a cokernel of u_{id} by $p_{id}p'_{i+1,d-1} = p'_{id}$, thus, we obtain the sequences (a), (b). Finally, the right upper square gives us the exact sequence

$$\text{(c')}\qquad 0 \longrightarrow E_i[d-1] \xrightarrow{[p_{i,d-1},u_{id}]} E_{i+1}[d-2] \oplus E_i[d] \xrightarrow{\begin{bmatrix} u_{i+1,d-1} \\ p_{i,d} \end{bmatrix}} E_{i+1}[d-1] \longrightarrow 0.$$

We claim that $E_i[d]$ has as only maximal subobject in $\mathcal{E}$ the image of u_{id}. Namely, let $f : E_i[d] \to E_t$ be some non-zero map. By induction, $u_{id}f : E_i[d-1] \longrightarrow E_t$ satisfies $u_{i,d-1}u_{id}f = 0$, thus $u_{id}f$ factors through p'_{id}, thus through p_{id}, say $u_{id}f = p_{i,d-1}f'$ for some f'. Therefore, $\begin{bmatrix} -f' \\ f \end{bmatrix}$ factors through the cokernel of $[u_{id}, p_{i,d-1}]$, thus $\begin{bmatrix} -f' \\ f \end{bmatrix} = \begin{bmatrix} u_{i+1,d-1} \\ p_{id} \end{bmatrix} f''$ for some $f'' : E_{i+1}[d-1] \longrightarrow E_t$. By induction $-f' = u_{i+1,d-1}f'' = 0$, thus $u_{id}f = p_{i,d-1}f' = 0$. Thus $E_i[d]$ has precisely one maximal subobject in $\mathcal{E}$, and this subobject is serial in $\mathcal{E}$, thus $E_i[d]$ is serial in $\mathcal{E}$.

Also, the sequence (c') is an Auslander-Reiten sequence. First of all, $E_{i+1}[d-1]$ is not projective, since (c') is not split. Thus $\tau E_{i+1}[d-1]$ is defined, and we claim that $\tau E_{i+1}[d-1] \approx E_i[d-1]$. Namely, there is the irreducible map $u_{i+1,d-1} : E_{i+1}[d-2] \longrightarrow E_{i+1}[d-1]$, using our knowledge of the Auslander-Reiten sequence starting in $E_{i+1}[d-1]$, thus, there has to exist an irreducible map $\tau E_{i+1}[d-1] \longrightarrow E_{i+1}[d-2]$, thus $\tau E_{i+1}[d-1]$ is a direct summand of the left term of the sink map $E_{i+1}[d-3] \oplus E_i[d-1] \longrightarrow E_{i+1}[d-2]$. Since by induction either $E_i[d-3] = 0$ (for $d = 3$) or else $\tau^- E_{i+1}[d-3] \approx E_{i+2}[d-3] \not\approx E_{i+1}[d-1]$, it follows that $\tau E_{i+1}[d-1] \approx E_i[d-1]$. Now, taking any non-invertible endomorphism φ of $E_i[d-1]$, we claim that the sequence induced from (c') by φ splits. Now φ has non-zero kernel, thus $u'_{i,d-1}\varphi = 0$ by the seriality of $E_i[d-1]$. Therefore, φ factors through $p_{i,d-1}$, say $\varphi = p_{i,d-1}\varphi'$. However, the sequence induced from (c') by $p_{i,d-1}$ obviously splits. This finishes the proof that (c') is an Auslander-Reiten sequence.

Now, let X be indecomposable in $\mathcal{E}$. Choose a monomorphism $f : E_i[j] \longrightarrow X$ with j largest possible, and some i. If f would not be an isomorphism, we can factor f through the source map $[u_{i,j+1}, p_{ij}]$ of $E_i[j]$, say $f = u_{i,j+1}f' + p_{ij}f''$. Now $u'_{ij}f \neq 0$, $u'_{ij}p_{ij} = 0$, thus $u'_{i,j+1}f' = u'_{ij}u_{i,j+1}f' \neq 0$. Since $E_i[j+1]$ is serial in $\mathcal{E}$ with socle given by $u'_{i,j+1}$, and X belongs to $\mathcal{E}$, it follows that f' is mono, contradicting the maximality of j. Thus f is an isomorphism. This shows that all indecomposable objects in $\mathcal{E}$ are of the form $E_i[j]$, for some i,j; in particular, they are serial.

We denote by $\mathcal{E}'$ the k-linear subcategory of $\mathcal{E}$ given by the objects $E_i[j]$ and the maps u_{ij}, p_{ij}. We claim that $\mathcal{E}'$ is, in fact, a full subcategory. Assume there is given a map $\varphi : E_i[j] \longrightarrow E_{i'}[j']$ in $\mathcal{E}$. We use induction on $j+j'$. For $j+j' = 2$, the map φ belongs to $\mathcal{E}'$ due to our assumption that we deal with orthogonal bricks $E_1,\ldots,E_n$. Now , let $j+j' > 2$. Since the image of φ belongs to $\mathcal{E}$, and any indecomposable object of $\mathcal{E}$ is isomorphic to an object in $\mathcal{E}'$, we can assume that φ is either mono or epi. Consider the case of φ being epi, the other case being dual. If φ is not an isomorphism, we can factor φ through p_{ij}, say $\varphi = p_{ij}\varphi'$, and by induction $\varphi' : E_{i+1}[j-1] \longrightarrow E_{i'}[j']$ belongs to $\mathcal{E}'$, thus also φ. Thus we can assume that φ is an automorphism of $E_i[j]$. Now $\varphi = c\cdot 1 + \varphi'$ for some $c \in k$ and some non-invertible endomorphism φ' of $E_i[j]$. Factoring φ' through its image and using induction, φ' belongs to $\mathcal{E}'$, thus also φ. This finishes the proof.

By construction, the set of all $E_i[j]$, $i \in \mathbb{Z}/n$, $j \in \mathbb{N}_1$, gives a component of $\Gamma(A)$, and this component is of the form $\mathbb{Z}\mathbb{A}_\infty/n$. Also, sending the arrow $[E_i[j-1]] \longrightarrow [E_i[j]]$ to u_{ij}, and the arrow $[E_i[j]] \longrightarrow [E_{i+1}[j-1]]$ to p_{ij}, we obtain a functor $F : k(\mathbb{Z}\mathbb{A}_\infty/n) \longrightarrow \mathcal{E}$, and by the previous considerations, this functor is full and dense. This functor is also faithful as we want to show. Let us call a non-zero map $E_i[j] \longrightarrow E_{i'}[j']$ standard, in case it is a composition of first a sequence of maps of the form p_{**} followed by a sequence of maps of the form u_{**}. Of course, the standard maps generate $\mathrm{Hom}(E_i[j],E_{i'}[j'])$ as a k-vectorspace, since F is full. On the other hand, we claim that the standard maps in $\mathrm{Hom}(E_i[j],E_{i'}[j'])$ are also linearly independent. Namely, given a standard map formed by first r maps of the form p_{**}, and then maps of the form u_{**}, then its image is isomorphic to $E_{i+1}[j-r]$, and we denote this map by φ_r. Now assume there is given a linear combination $\varphi = \sum_r c_r\varphi_r$ with $c_r \in k$, where r runs through some finite subset of $\mathbb{N}_0$, and take s maximal with $c_s \neq 0$. Then the image of $c_s\varphi_s$ is the unique subobject of $E_{i'}[j']$ of length $j-s$ in $\mathcal{E}$, whereas the image of $\sum_{r<s} c_r\varphi_r$ is properly contained in this subobject,thus $\varphi \neq 0$. Thus, the standard maps in $\mathrm{Hom}(E_i[j],E_{i'}[j'])$ form a k-basis, and this immediately shows that F is faithful. Thus, $\mathcal{E}$ is standard.

This finishes the proof.

Let A be a finite dimensional k-algebra. A component $\mathcal{C}$ of A-mod is said to be a tube provided $\Gamma(\mathcal{C})$ is a tube. Now, let I be some set, and let us consider families $\mathcal{T}(\rho)$, $\rho \in I$, of (pairwise different) tubes in A-mod, and the module class $\mathcal{T}$ generated by all $\mathcal{T}(\rho)$. Such a module class will be called a tubular family parametrized by I, or else a tubular I-family. In the cases we are interested in, the parameter set I will usually be the projective line $\mathbb{P}_1k$ over k. A tubular family $\mathcal{T}$ is said to be stable provided $\mathcal{T}$ does not contain non-zero projective or injective modules. Of course, the tubular I-family $\mathcal{T}$ is stable if and only if all tubes $\mathcal{T}(\rho)$, $\rho \in I$, are stable.

(3) <u>A standard stable tubular family in</u> A-mod <u>is an abelian category which is serial, and is closed under extensions in</u> A-mod.

<u>Proof.</u> Assume $\mathcal{T}$ is the module class generated by the family of pairwise different tubes $\mathcal{T}(\rho)$, $\rho \in I$, in A-mod. Since we assume that $\mathcal{T}$ is standard, $\mathrm{Hom}(\mathcal{T}(\rho),\mathcal{T}(\rho')) = 0$ for $\rho \neq \rho'$. As a consequence, it is sufficient to consider the case of $\mathcal{T}$ being a single tube. Let $\mathcal{T}$ be of rank n, thus $\Gamma(\mathcal{T}) = \mathbb{Z}\mathbb{A}_\infty/n$. Let $E_1,\ldots,E_n$ be the indecomposable modules of $\mathcal{T}$ which belong to the mouth, with $\tau E_i \approx E_{i-1}$, $1 \le i \le n$, where $E_0 = E_n$. Since $\mathcal{T}$ is standard, it is easy to see that $E_1,\ldots,E_n$ are pairwise orthogonal bricks. Thus $\mathcal{E} = \mathcal{E}(E_1,\ldots,E_n)$ is an abelian category which is serial and it is a component of $\Gamma(A)$. Since both $\mathcal{T}$ and $\mathcal{E}$ are components of $\Gamma(A)$ containing $E_1,\ldots,E_n$, it follows that $\mathcal{T} = \mathcal{E}$.

(3') Let $\mathcal{T}$ be a standard stable tubular I-family in A-mod, with $\mathrm{proj.dim.}M = 1$ for all indecomposable modules in $\mathcal{T}$. Let n_ρ be the rank of the tube $\mathcal{T}(\rho)$, and let $E_1^{(\rho)},\ldots,E_{n_\rho}^{(\rho)}$ be the indecomposable modules lying on the mouth of $\mathcal{T}(\rho)$, with $\tau E_i^{(\rho)} \approx E_{i-1}^{(\rho)}$, $1 \le i \le n_\rho$, and $E_0^{(\rho)} = E_{n_\rho}^{(\rho)}$. Then, in $K_0(A)$, we have

$$\langle \underline{\dim}\, E_i^{(\rho)}, \underline{\dim}\, E_j^{(\rho')}\rangle = \begin{cases} 1 & \\ -1 & \text{in case} \\ 0 & \end{cases} \begin{array}{l} \rho = \rho',\ i \equiv j \pmod{n_\rho}, \\ \rho = \rho',\ i \equiv j+1 \pmod{n_\rho}, \\ \text{otherwise.} \end{array}$$

The indecomposable modules in $\mathcal{T}$ are of the form $E_i^{(\rho)}[\ell]$, with $1 \le i \le n_\rho$, and $\rho \in I$, and

$$\chi_A(\underline{\dim}\, E_i^{(\rho)}[\ell]) = \begin{cases} 1 & \\ & \text{in case} \\ 0 & \end{cases} \begin{array}{l} \ell \not\equiv 0 \pmod{n_\rho}, \\ \\ \ell \equiv 0 \pmod{n_\rho}. \end{array}$$

<u>Proof.</u> We know that $\mathcal{T}$ is an abelian category, and that the modules $E_i^{(\rho)}$ with $1 \le i \le n_\rho$, $\rho \in I$, are simple objects in $\mathcal{T}$. Thus

$$\mathrm{Hom}(E_i^{(\rho)},E_j^{(\rho')}) = \begin{cases} k & \text{for } \rho = \rho',\ i = j \\ 0 & \text{otherwise.} \end{cases}$$

Also,

$$\mathrm{Ext}^1(E_i^{(\rho)},E_j^{(\rho')}) \approx D\,\mathrm{Hom}(E_j^{(\rho')},E_{i-1}^{(\rho)}) = \begin{cases} k & \text{for } \rho = \rho',\ j = i-1, \\ 0 & \text{otherwise,} \end{cases}$$

using 2.4.6. This gives the first assertion, using again that $\mathrm{proj.dim.}E_i^{(\rho)} = 1$. The second assertion follows from the first, taking into account that

$$\underline{\dim}\, E_i^{(\rho)}[\ell] = \sum_{j=i}^{i+\ell-1} \underline{\dim}\, E_j^{(\rho)},$$

where we assume that $E_j^{(\rho)}$ is defined for all $j \in \mathbb{Z}$, with $E_j^{(\rho)} = E_{j'}^{(\rho)}$ for $j \equiv j' \pmod{n_\rho}$.

The tubular I-family $\mathcal{T}$ in A-mod is said to be <u>separating</u>, provided the remaining indecomposable A-modules fall into two classes $\mathcal{P}$, $\mathcal{Q}$ such that the following conditions are satisfied:

(a) $\mathcal{T}$ is standard,

(b) $\mathrm{Hom}(\mathcal{Q},\mathcal{P}) = \mathrm{Hom}(\mathcal{Q},\mathcal{T}) = \mathrm{Hom}(\mathcal{T},\mathcal{P}) = o$,

(c) Given a map from $\mathcal{P}$ to $\mathcal{Q}$, and any $\rho \in I$, then this map can be factored through $\mathcal{T}(\rho)$.

In this case, we will say that $\mathcal{T}$ separates $\mathcal{P}$ from $\mathcal{Q}$.

(4) Let A be a connected algebra, and $\mathcal{T}$ a separating tubular family in A-mod, separating $\mathcal{P}$ from $\mathcal{Q}$. Then $\mathcal{P}$ and $\mathcal{Q}$ are uniquely determined by $\mathcal{T}$.

<u>Proof.</u> Define inductively

$$\mathcal{P}_o = \langle X \text{ indecomposable} \mid \mathrm{Hom}(\langle X\rangle,\mathcal{T}) \neq o,\ X \notin \mathcal{T}\rangle,$$

and, for $i \geq 1$,

$$\mathcal{P}_{2i-1} = \langle X \text{ indecomposable} \mid \mathrm{Hom}(\mathcal{P}_{2i-2},\langle X\rangle) \neq o,\ \mathrm{Hom}(\mathcal{T},\langle X\rangle) = o\rangle.$$

$$\mathcal{P}_{2i} = \langle X \text{ indecomposable } \mathrm{Hom}(\langle X\rangle,\mathcal{P}_{2i-1}) \neq o\rangle.$$

We claim that

$$\mathcal{P}_o \subseteq \mathcal{P}_1 \subseteq \mathcal{P}_2 \subseteq \dots \subseteq \mathcal{P}_j \subseteq \mathcal{P}_{j+1} \subseteq \dots \subseteq \mathcal{P}.$$

For, $\mathcal{P}_o \subseteq \mathcal{P}$, since $\mathrm{Hom}(\mathcal{Q},\mathcal{T}) = o$. For $i \geq 1$, we see that $\mathcal{P}_{2i-2} \subseteq \mathcal{P}_{2i-1}$ since, on the one hand, the identity map 1_X for X indecomposable and in $\mathcal{P}_{2i-2}$ is a non-zero homomorphism from $\mathcal{P}_{2i-2}$ to $\langle X\rangle$, and, on the other hand, by induction $\mathcal{P}_{2i-2} \subseteq \mathcal{P}$, and $\mathrm{Hom}(\mathcal{T},\mathcal{P}) = o$. Let us show that any indecomposable X in $\mathcal{P}_{2i-1}$ belongs to $\mathcal{P}$. Assume that X belongs to $\mathcal{T} \cup \mathcal{Q}$. There are non-zero maps from $\mathcal{P}_{2i-2} \subseteq \mathcal{P}$ to X, and we can factor such a map through $\mathcal{T}$, thus $\mathrm{Hom}(\mathcal{T},\langle X\rangle) \neq o$, contrary to assumption. This shows that $\mathcal{P}_{2i-1} \subseteq \mathcal{P}$. Similarly, $\mathcal{P}_{2i-1} \subseteq \mathcal{P}_{2i}$, since the identity map 1_X for X indecomposable and in $\mathcal{P}_{2i-1}$ is a non-zero homomorphism from $\langle X\rangle$ to $\mathcal{P}_{2i-1}$. Also $\mathcal{P}_{2i} \subseteq \mathcal{P}$, since by induction $\mathcal{P}_{2i-1} \subseteq \mathcal{P}$, and $\mathrm{Hom}(\mathcal{T},\mathcal{P}) = \mathrm{Hom}(\mathcal{Q},\mathcal{P}) = o$.

In order to see that $\mathcal{P}$ coincides with some $\mathcal{P}_j$, consider first an indecomposable projective module P belonging to $\mathcal{P}$. Since A is connected, there exists a chain $P_o,P_1,P_2,\dots,P_{2j-1},P_{2j} = P$, where $\mathrm{Hom}(\langle P_o\rangle,\mathcal{T}) \neq o$, and where $\mathrm{Hom}(P_{2i-2},P_{2i-1}) \neq o$ and $\mathrm{Hom}(P_{2i},P_{2i-1}) \neq o$ for all $1 \leq i \leq j$. Take such a chain with minimal j. We claim that all P_i belong to $\mathcal{P}$. Otherwise, let t be maximal with P_t not in $\mathcal{P}$. If t would be even, then $\operatorname{Hom}(P_t,P_{t+1}) \neq o$ implies that also $P_{t+1} \notin \mathcal{P}$, a contradiction. Thus t is odd. Now $P_{t+1} \in \mathcal{P}$, and $\operatorname{Hom}(P_{t+1},P_t) \neq o$. We can factor any non-zero map $P_{t+1} \longrightarrow P_t$ through $\mathcal{T}$ and obtain a non-zero map from P_{t+1} to $\mathcal{T}$. Thus $\operatorname{Hom}(\langle P_{t+1}\rangle,\mathcal{T}) \neq o$, and we can shorten the given chain by deleting $P_o,P_1,\dots,P_t$. This contradicts the minimality of j. Thus all P_i belong to $\mathcal{P}$, and it follows by induction that $P_i \in \mathcal{P}_i$ for $o \leq i \leq 2j$. In particular, $P \in \mathcal{P}_{2j}$.

Now let X be indecomposable and in $\mathcal{P}$. Take an indecomposable projective module P with $\operatorname{Hom}(P,X) \neq o$. Then also $P \in \mathcal{P}$, and, by previous considerations, $P \in \mathcal{P}_{2j}$ for some j. Thus $X \in \mathcal{P}_{2j+1}$.

This shows that $\mathcal{T}$ determines $\mathcal{P}$, and therefore also $\mathcal{Q}$.

(5) Let $\mathcal{T}$ be a sincere separating tubular family of A-modules, separating $\mathcal{P}$ from $\mathcal{Q}$. Then all projective modules belong to $\mathcal{P} \vee \mathcal{T}$, all injective modules to $\mathcal{T} \vee \mathcal{Q}$. If $X \in \mathcal{P}$, then proj.dim.$X \leq 1$, if $Y \in \mathcal{Q}$, then inj.dim.$Y \leq 1$. If, in addition, $\mathcal{T}$ contains no non-zero injective module, then proj.dim.$T \leq 1$ for any $T \in \mathcal{T}$, and gl.dim.$A \leq 2$. Similarly, if $\mathcal{T}$ contains no non-zero projective module, then inj.dim.$T \leq 1$ for any $T \in \mathcal{T}$, and again gl.dim.$A \leq 2$.

Proof. Since $\mathcal{T}$ contains a sincere module, and $\operatorname{Hom}(\mathcal{Q},\mathcal{T}) = 0$, any projective module belongs to $\mathcal{P} \vee \mathcal{T}$. Similarly, any injective module belongs to $\mathcal{T} \vee \mathcal{Q}$. Thus ${}_AA \in \mathcal{P} \vee \mathcal{T}$, and $D(A_A) \in \mathcal{T} \vee \mathcal{Q}$. If $X \in \mathcal{P}$, then also $\tau X \in \mathcal{P}$, since $\tau X \prec X$, thus $\operatorname{Hom}(D(A_A),\tau X) = 0$, and therefore proj.dim.$X \leq 1$, according to 2.4.1. Similarly, if $Y \in \mathcal{Q}$, then inj.dim.$Y \leq 1$. Now assume $\mathcal{T}$ contains no non-zero injective module. Then $D(A_A) \in \mathcal{Q}$. Thus, given $T \in \mathcal{T}$, then $\operatorname{Hom}(D(A_A),\tau T) = 0$, since with T also τT belongs to $\mathcal{T}$, thus proj.dim.$T \leq 1$. As a consequence, given P indecomposable projective, then with P also rad P belongs to $\mathcal{P} \vee \mathcal{T}$, thus proj.dim.rad $P \leq 1$, and therefore gl.dim.$A \leq 2$.

Now, let $\mathcal{T}(\rho)$, $\rho \in I$, be a family of stable tubes in A-mod. For $\rho \in I$, denote by n_ρ the rank of $\mathcal{T}(\rho)$; the function $n_* : I \longrightarrow \mathbb{N}_1$ will be called the type of this family. In the cases we will deal with, almost all $\mathcal{T}(\rho)$ will be homogeneous, thus, we usually will choose some finite subset I' of I such that $n_\rho = 1$ for all $\rho \in I \setminus I'$, say $I' = \{\rho_1,\dots,\rho_t\}$ with pairwise different ρ_s; we denote n_{ρ_s} just by n_s and $(n_1,\dots,n_t)$ will also be called the type of the family, or also the type of the module class $\mathcal{T}$ generated by all $\mathcal{T}(\rho)$, $\rho \in I$. For example, a family of stable tubes of type $(3,3,2)$ is given by two tubes of rank 3, one tube of rank 2, and, may be, additional homogeneous tubes.

3.2 Example: Kronecker modules

We are going to study one particular example, namely the category of representations of the quiver

 .

The representations of this quiver will be called Kronecker modules, they are of the form $(V_o, V_\omega, \gamma_1, \gamma_2)$ with V_o, V_ω k-vector spaces and $\gamma_1, \gamma_2 : V_\omega \longrightarrow V_o$ k-linear maps. Choosing bases in V_o and V_ω, we see that we deal with pairs (G_1, G_2) of (rectangular) matrices of the same size, and isomorphism of Kronecker modules means, in terms of matrices, multiplication of both G_1 and G_2 from the left by a regular matrix, and from the right by a regular matrix, simultaneously on G_1 and G_2. Thus, we consider the classification problem of matrix pencils, as presented, for example, in the book of Gantmacher [Gm], and used in the theory of differential equations. The classification of the indecomposable Kronecker modules is, as the name suggests, due to Kronecker. In fact, a partial solution of the classification problem was given by Weierstraß, and this work was completed by Kronecker. Note that the Kronecker modules are just the modules over the algebra

$$A = \begin{bmatrix} k & k^2 \\ o & k \end{bmatrix},$$

and A-mod is just $\mathcal{U}(\text{k-mod}, \text{Hom}(k^2, -))$.

We will use the following convention: there are two indecomposable projective A-modules $P(o)$ and $P(\omega)$, with $P(o)$ being simple. Ordering the indecomposable projective modules in this way, we have $\underline{\dim}\, P(o) = [10]$, $\underline{\dim}\, P(\omega) = [2,1]$. The Cartan matrix of A is given by $C = \begin{bmatrix} 1 & 2 \\ 0 & 1 \end{bmatrix}$, thus $C^{-T} = \begin{bmatrix} 1 & 0 \\ -2 & 1 \end{bmatrix}$, and $\chi_A = X_o^2 + X_\omega^2 - 2X_oX_\omega$ is the quadratic form of type $C(1)$. Also $\Phi = -C^{-T}C = \begin{bmatrix} -1 & -2 \\ 2 & 3 \end{bmatrix}$.
Let $w := [1,1]$, this is the minimal positive generator of $\text{rad}\,\chi_A$. For any $x = [x_o, x_\omega]$, we have

$$\iota_w(x) := \langle w,x \rangle = [1\ 1] \begin{bmatrix} 1 & 0 \\ -2 & 1 \end{bmatrix} \begin{bmatrix} x_o \\ x_\omega \end{bmatrix} = -x_o + x_\omega ,$$

and

$$x\Phi = x(I + \begin{bmatrix} -2 & -2 \\ 2 & 2 \end{bmatrix}) = x + 2\iota_w(x)w.$$

On the other hand, since A is hereditary,

$$\underline{\dim}\, \tau X = (\underline{\dim}\, X)\Phi$$

for X indecomposable and not projective, and

$$\underline{\dim}\, \tau^- Y = (\underline{\dim}\, Y)\Phi^{-1} ,$$

for Y indecomposable and not injective.

The preprojective modules : The two indecomposable projective modules $P(o)$, $P(\omega)$ have the dimension vectors $\underline{\dim}\, P(o) = [1,o]$, $\underline{\dim}\, P(\omega) = [2,1]$, thus the preprojective modules have the dimension vectors

$$\underline{\dim}\, \tau^{-t}P(o) = [2t+1,2t],$$
$$\underline{\dim}\, \tau^{-t}P(\omega) = [2t+2,2t+1];$$

we see that we obtain in this way all vectors $[n+1,n]$ with $n \in \mathbb{N}_o$, and that $\iota_w \underline{\dim}\, P = -1$ for any indecomposable preprojective module P. The preprojective component looks as follows:

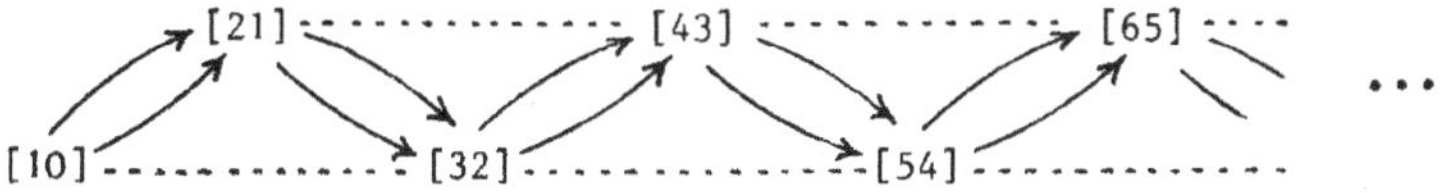

[Namely, rad $P(\omega) = P(o) \oplus P(o)$, thus $\mathrm{Irr}(P(o),P(\omega))$ is two-dimensional, and therefore, for all $t \in \mathbb{N}_o$, $\mathrm{Irr}(\tau^{-1}P(o),\tau^{-1}P(\omega))$ and $\mathrm{Irr}(\tau^{-1}P(\omega),\tau^{-t-1}P(o))$ are two-dimensional]. It follows that for P,P' indecomposable preprojective, with $\underline{\dim}\, P \leq \underline{\dim}\, P'$, there exists a non-zero map $P \longrightarrow P'$. [For , we can factorize any non-zero map $P(o) \longrightarrow P'$ through a sum of copies of P]. Since there are no cycles in the preprojective component, we have $\mathrm{Hom}(P,P') = o$ for P,P' indecomposable preprojective with $\underline{\dim}\, P' < \underline{\dim}\, P$, and therefore $\mathrm{Ext}^1(P,P') = o$ for P,P' indecomposable preprojective with $\underline{\dim}\, P < \underline{\dim}\, P' + [2,2]$.

Also, we note the following: if P,P' are indecomposable and preprojective, and $f : P \longrightarrow P'$ is a non-zero map, then f is mono. [For, let $X = \mathrm{Ker}\, f$, $Y = \mathrm{Im}\, f$, then $\iota_w \underline{\dim}\, X + \iota_w \underline{\dim}\, Y = \iota_w \underline{\dim}\, P = -1$. Since Y is a submodule of P', and P' is preprojective, all indecomposable summands of Y are preprojective, thus $\iota_w \underline{\dim}\, Y = -t$, with t the number of indecomposable direct summands in a direct decomposition of Y. Since Y is non-zero, $t \geq 1$, thus $\iota_w \underline{\dim}\, X \geq o$. However, also X is a submodule of a preprojective module, thus also all indecomposable summands of X are preprojective, thus $X = o$].

The only indecomposable Kronecker modules X with $\iota_w(\underline{\dim}\, X) < o$, are the preprojective ones. [For, if X is not preprojective, then no $\tau^t X$ with $t \in \mathbb{N}_o$ is projective, thus $\underline{\dim}\, \tau^{t+1}X = (\underline{\dim}\, \tau^t X)\Phi$. By induction, we see that $\underline{\dim}\, \tau^t X = (\underline{\dim}\, X)\Phi^t = \underline{\dim}\, X + 2t\, \iota_w(\underline{\dim}\, X)w$, for all $t \in \mathbb{N}_o$. However, if $\iota_w(\underline{\dim}\, X) < o$, then, for large t, this vector is no longer positive, contradicting the fact that it is a dimension vector.]

The preinjective modules. The two indecomposable injective modules $Q(o)$, $Q(\omega)$ have the dimension vectors $\underline{\dim}\, Q(o) = [1,2]$, $\underline{\dim}\, Q(\omega) = [o,1]$, thus, the preinjective modules have the dimension vectors

$$\underline{\dim}\, \tau^t Q(\omega) = [2t,2t+1]$$

$$\underline{\dim}\, \tau^t Q(o) = [2t+1,2t+2],$$

we obtain in this way all vectors $[n,n+1]$ with $n \in \mathbb{N}_o$, and we see that $\iota_w \underline{\dim}\, Q = 1$ for any indecomposable preinjective module Q. Conversely, if X is an indecomposable Kronecker module with $\iota_w \underline{\dim}\, X > o$, then X is preinjective. The preinjective component looks as follows:

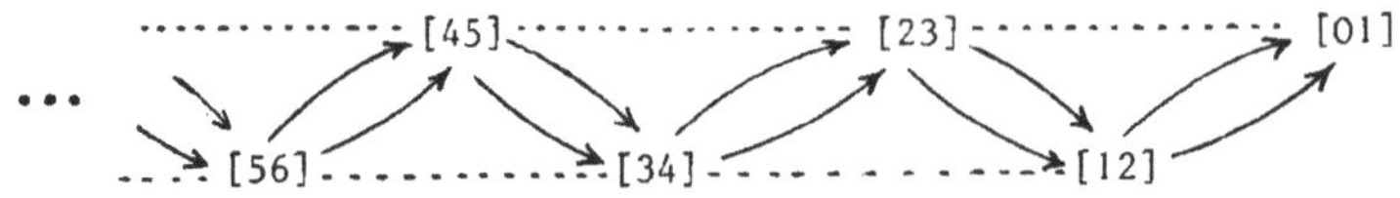

Note that if Q,Q' are indecomposable preinjective, and $\underline{\dim}\, Q \geq \underline{\dim}\, Q'$, then there exist non-zero maps $Q \longrightarrow Q'$, and all such maps are epi, whereas if $\underline{\dim}\, Q < \underline{\dim}\, Q'$, then $\mathrm{Hom}(Q,Q') = o$. As a consequence, $\mathrm{Ext}^1(Q,Q') = o$ for Q,Q' indecomposable preinjectives with $\underline{\dim}\, Q > \underline{\dim}\, Q' - [2,2]$.

The preprojective and the preinjective Kronecker modules are those indecomposable Kronecker modules X satisfying $\iota_w(\underline{\dim}\, X) \neq o$. We know that they are uniquely determined by their dimension vectors. The remaining indecomposable Kronecker modules and their direct sums will be said to be regular.

The regular modules. For any pair $\rho = (\rho_1,\rho_2) \in k^2 \setminus \{(o,o)\}$, let $k(\rho) = k(\rho_1,\rho_2)$ be the following Kronecker module

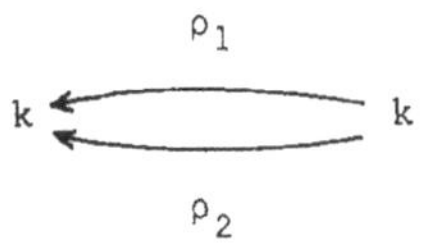

where ρ_i is the corresponding multiplication map. Given $\rho,\rho' \in k^2 \setminus \{(o,o)\}$, let $\rho \sim \rho'$ if and only if there is $c \in k$ with $\rho_1 = \rho_1' c$, $\rho_2 = \rho_2' c$, thus if and only if ρ and ρ' generate the same line through the origin in k^2. Of course, the quotient $(k^2 \setminus \{(o,o)\})/\sim$ is just the projective line $\mathbb{P}_1 k$ over k. Direct calculation shows

$$\mathrm{Hom}(k(\rho),\, k(\rho')) = \begin{cases} k & \text{if} \quad \rho \sim \rho' \\ o & \phantom{\text{if}} \quad \rho \not\sim \rho' \end{cases},$$

and any non-zero map $k(\rho) \longrightarrow k(\rho')$ (for $\rho \sim \rho'$) is an isomorphism. Fixing representatives ρ of the various equivalence classes with respect to $\sim$, we obtain a set of pairwise orthogonal bricks indexed over $\mathbb{P}_1 k$. By abuse of notation, we just will write $k(\rho)$, $\rho \in \mathbb{P}_1 k$, when considering such a set of representatives.

We claim that $\tau k(\rho) \approx k(\rho)$ for any $\rho \in \mathbb{P}_1 k$. [Namely, since $k(\rho)$ is not projective, $\underline{\dim}\, \tau k(\rho) = (\underline{\dim}\, k(\rho))\Phi = w\Phi = w$, thus $\tau k(\rho) \approx k(\rho')$ for some $\rho' \in \mathbb{P}_1 k$. Now

$$\mathrm{Ext}^1(k(\rho),k(\rho')) = \mathrm{Ext}^1(k(\rho),\tau k(\rho)) \neq o\ ,$$

thus

$$o = \langle w,w\rangle = \dim \mathrm{Hom}(k(\rho),k(\rho')) - \dim \mathrm{Ext}^1(k(\rho),k(\rho'))$$

shows that $\mathrm{Hom}(k(\rho),k(\rho')) \neq o$, thus $\rho \sim \rho'$.].

As a consequence, fixing any $\rho \in \mathbb{P}_1 k$, we see that the single module $k(\rho)$ is a τ-orbit, consisting of a brick, thus the component of A-mod containing $k(\rho)$ is a standard homogeneous tube, according to 3.1, and we will denote it by $\mathcal{T}(\rho)$. Since the various $k(\rho)$, with $\rho \in \mathbb{P}_1 k$, are pairwise orthogonal, we see that $\mathrm{Hom}(\mathcal{T}(\rho),\mathcal{T}(\rho')) = o$ for $\rho \not\sim \rho'$, and it follows that the module class $\mathcal{T}$ generated by all $\mathcal{T}(\rho)$, $\rho \in \mathbb{P}_1 k$, is standard.

Let us recall from 3.1 that $\mathcal{T}$ is a serial (abelian) category. In fact, the indecomposable objects in $\mathcal{T}$ are of the form $k(\rho)[d]$, with $\rho \in \mathbb{P}_1 k$, and $d \in \mathbb{N}_1$. Also note that $k(\rho)[d]$ has a unique chain of submodules belonging to $\mathcal{T}$, namely

$$o = k(\rho)[o] \subset k(\rho)[1] \subset \dots \subset k(\rho)[d],$$

with all factors $k(\rho)[j] \,/\, k(\rho)[j-1]$, $1 \le j \le d$, isomorphic to $k(\rho)$.

We want to show that any regular module belongs to $\mathcal{T}$. Let X be indecomposable and regular, say $X =$

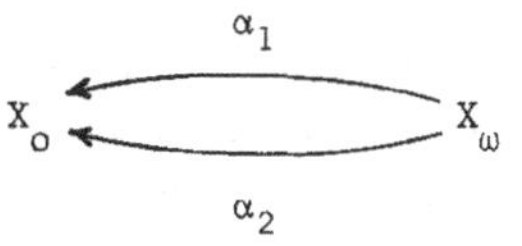

If $\mathrm{Ker}\ \alpha_1 \neq o$, let $o \neq x \in \mathrm{Ker}\ \alpha_1$, and define a map $f = (f_o,f_\omega) : k(o,1) \to X$ by $1f_o = \alpha_1(x)$, $1f_\omega = x$. Obviously, this is a monomorphism, since otherwise the injective module $Q(\omega)$ would embed into X. If $\mathrm{Ker}\ \alpha_1 = o$, then α_1 is bijective due to the assumption $\iota_w\, \underline{\dim}\, X = o$. In this case, let x be an eigenvector of $\alpha_1^{-1}\alpha_2$, say with eigenvalue μ, and define $f = (f_o,f_\omega) : k(1,\mu) \longrightarrow X$ again by

$1f_o = \alpha_1(x)$, $1f_\omega = x$. Again, this is a monomorphism. Let X' be the cokernel of f, say $X' = \bigoplus_{i=1}^{t} X'_i$, with indecomposable modules X'_i. Now no X'_i can be preprojective (since $\mathrm{Hom}(X,X'_i) \neq o$), thus $\iota_w \underline{\dim} X'_i \geq o$ for all i. Since also $\iota_w \underline{\dim} X' = o$, we see that all X'_i are regular again. By induction, we can assume that all X'_i belong to $\mathcal{T}$. Since $\mathcal{T}$ is closed under extensions, it follows that X belongs to $\mathcal{T}$.

Now, let P be indecomposable preprojective, Q indecomposable preinjective, and $f : P \longrightarrow Q$ any map. We claim that we can factor f through $k(\rho)[t]$, for any ρ and t suffiently large. Actually, if $\underline{\dim}\, P = [n+1,n]$, $\underline{\dim}\, Q = [m,m+1]$, then it is sufficient to choose $t \geq n+m+1$. For the proof, we first note that

$$\langle \underline{\dim}\, P, \underline{\dim}\, k(\rho)[t]\rangle = [n+1,n] \begin{bmatrix} 1 & 0 \\ -2 & 1 \end{bmatrix} \begin{bmatrix} t \\ t \end{bmatrix} = t ,$$

and that $\mathrm{Ext}^1(P, k(\rho)[t]) = o$, since P is preprojective, and $k(\rho)[t]$ not; thus, for $t \geq 1$

$$\dim \mathrm{Hom}(P,k(\rho)[t]) = t > t-1 = \dim \mathrm{Hom}(P,k(\rho)[t-1]).$$

Therefore, there exists $\varphi : P \longrightarrow k(\rho)[t]$ with image not contained in $k(\rho)[t-1]$. Assume $t \geq n+1$, then the image cannot be regular [since the image is not contained in $k(\rho)[t-1]$ and the only regular submodule of $k(\rho)[t]$ not contained in $k(\rho)[t-1]$ is $k(\rho)[t]$ itself, whereas $\underline{\dim}\, k(\rho)[t] > \underline{\dim}\, P$.] Also $k(\rho)[t]$ has no preinjective indecomposable submodule, thus, the image of φ has at least one indecomposable preprojective direct summand P'. However, the canonical map $P \longrightarrow P'$ has to be mono, thus P' is the image of φ and φ is mono. Let Q' be the cokernel of φ. Then Q' has no regular indecomposable direct summand, since $\varphi(P) \not\subseteq k(\rho)[t-1]$, thus all indecomposable direct summands of Q' are preinjective. However, $\iota_w(\underline{\dim}\, Q') = \iota_w(\underline{\dim}\, k(\rho)[t]) - \iota_w(\underline{\dim}\, P) = 1$, and therefore Q' is indecomposable. Thus, we have obtained an exact sequence

$$(*) \qquad o \longrightarrow P \longrightarrow k(\rho)[t] \longrightarrow Q' \longrightarrow o$$

with Q' being indecomposable preinjective, and having dimension vector $[t-n-1,t-n]$. Now assume $t \geq n+m+1$, then $[t-n-1,t-n] \geq [m,m+1]$, and therefore $\mathrm{Ext}^1(Q',Q) = o$. It follows that the sequence induced from $(*)$ by $f : P \longrightarrow Q$ splits, thus f factors through $k(\rho)[t]$. Thus, we have shown the following result:

$\mathcal{T}$ <u>is a stable separating tubular family, separating the preprojective component from the preinjective component.</u>

3.3 Wing modules

In order to construct separating tubular families, we will start with an algebra A_o, having a socalled sincere separating wing module which is dominated by some A_o-module R, and consider the one-point extension $A_o[R]$. Let us first give the relevant definitions.

The Auslander-Reiten quiver of the linearly oriented quiver of type $\mathbb{A}_n$ will be denoted by $\Theta(n)$, its vertices are of the form w_{ij}, with $1 \leq i \leq j \leq n$, there are arrows $w_{ij} \to w_{i+1,j}$ for $1 \leq i < j \leq n$ and $w_{ij} \to w_{i,j+1}$ for $1 \leq i \leq j < n$, and there are extensions $w_{i-1,j-1} \rceil w_{ij}$ for $1 < i \leq j \leq n$. Note that the vertices w_{1j} with $1 \leq j \leq n$, are the projective vertices, the vertices w_{in}, $1 \leq i \leq n$, are the injective ones; thus w_{1n} is the only projective-injective vertex of $\Theta(n)$. The vertices which are neither projective nor injective will be said to belong to the <u>interior</u> of $\Theta(n)$.

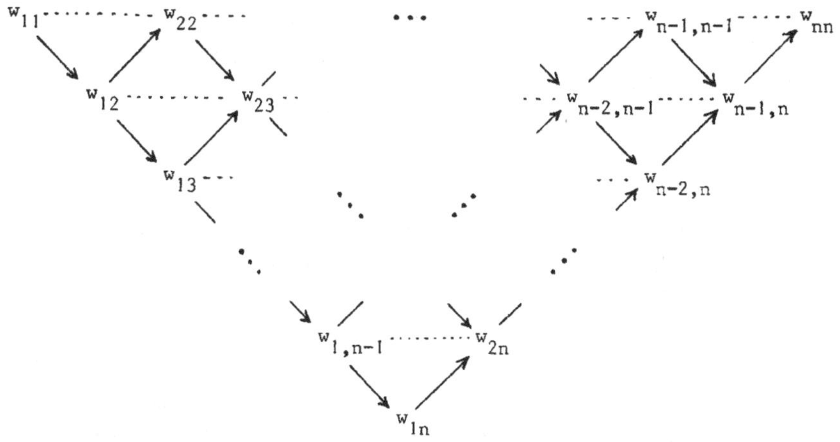

Let w_o be a vertex of a translation quiver Γ. A mesh-complete subquiver Θ of Γ will be called a <u>wing</u> of w_o, provided Θ is of the form $\Theta(n)$ for some $n \geq 2$, with w_o being the projective-injective vertex of Θ. The number n will be called the <u>length</u> of the wing. The vertex w_o is said to be a <u>wing vertex</u> provided any neighbor of w_o belongs to a wing of w_o. Let w_o be a wing vertex with wings $\Theta_1,\ldots,\Theta_t$ of lengths $n_1,\ldots,n_t$, respectively. Then $(n_1,\ldots,n_t)$ will be called the <u>type</u> of the wing vertex w_o.

For example, a translation quiver with a wing vertex w_o of type $(5,3,2)$ has locally (in the vicinity of w_o) the following shape

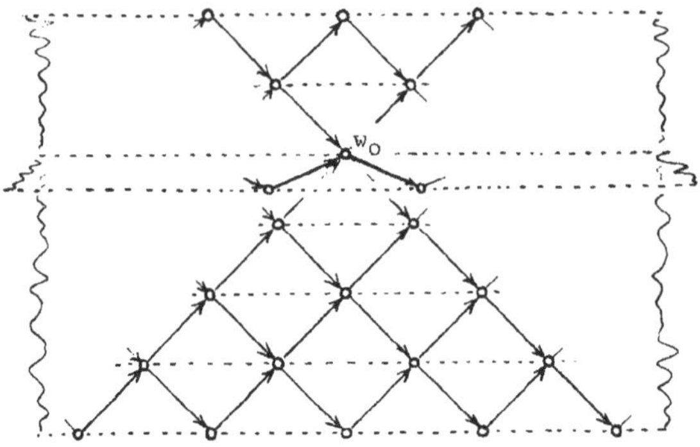

Let A_o be a finite dimensional algebra. An indecomposable A_o-module W_o belonging to a component Γ is said to be a <u>wing module</u> provided $[W_o]$ is a wing vertex in Γ.

Assume that the A_o-module W_o is a sincere, directing wing module. We recall that the existence of a sincere directing module implies that $\mathrm{gl.dim.}A_o \leq 2$, thus the Coxeter transformation Φ_o for A_o is defined. We will consider a certain one-point extension $A = A_o[R]$ of A_o using a suitable A_o-module R ; given $0 \neq \rho : R \to W_o$, we will be interested in the A-module $W_o(\rho) := (W_o,k,\rho)$. We will say that W_o is <u>dominated</u> by the A_o-module R, provided

$$\underline{\dim}\, R = (\underline{\dim}\, W_o)(I - \Phi_o^{-1}) ,$$

and

$$\mathrm{proj.dim.}W_o(\rho) \leq 1 \quad \text{for all} \quad 0 \neq \rho : R \to W_o .$$

Examples of modules R dominating a sincere directing wing module will be considered in great detail, later. The condition $\mathrm{proj.dim.}W_o(\rho) \leq 1$ for $0 \neq \rho : R \to W_o$, is not always easy to check. However, in some cases, it will be straight forward to see that this condition is satisfied. First of all, if R is projective, then $\mathrm{proj.dim.}W_o \leq 1$ implies that $\mathrm{proj.dim.}W_o(\rho) \leq 1$ for all $\rho : R \to W_o$. [For, $W_o(\rho)$ is an extension of W_o by the simple module $E(\omega) := P(\omega)/R$, where $P(\omega) = (R,k,1_R)$.] Also, given $0 \neq \rho : R \to W_o$ with $\mathrm{Ker}(\rho)$ being projective and $\mathrm{proj.dim.Cok}(\rho) \leq 1$, then $\mathrm{proj.dim.}W_o(\rho) \leq 1$. [Indeed, ρ induces an exact sequence

$$0 \to \mathrm{Ker}(\rho) \to P(\omega) \to W_o(\rho) \to \mathrm{Cok}(\rho) \to 0 ,$$

and thus $W_o(\rho)$ is an extension of $P(\omega)/\mathrm{Ker}(\rho)$ by $\mathrm{Cok}(\rho)$.] In particular, if A_o is hereditary, and all proper submodules of R are projective, then $\mathrm{proj.dim.}W_o(\rho) \leq 1$ for all $0 \neq \rho : R \to W_o$.

<u>Lemma</u>. <u>Let W_o be a sincere directing wing A_o-module. Then W_o is separating, and any indecomposable A_o-module not belonging to the interior of a wing of W_o is either generated or cogenerated by W_o.</u>

The <u>proof</u> of the lemma for a general wing module W_o will be given in section 4.2, it will be based on tilting techniques (see 4.2.7'). Note however that we will use the lemma in the course of the proof of theorem 3.4. Since our main applications of this theorem deal with wing modules which belong to preprojective or preinjective components, let us give a direct proof of the lemma in this case.

Thus, assume W_0 is a sincere wing A_0-module which belongs to a preprojective component of A_0-mod. Of course, in this case, W_0 is necessarily directing, according to 2.2.4 and 2.1.4.

Let $\mathcal{X}$ be the module class given by the indecomposable A_0-modules X with $X \prec W_0$, let $\mathcal{Y}$ be the module class given by the indecomposable A_0-modules Y with $W_0 \prec Y$. Let M be an indecomposable A_0-module either in $\mathcal{Y}$, or belonging neither to $\mathcal{X}$ nor to the interior of a wing of W_0, nor being isomorphic to W_0. We want to show the following: Given $X \in \mathcal{X}$, then every map $h : X \to M$ factors through a direct sum of copies of W_0. Now h cannot be an isomorphism, since W_0 is directing. Thus, we can factor h through the source map $f : X \longrightarrow X'$ of X, say $h = fh'$. If all indecomposable direct summands X'' of X' again satisfy $X'' \prec W_0$, then we can use induction in order to conclude that h' factors through a direct sum of copies of h'. [Note that the length of all paths from X to W_0 is bounded, and denoting by $\ell(X,W_0)$ the maximal length of such a path, we must have $\ell(X,W_0) > \ell(X'',W_0)$.]. However, if at least one indecomposable direct summand X'' of X' does not satisfy $X'' \prec W_0$, then it is easy to see that there exists a wing θ of $[W_0]$, such that $[X]$ belongs to θ. [Note that since $X \prec W_0$, and W_0 is sincere and directing, X cannot be injective.] Let θ be a wing of length n, denote its vertices as above by w_{ij}, $1 \leq i \leq j \leq n$, and let W_{ij} be an indecomposable module with $[W_{ij}] = w_{ij}$. Thus $X = W_{1t}$ for some $1 \leq t < n$ (and $X' = W_{12}$ if $t = 1$, and $X' = W_{2,t-1} \oplus W_{1,t+1}$ if $t > 1$).
Consider now a map $g : W_{ij} \longrightarrow M$, where $i \leq j \leq n$. Note that we can factor g through a direct sum of copies of the various W_{sn} with $1 \leq s \leq n$. [Namely, we use going down induction on $i+j$. In case $j = n$, nothing has to be shown, whereas in case $j < n$, we factor g through the source map $W_{ij} \longrightarrow W_{i+1,j} \oplus W_{i,j+1}$ (with $W_{i+1,i} = 0$), and use induction.] Thus we see that it is sufficient to show that any map $W_{1t} \longrightarrow W_{sn}$ factors through $W = W_{1n}$. However, a preprojective component of A_0-mod is standard, see 2.3, Lemma 3, and in $k(\theta(n))$ any map from W_{1t} to W_{sn} obviously factors through W_{1n}. Thus, h factors through a direct sum of copies of W_0. It follows that W_0 is separating. Also, we see that the module M considered above actually is generated by W_0. In particular, it follows then that $M \in \mathcal{Y}$. [For, any indecomposable projective A_0-module belongs to $\mathcal{X} \vee \langle W_o \rangle$, since W_0 is sincere. Let $p : P \to M$ be a projective cover of M. We apply the previous considerations to $h = p$ and obtain a factorization of p, thus a surjective map from a direct sum of copies of W_0 to M.] It remains to see that the modules in $\mathcal{X}$ are cogenerated by W_0. We use the dual argument: Since W_0 is sincere, any indecomposable injective A_0-module belongs to $\mathcal{Y}$. Given $X \in \mathcal{X}$, let $X \to Q$ be an injective envelope. Since $Q \in \mathcal{Y}$, we can factor it through a direct sum of copies of W_0 and obtain an embedding of X into such a direct sum. This finishes the proof.

3.4 The main theorem

Theorem. *Let k be an algebraically closed field. Let A_o be a finite dimensional k-algebra with a sincere, directing wing module W_o of type $(n_1,\dots,n_t)$, dominated by R. Let $A = A_o[R]$, and $w = e(\omega) + \underline{\dim}\, W_o \in K_o(A)$. Denote by ι_w the linear form $\iota_w = \langle w,-\rangle$ on $K_o(A)$. Let $\mathcal{P}_w, \mathcal{T}_w, \mathcal{Q}_w$ be the classes of A-modules given by the indecomposable A-modules M satisfying $\iota_w(\underline{\dim}\, M) < 0$, $= 0$, or > 0, respectively. Then $\mathcal{T}_w$ is an abelian category, and a stable tubular $\mathbb{P}_1$k-family of type $(n_1,\dots,n_t)$, separating $\mathcal{P}_w$ from $\mathcal{Q}_w$ and being controlled by the restriction of χ_A to Ker ι_w.*

The proof of the theorem will be given in several steps; it will cover all of this section. First, we consider the category A_o-mod in more detail. The assertions (1) - (4) are concerned with properties of W_o, R, and the wings of W_o inside A_o-mod. We need the following notation. By assumption, W_o is a directing module, thus, let

$$\begin{aligned} \mathcal{X} &= \langle X \mid X \prec W_o\rangle \\ \mathcal{Y} &= \langle Y \mid W_o \prec Y\rangle \\ \mathcal{W}^o &= \langle M \mid M \not\preceq W_o \not\preceq M\rangle . \end{aligned}$$

In order to relate these classes to the wings of W_o, let us denote by $\Theta(n_1,\dots,n_t)$ the translation quiver obtained from the disjoint union of copies of the various $\Theta(n_s)$, $1 \le s \le t$, by identifying all the projective-injective vertices.

(1) The full subcategory given by the indecomposable modules belonging to the wings of W_o is of the form $k(\Theta(n_1,\dots,n_t))$.

Proof. Let $\Theta_1,\dots,\Theta_t$ be the wings of W_o, with Θ_s of length n_s, for $1 \le s \le t$, and denote the vertices in Θ_s in the obvious way by $w_{ij}^{(s)}$, $1 \le i \le j \le n_s$. We are going to choose indecomposable modules $W_{ij}^{(s)}$ with $[W_{ij}^{(s)}] = w_{ij}^{(s)}$ in an appropriate way: For $i = 1$, and $1 \le j < n_s$, an irreducible map $W_{1j}^{(s)} \to W_{1,j+1}^{(s)}$, is a monomorphism, thus we may assume an inclusion map, and we denote it by $\mu_{1j}^{(s)}$. In this way, all $W_{1j}^{(s)}$, $1 \le j < n_s$, are chosen as submodules of W_o. For $i > 1$, let $W_{ij}^{(s)} = W_{1j}^{(s)}/W_{1,i-1}^{(s)}$, and let $\mu_{ij}^{(s)}: W_{ij}^{(s)} \to W_{i,j+1}^{(s)}$ be the canonical inclusion maps. For $1 \le i < j \le n_s$, let $\varepsilon_{ij}^{(s)}: W_{ij}^{(s)} \to W_{i+1,j}^{(s)}$ be the canonical projection map multiplied by $(-1)^j$.

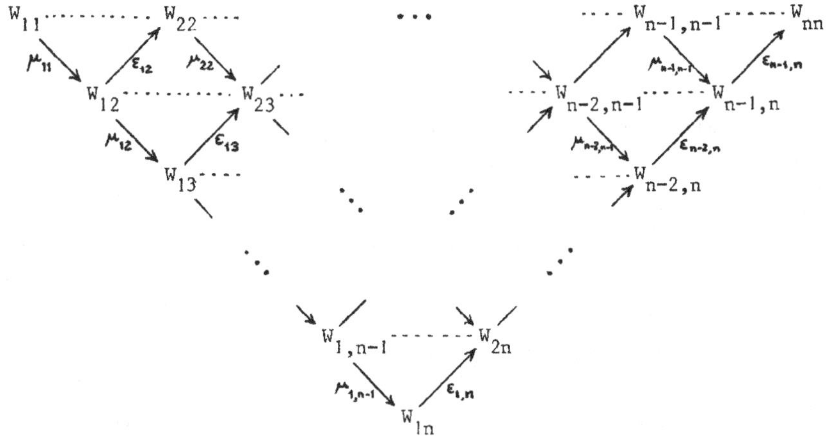

(we have deleted the index s referring to the index of the wing).

Let N be the k-subcategory of A_o-mod having as objects the indecomposable modules $W_{ij}^{(s)}$, $1 \leq i \leq j \leq n_s$, $1 \leq s \leq t$, and with maps generated by the various $\mu_{ij}^{(s)}$, $\varepsilon_{ij}^{(s)}$. First, we want to show that N is a full subcategory of A_o-mod. Thus, let W', W'' be indecomposable modules in N, and assume there exists $\alpha : W' \to W''$, not belonging to N. If $W' = W''$, then we may suppose that α is not invertible [subtracting, if necessary, a scalar multiple of the identity map of W']. If $W'' = W_o$, the sink map for W'' is of the form $(\mu_{1,n_s-1}^{(s)})_s : \bigoplus_s W_{1,n_s-1}^{(s)} \to W_o = W''$, and we obtain a factorization $\alpha = \sum_s \alpha_s \mu_{1,n_s-1}^{(s)}$, with at least one α_s not in N. Similarly, if $W'' = W_{ij}^{(s)}$ with $2 \leq i$, the sink map for W'' is of the form (μ,ε), thus we can factor $\alpha = \alpha_1\mu + \alpha_2\varepsilon$, and at least one of α_1,α_2 is not in N. Using going down induction on $i+j$, we see that we may assume $W'' = W_{1j}^{(s)}$ with $j < n_s$. In particular, W'' belongs to X. Dually, using source maps and going up induction on $i+j$, we see that we may assume $W' = W_{in_s}^{(s)}$ with $1 < i$; in particular, W' belongs to Y. However, since W_o is a directing module, we have $\mathrm{Hom}(Y,X) = 0$, thus $\alpha = 0$ belongs to N, a contradiction. This shows that N is a full subcategory of A_o-mod. In order to see that N is of the form $k(\Theta(n_1,\ldots,n_t))$, we first note that the full subcategory of N given by the indecomposable modules in the fixed wing Θ_s obviously is of the form $k(\Theta(n_s))$. It only remains to see that the compositions $\mu_{11}^{(s)}\cdots\mu_{1,n_s-1}^{(s)}\varepsilon_{1,n_{s'}}^{(s')}\cdots\varepsilon_{n_{s'}-1,n_{s'}}^{(s')}$, for $s \neq s'$, are non-zero. Assume, on the contrary, this composition is zero, thus $\varepsilon' = \varepsilon_{1,n_{s'}}^{(s')}\cdots\varepsilon_{n_{s'}-1,n_{s'}}^{(s')}$ factors through the cokernel $W_{2,n_s}^{(s)}$ of $\mu_{11}^{(s)}\cdots\mu_{1,n_s-1}^{(s)}$. However, in $\Theta(n_1,\ldots,n_t)$, there is no path from $w_{2,n_s}^{(s)}$ to $w_{n_{s'},n_{s'}}^{(s')}$, thus $\mathrm{Hom}(W_{2,n_s}^{(s)}, W_{n_{s'},n_{s'}}^{(s')}) = 0$, whereas $\varepsilon' \neq 0$. This finishes the proof of (1).

For the proof of (1), we only used that W_0 is a directing wing module, and did not refer to lemma 3.3. From now on, let us assume that W_0 is, in addition, separating, and that any A_0-module not belonging to a wing of W_0 is generated or cogenerated by W_0.

It follows that $\mathcal{W}^0$ is given by the modules of the form $W_{ij}^{(s)}$, with $1 < i \leq j < n_s$. Also, let $\mathcal{X}^0$ be given by the modules in $\mathcal{X}$ which belong to a wing of W_0, and $\mathcal{Y}^0$ the modules in $\mathcal{Y}$ which belong to a wing of W_0. The indecomposable modules in $\mathcal{X}^0$ are just those of the form $W_{1j}^{(s)}$, with $1 \leq j < n_s$, those in $\mathcal{Y}^0$ are those of the form $W_{in_s}^{(s)}$ with $1 < i \leq n_s$.

(2) If M is an indecomposable module belonging to a wing of W_0, then $\chi(\underline{\dim} M) = 1$, $\underline{\dim}\, M$ determines M uniquely, $\text{proj.dim.} M \leq 1$, and $\text{inj.dim}\, M. \leq 1$.

Proof. First we show that M itself is a directing module. Thus, let $M = M_0 \to M_1 \to \dots \to M_n = M$ be a cycle. According to (1), not all M_i belong to the wings of W_0. Take i minimal and j maximal, with M_i, M_j not in the wings. It follows that $M_i \in \mathcal{Y}$, and $M_j \in \mathcal{X}$, thus then is given a path from $\mathcal{Y}$ to $\mathcal{X}$, a contradiction. Now it follows from 2.4 (8) that $\underline{\dim}\, M$ determines M uniquely, and $\chi(\underline{\dim}\, M) = 1$. Also, it is obvious that τM belongs to $\mathcal{X} \vee \mathcal{W}^0$, whereas $\tau^- M$ belongs to $\mathcal{W}^0 \vee \mathcal{Y}$. Now, since W_0 is sincere, all indecomposable projective A_0-modules P belong to $\mathcal{X} \vee \langle W_0 \rangle$, whereas all indecomposable injective A_0-modules I belong to $\langle W_0 \rangle \vee \mathcal{Y}$. It follows that $\text{Hom}(\tau^- M, P) = 0$ and $\text{Hom}(I, \tau M) = 0$, thus $\text{inj.dim.} M \leq 1$, and $\text{proj.dim.} M \leq 1$, according to 2.4 (1), (1)*.

In dealing with $A_0[R]$-modules, we usually will consider them as triples $V = (V_0, V_\omega, \gamma : R \otimes V_\omega \to V_0)$, where V_0 is an A_0-module, V_ω a k-space, and γ is A_0-linear, or also as given by $(V_0, V_\omega, \overline{\gamma})$, where $\overline{\gamma} : V_\omega \to \text{Hom}_{A_0}(R, V_0)$ is the adjoint map of γ. If $\dim V_\omega = n$, we just will abbreviate V as being given by a map $R^n \to V_0$.

(3) The module R belongs to $\mathcal{X} \vee \langle W_0 \rangle$. It is the direct sum of at most two indecomposable direct summands. If R decomposes, then R is projective. If R is indecomposable, then $R \in \mathcal{X} \setminus \mathcal{X}^0$.

Proof. Assume $R = R' \oplus R''$ with R' indecomposable and not projective, and with $R'' \neq 0$. We may assume that $\text{Hom}(R'', W_0) \neq 0$. [For , either R'' has an indecomposable projective direct summand, and then $\text{Hom}(R'', W_0) \neq 0$ due to the fact that W_0 is sincere. Or else we may, if necessary, exchange R' with an indecomposable direct summand of R'', since we know that at least $\text{Hom}(R, W_0) \neq 0$, due to the fact that

(*) $$\langle \underline{\dim}\, R, \underline{\dim}\, W_o \rangle = 2(\underline{\dim}\, W_o, \underline{\dim}\, W_o) = 2 ,$$

and that $\operatorname{inj.dim.} W_o \le 1]$. Thus, let $0 \neq \rho : R'' \to W_o$, and consider the following map of A-modules:

$$\begin{array}{ccc} R' \oplus R'' & \xrightarrow{\quad 1 \quad} & R' \oplus R'' \\ \downarrow \begin{bmatrix} 0 & 1 & 0 \\ 0 & 0 & 1 \end{bmatrix} & \begin{bmatrix} \pi \\ 0 \\ \rho \end{bmatrix} & \downarrow \begin{bmatrix} 0 \\ \rho \end{bmatrix} \\ P \oplus R' \oplus R'' & \xrightarrow{\qquad\quad} & W_o \end{array}$$

where $\pi : P \to W_o$ is a projective cover of W_o . The A-module on the left is projective, the map is surjective, and its kernel has $(R',0,0)$ as a direct summand. Since $(R',0,0)$ is not a projective A-module, it follows that $\operatorname{proj.dim.} W_o([{}^{o}_{\rho}]) \ge 2$, contrary to one of the conditions in the definition of domination. This shows that R is projective or indecomposable. - If R is indecomposable, then the equality (*) together with the fact that $\operatorname{inj.dim.} W_o \le 1$ shows that $\operatorname{Hom}(R,W_o) \neq 0$. If R is projective, and R' is any indecomposable direct summand of R , then $\operatorname{Hom}(R',W_o) \neq 0$, since W_o is sincere. This proves the first assertion. - Let us show that if R decomposes, then R is the direct sum of two indecomposable direct summands. In this case, R is projective, thus

$$2 = \langle \underline{\dim}\, R, \underline{\dim}\, W_o \rangle = \dim \operatorname{Hom}(R,W_o) .$$

The assertion now follows from the fact that any indecomposable direct summand R' of R satisfies $\operatorname{Hom}(R',W_o) \neq 0$. - Finally, if R is indecomposable, then $\operatorname{Hom}(R,W_o) \neq 0$ shows that R is in $\mathcal{X} \cup \langle W_o \rangle$. However, $R \neq W_o$ and $R \notin \mathcal{X}^o$, since $\dim \operatorname{Hom}(R,W_o) \ge 2$ according to (*).

(4) Let M be an indecomposable module belonging to a wing of W_o . Then

$$\dim \operatorname{Hom}(R,M) = \begin{cases} 2 & M = W_o \\ 1 \quad \text{if} & M \in \mathcal{X}^o \vee \mathcal{Y}^o \\ 0 & M \in \mathcal{W}^o \end{cases} .$$

<u>Proof</u>. Since $\operatorname{inj.dim.} M \le 1$, we have $\dim \operatorname{Hom}(R,M) = \langle \underline{\dim}\, R, \underline{\dim}\, M \rangle + \dim \operatorname{Ext}^1(R,M)$. However, $\operatorname{Ext}^1(R,M) = D\operatorname{Hom}(\tau^- M,R)$, and $\tau^- M$ belongs to $\mathcal{W}^o \cup \mathcal{Y}$, whereas R belongs to $\mathcal{X} \vee \{W_o\}$, thus $\operatorname{Ext}^1(R,M) = 0$. Thus

$$\dim \operatorname{Hom}(R,M) = \langle \underline{\dim}\, R, \underline{\dim}\, M \rangle = 2(\underline{\dim}\, W_o, \underline{\dim}\, M) ,$$

since R dominates W_o . Now $\operatorname{proj.dim.} W_o \le 1$, $\operatorname{inj.dim.} W_o \le 1$, and $\operatorname{Ext}^1(W_o,M) = D\operatorname{Hom}(M,\tau W_o) = 0$, $\operatorname{Ext}^1(M,W_o) = D\operatorname{Hom}(\tau^- W_o,M) = 0$, since $\tau W_o \in \mathcal{X} \setminus \mathcal{X}^o$, $\tau^- W_o \in \mathcal{Y} \setminus \mathcal{Y}^o$, thus

$$\begin{aligned} 2(\underline{\dim}\, W_o, \underline{\dim}\, M) &= \langle \underline{\dim}\, W_o, \underline{\dim}\, M \rangle + \langle \underline{\dim}\, M, \underline{\dim}\, W_o \rangle \\ &= \dim \operatorname{Hom}(W_o,M) + \dim \operatorname{Hom}(W_o,M) . \end{aligned}$$

The assertion now follows directly from (1).

Now we can consider A-modules in more detail. Recall that A-mod is equivalent to $\check{\mathcal{U}}(A_o\text{-mod}, \operatorname{Hom}_{A_o}(R,-))$. Since the functor $\operatorname{Hom}_{A_o}(R,-)$ will not be changed throughout the discussion, we will omit the reference to it, thus given a module class N in A_o-mod , we will denote $\check{U}(N,\mathrm{Hom}_{A_o}(R,-))$ just by $\check{U}(N)$, and similarly for U . Since $\mathrm{End}(W_o) = k$, $\dim \mathrm{Hom}(R,W_o) = 2$, it follows that the category $\check{U}(<W_o>)$ is equivalent to the category of Kronecker modules, thus we may speak of preprojective, regular, and preinjective objects in $\check{U}(<W_o>)$. Actually, we only are interested in $U(<W_o>)$ and we denote by P^o the class of preprojective, by T^o the class of regular, and by Q^o the class of preinjective objects in $U(<W_o>)$. [To be precise: an indecomposable A-module (V_o,V_ω,γ) belongs to P^o , or T^o , or Q^o , if and only if V_o is the direct sum of copies of W_o , say n copies, with $n \geq 1$, and $\dim V_\omega = n-1$, or n , or $n+1$, respectively.] Also, let

$$P^* = U(X) \int P^o \ , \ Q^* = Q^o \int \check{U}(Y) \ ,$$

and finally

$$\begin{aligned} P &= P^* \setminus \overline{X^o} \\ T &= (\overline{X^o} \int T^o \int Y^o) \vee W^o \\ Q &= Q^* \setminus Y^o \ , \end{aligned}$$

these three classes are the module classes in A-mod which we are interested in.

Given a wing Θ_s of W_o , there is given a chain of inclusion maps

$$W_{11}^{(s)} \subset W_{12}^{(s)} \subset \dots \subset W_{1,n_s-1}^{(s)} \subset W_{1,n_s}^{(s)} = W_o$$

all being denoted by μ . If we denote by ρ_s' a fixed non-zero map $R \to W_{1,n_s-1}^{(s)}$ and let $\rho_s = \rho_s'\mu : R \to W_o$, then $W_o(\rho_s) = (W_o,k,\rho_s)$ may be identified with $(W_o,\mathrm{Hom}(R,W_{1,n_s-1}^{(s)}),e\cdot\mathrm{Hom}(1,\mu))$, where $e : R \otimes \mathrm{Hom}(R,W_{1,n_s-1}^{(s)}) \to W_{1,n_s-1}^{(s)}$ is the evaluation map. In this way, we obtain a chain of inclusions of A-modules

$$\overline{W_{11}^{(s)}} \subset \overline{W_{12}^{(s)}} \subset \dots \subset \overline{W_{1,n_s-1}^{(s)}} \subset W_o(\rho_s) \ ,$$

where all but the last maps are of the form $(\mu,\mathrm{Hom}(1,\mu))$, whereas the last map is of the form $(\mu,1)$. Again, we will denote these maps just by μ . Also the maps $\varepsilon : W_{1j}^{(s)} \to W_{2j}^{(s)}$, $1 < j < n_s$ induce zero maps $\mathrm{Hom}(R,\varepsilon) = 0$, thus we obtain maps $(\varepsilon,0) : \overline{W_{1j}^{(s)}} \to W_{2j}^{(s)}$ which will be denoted by ε . Finally, the map $\varepsilon : W_o = W_{1,n_s}^{(s)} \to W_{2,n_s}^{(s)}$ induces a map $\mathrm{Hom}(1_R,\varepsilon)$ which has zero composition with $\mathrm{Hom}(1_R,\mu_{1,n_s-1}^{(s)})$, since the composition factors through $\mathrm{Hom}(R,W_{2,n_s-1}^{(s)}) = 0$. Thus we obtain a map $(\varepsilon,0) : W_o(\rho_s) \to W_{2,n_s}^{(s)}$ which also will be denoted by ε . Thus, we have constructed the following diagram:

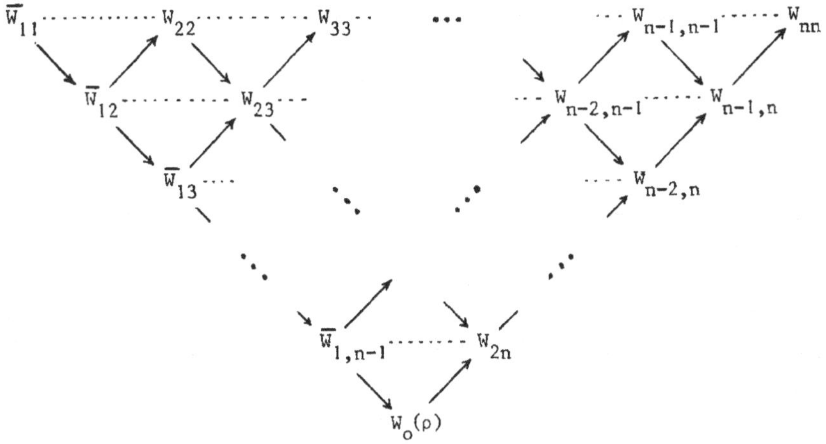

(again, we have deleted the index s).

Let us recall that for $1 < i \leq j < n_s$, we have $\mathrm{Hom}(R, W_{ij}^{(s)}) = 0$, thus $\overline{W_{ij}^{(s)}} = W_{ij}^{(s)}$. We claim that the meshes encountered in the diagram above are actually given by Auslander-Reiten sequences in A-mod. Namely, we have

(5) For $1 \leq i \leq j < n$, the source map for $\overline{W_{ij}^{(s)}}$ is of the form $[\varepsilon, \mu]$ with ε, μ given by the diagram above. For $1 < i \leq j \leq n$, the sink map for $W_{ij}^{(s)}$ is of the form $\begin{bmatrix}\mu\\ \varepsilon\end{bmatrix}$ with ε, μ given by the diagram above.

Proof. This is a direct consequence of 2.5 (5).

(6) $\mathrm{Hom}(\mathcal{Q}, \mathcal{P}) = \mathrm{Hom}(\mathcal{Q}, \mathcal{T}) = \mathrm{Hom}(\mathcal{T}, \mathcal{P}) = 0$.

Proof. Let $f = (f_0, f_\omega) : U \longrightarrow V$ be a map. If $\gamma_V : V_\omega \longrightarrow \mathrm{Hom}(R, V_0)$ is mono (thus, if V is in $\mathcal{U}(A_0\text{-mod})$), then $f_0 = 0$ implies that $f = 0$. Since $\mathrm{Hom}(\mathcal{Y}, \mathcal{X}) = \mathrm{Hom}(\mathcal{Y}, \mathcal{W}^o \vee \langle W_o\rangle) = \mathrm{Hom}(\mathcal{W}^o \vee \langle W_o\rangle, \mathcal{X}) = 0$, it follows that $\mathrm{Hom}(\check{\mathcal{U}}(\mathcal{Y}), \mathcal{U}(\mathcal{X})) = \mathrm{Hom}(\check{\mathcal{U}}(\mathcal{Y}), \mathcal{W}^0 \vee \mathcal{U}(\langle W_o\rangle)) = \mathrm{Hom}(\mathcal{W}^0 \vee \mathcal{U}(\langle W_o\rangle), \mathcal{U}(\mathcal{X})) = 0$. Since $\mathrm{Hom}(\mathcal{W}^0, W_0) = \mathrm{Hom}(W_0, \mathcal{W}^0) = 0$, it follows that $\mathrm{Hom}(\mathcal{W}^0, \mathcal{U}(\langle W_o\rangle)) = \mathrm{Hom}(\mathcal{U}(\langle W_o\rangle), \mathcal{W}^o) = 0$. From the investigations of Kronecker modules we know that $\mathrm{Hom}(\mathcal{Q}^0, \mathcal{P}^0) = \mathrm{Hom}(\mathcal{Q}^0, \mathcal{T}^0) = \mathrm{Hom}(\mathcal{T}^0, \mathcal{P}^0) = 0$. Thus, we only have to show that $\mathrm{Hom}(\mathcal{Q}, \mathcal{Y}^o) = 0$ and $\mathrm{Hom}(\overline{\mathcal{X}^0}, \mathcal{P}) = 0$.

Let $X = W_{1j}^{(s)}$ be in $\mathcal{X}^0$, with $1 \leq j \leq n_s - 1$, and $V \in \mathcal{P}$. By decreasing induction on j, we show that $\mathrm{Hom}(\overline{X}, V) = 0$. Given a map $f : \overline{X} \longrightarrow V$, we factor it through the source map $\overline{X} \longrightarrow M$ in A-mod. If $j = n_s - 1$, then $M = W_0(\rho_s) \oplus W_{2,n_s-1}^{(s)}$, thus M_0 is in $\langle W_o\rangle \vee \mathcal{W}^o$, and therefore $\mathrm{Hom}(M, V) = 0$. If $j < n_s - 1$, then $M = \overline{W}_{1,j+1} \oplus W_{2,j}$. Since $W_{2,j} \in \mathcal{W}^0$, we have $\mathrm{Hom}(W_{2,j}, V) = 0$ and by induction, also

$\mathrm{Hom}(\overline{W_{1,j+1}^{(s)}},V) = 0$. Thus again $\mathrm{Hom}(M,V) = 0$. Therefore $\mathrm{Hom}(X,V) = 0$.

Similarly, for $Y = W_{in_s}^{(s)}$ in $\mathcal{Y}^o$, with $1 \leq i \leq n_s-1$, and $U \in \mathcal{Q}$ one uses induction on i and the sink map $N \longrightarrow Y$ in A-mod, in order to show that $\mathrm{Hom}(U,Y) = 0$.

(7) The projective A-modules belong to $\mathcal{P}$, the injective A-modules to $\mathcal{Q}$.

Proof. Let P be an indecomposable projective A-module. First, consider the case of P being actually a projective A_o-module. Since W_o is a sincere A_o-module, it follows that P is in $\mathcal{X} \vee \langle W_o \rangle$. If P is in $\mathcal{X}$, then in $\mathcal{U}(\mathcal{X}) \setminus \overline{\mathcal{X}^o}$, thus in $\mathcal{P}$. If $P = W_o$, then $P \in \mathcal{P}^o$, thus in $\mathcal{P}$. Now, consider the remaining case of $P = (R,k,1_R)$. If R is indecomposable, then R is in $\mathcal{X} \setminus \mathcal{X}^o$, thus $P \in \mathcal{U}(\mathcal{X}) \setminus \overline{\mathcal{X}^o} \subseteq \mathcal{P}$. Thus consider the case of R being decomposable, say $R = R' \oplus R''$. Then both R',R'' are in $\mathcal{X} \vee \langle W_o \rangle$. If both R',R'' are in $\mathcal{X}$, then P is in $\mathcal{U}(\mathcal{X})$. If both R',R'' are equal to W_o, then P is in $\mathcal{U}(\langle W_o \rangle)$, and actually in $\mathcal{P}^o$. If, finally, R' is in $\mathcal{X}$, $R'' = W_o$, then P is an extension of W_o by P/W_o, and W_o is in $\mathcal{P}^o$, whereas P/W is in $\mathcal{U}(\mathcal{X})$. We see that for R being decomposable, always P belongs to $\mathcal{P}^*$. However, the indecomposable modules in $\mathcal{P}^* \setminus \mathcal{P}$ are those in $\overline{\mathcal{X}^o}$, and they have the property that their restrictions to A_o are indecomposable. This shows that $P \in \mathcal{P}$.

Next, let Q be an indecomposable injective A-module. If $Q = (0,k,o)$, then $Q \in \check{\mathcal{U}}(\mathcal{Y}) \setminus \mathcal{Y}^o$, thus $Q \in \mathcal{Q}$. Otherwise, the restriction Q_o of Q to A_o is an indecomposable injective A_o-module, and $Q = (Q_o,\mathrm{Hom}(R,Q_o),e)$, with e the evaluation map. Since W_o is sincere, $Q_o \in \langle W_o \rangle \vee \mathcal{Y}$. If $Q_o = W_o$, then $\dim \mathrm{Hom}(R,Q_o) = 2$ shows that Q belongs to $\mathcal{Q}^o$. If $Q_o \in \mathcal{Y}^o$, then $\dim \mathrm{Hom}(R,Q_o) = 1$ shows that $Q \in \mathcal{U}(\mathcal{Y}) \setminus \mathcal{Y}^o$. Of course, if $Q_o \in \mathcal{Y} \setminus \mathcal{Y}^o$, then also $Q \in \mathcal{U}(\mathcal{Y}) \setminus \mathcal{Y}^o$. Thus, always $Q \in \mathcal{Q}$.

(8) For $T \in \mathcal{T}$, $\mathrm{proj.dim.} T \leq 1$.

Proof. The modules in $\mathcal{Y}^o$ and $\mathcal{W}^o$ have as A_o-modules projective dimension ≤ 1, thus also as A-modules. By assumption the A-modules $W_o(\rho)$, with $0 \neq \rho : R \longrightarrow W_o$, have projective dimension ≤ 1, thus all modules in $\mathcal{T}^o$. Since the class of modules of projective dimension ≤ 1 is closed under extensions, it remains to be seen that $\mathrm{proj.dim.} T \leq 1$ for T in $\overline{\mathcal{X}^o}$. Thus, let $T = \overline{W_{1j}^{(s)}}$ for some $1 \leq j \leq n_s-1$. Then we have an exact sequence

$$0 \longrightarrow \overline{W_{1j}^{(s)}} \longrightarrow W_o(\rho_s) \longrightarrow W_{j+1,n_s}^{(s)} \longrightarrow 0,$$

and $\mathrm{proj.dim.}\, W_o(\rho_s) \leq 1$, $\mathrm{proj.dim.}\, W_{j+1,n_s}^{(s)} \leq 1$ imply $\mathrm{proj.dim.}\, \overline{W_{1j}^{(s)}} \leq 1$. This finishes the proof.

(9) For $T \in \mathcal{T}$, $\underline{\dim}\, \tau T = (\underline{\dim}\, T)\Phi$.

Proof. This a direct consequence of 2.4 (4) using (6),(7), and (8).

(10) For $3 \leq i \leq n_s$, $\tau_A W^{(s)}_{ii} = W^{(s)}_{i-1,i-1}$, $\tau_A W^{(s)}_{22} = \overline{W^{(s)}_{11}}$, and $\tau\overline{W^{(s)}_{11}} = W^{(s)}_{n_s n_s}$.

Proof. All but the last assertions are direct consequences of (5). The diagram before (5) shows that

$$w = \underline{\dim}\, W_o(\rho_s) = \underline{\dim}\, \overline{W^{(s)}_{11}} + \sum_{i=2}^{n_s} \underline{\dim}\, W^{(s)}_{ii}.$$

Applying Φ to this equality and noting that w is Φ-invariant, according to 2.5 (11), it follows that

$$w = (\underline{\dim}\, \overline{W^{(s)}_{11}})\Phi + \sum_{i=2}^{n_s} (\underline{\dim}\, W^{(s)}_{ii})\Phi$$
$$= \underline{\dim}\, \tau\overline{W^{(s)}_{11}} + \underline{\dim}\, \overline{W^{(s)}_{11}} + \sum_{i=2}^{n_s-1} \underline{\dim}\, W^{(s)}_{ii},$$

using (9) and the assertions of (10) which already are established. Thus

$$\underline{\dim}\, \tau\overline{W^{(s)}_{11}} = \underline{\dim}\, W^{(s)}_{n_s n_s},$$

thus $\tau\overline{W^{(s)}_{11}}$ is in fact an A_0-module, indecomposable and with dimension vector $W^{(s)}_{n_s n_s}$. According to (2), $W^{(s)}_{n_s n_s}$ is uniquely determined by its dimension vector.

We will say that a map $\eta : M \longrightarrow W_o$ factors through a wing of W_o if the image of η is contained in $W^{(s)}_{1,n_s-1}$, for some s. (Recall that the modules $W^{(s)}_{1j}$ may be considered as submodules of W_o).

(11) If $\rho : R \longrightarrow W_o$ does not factor through a wing of W_o, then $\tau W_o(\rho) \approx W_o(\rho)$.

Proof. We have seen above that $\underline{\dim}\, \tau W_o(\rho) = (\underline{\dim}\, W_o(\rho))\Phi = w\Phi = w$. We consider the restriction of $\tau W_o(\rho)$ to A_o, say $\tau W_o(\rho)\big|_{A_o} = \bigoplus_{i=1}^{t} M_i$ with indecomposable A_o-modules M_i. First, consider the case $t > 1$. Now

$$1 = \langle w_o, w_o\rangle = \sum_{i=1}^{t} \langle \underline{\dim}\, M_i, w_o\rangle,$$

thus there is some j with $\langle \underline{\dim}\, M_j, \underline{\dim}\, W_o \rangle > 0$, and consequently $\mathrm{Hom}(M_j, W_o) \neq 0$ because $\mathrm{inj.dim.} W_o \leq 1$. It follows that M_j is cogenerated by W_o. [Indeed, $M_j \in X \vee \langle W_o \rangle$, and is cogenerated by an injective module, thus a module in $\langle W_o \rangle \vee Y$. Since we can factor any map from X to Y through W_o, we see that M_j is cogenerated by W_o]. Denoting M_j by M, and the direct sum of the remaining M_i by M', we have $\tau W_o(\rho) = ([\xi, \xi'] : R \longrightarrow M \oplus M')$, with M indecomposable, and cogenerated by W_o. Now $\xi \neq 0$, since $\tau W_o(\rho)$ is indecomposable, thus there exists $\eta : M \longrightarrow W_o$ with $\xi\eta \neq 0$. We can assume that η factors through a wing of W_o. [For, since $\underline{\dim}\, M < w_o$, any $\eta : M \longrightarrow W_o$ factors through the sink map $(\mu^{(s)})_s : \bigoplus_s W^{(s)}_{1,n_s-1} \longrightarrow W_o$, say $\eta = \sum_s \eta_s \mu^{(s)}$. Thus, if $\xi\eta \neq 0$, then $\xi\eta_s\mu^{(s)} \neq 0$ for some s, so we may replace η by $\eta_s\mu^{(s)}$]. The following diagram

$$\begin{array}{ccccccccc}
0 & \longrightarrow & R & \xrightarrow{[1\ 1]} & R \oplus R & \xrightarrow{\left[\begin{smallmatrix}1\\-1\end{smallmatrix}\right]} & R & \longrightarrow & 0 \\
 & & \downarrow {\scriptstyle [\xi,\xi']} & & \downarrow {\scriptstyle \left[\begin{smallmatrix}\xi & 0 & 0\\ 0 & \xi' & \xi\eta\end{smallmatrix}\right]} & & \downarrow {\scriptstyle \xi\eta} & & \\
0 & \longrightarrow & M \oplus M' & \xrightarrow{\left[\begin{smallmatrix}1 & 0 & \eta\\ 0 & 1 & 0\end{smallmatrix}\right]} & M \oplus M' \oplus W_o & \xrightarrow{\left[\begin{smallmatrix}\eta\\0\\-1\end{smallmatrix}\right]} & W_o & \longrightarrow & 0
\end{array}$$

is commutative, with exact rows, thus it is an exact sequence

$$0 \longrightarrow \tau W_o(\rho) \longrightarrow T \longrightarrow W_o(\xi\eta) \longrightarrow 0$$

of A-modules. The sequence does not split, since T has $(\xi : R \longrightarrow M)$ as a direct summand. Thus the sequence induces the Auslander-Reiten sequence starting with $\tau W_o(\rho)$, and therefore $\mathrm{Hom}(W_o(\rho), W_o(\xi\eta)) \neq 0$. However, this implies that ρ is a scalar multiple of $\xi\eta$, since $\mathrm{End}(W_o) = k$. Thus ρ factors through η, and therefore through a wing.

It remains to consider the case $t = 1$. In this case, $\tau W_o(\rho)|_{A_o} = W_o$, since W_o is determined by its dimension vector. Therefore $\tau W_o(\rho) = W(\rho')$ for some $0 \neq \rho' : R \longrightarrow W_o$. However, then $\mathrm{Ext}^1(W_o(\rho), W_o(\rho')) \neq 0$, the Auslander-Reiten sequence being a non-zero element in this Ext-group. Since $W_o(\rho), W_o(\rho')$ belong to T^o, the category of regular objects in a category equivalent to the category of Kronecker modules, $\mathrm{Ext}^1(W_o(\rho), W_o(\rho')) \neq 0$ implies that $W_o(\rho)$ and $W_o(\rho')$ are isomorphic. Thus $\tau W_o(\rho) \approx W_o(\rho)$.

(12) For any s, the modules $\overline{W^{(s)}_{11}}$, $W^{(s)}_{ii}$, $2 \leq i \leq n_s$, are orthogonal bricks and they lie on the mouth of a stable tube of rank n_s. If $\rho : R \longrightarrow W_o$ does not factor through a wing, then $W_o(\rho)$ lies on the mouth of a stable tube of rank 1.

Proof. Since $W_{ii}^{(s)}$, $1 \leq i \leq n_s$, are orthogonal bricks, the same is true for $\overline{W_{11}^{(s)}}$, $W_{ii}^{(s)}$, $2 \leq i \leq n_s$. According to (10), these latter A-modules form a τ-orbit. Similarly, $W_o(\rho)$ for $0 \neq \rho : R \longrightarrow W_o$, is a brick, and it is a τ-orbit provided ρ does not factor through a wing. Since the projective dimension of all modules in T is ≤ 1, the assertion follows from 3.1 (2).

(13) Any non-zero map $\rho : R \longrightarrow W_o$ factors through at most one wing.

Proof. If ρ factors through the wing θ_s, then there is an irreducible monomorphism $\overline{W_{1,n_s-1}^{(s)}} \longrightarrow W_o(\rho)$. In a regular tube, any module is the target of at most one irreducible monomorphism. Since for $s \neq s'$, the modules $\overline{W_{1,n_s-1}^{(s)}}$ and $\overline{W_{1,n_{s'}-1}^{(s')}}$ are non-isomorphic, ρ cannot factor both through θ_s and $\theta_{s'}$.

For any $\rho \in \mathbb{P}\,\mathrm{Hom}(R,W_o)$, let us define a module class $T(\rho)$. If ρ factors through the wing θ_s, let $T(\rho) = E(\overline{W_{11}^{(s)}}, W_{22}^{(s)},\ldots,W_{n_s n_s}^{(s)})$. If ρ does not factor through any wing, let $T(\rho) = E(W_o(\rho))$. We set $T^o(\rho) = T(\rho) \cap T^o$.

(14) If ρ,ρ' are linearly independent maps $R \longrightarrow W_o$, then $\mathrm{Hom}(T(\rho),T(\rho')) = 0$.

Proof. It is sufficient to show that $\mathrm{Hom}(T,T') = 0$ for $T \in T(\rho), T' \in T(\rho')$ and both T,T' being of the form $\overline{W_{11}^{(s)}}$, $W_{ii}^{(s)}$, $2 \leq i \leq n_s$, or $W_o(\rho'')$ with ρ'' not factoring through a wing. Of course, for ρ,ρ' linearly independent, we have $\mathrm{Hom}(W_o(\rho),W_o(\rho')) = 0$. Now assume, ρ factors through the wing θ_s. There is a non-zero homomorphism $W_{ii}^{(s)} \longrightarrow W_o$ only for $i = 1$, thus we consider $\mathrm{Hom}(\overline{W_{11}^{(s)}},W_o(\rho'))$. Assume there is some non-zero $f = (f_o,f_\omega) : \overline{W_{11}^{(s)}} \longrightarrow W_o(\rho')$. Then f_ω defines a scalar multiplication on R, f_o is a scalar multiple of the inclusion map $\mu : W_{11} \longrightarrow W_o$, thus, the commuting diagram

$$\begin{array}{ccc} R & \xrightarrow{\ f_\omega\ } & R \\ \big\downarrow & & \big\downarrow \rho' \\ W_{11} & \xrightarrow{\ f_o\ } & W_o \end{array}$$

shows that ρ' also factors through the wing θ_s. Similarly, assume ρ' factors through θ_s. There is a non-zero homomorphism $W_o \longrightarrow W_{ii}^{(s)}$ only for $i = n_s$, thus assume there exists a non-zero $g = (g_o,g_\omega) : W_o(\rho) \longrightarrow W_{n_s n_s}^{(s)}$, hence a commutative diagram

$$\begin{array}{ccc} R & \longrightarrow & 0 \\ {\scriptstyle\rho}\downarrow & & \downarrow \\ W_o & \xrightarrow{g_\omega} & W^{(s)}_{n_s n_s} \end{array} .$$

Now g_ω is a non-zero scalar multiple of the projection $W_o \longrightarrow W_o/W^{(s)}_{1,n_s-1} \approx W^{(s)}_{n_s n_s}$, and ρ factors through its kernel $W^{(s)}_{1,n_s-1}$. Thus ρ also factors through the wing θ_s. Finally, assume ρ factors through θ_s, and ρ' through $\theta_{s'}$. There is a non-zero map $W^{(s)}_{ii} \longrightarrow W^{(s')}_{jj}$ only for $i = 1$, $j = n_{s'}$, and it factors through W_o. Thus, assume there exists a non-zero $h = (h_o, h_\omega) : \overline{W^{(s)}_{11}} \longrightarrow W^{(s')}_{n_{s'},n_{s'}}$. Note that $h_\omega = 0$, and h_o factors through W_o, say $h_o = h'_o h''_o$. Thus we have the following diagram

$$\begin{array}{ccccc} R & \xrightarrow{1} & R & \xrightarrow{o} & 0 \\ \downarrow & & \downarrow{\scriptstyle\rho'''} & & \downarrow \\ \overline{W^{(s)}_{11}} & \xrightarrow{h'_o} & W_o & \xrightarrow{h''_o} & W^{(s')}_{n_{s'},n_{s'}} \end{array}$$

with ρ''' being a non-zero map. It follows that $\mathrm{Hom}(\overline{W^{(s)}_{11}}, W_o(\rho''')) \neq 0$, therefore ρ''' factors through θ_s, and $\mathrm{Hom}(W_o(\rho'''), W^{(s')}_{n_{s'},n_{s'}}) \neq 0$, therefore ρ''' factors also through $\theta_{s'}$, thus $s = s'$.

(15) $$\mathcal{T} = \bigvee_{\rho \in \mathbb{P}\,\mathrm{Hom}(R,W_o)} \mathcal{T}(\rho) .$$

<u>Proof.</u> First, let us show that any $\mathcal{T}(\rho)$ is contained in $\mathcal{T}$. If ρ does not factor through a wing, then $\mathcal{T}(\rho) \subseteq \mathcal{T}^o$. Thus, assume ρ factors through the wing θ_s. Denote $\overline{W^{(s)}_{11}}$ by E_1, and $W^{(s)}_{ii}$ by E_i, for $2 \le i \le n_s$. Also, for $i \in \mathbb{Z}$, $\ell \in \mathbb{N}_1$, define $E_i[\ell]$ as follows: $E_i[1] = E_{i'}$ for $i \equiv i' \pmod{n_s}$, and for $\ell \ge 2$, let $E_i[\ell]$ be the indecomposable module in $\mathcal{T}$ with $E_i[1]$ as submodule such that $E_i[\ell] / E_i[1] \approx E_{i+1}[\ell-1]$. Of course, $E_i[\ell] \approx E_{i'}[\ell']$ if and only if $i \equiv i' \pmod{n_s}$ and $\ell = \ell'$. We have to show that all $E_i[\ell]$, $1 \le i \le n_s$, $\ell \in \mathbb{N}_1$, belong to $\mathcal{T}$. Also, let $E_i[0] = 0$. Consider first the case $i = 1$. Let $\ell = \ell' n_s + \ell''$ with $0 \le \ell'$, and $0 \le \ell'' < n_s$, then there is an exact sequence

$$0 \longrightarrow E_1[\ell' n_s] \longrightarrow E_1[\ell] \longrightarrow E_1[\ell''] \longrightarrow 0 ,$$

$E_1[\ell' n_s]$ is ℓ'-fold extension of $E_1[n_s] = W_0(\rho_s)$, thus belongs to $\mathcal{T}^0$, and $E_1[\ell''] = \overline{W^{(s)}_{1\ell''}}$ belongs to $\overline{\mathcal{X}^0}$, thus $E_1[\ell]$ is in $\mathcal{T}$. Now consider the case $1 < i \leq n_s$. If $\ell \leq n_s - i$, then $E_i[\ell] = W^{(s)}_{i,i+\ell-1} \in \mathcal{W}^0$. If $\ell \geq n_s - i + 1$, then there is an exact sequence

$$0 \longrightarrow E_i[n_s-i+1] \longrightarrow E_i[\ell] \to E_1[\ell-n_s+i-1] \longrightarrow 0.$$

Now $E_i[n_s-i+1] = W^{(s)}_{i,n_s} \in \mathcal{Y}^0$, and, as we have seen above, $E_1[\ell-n_s+i-1]$ belongs to $\overline{\mathcal{X}^0} \int \mathcal{T}^0$; therefore $E_i[\ell] \in \overline{\mathcal{X}^0} \int \mathcal{T}^0 \int \mathcal{Y}^0$. Thus, for all $1 \leq i \leq n_s$, and all $\ell \in \mathbb{N}_1$, $E_i[\ell] \in \mathcal{T}$.

On the other hand, any $\mathcal{T}(\rho)$ is closed under extensions. If $T \in \mathcal{T}(\rho)$, and $T' \in \mathcal{T}(\rho')$ with ρ,ρ' different elements in $\mathbb{P}\,\mathrm{Hom}(R,W_0)$, then $\mathrm{Ext}^1(T,T') = 0$, since $\mathrm{Hom}(T',\tau T) = 0$ (with T also τT belongs to $\mathcal{T}(\rho)$). Thus $\bigvee_{\rho \in \mathbb{P}\,\mathrm{Hom}(R,W_0)} \mathcal{T}(\rho)$ is closed under extensions. Now an indecomposable module in $\overline{\mathcal{X}^0}$ is of the form $\overline{W^{(s)}_{1j}}$ with $1 \leq j < n_s$, an indecomposable module in $\mathcal{Y}^0$ is of the form $W^{(s)}_{in_s}$, $2 \leq i \leq n_s$, and the indecomposable modules in $\mathcal{W}^0$ are of the form $W^{(s)}_{ij}$ with $1 < i < j < n_s$, where for a fixed s, all these modules belong to $\mathcal{E}(\overline{W^{(s)}_{11}}, W^{(s)}_{22},\ldots,W^{(s)}_{n_s n_s}) = \mathcal{T}(\rho_s)$, where $0 \neq \rho_s : R \longrightarrow W_0$ factors through the wing θ_s. Finally, any indecomposable object in $\mathcal{T}^{\sim}$ is an n-fold extension of some $W_0(\rho)$, thus it belongs to $\mathcal{T}(\rho)$. This finishes the proof.

(16) $\mathrm{Ext}^1(\mathcal{P},\mathcal{T}) = 0 = \mathrm{Ext}^1(\mathcal{T},\mathcal{Q})$.

<u>Proof.</u> Let $P \in \mathcal{P}$, $T \in \mathcal{T}$, $Q \in \mathcal{Q}$. Since inj.dim.$T \leq 1$ and proj.dim.$T \leq 1$, we have

$$\mathrm{Ext}^1(P,T) = D\,\mathrm{Hom}(\tau^- T,P),$$
$$\mathrm{Ext}^1(T,Q) = D\,\mathrm{Hom}(Q,\tau T).$$

However, $\mathcal{T}$ is closed unter τ,τ^-, and $\mathrm{Hom}(\mathcal{T},\mathcal{P}) = \mathrm{Hom}(\mathcal{Q},\mathcal{T}) = 0$.

Before we can show that also $\mathrm{Ext}^1(\mathcal{P},\mathcal{Q}) = 0$, we show that any map from $\mathcal{P}$ to $\mathcal{Q}$ factors through $\mathcal{T}$, or better, through any $\mathcal{T}(\rho)$. In order to do so, we consider the universal maps to and from modules in $\mathcal{U}(W_0)$. Given an A_0-module M, let $\eta_M : M \longrightarrow W_0 \otimes \mathrm{Hom}(M,W_0)$ be the universal map from M to $\langle W_0 \rangle$; it is defined as follows: Let $\alpha_1,\ldots,\alpha_r$ be a k-basis of $\mathrm{Hom}(M,W_0)$, then $m\eta_M = \sum_{i=1}^{r} m\alpha_i \otimes \alpha_i$. For U in $\mathcal{P}$, we consider $\vec{U} = (W_0 \otimes \mathrm{Hom}(U_\sigma,W_0),\ U_\omega,\ \gamma_U \mathrm{Hom}(R,\eta_{U_0}))$, and the map $\eta_U = (1,\eta_{U_0}) : U \longrightarrow \vec{U}$. The dual construction is given as follows:

Given V in $\mathcal{Q}$, let $\overleftarrow{V}_0 = W_0 \otimes \mathrm{Hom}(R,W_0)$, let $\zeta_0 : \overleftarrow{V} \longrightarrow V_o$ be the evaluation map, with $(x \otimes f)\zeta_0 = xf$ for $x \in W_0$, $f \in \mathrm{Hom}(W_0,M)$ and let $\overleftarrow{V}_\omega, \gamma_{\overleftarrow{V}}$, and ζ_ω be defined by the following pullback diagram

$$\begin{array}{ccc}
\overleftarrow{V}_\omega & \xrightarrow{\ \zeta_\omega\ } & V_\omega \\
\Big\downarrow{\scriptstyle \gamma_{\overleftarrow{V}}} & & \Big\downarrow{\scriptstyle \gamma_V} \\
\mathrm{Hom}(R,\overleftarrow{V}_\omega) & \xrightarrow{\ \mathrm{Hom}(R,\zeta_o)\ } & \mathrm{Hom}(R,V_o)
\end{array}$$

In this way, we obtain $\overleftarrow{V} = (\overleftarrow{V}_0, \overleftarrow{V}_\omega, \gamma_{\overleftarrow{V}})$ and a map $\zeta = (\zeta_0, \zeta_\omega) : \overleftarrow{V} \longrightarrow V$.

(17) For U in $\mathcal{P}$, the module $\vec{U}$ belongs to $\mathcal{P}^0$. For V in $\mathcal{Q}$, the module $\overleftarrow{V}$ belongs to $\mathcal{Q}^0$.

<u>Proof.</u> For $1 \leq s \leq t$, let $\rho_s : R \longrightarrow W_0$ be a non-zero map factoring through the wing θ_s. We fix some j (with $1 \leq j \leq t$), and some $\rho : R \longrightarrow W_0$ not factoring through a wing. Given U in $\mathcal{P}$, we want to show that $\vec{U}$ is cogenerated by $\mathcal{T}' = \mathcal{T}(\rho) \vee \bigvee_{s \neq j} \mathcal{T}(\rho_s)$. From this it follows that $\vec{U}$ is the direct sum of a module in $\mathcal{P}^0$ and a module in $\mathcal{T}'$. Since we may vary both j and ρ, the module $\vec{U}$ actually belongs to $\mathcal{P}^0$.

Consider first the case of U being in $\mathcal{U}(\mathcal{X}) \setminus \overline{\mathcal{X}^0}$. We want to show that there exists a non-zero homomorphism $f : U \longrightarrow T$ with $T \in \mathcal{T}'$. If there exists a non-zero homomorphism $f_0 : U_0 \longrightarrow W^{(s)}_{1,n_s-1}$, for some $s \neq j$, let $f_\omega = \gamma_U \mathrm{Hom}(R,f_0)$. We obtain, in this way, a non-zero map $f : U \longrightarrow \overline{W^{(s)}_{1,n_s-1}}$. Thus, assume $\mathrm{Hom}(U_0, W^{(s)}_{1,n_s-1}) = 0$ for all $s \neq j$. Then the indecomposable direct summands of U_0 belong to the wing θ_s (and to $\mathcal{X}$). Since U has no submodule belonging to $\mathcal{X}^0$, it follows that $U_\omega = 0$, therefore there is a non-zero homomorphism $f_0 : U_0 \longrightarrow W_0$, and it induces a non-zero map $f : U \longrightarrow W_0(\rho)$.

By induction on the length of U, where $U \in \mathcal{U}(\mathcal{X}) \setminus \overline{\mathcal{X}^0}$, we want to show that U is cogenerated by $\mathcal{T}'$. We have seen above, that there exists $0 \neq f : U \longrightarrow T$ with $T \in \mathcal{T}'$. Let $U' = \mathrm{Ker}(f)$. Then again $U' \in \mathcal{U}(\mathcal{X}) \setminus \overline{\mathcal{X}^0}$, then, by induction U' is cogenerated by $\mathcal{T}'$, say there exists a monomorphism $f' : U' \longrightarrow T'$ with $T' \in \mathcal{T}'$. Forming the induced exact sequence and using $\mathrm{inj.dim.}T' \leq 1$, we obtain the following commutative diagram with exact rows:

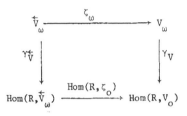

$$\begin{array}{ccccccccc}
0 & \longrightarrow & U' & \longrightarrow & U & \longrightarrow & U/U' & \longrightarrow & 0 \\
 & & \downarrow & & \downarrow & & \| & & \\
0 & \longrightarrow & T' & \longrightarrow & U'' & \longrightarrow & U/U' & \longrightarrow & 0 \\
 & & \| & & \downarrow & & \downarrow & & \\
0 & \longrightarrow & T' & \longrightarrow & T'' & \longrightarrow & T & \longrightarrow & 0
\end{array}$$

Since $\mathcal{T}'$ is closed under extensions, $T'' \in \mathcal{T}'$.

Now, consider the general case of $U \in \mathcal{P}$. Since $U \in \mathcal{P}^*$, there is a submodule P of U with $P \in \mathcal{P}^o$ and $U/P \in \mathcal{U}(\mathcal{X})$. Since $\mathcal{T}^o$ separates $\mathcal{P}^o$ from $\mathcal{Q}^o$, there exists a monomorphism $g' : P \longrightarrow T$ with $T \in \mathcal{T}(\rho)$. Since $U/P \in \mathcal{U}(\mathcal{X}) \subseteq \mathcal{P} \vee \overline{\mathcal{X}^o}$, and $\mathrm{Ext}^1(\mathcal{P},\mathcal{T}) = 0$, $\mathrm{Ext}^1(\overline{\mathcal{X}^o},\mathcal{T}(\rho)) = 0$, it follows that g' can be extended to U, thus there is $g : U \longrightarrow T$ with $\mathrm{Ker}(g) \cap P = 0$. Now $\mathrm{Ker}(g)$ belongs to $\mathcal{U}(\mathcal{X}) \setminus \mathcal{X}^o$, thus by the previous considerations, $\mathrm{Ker}(g)$ is cogenerated by $\mathcal{T}'$. Using again that $\mathrm{inj.dim.}T \leq 1$ for $T \in \mathcal{T}'$, we see that U is cogenerated by $\mathcal{T}'$.

This shows that $\vec{U}$ belongs to $\mathcal{P}^o$. A similar proof shows that for $V \in \mathcal{Q}$, the module $\overleftarrow{V}$ belongs to $\mathcal{Q}^o$.

(18) Any homomorphism $\mathcal{P} \longrightarrow \mathcal{Q}$ factors through $\mathcal{T}^o(\rho)$, for any $\rho \in \mathbb{P}\,\mathrm{Hom}(R,W_o)$.

<u>Proof.</u> Let $U \in \mathcal{P}$, $V \in \mathcal{Q}$, $f : U \longrightarrow V$. We can factor $f_o : U_o \longrightarrow V_o$ through a direct sum of copies of W_o, say $f_o = g_o h_o$. Now g_o factors through $\eta_o : U_o \longrightarrow W_o \otimes \mathrm{Hom}(U_o,W_o)$, and h_o factors through $\zeta_o : W_o \otimes \mathrm{Hom}(W_o,V_o) \longrightarrow V_o$, thus $f_o = \eta_o f'_o \zeta_o$ for some $f'_o : \vec{U}_o = W_o \otimes \mathrm{Hom}(U_o,W_o) \longrightarrow W_o \otimes \mathrm{Hom}(W_o,V_o) = \overleftarrow{V}_o$. Since

$$\gamma_U \mathrm{Hom}(R,\eta_o)\mathrm{Hom}(R,f'_o)\mathrm{Hom}(R,\zeta_o) = \gamma_U \mathrm{Hom}(R,\eta_o f'_o \zeta_o)$$

$$= \gamma_U \mathrm{Hom}(R,f_o) = f_\omega \gamma_V,$$

we can factor $\gamma_U \mathrm{Hom}(R,\eta_o)\mathrm{Hom}(R,f'_o)$ and f_ω through the pullback $\overleftarrow{V}_\omega$ of $\mathrm{Hom}(R,\zeta_o)$ and γ_V, thus we obtain f'_ω such that the following diagram commutes

$$\begin{array}{ccccccc}
U_\omega & \xrightarrow{\;1\;} & \vec{U}_\omega & \xrightarrow{\;f'_\omega\;} & \overleftarrow{V}_\omega & \xrightarrow{\;\zeta_\omega\;} & V_\omega \\
\downarrow{\scriptstyle \gamma_U} & & \downarrow{\scriptstyle \gamma_U \mathrm{Hom}(R,\eta_o)} & & \downarrow{\scriptstyle \gamma_{\overleftarrow{V}}} & & \downarrow{\scriptstyle \gamma_V} \\
\mathrm{Hom}(R,U_o) & \xrightarrow{\mathrm{Hom}(R,\eta_o)} & \mathrm{Hom}(R,\vec{U}_o) & \xrightarrow{\mathrm{Hom}(R,f'_o)} & \mathrm{Hom}(R,\overleftarrow{V}_o) & \xrightarrow{\mathrm{Hom}(R,\zeta_o)} & \mathrm{Hom}(R,V_o)
\end{array}$$

Thus we have factored $f : U \longrightarrow V$ as $f = \eta f' \zeta$ with $f' : \vec{U} \longrightarrow \overleftarrow{V}$. Since $\vec{U} \in \mathcal{P}^o$, $\overleftarrow{V} \in \mathcal{Q}^o$, we can factor f' through $\mathcal{T}^o(\rho)$, for any $\rho \in \mathbb{P}\,\mathrm{Hom}(R,W_o)$.

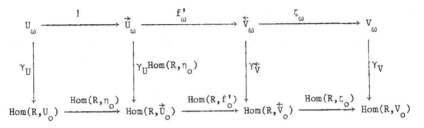

(19) Let $\rho \in \mathbb{P}\,\mathrm{Hom}(R,W_o)$. Any module in $\mathcal{P}$ is cogenerated by $\mathcal{T}^o(\rho)$, any module in $\mathcal{Q}$ is generated by $\mathcal{T}^o(\rho)$.

Proof. Let U in $\mathcal{P}$. Let $f : U \longrightarrow Q$ be an embedding of U into an injective module. According to (7), Q belongs to $\mathcal{Q}$. Factorizing f through $\mathcal{T}^o(\rho)$, we obtain an embedding of U into a module in $\mathcal{T}^o(\rho)$. Similarly, for V in $\mathcal{Q}$, let $g : P \longrightarrow V$ be an epimorphism with P projective. Then $P \in \mathcal{P}$, so we can factor g through $\mathcal{T}^o(\rho)$.

(20) $\mathrm{Ext}^1(\mathcal{P},\mathcal{Q}) = 0$.

Proof. Let $U \in \mathcal{P}$, $V \in \mathcal{Q}$. Choose a monomorphism $f : U \longrightarrow T$ with $T \in \mathcal{T}$. Since $U_o \in \mathcal{X} \vee \langle W_o \rangle$, and $V_o \in \langle W_o \rangle \vee \mathcal{Y}$, we have $\mathrm{Ext}^1(U_o,V_o) = 0$, thus $\mathrm{Ext}^1(U,V) = \mathrm{Ext}^1_{\mathcal{K}}(U,V)$. According to 2.5 (3'), the induced map $\mathrm{Ext}^1_{\mathcal{K}}(f,V) : \mathrm{Ext}^1_{\mathcal{K}}(T,V) \longrightarrow \mathrm{Ext}^1_{\mathcal{K}}(U,V)$ is surjective. However, $\mathrm{Ext}^1_{\mathcal{K}}(T,V) \subseteq \mathrm{Ext}^1(T,V) = 0$, since $T \in \mathcal{T}$, $V \in \mathcal{Q}$.

(21) $A\text{-mod} = \mathcal{P} \vee \mathcal{T} \vee \mathcal{Q}$.

Proof. We show that $A\text{-mod} = \mathcal{P} \int \mathcal{T} \int \mathcal{Q}$, the assertion then follows from (15) and (18). Clearly, $A\text{-mod} = \mathcal{U}(\mathcal{X}) \int \mathcal{U}(\langle W_o \rangle \vee \mathcal{W}^o) \int \check{\mathcal{U}}(\mathcal{Y})$, and $\mathcal{U}(\mathcal{W}^o) = \mathcal{W}^o$, since $\mathrm{Hom}(R,\mathcal{W}^o) = 0$. Also $\mathcal{U}(\langle W_o \rangle) = \mathcal{P}^o \vee \mathcal{T}^o \vee \mathcal{Q}^o$, thus

$$A\text{-mod} = \mathcal{U}(\mathcal{X}) \int \mathcal{P}^o \int (\mathcal{T}^o \vee \mathcal{W}^o) \int \mathcal{Q}^o \int \check{\mathcal{U}}(\mathcal{Y}).$$
$$= \mathcal{P}^* \int (\mathcal{T}^o \vee \mathcal{W}^o) \int \mathcal{Q}^*.$$

Now $\mathcal{P}^* = \mathcal{P} \vee \overline{\mathcal{X}^o}$, $\mathcal{Q}^* = \mathcal{Q} \vee \mathcal{Y}^o$, thus

$$A\text{-mod} = \mathcal{P} \int \overline{\mathcal{X}^o} \int (\mathcal{T}^o \vee \mathcal{W}^o) \int \mathcal{Y}^o \int \mathcal{Q},$$

and $\overline{\mathcal{X}^o} \int (\mathcal{T}^o \vee \mathcal{W}^o) \int \mathcal{Y}^o = (\overline{\mathcal{X}^o} \int \mathcal{T}^o \int \mathcal{Y}^o) \vee \mathcal{W}^o = \mathcal{T}$.

(22) $\mathcal{P} = \mathcal{P}_w$, $\mathcal{T} = \mathcal{T}_w$, $\mathcal{Q} = \mathcal{Q}_w$.

Proof. We only have to show that $\mathcal{P} \subseteq \mathcal{P}_w$, $\mathcal{T} \subseteq \mathcal{T}_w$, $\mathcal{Q} \subseteq \mathcal{Q}_w$, since we already know that $A\text{-mod} = \mathcal{P} \vee \mathcal{T} \vee \mathcal{Q}$. Let $0 \neq P \in \mathcal{P}$. Now $\mathcal{P}$ is cogenerated by any $\mathcal{T}^o(\rho)$, thus take a non-zero map $\varphi : P \longrightarrow T'$ with $T' \in \mathcal{T}^o(\rho)$. We have $\mathrm{Hom}(P,T') \neq 0$, whereas $\mathrm{Ext}^1(P,T') = 0$ according to (16). Note that $\underline{\dim}\, T' = tw$, for some $t \geq 1$. Since proj.dim. $P \leq 1$.

$$\begin{aligned}\iota_w(\underline{\dim}\, P) &= \langle w,\underline{\dim}\, P\rangle = -\langle \underline{\dim}\, P,w\rangle \\ &= -\frac{1}{t}\langle \underline{\dim}\, P,\underline{\dim}\, T'\rangle \\ &= -\frac{1}{t}(\dim \mathrm{Hom}(P,T') - \dim \mathrm{Ext}^1(P,T)) < 0.\end{aligned}$$

Similarly, for $Q \in \mathcal{Q}$, we use that Q is generated by some $T'' \in \mathcal{T}^o(\rho)$, that $\mathrm{Ext}^1(T'',Q) = 0$ and that inj.dim. $Q \leq 1$, in order to see that $\iota_w(\underline{\dim}\, Q) > 0$.

Finally, for showing $\iota_w(\underline{\dim}\, T) = 0$ for $T \in \mathcal{T}$, choose some $\rho \in \mathbb{P}\,\mathrm{Hom}(R,W_o)$ with no indecomposable direct summand of T belonging to $\mathcal{T}(\rho)$. Then $\mathrm{Hom}(W_o(\rho),T) = 0 = \mathrm{Ext}^1(W_o(\rho),T)$. Using again that inj.dim. $T \leq 1$, the assertion follows.

Recall that we have denoted by $\eta_M : M \longrightarrow W_o \otimes \mathrm{Hom}(M,W_o)$ the universal map from M to $\langle W_o\rangle$.

(23) For $M \in \mathcal{X}$, the map η_M is a monomorphism with cokernel in $\mathcal{Y}^o$.

Proof. Since the modules M in $\mathcal{X}$ are cogenerated by W_o, η_M is mono. Let $Y = \mathrm{Cok}\, \eta_M$. Let $n = \dim \mathrm{Hom}(M,W_o)$, thus $W_o \otimes \mathrm{Hom}(M,W_o) = W_o^n$. The exact sequence

$$0 \longrightarrow M \xrightarrow{\eta_M} W_o^n \longrightarrow Y \longrightarrow 0$$

gives rise to an exact sequence

$$\mathrm{Hom}(W_o^n,W_o) \xrightarrow{\mathrm{Hom}(\eta_M,W_o)} \mathrm{Hom}(M,W_o) \longrightarrow \mathrm{Ext}^1(Y,W_o) \longrightarrow \mathrm{Ext}^1(W_o^n,W_o).$$

The last term is zero, since $\mathrm{Ext}^1(W_o,W_o) = 0$, and $\mathrm{Hom}(\eta_M,W_o)$ is surjective, due to the universality of the map η_M. Thus $\mathrm{Ext}^1(Y,W_o) = 0$, and therefore $\mathrm{Hom}(\tau^- W_o,Y) = 0$. On the other hand, Y is generated by W_o, thus $Y \in \langle W_o\rangle \vee \mathcal{Y}$. Also, Y does not split off a copy of W_o, therefore $Y \in \mathcal{Y}$. However, the only modules in $\mathcal{Y}$ satisfying $\mathrm{Hom}(\tau^- W_o,Y) = 0$ are those in $\mathcal{Y}^o$.

Let $e_1^{(s)} = \underline{\dim}\, \overline{W_{11}^{(s)}}$, and, for $2 \leq i \leq n_s$, let $e_i^{(s)} = \underline{\dim}\, W_{ii}^{(s)}$. Actually, define $e_i^{(s)}$ for all $i \in \mathbb{Z}$ according to $e_i^{(s)} = e_{i'}^{(s)}$ provided $i \equiv i' \pmod{n_s}$. Also, for $1 \leq i \leq n_s$, and all $\ell \geq 0$, let $e_i^{(s)}[\ell] = \sum_{r=i}^{i+\ell-1} e_r^{(s)}$.

(24) The elements w_o and $e_i^{(s)}$, $2 \leq i \leq n_s$, $1 \leq s \leq t$, form a $\mathbb{Z}$-basis of $K_o(A_o)$.

Proof. If P is an indecomposable projective A_o-module, then P belongs to $\mathcal{X} \vee \langle W_o \rangle$, thus there is an exact sequence

$$0 \longrightarrow P \longrightarrow W_o^n \longrightarrow Y \longrightarrow 0$$

with $Y \in \mathcal{Y}^o$, thus $\underline{\dim}\, P = nw_o - \underline{\dim}\, Y$. However, $\underline{\dim}\, Y$ is an integral linear combination of dimension vectors $e_i^{(s)}$, with $2 \le i \le n_s$, $1 \le s \le t$ [note that

$\underline{\dim}\, W_{in_s}^{(s)} = e_i^{(s)}[n_s - i + 1] = \sum_{r=i}^{n_s} e_r^{(s)}$ for $2 \le i \le n_s$]. Thus the elements $e_i^{(s)}$, $2 \le i \le n_s$, $1 \le s \le t$, together with w_o, generate $K_o(A_o)$ as an abelian group. On the other hand, the bilinear form $\langle -,- \rangle$ evaluated on these elements, gives a lower triangular matrix with main diagonal entries equal to 1; thus, these elements are linearly independent.

(25) The elements w and $e_i^{(s)}$, $2 \le i \le n_s$, $1 \le s \le t$, form a $\mathbb{Z}$-basis of $\operatorname{Ker} \iota_w$.

Proof. Clearly, these elements are linearly independent and belong to $\operatorname{Ker} \iota_w$. Since these elements together with $e(\omega)$ generate all of $K_o(A)$, and ι_w is a non-zero homomorphism from $K_o(A)$ to $\mathbb{Z}$, the assertion follows.

Let K_s be the subspace of $K_o(A)$ generated by the elements $e_i^{(s)}$, $2 \le i \le n_s$. The previous assertion asserts

$$\operatorname{Ker} \iota_w = \langle w \rangle \oplus \bigoplus_{1 \le s \le t} K_s.$$

This decomposition is compatible with the quadratic form χ_A. In fact:

(26) The subspaces $\langle w \rangle$, K_s, $1 \le s \le t$, are pairwise orthogonal with respect to χ_A. The subspace $\langle w \rangle$ is the radical of $\chi_A | \operatorname{Ker} \iota_w$, and $\chi_A | K_s$ is of the form $\mathbb{A}_{n_s - 1}$, using the basis $e_2^{(s)}, \ldots, e_{n_s}^{(s)}$.

Proof. We know that w belongs to the radical of χ_A, thus $\langle w \rangle$ is orthogonal to all of $K_o(A)$. Also, it is easy to check that for $2 \le i \le n_s$, $2 \le j \le n_t$,

$$(e_i^{(s)}, e_j^{(t)}) = \begin{cases} 1 & s = t,\ i = j \\ -\frac{1}{2} & s = t,\ |i-j| = 1 \\ 0 & \text{otherwise,} \end{cases}$$

using (1), (2), and (10) [calculating, in fact, $\langle e_i^{(s)}, e_j^{(t)} \rangle$.] It follows that the subspaces K_s are pairwise orthogonal, and that the restriction of χ_A to K_s is of the form $\mathbb{A}_{n_s-1}$. As a consequence, $\bigoplus_s K_s$ contains no non-zero radical element for χ_A, thus $\langle w\rangle$ is the radical of $\chi_A|\mathrm{Ker}\,\iota_w$.

(27) T_w is controlled by the restriction of χ_A to $\mathrm{Ker}\,\iota_w$.

Proof. Let $x \in \mathrm{Ker}\,\iota_w$ be a positive root of $\chi = \chi_A$. Subtracting a suitable multiple of w from x, we obtain an element y in $\bigoplus_s K_s$; in fact, $y = x - x_\omega w$. Since w belongs to the radical of χ, we have $\chi(y) = \chi(x)$, thus y is again a root. However, $\chi \mid \bigoplus_s K_s$ is positive definite, thus either $y = 0$, or else $\chi(y) = 1$. For $y = 0$, x is a multiple of w, and is the dimension vector of the (indecomposable) representations $W_0(\rho)[x_\omega]$, with $\rho \in \mathbb{P}\,\mathrm{Hom}(R,W_0)$, and also of the (indecomposable) representations $W_{ii}^{(s)}[n_s x_\omega]$, $2 \leq i \leq n_s$, $1 \leq s \leq t$, (of course, also $\underline{\dim}\,\overline{W_{11}^{(s)}}[n_s x_\omega] = x$, however these modules $\overline{W_{11}^{(s)}}[n_s x_\omega]$ are of the form $W_0(\rho)[x_\omega]$, so we have already taken into account these modules).

Thus, it remains to consider the case $\chi(x) = 1$. In this case, y belongs to some K_s, and is of the form $y = \pm e_i^{(s)}[\ell']$, for some $2 \leq i \leq n_s$, and $1 \leq \ell' \leq n_s-i+1$ (since these are the roots in $\bigoplus_s K_s$). Now, if y is a positive root, then

$$x = x_\omega w + y = e_i^{(s)}[\ell'+n_s x_\omega] = \underline{\dim}\,W_{ii}^{(s)}[\ell'+n_s x_\omega].$$

On the other hand, let y be a negative root, say $y = -e_i^{(s)}[\ell']$ with $2 \leq i \leq n_s$, then $x_\omega \geq 1$ (since otherwise x is not positive), and $x = x_\omega w - e_i^{(s)}[\ell'] = e_{i+\ell'}^{(s)}[n_s x_\omega - \ell']$. Thus, for $\ell' \leq n_s - i$, we see that $x = \underline{\dim}\,W_{ii}^{(s)}[n_s x_\omega - \ell']$, whereas for $\ell' = n_s - i + 1$, we have $x = \underline{\dim}\,\overline{W_{11}^{(s)}}[n_s x_\omega - \ell']$. Altogether, we see that $\underline{\dim}$ induces a bijection between the positive elements x in $\mathrm{Ker}\,\iota_w$ with $\chi(x) = 1$, and the indecomposable modules of the form $\overline{W_{11}^{(s)}}[\ell]$ or $W_{ii}^{(s)}[\ell]$, with $2 \leq i \leq n_s$, $1 \leq s \leq t$, and $\ell \not\equiv 0 \pmod{n_s}$.

This finishes the proof of the theorem.

Remark. We keep the assumptions of the theorem. Assume, in addition, that $\mathbb{T}_{n_1,\dots n_t}$ is neither a Dynkin nor a Euclidean diagram. Then both $\mathcal{P}_w$ and $\mathcal{Q}_w$ are strictly wild.

Proof. It is well-known (and easy to see) that $\mathcal{U}(X^0)$ is strictly wild, and the image of any full exact embedding

$$k\langle T_1,T_2\rangle\text{-mod} \longrightarrow \mathcal{U}(X^0)$$

lies in $\mathcal{U}(\mathcal{X}^{o}) \setminus \overline{\mathcal{X}^{o}}$ (namely, any indecomposable $k\langle T_1,T_2\rangle$-module M satisfies $\mathrm{Ext}^1(M,M) \neq 0$). Thus $\mathcal{P}_w$ is strictly wild. Similarly, $\mathcal{U}(\mathcal{Y}^{o}) \setminus \mathcal{Y}^{o}$ is strictly wild.

3.5 The operation of Φ_A on $K_o(A)$.

Let $\Phi = \Phi_A$. We will use the notation and results of the previous section. Recall that $\text{Ker}\ \iota_w$ has the following generating set: $w, e_i^{(s)}$, $1 \leq i \leq n_s$, $1 \leq s \leq t$, with relations $\sum_{i=1}^{n_s} e_i^{(s)} = w$, for all $1 \leq s \leq t$, and that Φ operates on these elements as follows:

$$w\Phi = w, \quad e_i^{(s)}\Phi = e_{i-1}^{(s)} \quad \text{for all } i,s \quad (\text{with } e_o^{(s)} = e_{n_s}^{(s)}),$$

thus, $\langle w \rangle$ and $\text{Ker}\ \iota_w$ (and also the subgroups $\langle e_1^{(s)},\ldots,e_{n_s}^{(s)}\rangle = \langle w\rangle \oplus K_s$) are Φ-invariant. We denote by d the lowest common multiple of $n_1,\ldots n_t$. Then we have

(a) Φ operates trivially on $\langle w\rangle$, and periodically, with minimal period d, on $\text{Ker}\ \iota_w$.

Similarly, we show:

(b) Φ operates trivially on $K_o(A)/\text{Ker}\ \iota_w$, and periodically, with minimal period d, on $K_o(A)/\langle w\rangle$.

Proof. For any $x \in K_o(A)$, we have

$$\langle w,x\Phi\rangle = \langle w\Phi,x\Phi\rangle = \langle w,x\rangle,$$

using that w is Φ-invariant and that the bilinear form $\langle -,-\rangle$ is Φ-invariant. Also, let $y = x\Phi - x$. Then

$$x\Phi^d - x = \sum_{j=o}^{d-1} (x\Phi - x)\Phi^j = \sum_{j=o}^{d-1} y\Phi^j$$

is an element of $\text{Ker}\ \iota_w$ which is Φ-invariant, thus in $\langle w\rangle$. This shows that Φ operates periodically on $K_o(A)/\langle w\rangle$ with period d. This period is minimal, since the minimal period of Φ even on $\text{Ker}\ \iota_w/\langle w\rangle$ is d.

We define the w-defect ∂_w on $K_o(A)$ by $\partial_w(x)\circ w = x\Phi^d - x$, for $x \in K_o(A)$. In this way, we obtain a linear form $\partial_w : K_o(A) \longrightarrow \mathbb{Z}$.

Proposition. $\partial_w = (2 - \sum_{s=1}^{t} (1 - \frac{1}{n_s}))d\cdot\iota_w$

Proof. Since Φ^d is the identity on $\operatorname{Ker} \iota_w$, we see that $\operatorname{Ker} \iota_w \subseteq \operatorname{Ker} \partial_w$, thus ∂_w is a scalar multiple of ι_w, say $\partial_w = a\iota_w$, and we have to determine a. We evaluate ι_w and ∂_w on $p(\omega) = \dim P(\omega)$. For ι_w, we obtain:

$$\iota_w(p(\omega)) = \langle w, p(\omega)\rangle = -\langle p(\omega), w\rangle = -1 .$$

In order to determine $\partial_w(p(\omega))$, we first show:

$$\underline{\dim}\, R = 2w_0 - e ,$$

where
$$e = \sum_{s=1}^{t} \sum_{i=2}^{n_s} e_i^{(s)} .$$

For the proof, we note that the source map for W_0 in A_0-mod is of the form $W_0 \longrightarrow \bigoplus_s W_{2n_s}^{(s)}$, and $e = \underline{\dim} \bigoplus_s W_{2n_s}^{(s)}$. Now, if W_0 is not injective, there is an exact sequence

$$0 \longrightarrow W_0 \longrightarrow \bigoplus_s W_{2n_s}^{(s)} \longrightarrow \tau_{A_0}^- W_0 \longrightarrow 0,$$

and, since $\text{inj.dim.}_{A_0} W_0 \leq 1$, and $\operatorname{Hom}(I, W_0) = 0$, for any indecomposable injective A_0-module I, we have

$$w_0 \Phi_0^{-1} = \underline{\dim}\, \tau_{A_0}^- W_0 = -w_0 + e .$$

On the other hand, if W_0 is injective, then the socle of W_0 is a simple projective A_0-module P, since $\operatorname{Hom}(I, W_0) = 0$ for any indecomposable injective A_0-module $I \neq W_0$, and $\operatorname{End}(W_0) = k$, thus we have the exact sequence

$$0 \longrightarrow P \longrightarrow W_0 \longrightarrow \bigoplus_s W_{2n_s}^{(s)} \longrightarrow 0,$$

and $\nu P = W_0$, therefore

$$w_0 \Phi_0^{-1} = -\underline{\dim}\, P = -w_0 + e.$$

In any case, we have

$$\underline{\dim}\, R = w_0(I - \Phi_0^{-1}) = 2w_0 + e.$$

Now $\nu P(\omega)$ is the simple injective module $E(\omega)$, thus $p(\omega)\Phi = -e(\omega)$, whereas $p(\omega) = \underline{\dim}\, R + e(\omega)$, thus

$$p(\omega)\Phi - p(\omega) = - \dim R - 2e(\omega)$$

$$= - 2w_0 + e - 2e(\omega) = - 2w + e .$$

Let us determine $\sum_{j=o}^{d-1} e\Phi^j$. For any $e_i^{(s)}$, we have

$$\sum_{j=o}^{d-1} e_i^{(s)}\Phi^j = \frac{d}{n_s} \sum_{i=1}^{n_s} e_i^{(s)} = \frac{d}{n_s} w ,$$

thus

$$\sum_{j=o}^{d-1} e\Phi^j = \sum_{j=o}^{d-1} \sum_{s=1}^{t} \sum_{i=2}^{n_s} e_i^{(s)}\Phi^j$$

$$= \sum_{s=1}^{t} \sum_{i=2}^{n_s} \frac{d}{n_s} w = \sum_{s=1}^{t} \frac{n_s - 1}{n_s} dw.$$

Altogether, we see

$$\partial_w(p(\omega)) = \sum_{j=o}^{d-1} (p(\omega)\Phi - p(\omega))\Phi^j =$$

$$= \sum_{j=o}^{d-1} (-2w + e)\Phi^j = - 2dw + \sum_{s=1}^{t} \frac{n_s - 1}{n_s} dw$$

$$= -(2 - \sum_{s=1}^{t} \frac{n_s - 1}{n_s}) dw ,$$

and therefore $$a = \frac{\partial_w(p(\omega))}{\iota_w(p(\omega))} = (2 - \sum_{s=1}^{t} \frac{n_s - 1}{n_s}) d .$$

Always, we will assume that the wings are ordered in such a way that $n_1 \geq n_2 \geq \dots \geq n_t$. We note the following:
The factor $2 - \sum_{s=1}^{t} \frac{n_s - 1}{n_s}$ is positive if and only if $t \leq 2$, or $t = 3$ and $n_1 = 3,4$, or 5, and $n_2 = 3$, $n_3 = 2$ (the case of $\mathbb{T}_{n_1,\dots,n_t}$ being a Dynkin diagram, namely $\mathbb{A}_n, \mathbb{D}_n, \mathbb{E}_6, \mathbb{E}_7$, or $\mathbb{E}_8$). The factor $2 - \sum_{s=1}^{t} \frac{n_s - 1}{n_s}$ is zero if an only if $(n_1,\dots,n_t)$ is one of $(2,2,2,2)$, $(3,3,3)$, $(4,4,2)$, or $(6,3,2)$ (the case of $\mathbb{T}_{n_1,\dots,n_t}$ being a Euclidean diagram, namely $\tilde{\mathbb{D}}_4, \tilde{\mathbb{E}}_6, \tilde{\mathbb{E}}_7$ or $\tilde{\mathbb{E}}_8$). Thus, we have the following corollaries:

Corollary 1. Φ operates periodically on $K_o(A)$ if and only if $\mathbb{T}_{n_1,\ldots,n_t}$ is one of $\widetilde{\mathbb{D}}_4$, $\widetilde{\mathbb{E}}_6$, $\widetilde{\mathbb{E}}_7$ or $\widetilde{\mathbb{E}}_8$, and in these cases, the minimal period is $d = 2,3,4$, or 6, respectively.

Corollary 2. The radical of χ is one-dimensional except in the cases of $\mathbb{T}_{n_1,\ldots,n_t}$ being of the form $\widetilde{\mathbb{D}}_4$, $\widetilde{\mathbb{E}}_6$, $\widetilde{\mathbb{E}}_7$, $\widetilde{\mathbb{E}}_8$. In these exceptional cases, the radical of χ is two-dimensional.

3.6 Tame hereditary algebras

As a first application of the main theorem, we are going to determine the Auslander-Reiten quiver of the path algebras A of quivers with underlying graph $\tilde{\mathbb{A}}_n$, $\tilde{\mathbb{D}}_n$, $\tilde{\mathbb{E}}_6$, $\tilde{\mathbb{E}}_7$, $\tilde{\mathbb{E}}_8$. Almost always, such an algebra or its opposite is obtained from the path algebra A_o of a quiver with underlying graph $\mathbb{A}_n$, $\mathbb{D}_n$, $\mathbb{E}_6$, $\mathbb{E}_7$, $\mathbb{E}_8$ as a one-point extension $A = A_o[R]$, where R is a projective module, and such that R dominates the (unique) indecomposable A_o-module M with $\underline{\dim}\, M$ being maximal. Actually, this module M turns out to be a sincere directing wing module, and, in this way, we can apply the main theorem. Thus, let us consider the A_o-module M with $\underline{\dim}\, M$ being maximal in more detail. Actually, we may consider in the same way a more general case.

Let B be a sincere, directed algebra. We recall from 2.4.9 that $\underline{\dim}$ furnishes a bijection between the indecomposable B-modules and the positive roots of χ_B. An indecomposable B-module will be said to be <u>maximal</u> provided $\underline{\dim}$ B is a maximal root. In dealing with a maximal module M, it will be of interest to know whether M is the only sincere indecomposable module, whether $\underline{\dim}\, M$ has one or two exceptional vertices (note that they correspond to composition factors in M) and to determine the position of the τ-orbit of M in the orbit quiver of B. The general investigation will be done in section 6.1, here we content ourself with two results which are sufficient for the consideration of the tame hereditary algebras. First, we show that a sincere maximal module always is dominated by a projective module.

(1) <u>Let</u> B <u>be a directed algebra,</u> ${}_BM$ <u>a sincere maximal module. Then</u> M <u>is dominated by the projective</u> B-module $R = \bigoplus_a P(a)^{d_a}$, <u>with</u> $d_a = D_a\chi\,(\underline{\dim}\, M)$.

<u>Proof.</u> First we note that $d_a \geq 0$ for all a, since $\underline{\dim}\, M$ is a maximal positive root, see 1.1.6, thus R is defined. Let $m = \underline{\dim}\, M$, $r = \underline{\dim}\, R = \sum_a d_a p(a)$. We have

$$\langle r, e(b)\rangle = \sum_a d_a \langle p(a), e(b)\rangle = d_b = 2(m, e(b)),$$

for all b, thus the linear forms $\langle r, -\rangle$ and $2(m,-)$ coincide. According to 2.5.11, we see that $r = m(I - \Phi_B^{-1})$. Since R is projective and proj.dim $M \leq 1$, we see that also proj.dim.$M(\rho) \leq 1$ for any $\rho : R \longrightarrow M$. Thus R dominates M.

Let $I = (I_o, I_1)$ be a graph. The vertex a of I is called an <u>end-point</u> provided a has precisely one neighbor b, there is precisely one edge a — b, and $a \neq b$. The vertex a of I is said <u>to cut off a branch</u> I' <u>of length</u> n from I provided I is obtained from $I' \sqcup I''$, where $I' = \mathbb{A}_n$ with a being one of its endpoints, and I" an arbitrary graph, by adding a single edge connecting a with some vertex in I". Thus, in this case, I has the form

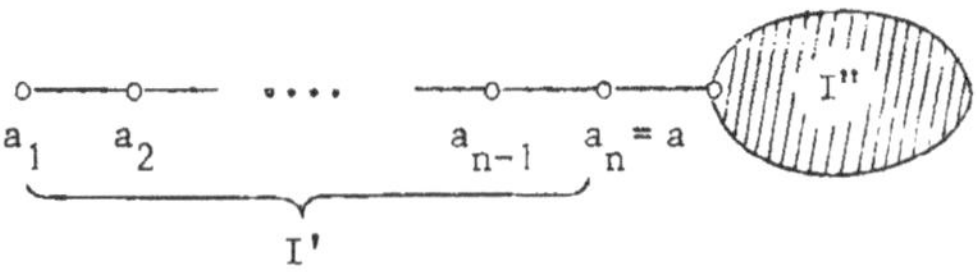

Of course, a vertex a of I is an endpoint of I if and only if it cuts off a branch of length 1.

(2) Let B be a directed algebra, and M a sincere maximal B-module. Assume the τ-orbit of M cuts off a branch $\mathcal{O}'$ of length n from the orbit graph of B. If n = 1, then M is projective-injective, say M = P(a) = Q(b). If $n \geq 2$, then [M] has a wing θ of length n, such that the τ-orbits of the modules in θ are the vertices of $\mathcal{O}'$. Also, the source of θ is of the form [P(a)], the sink of θ is of the form [Q(b)]. Always, $a \neq b$, and these are the exceptional vertices of **dim** M. Also, M is the only indecomposable module X with $\mathrm{Hom}(P(a),X) \neq 0 \neq \mathrm{Hom}(X,Q(b))$, in particular, M is the only sincere indecomposable module.

Proof. Let m = **dim** M. Consider first the case n = 1. In this case, the τ-orbit of M is an endpoint of the orbit quiver of B. In case there would not exist an irreducible map $X \longrightarrow M$ with X indecomposable, M would be simple projective, but then M can only be sincere in case B being a simple algebra, against our assumption that the orbit graph of B contains at least one edge. Thus, choose an irreducible map $f : X \longrightarrow M$ with X indecomposable. Now f cannot be an epimorphism, since **dim** X is a root and **dim** M is a maximal root. Thus f is mono. Since $f : X \longrightarrow M$ is the sink map for M, we conclude that M is projective, say M = P(a). Dually, M is also injective, say M = Q(b). If X is an indecomposable module with $\mathrm{Hom}(P(a),X) \neq 0$, $\mathrm{Hom}(X,Q(b)) \neq 0$, and $X \not\approx M$, then we obtain a cycle of length 2, impossible. Thus M is the only indecomposable module X with $\mathrm{Hom}(P(a),X) \neq 0$, $\mathrm{Hom}(X,Q(b)) \neq 0$.

Now, assume $n \geq 2$. Since M is a sincere indecomposable module, we conclude that $\Gamma(B)$ contains a full convex subquiver of the form

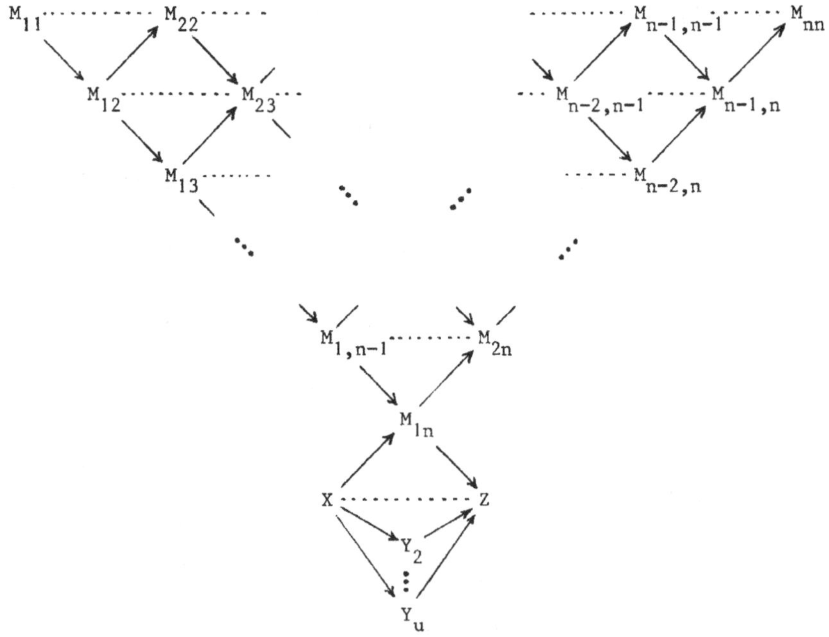

where $M_{1n} \approx M$, where u is the number of neighbors of the τ-orbit of $[X]$ in $\overline{\mathcal{O}(B)}$, and where the τ-orbit of the modules M_{ij}, $1 \leq i \leq j \leq n$, are the vertices of $\mathcal{O}'$. Of course, we see that in this way, we obtain a wing for $[M]$ of length n.
We denote, for $2 \leq i \leq n$, by $\mu_i : M_{1,i-1} \longrightarrow M_{1i}$ some irreducible map. First, assume some M_{1t}, $2 \leq t \leq n$, is projective. Then also $M_{1,t-1}$, and therefore, by induction, also M_{11}, is projective. [Namely, if M_{1t} is projective, then there is at most one arrow in $\Gamma(B)$ ending in $[M_{1,t-1}]$, and none in case $t = 2$, due to the fact that the τ-orbit of $[M_{1t-1}]$ has precisely two neighbors in $\mathcal{O}(B)$ in case $t > 2$, and one in case $t = 2$, and one of these neighbors is the τ-orbit of $[M_{1t}]$. However, for $t > 2$, there is the irreducible map $\mu_{t-1} : M_{1,t-2} \longrightarrow M_{1,t-1}$ which therefore has to be a sink map for $M_{1,t-1}$. Since μ_{t-1} is mono, we see that $M_{1,t-1}$ is projective. Similarly, for $t = 2$, we conclude that $0 \longrightarrow M_{11}$ is a sink map, thus M_{11} is (simple) projective.] Now assume, no M_{1j}, $2 \leq j \leq n$, is projective. Also in this case, we want to show that M_{11} is projective. Denote an irreducible map $X \longrightarrow M_{1n}$ by f_n. Note that f_n has to be mono, since $\underline{\dim}\, M$ is a maximal root, and $\underline{\dim}\, X$ is some root, comparable with $\underline{\dim}\, M$. Using going down induction, we construct an irreducible monomorphism $f_i : \tau M_{1,i+1} \longrightarrow M_{1i}$ for all $1 \leq i \leq n$. Namely, assume there is given an irreducible monomorphism $f_i : \tau M_{1,i+1} \longrightarrow M_{1i}$, $2 \leq i \leq n$. Now, the τ-orbit of $[M_{1i}]$ has precisely two neighbors in $\mathcal{O}(B)$, both connected to it by a single arrow, thus the Auslander-Reiten

sequence ending in M_{1i} is given by

$$0 \longrightarrow \tau M_{1i} \xrightarrow{[f_{i-1}\ *]} M_{1,i-1} \oplus \tau M_{1,i+1} \xrightarrow{\begin{bmatrix}\mu_i\\ f_i\end{bmatrix}} M_{1i} \longrightarrow 0,$$

and with f_i also f_{i-1} is mono. In this way, we obtain an irreducible monomorphism $f_1 : \tau M_{12} \longrightarrow M_{11}$, and, since the τ-orbit of $[M_{11}]$ in $\mathcal{O}(B)$ is an endpoint, we see that f_1 is the sink map for M_{11}. In particular, M_{11} is projective. Let $M_{11} = P(a)$. Dually, M_{nn} is injective, say $M_{nn} = Q(b)$. Since $\mathrm{Hom}(P(a),Q(b)) = 0$, we must have $a \neq b$. Also, we claim that if X is indecomposable and $\mathrm{Hom}(P(a),X) \neq 0$, then either $X = M_{1j}$ for some $1 \le j \le n$, or else $M \prec X$. Namely, given a map $0 \neq g_1 : P(a) = M_{11} \longrightarrow X$, and assume $X \not\approx M_{11}$, then we factor g through the source map μ_2 of M_{11}, say $g = \mu_2 g_2$, with $g_2 : M_{22} \longrightarrow X$. Assume, $X \not\approx M_{1j}$ for $1 \le j \le t$, where $t < n$, and we have factored g through $\mu_2 \ldots \mu_t$, say $g = \mu_1 \ldots \mu_t\, g_t$. Then we can factor g_t through the source map $[\varepsilon_t, \mu_{t+1}] : M_{1t} \longrightarrow M_{2,t-1} \oplus M_{1,t+1}$, where $\varepsilon_t : M_{1t} \longrightarrow M_{2,t-1}$ is an irreducible map with $\mu_2 \ldots \mu_t \varepsilon_t = 0$. Thus $g_t = \varepsilon_t h_t + \mu_{t+1}\, g_{t+1}$, and therefore

$$g = \mu_2 \ldots \mu_t\, g_t = \mu_2 \ldots \mu_t \mu_{t+1}\, g_{t+1}\,.$$

Finally, if also $X \not\approx M_{1n} = M$, and we have $g = \mu_2 \ldots \mu_n g_n$, then we factor g_n through the source map of M, and obtain $M \prec X$. - Dually, if X is indecomposable, and $\mathrm{Hom}(X,Q(b)) \approx 0$, then $X = M_{in}$ with $1 \le i \le n$ or else $X \prec M$. As a consequence, we see that $X = M_{1n} = M$ is the only indecomposable module X with $\mathrm{Hom}(P(a),X) \neq 0$, $\mathrm{Hom}(X,Q(b)) \neq 0$.

In both cases $n = 1$ and $n > 1$, we have seen that m is the only positive root with both $m_a \neq 0$ and $m_b \neq 0$. Now, if c is an exceptional vertex for m, then $\sigma_c(m) = m - e(c)$ is a positive root. If we would have $c \neq a,b$, then we would obtain an additional positive root $x = \sigma_c(m)$ with $x_a \neq 0$, $x_b \neq 0$, impossible. Also, if there would be only one exceptional vertex c, then $\sigma_c(m)$ would be sincere, thus also in this case, we would obtain an additional positive root $x = \sigma_c(m)$ with $x_a \neq 0$, $x_b \neq 0$, again impossible. This shows that both a,b are exceptional vertices for m. This finishes the proof.

(3) <u>Let</u> B <u>be directed</u>, ${}_BM$ <u>a sincere maximal module. Assume that either the orbit graph</u> $\overline{\mathcal{O}(B)}$ <u>is</u> $\mathbb{A}_n$ <u>with</u> $n \ge 2$, <u>or else that</u> $\overline{\mathcal{O}(B)}$ <u>is a star and</u> <u>dim</u> M <u>has a unique exceptional vertex. Then</u> M <u>is a separating wing module dominated by</u> $R = \oplus P(a)^{d_a}$, <u>with</u> $d_a = D_a \chi$ (<u>dim</u> M).

<u>Proof.</u> First, assume the orbit graph $\overline{\mathcal{O}(B)}$ is $\mathbb{A}_n$ with $n \ge 2$, say

$$x_1 —— x_2 —— \ \ldots\ —— x_n\,,$$

and the τ-orbit of M is x_p. Then this τ-orbit cuts off both a branch of length p, as well as a (different) branch of length $n-p+1$. In this way, we obtain two wings of length p, and $n-p+1$, respectively (or one wing, in case $p = 1$ or $p = n$).

Second, we consider the case of $\overline{\mathcal{O}(B)}$ being a star, say with center z, and $\underline{\dim}\, M$ having a unique exceptional vertex. Now, the τ-orbit of M cannot cut off a branch from $\overline{\mathcal{O}(B)}$, since otherwise $\underline{\dim}\, M$ would have two different exceptional vertices. Thus, the τ-orbit of M has to be z, and it follows that M is a wing module (using that M is sincere). Since B-mod is a preprojective component, we see that M is also directing. Also, we have seen in (1) that M is dominated by the module R as defined above.

Now, we consider the special case of A_o (= B) being the path algebra of a quiver Δ with underlying graph of the form $\mathbb{A}_n, \mathbb{D}_n, \mathbb{E}_6, \mathbb{E}_7, \mathbb{E}_8$. We define its type as follows: In case $\mathbb{D}_n$, the type of Δ is $(n-2,2,2)$, in case $\mathbb{E}_m$ $(m = 6,7,8)$, the type of Δ is $(m-3,3,2)$. Only in case $\mathbb{A}_n$, the type of Δ depends on the orientation of Δ. Since the case $\mathbb{A}_1$ leads to the case of Kronecker modules, already discussed, we only consider $\mathbb{A}_n$ with $n \geq 2$. We assume that $\Delta_o = \{1,2\ldots,n\}$, with edges $i — i+1$ in $\bar{\Delta}$, we denote by p the number of arrows $i \leftarrow i+1$, by q the number of arrows $i \rightarrow i+1$, and we assume that $p \geq q$; then, by definition, the type of Δ is $(p+1,q+1)$ in case $q \geq 1$, and (n) in case $q = 0$. Thus, in all cases the type $(n_1,\ldots,n_r)$ of Δ is given by a sequence of integers $n_1 \geq n_2 \geq \ldots \geq n_r \geq 2$ such that $\mathbb{T}_{n_1,\ldots,n_r} = \bar{\Delta}$.

(4) <u>Let A_o be the path algebra of a quiver Δ with underlying graph $\bar{\Delta} = \mathbb{A}_n$ $(n \geq 2)$, $\mathbb{D}_n$ $(n \geq 4)$, $\mathbb{E}_6, \mathbb{E}_7$, or $\mathbb{E}_8$. Let $(n_1,\ldots,n_r)$ be the type of Δ. Let M be the unique maximal module of A_o. Then M is a sincere, directing wing module of type $(n_1,\ldots,n_r)$, and M is dominated by $R = \oplus P(a)^{d_a}$, where $d_a = D_a\chi(\underline{\dim}\, M)$. The algebra $A_o[R]$ is the path algebra of a quiver of type $\tilde{\bar{\Delta}}$.</u>

<u>Proof.</u> We note from 1.2 that for $\bar{\Delta} = \mathbb{A}_n, \mathbb{D}_n, \mathbb{E}_6, \mathbb{E}_7, \mathbb{E}_8$, the quadratic form $\chi(\bar{\Delta})$ has a unique maximal root, and this root is sincere and positive, thus, there is a unique maximal module M (see 2.4 (12)), and M is sincere. In case $\Delta = \mathbb{A}_n$, we have $\underline{\dim}\, M = (1,1,\ldots,1)$, and one easily sees that the τ-orbit of M divides the orbit graph of A_o into a branch of length $p+1$, and a branch of length $q+1$, thus M is a sincere wing module of the mentioned type. In the cases $\bar{\Delta} = \mathbb{D}_n, \mathbb{E}_6, \mathbb{E}_7, \mathbb{E}_8$, the maximal root $\underline{\dim}\, M$ has a unique exceptional vertex, thus the τ-orbit of M cannot cut off a branch of the orbit quiver of A_o, thus it is the center of the star Δ. Since M is sincere, it follows that M is a wing module of the corresponding type. As we have noted above, M is directing and dominated by the given R. Since R is projective, the algebra $A_o[R]$ is again hereditary, thus the path algebra of some quiver Δ', and we obtain Δ' from Δ by adding a new vertex ω and

d_a arrows from a to ω, for any $a \in Q$. We have noted the exceptional vertices of the unique maximal root $\underline{\dim}\, M$ of $\chi(\bar{\Delta})$, and we conclude that $\overline{\Delta'} = \widetilde{\bar{\Delta}}$.

It follows that we can apply in this situation our main theorem. We therefore obtain the following structure theorem for the module category of a path algebra of a quiver Δ with underlying graph $\tilde{\mathbb{A}}_n$, $\tilde{\mathbb{D}}_n$, $\tilde{\mathbb{E}}_6$, $\tilde{\mathbb{E}}_7$, $\tilde{\mathbb{E}}_8$, where for $\tilde{\mathbb{A}}_n$, we have to exclude the cyclic orientation. Thus, in case $\bar{\Delta} = \tilde{\mathbb{A}}_n$, assume that Δ has at least one source. We define the <u>tubular type</u> of Δ as follows: for $\bar{\Delta} = \tilde{\mathbb{D}}_n$, $(n \geq 4)$ the tubular type of Δ is $(n-2,2,2)$, for $\bar{\Delta} = \tilde{\mathbb{E}}_m$ $(m = 6,7,8)$, the tubular type Δ is $(m-3,3,2)$, whereas for $\bar{\Delta} = \tilde{\mathbb{A}}_n$, it depends on the orientation of Δ. Let $\bar{\Delta}_o$ have as set of vertices $\{0,1,\dots,n\}$, with edges $i \text{ --- } i+1$ for $0 \leq i \leq n$, (modulo $n+1$), denote by p the number of arrows $i \longleftarrow i+1$, by q the number of arrows $i \longrightarrow i+1$, and we may assume $p \geq q$; then the tubular type of Δ is given by (p,q).

(5) <u>Theorem.</u> <u>Let A be the path algebra of a quiver with underlying graph $\tilde{\mathbb{A}}_n$, $\tilde{\mathbb{D}}_n$, $\tilde{\mathbb{E}}_6$, $\tilde{\mathbb{E}}_7$, $\tilde{\mathbb{E}}_8$, and having at least one sink, let $(n_1,\dots,n_r)$ be its tubular type. Then A has a preprojective component $\mathcal{P}$, a preinjective component $\mathcal{Q}$, and a stable separating tubular $\mathbb{P}_1k$-family $\mathcal{T}$ of type $(n_1,\dots,n_r)$, separating $\mathcal{P}$ from $\mathcal{Q}$, and in this way, we obtain all indecomposable A-modules. Also, A-mod is controlled by χ_A. If we denote by w the minimal positive radical vector of χ_A in $K_o(A)$, then $\mathcal{P} = \mathcal{P}_w$, $\mathcal{T} = \mathcal{T}_w$, $\mathcal{Q} = \mathcal{Q}_w$.</u>

We may visualize the structure of A-mod as follows (again, with non-zero maps being possible only from left to right):

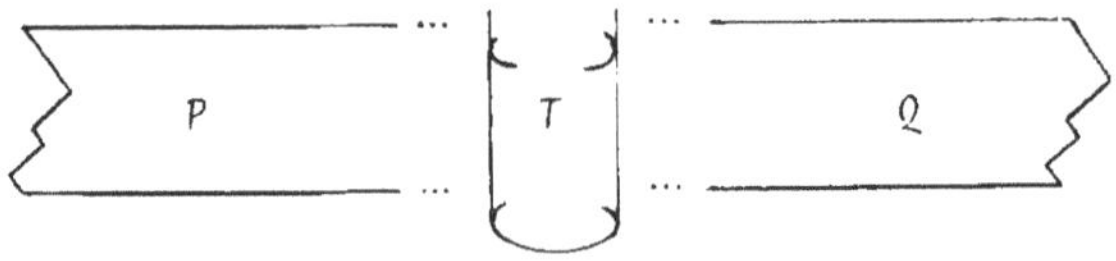

<u>Proof.</u> In case $\tilde{\mathbb{A}}_1$, we deal with the Kronecker quiver, thus we can refer to 3.2. Consider now the case of a quiver Δ' with $\bar{\Delta}' = \tilde{\mathbb{A}}_n$, $n \geq 2$, and let ω be a sink of Δ'. Similarly, in case Δ' is a quiver with $\bar{\Delta}' = \tilde{\mathbb{D}}_n$, $\tilde{\mathbb{E}}_6$, $\tilde{\mathbb{E}}_7$, or $\tilde{\mathbb{E}}_8$, then $\overline{\Delta'}$ is obtained from a graph $\mathbb{D}_n$, $\mathbb{E}_6$, $\mathbb{E}_7$, or $\mathbb{E}_8$ (respectively), by adding some vertex ω. Fix such a vertex ω, and assume first that ω is a sink. Let Δ be obtained from Δ' by deleting ω. Let A_o be the path algebra of Δ, and $R = \operatorname{rad} P(\omega)$, considered as an A_o-module. Then $A = A_o[R]$, and R dominates the maximal A_o-module M. Note that by definition, the type of A_o is just the tubular type of A. The main theorem asserts that $\mathcal{T}_w$ is a stable separating tubular $\mathbb{P}_1k$-family with type that of A_o, separating $\mathcal{P}_w$ from $\mathcal{Q}_w$ and being controlled by $\chi_{A_o}|\operatorname{Ker} \iota_w$. Also, we know that A (being hereditary) has a preprojective component $\mathcal{P}$. Since the projective A-modules belong to $\mathcal{P}_w$, we must have $\mathcal{P} \subseteq \mathcal{P}_w$; similarly, there is a preinjective component $\mathcal{Q}$,

and $\mathcal{Q} \subseteq \mathcal{Q}_w$. Also, we know that for X indecomposable in $\mathcal{P}$ (or in $\mathcal{Q}$), $\underline{\dim}\, X$ is a positive root for χ_A. Since χ_A is positive semi-definite with $\operatorname{rad} \chi_A \subseteq \operatorname{Ker} \iota_w$, any $x \in K_o(A)$ outside $\operatorname{Ker} \iota_w$ satisfies $\chi_A(x) > 0$. Thus, in order to see that A-mod is controlled by χ_A, we only have to show that any positive root of χ_A outside $\operatorname{Ker} \iota_w$ is the dimension vector of an indecomposable module.

Let x be a positive element of $K_o(A)$, with $\iota_w(x) < 0$. According to 3.5, also $\partial_w(x) < 0$, and $x\Phi^d - x = \partial_w(x)w$, where d is the lowest common multiple of $n_1,\ldots,n_r$. Thus $x\Phi^d < x$. Since $\partial_w(x) = \partial_w(x\Phi)$, we obtain a proper descending sequence

$$x > x\Phi^d > x\Phi^{2d} > \ldots \qquad > x\Phi^{id} > \ldots ,$$

thus, there is some $j \geq 1$ such that $x\Phi^{jd}$ is neither positive nor zero. Let X be a module with $\underline{\dim}\, X = x$, and assume that X has no indecomposable direct summand belonging to $\mathcal{P}$. Then $\operatorname{Hom}(\tau^t X, {}_A A) = 0$ for all $\tau \in \mathbb{N}_o$. [Namely, a non-zero map $\varphi : \tau^t X \longrightarrow {}_A A$ has as image a non-zero submodule Y, and Y has to be projective, since A is hereditary, thus $\tau^t X$ has a direct summand isomorphic to Y, and therefore X has a direct summand of the form $\tau^{-t} Y$, and $\tau^{-t} Y$ is a (non-zero) direct sum of modules in $\mathcal{P}$. This shows that $\operatorname{Hom}(\tau^t X, {}_A A) = 0$.] Since also $\operatorname{proj.dim} \tau^t X \leq 1$ for all $t \in \mathbb{N}_o$, we conclude from 2.4 (4) that

$$\underline{\dim}\, \tau^t X = (\underline{\dim}\, X)\Phi^t = x\Phi^t .$$

However, for $t = jd$, we obtain a contradiction, since the dimension vector of any module is positive or zero. This shows that any module X with $\underline{\dim}\, X = x$ has an indecomposable direct summand in $\mathcal{P}$.

As a first application, let X be an indecomposable module in $\mathcal{P}_w$ and $x \doteq \underline{\dim}\, X$. It follows that X belongs to $\mathcal{P}$, thus $\mathcal{P} = \mathcal{P}_w$. By duality, we also obtain $\mathcal{Q} = \mathcal{Q}_w$, thus A-mod $= \mathcal{P} \vee \mathcal{T} \vee \mathcal{Q}$.

As a second application, start with some positive root x of χ_A, and choose X with $\underline{\dim}\, X = x$ and $\dim_k \operatorname{End}(X)$ smallest possible. First, assume $\iota_w(x) < 0$. Let $X = \bigoplus_{i=1}^{s} X_i$, with all X_i being indecomposable. By the considerations above, some X_j, say X_1 belongs to $\mathcal{P}$, in particular, $\iota_w(\underline{\dim}\, X_1) < 0$. Note that Lemma 1 of 2.3 asserts that $\operatorname{Ext}^1(X_i,X_j) = 0$ for all $i \neq j$. Assume $\operatorname{Ext}^1(X_i,X_i) \neq 0$ for some i, then $i > 1$ and X_i has a submodule W in $\mathcal{T}$ with $\underline{\dim}\, W = w$, let $\mu : W \hookrightarrow X_i$ denote the embedding. Now μ gives rise to the induced map

$$\operatorname{Ext}^1(\mu,X_1) : \operatorname{Ext}^1(X_i,X_1) \longrightarrow \operatorname{Ext}^1(W,X_1)$$

which is surjective, since A is hereditary. Now

$$\iota_w(\underline{\dim}\, X_1) = \langle w, \underline{\dim}\, X_1\rangle = \dim \mathrm{Hom}(W,X_1) - \dim \mathrm{Ext}^1(W,X_1)$$

is negative, thus $\mathrm{Ext}^1(W,X_1) \neq 0$, and therefore also $\mathrm{Ext}^1(X_i,X_1) \neq 0$. This contradiction shows that $\mathrm{Ext}^1(X_i,X_i) = 0$, for all i. Consequently

$$1 = \chi_A(x) = \dim \mathrm{End}(X),$$

and therefore X is indecomposable. Dually, we also conclude from $\iota_w(x) > 0$, that X is indecomposable. This shows that A-mod is controlled by χ_A.

We have fixed above some vertex ω of the quiver Δ' and have assumed, in the cases $\overline{\Delta'} = \widetilde{\mathbb{D}}_n, \widetilde{\mathbb{E}}_6, \widetilde{\mathbb{E}}_7, \widetilde{\mathbb{E}}_8$, that ω is a sink. Now in these cases, ω is an endpoint of $\overline{\Delta'}$, thus ω is either a sink or a source. If ω is a source, we consider instead of A the opposite algebra A^{op}, and use the previous results and note that A-mod is isomorphic to the dual of the category A^{op}-mod.

We have excluded above the cyclic orientation on the graph $\widetilde{\mathbb{A}}_n$. However, we may use these considerations in order to deal also with this case. Note however, that in this case the path algebra no longer is finite dimensional.

(6) Let A be the path algebra of the quiver given by the graph $\widetilde{\mathbb{A}}_n$ with cyclic orientation. Then A-mod is a standard stable tubular $\mathbb{A}^1(k)$-family of type $(n+1)$, and is controlled by χ_A.

Proof. We embed A-mod in the category of representations of the quiver Δ''

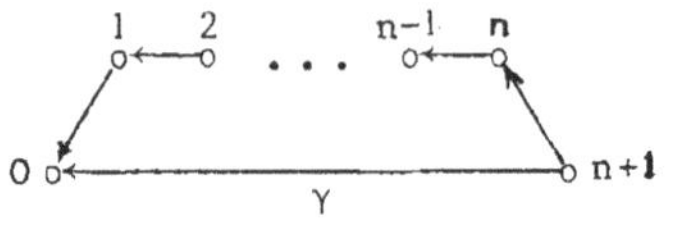

as the full subcategory $\mathcal{U}$ of all representations (V_i, V_α) with V_γ being an identity map. Let $\mathcal{T}(\infty)$ be the component of Δ'' containing the representation

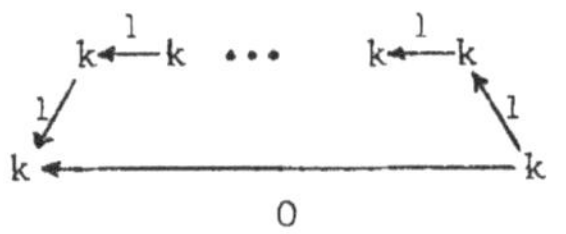

and let $\mathcal{T}$ be the tubular $\mathbb{P}_1(k)$-family of representations of Δ''. Then $\mathcal{T} = \mathcal{U} \vee \mathcal{T}(\infty)$. This finishes the proof.

3.7 Examples: The canonical algebras

We are going to construct sincere stable separating $\mathbb{P}_1k$-families of arbitrary type $(n_1,\ldots,n_t)$. Given $n_1,\ldots,n_t \in \mathbb{N}_1$, with $t \geq 2$ let $\Delta(n_1,\ldots,n_t)$ be the quiver obtained from the disjoint union of linearly ordered quivers of types $\mathbf{A}_{n_1+1},\ldots,\mathbf{A}_{n_t+1}$, by identifying all sinks to a single vertex 0, and all sources to a single vertex ω. Thus, $\Delta(n_1,\ldots,n_t)$ has the following form

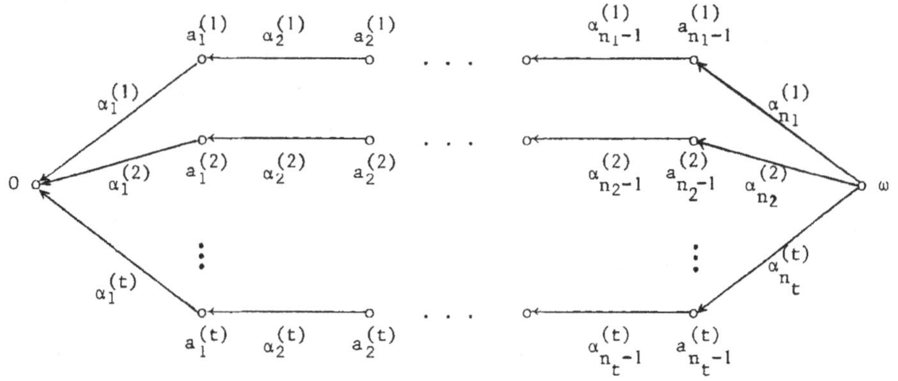

Let us denote by $\alpha^{(s)}$ the path $\alpha^{(s)} = \alpha^{(s)}_{n_s}\ldots\alpha^{(s)}_2\alpha^{(s)}_1$, and let I be the vector-space with basis $\alpha^{(1)},\ldots,\alpha^{(t)}$. A subspace J of I will be said to be generic provided $\dim J = t-2$ and J intersects any 2-dimensional coordinate subspace $\langle\alpha^{(s)},\alpha^{(s')}\rangle$ (where $s \neq s'$) in zero. The algebras defined by the quiver $\Delta(n_1,\ldots,n_t)$ with generic relations J will be said to be the canonical algebras of type $(n_1,\ldots,n_t)$. Of course, we can assume $n_1 \geq n_2 \geq \ldots \geq n_t$. If $t \geq 3$, we can assume, in addition, that $n_t \geq 2$ [namely, if $t \geq 3$, and $n_t = 1$, then we delete the arrow $\alpha^{(t)}_1$, and replace J by $J \cap \langle\alpha^{(1)},\ldots,\alpha^{(t-1)}\rangle$; the algebra we obtain will be isomorphic to the given one].

Note that for $t = 2$, we have $J = 0$, thus a canonical algebra of type (n_1,n_2) is just a hereditary algebra of tubular type (n_1,n_2). In particular, the canonical algebra of type $(1,1)$ is the Kronecker algebra. Similarly, for $t = 3$, there is up to isomorphism only one canonical algebra of type (n_1,n_2,n_3); namely given by $J = \langle\alpha^{(1)} + \alpha^{(2)} + \alpha^{(3)}\rangle$. Always, we may consider a canonical algebra C as a one-point extension $C = C_0[M]$, where C_0 is obtained from C by deleting the vertex ω. Of course, $C = C_0[M]$ is canonical of type $\mathbf{T}_{n_1,\ldots,n_t}$, if and only if the algebra C_0 is the hereditary algebra given by the star $\mathbf{T}_{n_1,\ldots,n_t}$ with subspace orientation, and M is of the following form

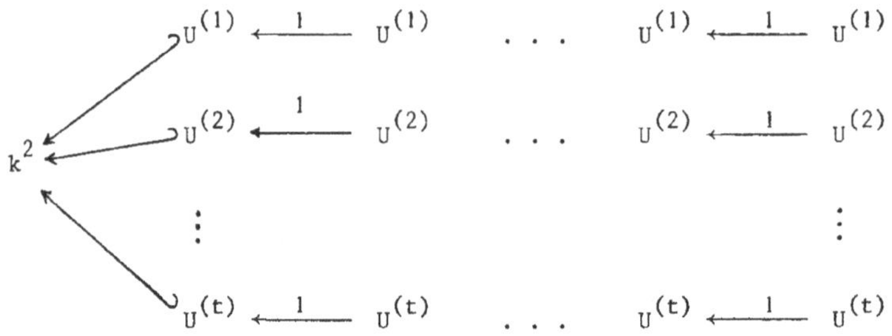

where $U^{(1)}, U^{(2)}, \ldots U^{(t)}$ are pairwise different one-dimensional subspaces of k^2 (such a C_o-module will be called a coordinate module). Thus, the canonical algebras of a fixed type are determined by the various possibilities of choosing t one-dimensional subspaces in a two-dimensional vectorspace, or, equivalently, by choosing t pairwise different points on the projective line $\mathbb{P}_1 k$. In particular, for $t = 4$, we have to consider four points on a projective line, and, in this way, we obtain a one-parameter family of algebras, distinguished by the cross ratio of these points.

Given a canonical algebra C of type $(n_1,\ldots,n_t)$, let us consider the following module classes. C-modules M are given in the form $M = (M_x, \alpha_i^{(s)})$, where x runs through the vertices and $\alpha_i^{(s)}$ through the arrows of $\Delta(n_1,\ldots,n_t)$, thus $1 \leq i \leq n_s$, $1 \leq s \leq t$. Let $\mathcal{P}$ be the module class given by all indecomposable modules $M = (M_x, \alpha_i^{(s)})$, with all $\alpha_i^{(s)}$ being mono, and not all $\alpha_i^{(s)}$ being isomorphisms. Dually, let $\mathcal{Q}$ be the module class given by all indecomposable modules M with all $\alpha_i^{(s)}$ being epi, and not all $\alpha_i^{(s)}$ being isomorphisms. Finally, let $\mathcal{T}$ be the module class given by the remaining indecomposable modules M. Thus $\mathcal{T} = \mathcal{T}' \vee \mathcal{T}''$, where $\mathcal{T}'$ is the module class given by all indecomposable modules M such that not all $\alpha_i^{(s)}$ are mono and also not all $\alpha_i^{(s)}$ are epi, and $\mathcal{T}''$ is the class of all modules M such that all $\alpha_i^{(s)}$ are isomorphisms (note that a module has this last property if and only if every indecomposable direct summand has this property). By definition, we have

$$A\text{-mod} = \mathcal{P} \vee \mathcal{T} \vee \mathcal{Q}$$

Also, let us denote by $\iota : K_o(A) \longrightarrow \mathbb{Z}$ the linear form given by $-X_0 + X_\omega$, thus $\iota(\underline{\dim}\, M)$ is just the difference of the dimensions of M_ω and M_0. There is the following result which is a direct application of the main theorem:

<u>Theorem.</u> <u>$\mathcal{T}$ is a sincere stable tubular $\mathbb{P}_1 k$-family of type</u> $(n_1,\ldots,n_t)$, <u>separating</u> $\mathcal{P}$ <u>from</u> $\mathcal{Q}$. <u>An indecomposable module</u> M <u>belongs to</u> $\mathcal{P},\mathcal{T}$, <u>or</u> $\mathcal{Q}$, <u>if and only if</u> $\iota(\underline{\dim}\, M) < 0$, $= 0$, <u>or</u> > 0, <u>respectively, and</u> $\mathcal{T}$ <u>is controlled by the restriction of</u> χ_C <u>to</u> Ker ι.

Proof. In order to show that we can apply the main theorem, let C_o be the restriction of C to the quiver obtained from $\Delta(n_1,\dots n_t)$ by deleting ω, and let R = rad $P(\omega)$, considered as an C_o-module. Thus $C = C_o[R]$. Let W_o be the indecomposable injective C_o-module $W_o = Q_o(0)$ corresponding to the vertex 0. Note that C_o is hereditary. The orbit graph of its preinjective component is $\mathbb{T}_{n_1,\dots,n_t}$, and the τ-orbit of W_o is the center of $\mathbb{T}_{n_1,\dots,n_t}$. Since W_o is a sincere C_o-module, it is a wing module of type $(n_1,\dots,n_t)$, and also directing. It remains to be shown that W_o is dominated by R. We denote by Φ_o the Coxeter transformation for C_o; according to 2.4.b , we have

$$(\underline{\dim}\, W_o)\Phi_o^{-1} = \underline{\dim}\, Q_o(0)\Phi_o^{-1} = -\underline{\dim}\, P_o(0),$$

where $P_o(0)$ is the indecomposable projective module corresponding to the vertex 0, but this is just the simple module $E(0)$, therefore

$$(\underline{\dim}\, W_o)(I - \Phi_o^{-1}) = \underline{\dim}\, W_o + \underline{\dim}\, E(0)$$

$$= \begin{smallmatrix} 1\dots1 \\ 1 \\ 1\dots1 \end{smallmatrix} + \begin{smallmatrix} 0\dots0 \\ 1 \\ 0\dots0 \end{smallmatrix} = \begin{smallmatrix} 1\dots1 \\ 2 \\ 1\dots1 \end{smallmatrix} = \underline{\dim}\, R.$$

Now, consider any non-zero map $\rho : R \longrightarrow W_o$. We claim that the kernel of ρ is projective. Namely, consider the (non-zero) k-linear map $\rho_0 : R_0 = I/J \longrightarrow k = (W_o)_0$ (note that, as above, given a C_o-module M, we denote by M_0 its component at the vertex 0), and correspondingly, given a C_o-map ρ, then $\rho = (\rho_x)_x$, where x runs through all vertices, thus we may consider ρ_0). Thus, the kernel of ρ_0 gives a hyperplane J' in I, with $J \subset J' \subset I$. Since J is generic, there is at most one s with $\alpha^{(s)} \in J'$, thus the kernel of ρ is either $P_o(0)$ or else some $P_o(x^{(s)}_{n_s-1})$, $1 \le s \le t$. Since for any non-zero $\rho : R \longrightarrow W_o$, the kernel is projective and the cokernel has projective dimension ≤ 1 (recall that C_o is hereditary), it follows that proj.dim.$W_o(\rho) \le 1$ (see 3.3). Thus, R dominates W_o.

Now, $w = 1\begin{smallmatrix} 1\dots1 \\ \\ 1\dots1 \end{smallmatrix}1$, and $\iota_w = \langle w,-\rangle$ coincides with $\iota = -X_o + X_\omega$. Thus, according to the main theorem, $\mathcal{T}_w$ is a sincere, stable tubular $\mathbb{P}_1k$-family of type $(n_1,\dots,n_t)$, separating $\mathcal{P}_w$ from $\mathcal{Q}_w$ and controlled by $\chi_C|\text{Ker}\,\iota$. We want to identify $\mathcal{P}_w,\mathcal{T}_w,\mathcal{Q}_w$ with $\mathcal{P},\mathcal{T},\mathcal{Q}$. Of course, $\mathcal{P} \subseteq \mathcal{P}_w$, $\mathcal{Q} \subseteq \mathcal{Q}_w$ [namely, if M is in $\mathcal{P}$, then all maps $\alpha_i^{(s)}$ are monos, thus $\dim M_0 \ge \dim M_\omega$, and since not all maps $\alpha_i^{(s)}$ are isomorphisms, even $\dim M_0 > \dim M_\omega$, thus $\iota(\underline{\dim}\, M) < o$, and similarly for $\mathcal{Q}$].

For all $1 \le i \le n_s$, $1 \le s \le t$, let us consider the map

$$Q(\alpha_i^{(s)}) : Q(a_{i-1}^{(s)}) \longrightarrow Q(a_i^{(s)})$$

induced by the arrow $\alpha_i^{(s)}$, where $a_o^{(s)} = 0$, and $a_{n_s}^{(s)} = \omega$. We denote the kernel of this map by $E_i^{(s)}$. As a non-zero submodule of an indecomposable injective module, $E_i^{(s)}$ is indecomposable, and, using that $Q(\alpha_i^{(s)})$ is surjective, we easily calculate the dimension vector of $E_i^{(s)}$ as follows:

$$\underline{\dim}\, E_1^{(s)} = w - \sum_{i=1}^{n_s-1} e(a_i^{(s)}), \quad \text{and}$$

$$\underline{\dim}\, E_i^{(s)} = e(a_{i-1}^{(s)}) , \quad \text{for } 2 \le i \le n_s.$$

Consequently, $E_i^{(s)}$ belongs to $\mathcal{T}_w$. According to 2.1, for any C-module $M = (M_x, \alpha_i^{(s)})$, we have the following property:

(a) $\quad \mathrm{Hom}(M, E_i^{(s)}) = o$ if and only if $\alpha_i^{(s)}$ is epi on M.

As a consequence, let Q be indecomposable in $\mathcal{Q}_w$. We have $\mathrm{Hom}(Q, E_i^{(s)}) = o$ since $Q \in \mathcal{Q}_w$, $E_i^{(s)} \in \mathcal{T}_w$, thus using (a), it follows that $Q \in \mathcal{Q}$. Since we already know that $\mathcal{Q} \subseteq \mathcal{Q}_w$, it follows that $\mathcal{Q} = \mathcal{Q}_w$.

Similarly, denote by $F_i^{(s)}$ the cokernel of the canonical map

$$P(\alpha_i^{(s)}) : P(a_{i-1}^{(s)}) \longrightarrow P(a_i^{(s)}).$$

Again, $F_i^{(s)}$ is indecomposable, and

$$\underline{\dim}\, F_i^{(s)} = e(a_i^{(s)}) , \quad \text{for } 1 \le i \le n_s - 1, \quad \text{and}$$

$$\underline{\dim}\, F_{n_s}^{(s)} = w - \sum_{i-1}^{n_s-1} e(a_i^{(s)}) ,$$

thus, also all $F_i^{(s)}$ belong to $\mathcal{T}_w$ (actually, up to a shift of the indices, they coincide with the $E_i^{(s)}$). Since for any C-module $M = (M_x, \alpha_i^{(s)})$, we have

(a*) $\quad \mathrm{Hom}(F_i^{(s)}, M) = o$ if and only if $\alpha_i^{(s)}$ is mono on M.

Similarly to the previous case, it follows that $\mathcal{P}_w \subseteq \mathcal{P}$, thus $\mathcal{P}_w = \mathcal{P}$. Since, by definition, an indecomposable module belongs to $\mathcal{T}$ if and only if it does not belong to $\mathcal{P} \vee \mathcal{Q}$, it follows that $\mathcal{T} = \mathcal{T}_w$. This finishes the proof.

Remark 1. Let us identify in this case the various non-homogeneous tubes. Fix some s with $n_s \ge 2$. For any $T \in \mathcal{T}$, we have $\underline{\dim}\, \tau T = (\underline{\dim}\, T)\Phi$, where Φ is the Coxeter transformation for C, thus $\underline{\dim}\, \tau E_i^{(s)} = (\underline{\dim}\, E_i^{(s)})\Phi = \underline{\dim}\, E_{i-1}^{(s)}$, where $E_o^{(s)} = E_{n_s}^{(s)}$. Since (under our assumption $n_s \geq 2$) all $\underline{\dim}\, E_i^{(s)}$ are roots, they determine uniquely the corresponding modules, thus $E_1^{(s)},\ldots,E_{n_s}^{(s)}$ is a full τ-orbit. Also, since for example $E_{n_s}^{(s)}$ is a simple C-module, $E_{n_s}^{(s)}$, and therefore all $E_i^{(s)}$, lie on the mouth of a tube $T^{(s)}$. The indecomposable modules M in $T^{(s)}$ are characterized as those indecomposable modules in T which satisfy $\mathrm{Hom}(M,E_i^{(s)}) \neq o$, thus they are those indecomposable modules M in T for which some map $\alpha_i^{(s)}$ is not epi.

Similarly, we see that $F_1^{(s)},\ldots,F_{n_s}^{(s)}$ is a full τ-orbit, and since $E_{n_s}^{(s)} = F_{n_s-1}^{(s)}$ $(= E(a_{n_s-1}^{(s)}))$, we conclude that

(b) $$E_i^{(s)} = F_{i-1}^{(s)} \quad \text{for all } 2 \leq i \leq n_s, \text{ and } E_1^{(s)} = F_{n_s}^{(s)}.$$

It follows that the indecomposable modules in $T^{(s)}$ are just those indecomposable modules M in T for which some map $\alpha_i^{(s)}$ is not mono.

As a consequence, <u>an indecomposable module belongs to</u> $T^{(s)}$ <u>if and only if some map</u> $\alpha_i^{(s)}$ <u>is not mono, and some map</u> $\alpha_j^{(s)}$ <u>is not epi</u>.

In this way, we obtain for the various s with $n_s \geq 2$, non-homogeneous tubes belonging to T and of rank n_s. All remaining tubes belonging to T are homogeneous (since we know the type of T). In particular, if $n_s = 1$ for some s, then $\tau E_1^{(s)} = E_1^{(s)}$, and consequently $E_1^{(s)}$ belongs to the mouth of some tube. For $F_1^{(s)}$, the map $\alpha_1^{(s)}$ is not mono, thus zero, thus not epi, therefore $\mathrm{Hom}(F_1^{(s)},E_1^{(s)}) \neq o$. The image of a non-zero map $f : F_1^{(s)} \to E_1^{(s)}$ has to be all of $E_1^{(s)}$, thus f is an isomorphism. Consequently, the formula (b) is valid for all s. We see that T' decomposes into t tubes of rank $n_1,\ldots,n_t$, namely $T' = \bigvee_{s=1}^{t} T^{(s)}$, and T'' consists of homogeneous tubes.

It follows that <u>an indecomposable</u> C-<u>module</u> X <u>belongs to a homogeneous tube in</u> T <u>if and only if all the maps</u> $\alpha_i^{(s)}$, $1 \leq i \leq n_s$, $1 \leq s \leq t$, <u>are isomorphisms</u>. Also, <u>in this case</u>, X <u>belongs to the mouth of a homogeneous tube if and only if</u> $\underline{\dim}\, X = 1\,{}^{1\ldots 1}_{1\ldots 1}\,1$.

The proof of the last assertion is as follows: we embed the category of all Kronecker modules into C-mod, by sending the Kronecker module $V = (V_o, V_\omega, \gamma_1, \gamma_2)$ to the C-module ιV, with $(\iota V)_o = V_o$, $(\iota V)_\omega = V_\omega = (\iota V)_{a_i^{(s)}}$, for all $1 \leq i < n_s$, $1 \leq s \leq t$, and with maps

$$(\iota V)_{\alpha_1^{(s)}} = \gamma_s, \qquad \text{for } s = 1,2,$$

$$(\iota V)_{\alpha_1^{(s)}} = \lambda_s\gamma_1 + \mu_s\gamma_2, \qquad \text{for } s \geq 3$$

where $\alpha^{(s)} = \lambda_s\gamma_1 + \mu_s\gamma_2$, with $\lambda_s, \mu_s \in k$, and

$$(\iota V)_{\alpha_i^{(s)}} = 1_{V_o}, \qquad \text{for } 1 < i \leq n_s,\ 1 \leq s \leq t.$$

Of course, we obtain in this way a full exact embedding. Also, if X is a C-module with all $\alpha_i^{(s)}$ isomorphisms, then X is isomorphic to an image under ι. Then, assume X belongs to the mouth of a homogeneous tube, therefore $X = \iota V$ for some Kronecker module V. Now, V is indecomposable, and $\underline{\dim}\, V$ is of the form $[m,m]$, therefore V is regular. But $\mathrm{Ext}^1(V,V) = 0$ implies $m = 1$. Therefore, $\underline{\dim}\, X = 1{}^{1\ldots1}_{1\ldots1}1$.

Remark 2. It seems curious that for a canonical algebra C, the module class $\mathcal{T}$ is always very well behaved, whereas the two classes $\mathcal{P}$ and $\mathcal{Q}$ may be wild. The representation type of both $\mathcal{P}$ and $\mathcal{Q}$ very much depend on the type of C. Assume C is a canonical algebra of type $(n_1,\ldots,n_t)$. We later will see that for $\mathbb{T}_{n_1,\ldots,n_t}$ being a Dynkin diagram, C is concealed, thus $\mathcal{P}$ is a preprojective component, and $\mathcal{Q}$ is a preinjective component. In case $\mathbb{T}_{n_1,\ldots,n_t}$ is a Euclidean diagram, the structure of the module classes $\mathcal{P}$ and $\mathcal{Q}$ completely will be determined in part 5, the algebra C being a tubular algebra. Finally, for $\mathbb{T}_{n_1,\ldots,n_t}$ being neither Dynkin nor Euclidean, $\mathcal{P}$ and $\mathcal{Q}$ are strictly wild, according to 3.4.

References

The representation theory of the tame hereditary algebras was developed in several steps. At the beginning of section 3.2, we have mentioned that the representations of the quiver $\circ \rightrightarrows \circ$ have been determined by Kronecker [K], this was in 1890. The next special case to be considered was the four-subspace quiver: its indecomposable representations were listed by Nazarova [N1] in 1967 (and later also by Gelfand-Ponomarev [GP]). After the introduction of the notion of a quiver by Gabriel, the problem of determining the indecomposable representations of the connected quivers with semidefinite quadratic form was apparent. It was solved independently in 1973 by Donovan-Freislich [DF] and Nazarova [N2]. On the other hand, as we have outlined in the references to chapter 2, Bernstein-Gelfand-Ponomarev [BGP] have shown that any connected representation infinite quiver has a preprojective component and a preinjective component. In the tame case, the module class given by the remaining indecomposable modules is an abelian serial category $\mathcal{T}$; this was shown in our joint work [DR] with Dlab (there, the more general case of an arbitrary basefield, thus of representations of species, was considered). Note that the tables of [DR] exhibit that the type of the tubular family $\mathcal{T}$ is given by the Dynkin diagram corresponding to the given Euclidean diagram.

The introduction of the w-defect in 3.5 is copied from [DR], the notion of defect was in fact a main working tool already in [GP] and [DF]. Given pairwise orthogonal bricks $E_1,\ldots,E_n$, the consideration of the full subcategory $\mathcal{E}(E_1,\ldots,E_n)$ has been called the process of "simplification" in [Ri1], lemma 3.1.2 is taken from this paper. The general notion of a tube was introduced in the joint paper [ER] with d'Este; in particular, this paper also contains lemma 3.1.1.

4. Tilting functors and tubular extensions (Notation, results, some proofs)

The aim of this chapter is to outline two techniques of the representation theory of finite-dimensional algebras which are very useful for studying several classes of algebras closely related to hereditary ones.

4.1 Tilting modules

Let A be a finite-dimensional algebra. A module ${}_AT$ is called a tilting module provided it satisfies the following properties

(α) $\text{proj.dim }{}_AT \leq 1$.

(β) $\text{Ext}^1_A({}_AT, {}_AT) = 0$.

(γ) The number of isomorphism classes of indecomposable direct summands of ${}_AT$ is equal to the rank of $K_o(A)$.

Under the conditions (α) and (β), the condition (γ) is equivalent to any one of the following conditions:

(γ') There exists an exact sequence $o \longrightarrow {}_AA \longrightarrow T' \longrightarrow T'' \longrightarrow o$ with $T', T'' \in \langle {}_AT \rangle$.

(γ") For any indecomposable projective A-module $P(a)$, there exists an exact sequence $o \longrightarrow P(a) \longrightarrow T'_a \longrightarrow T''_a \longrightarrow o$, with $T'_a, T''_a \in \langle {}_AT \rangle$.

Any module ${}_AT$ satisfying the conditions (α) and (β) may be called a partial tilting module, namely, as it turns out, any partial tilting module is a direct summand of a tilting module.

We note the following: if ${}_AT$ is a tilting module, and $T \oplus S$ is a partial tilting module, then $S \in \langle {}_AT \rangle$. Namely, $S \oplus T$ satisfies the axioms (α), (β), (γ'), thus it is a tilting module. However, according to (γ), the indecomposable direct summands of ${}_AS$ must have been already direct summands of ${}_AT$.

The dual notion of a tilting module is that of a cotilting module; note that the conditions (β) and (γ) are self-dual, thus a module ${}_AT$ is cotilting if and only if it satisfies (β), (γ), and

(α*) $\text{inj.dim }{}_AT \leq 1$.

Let us show that any partial tilting module is a direct summand of a tilting module; in fact, there is the following lemma due to Bongartz:

(1) Let ${}_AT$ be a partial tilting module. Then there exists a tilting module ${}_AT \oplus {}_AT'$ such that any indecomposable direct summand T'' of T' is projective

<u>or satisfies</u> $\mathrm{Hom}_A(T'',T) \neq 0$.

<u>Proof.</u> Let $E_1,\dots,E_m$ be a basis of $\mathrm{Ext}^1({}_AT,{}_AA)$, and let

$$(E_i)_i : 0 \longrightarrow {}_AA \xrightarrow{\mu} {}_AT' \xrightarrow{\varepsilon} \bigoplus_{i=1}^{m} {}_AT \longrightarrow 0$$

be the pushout along the diagonal map $({}_AA)^m \to A$ of the exact sequence $\oplus E_i$. By (α), $\mathrm{proj.dim.}_AT' \leq 1$. Applying $\mathrm{Hom}({}_AT,-)$ to the sequence above, we obtain

$$\mathrm{Hom}({}_AT, \bigoplus_{i=1}^{m} {}_AT) \longrightarrow \mathrm{Ext}^1({}_AT,{}_AA) \longrightarrow \mathrm{Ext}^1({}_AT,{}_AT') \longrightarrow 0,$$

and by construction, the first map is surjective, thus $\mathrm{Ext}^1({}_AT,{}_AT') = 0$. Also, there are exact sequences

$$0 = \mathrm{Ext}^1_A({}_AA,{}_AT) \longleftarrow \mathrm{Ext}^1_A({}_AT',{}_AT) \longleftarrow \mathrm{Ext}^1_A(\oplus\, {}_AT,{}_AT\ ) = 0,$$

$$0 = \mathrm{Ext}^1_A({}_AA,{}_AT') \longleftarrow \mathrm{Ext}^1_A({}_AT',{}_AT') \longleftarrow \mathrm{Ext}^1_A(\oplus\, {}_AT,{}_AT') = 0,$$

thus $\mathrm{Ext}^1_A(T \oplus T',T \oplus T') = 0$. Altogether we see that the module $T \oplus T'$ satisfies the conditions (α), (β), (γ'). In case there is a direct summand T'' of T' satisfying $\mathrm{Hom}(T'',T) = 0$, it follows that T'' lies in the kernel of ε, thus it is a direct summand of the image of μ, and therefore projective.

Given a module ${}_AM$, with endomorphism ring B, we may consider M as a right B-module, and denote it by M_B. Dualizing M_B, we obtain a (left) B-module $D(M_B)$. Now $D(M_B)$ is in fact a B-A-bimodule, thus there is a canonical ring homomorphism $A \longrightarrow \mathrm{End}(D(M_B))$. In case this is an isomorphism, we will identify A and $\mathrm{End}(D(M_B))$.

(2) <u>If</u> ${}_AT$ <u>is a tilting</u> A-<u>module, and</u> $B = \mathrm{End}({}_AT)$, <u>then</u> $D(T_B)$ <u>is a cotilting</u> B-<u>module, and</u> $A = \mathrm{End}(D(T_B))$ <u>canonically</u>.

Or, we may formulate this as follows:

If ${}_AT$ is a tilting module, with $B = \mathrm{End}({}_AT)$, then also ${}_{B^{op}}T$ is a tilting module, and $A^{op} = \mathrm{End}({}_{B^{op}}T)$.

(3) Given a finite-dimensional algebra A, a pair $(\mathcal{F},\mathcal{G})$ of module classes in A-mod is said to be a <u>torsion pair</u> provided $\mathcal{F}$ is the set of all modules F in A-mod with $\mathrm{Hom}(\mathcal{G},F) = 0$, and $\mathcal{G}$ is the set of all modules G with $\mathrm{Hom}(G,\mathcal{F}) = 0$. In this case, the modules in $\mathcal{G}$ are called the <u>torsion</u> modules, those in $\mathcal{F}$ the <u>torsionfree</u> modules (with respect to the given torsion pair $(\mathcal{F},\mathcal{G})$).

[Note that in a torsion pair $(\mathcal{F},\mathcal{G})$, we first specify the torsionfree class, and then the torsion class, contrary to the usual convention. Also, we had to refrain from denoting the torsion class by $\mathcal{T}$, since we cannot denote all module classes by $\mathcal{T}$; the letter $\mathcal{G}$ stands for "generated", the reason being that most torsion classes we are interested in will be classes of modules generated by a fixed module G]. Given a torsion pair $(\mathcal{F},\mathcal{G})$, the class $\mathcal{F}$ is closed under submodules, and extensions, the class $\mathcal{G}$ is closed under factor modules and extensions. Also

$$\text{A-mod} = \mathcal{F} \,\rfloor\!\lceil\, \mathcal{G} ,$$

and actually, given a module ${}_AM$, there is a unique submodule M' of M belonging to $\mathcal{G}$ such that M/M' is in $\mathcal{F}$; this submodule M' is the largest submodule of M belonging to $\mathcal{G}$, and M/M' is the largest factor module of M belonging to $\mathcal{F}$; the submodule M' is called the <u>torsion submodule</u> of M, and denoted by $t(M)$ (or $t_{\mathcal{G}}(M)$). In case $(\mathcal{F},\mathcal{G})$ is a torsion pair with A-mod $= \mathcal{F} \vee \mathcal{G}$, or, equivalently, with $t(M)$ being a direct summand of M, for every A-module M, then $(\mathcal{F},\mathcal{G})$ is said to <u>split</u>.

<u>Example</u>. Let ${}_AG$ be a module with $\operatorname{Ext}^1(G,G) = 0$ and $\operatorname{proj.dim} {}_AG \leq 1$. Let $\mathcal{F} = \mathcal{F}({}_AG) := \{M \mid \operatorname{Hom}(G,M) = 0\}$, $\mathcal{G} = \mathcal{G}({}_AG) := \{M \mid M \text{ generated by } G\}$. Then $(\mathcal{F},\mathcal{G})$ is a torsion pair.

<u>Proof.</u> Of course, $\operatorname{Hom}(\mathcal{G},\mathcal{F}) = 0$. It only remains to be seen that any M with $\operatorname{Hom}(M,\mathcal{F}) = 0$ is generated by G. Let $\varepsilon : \oplus G \longrightarrow M$ be a universal map from a direct sum of copies of G to M, let I be the image of ε, and C its cokernel with projection map $\pi : M \longrightarrow C$. We claim that $C \in \mathcal{F}$. Namely, given $\varphi : G \longrightarrow C$, we can construct the following commutative diagram with exact rows:

$$\begin{array}{ccccccccc}
0 & \longrightarrow & I & \longrightarrow & M & \xrightarrow{\pi} & C & \longrightarrow & 0 \\
 & & \| & & \uparrow & & \uparrow{\scriptstyle\varphi} & & \\
0 & \longrightarrow & I & \longrightarrow & M' & \longrightarrow & G & \longrightarrow & 0 \\
 & & \uparrow{\scriptstyle\varepsilon} & & \uparrow & & \| & & \\
0 & \longrightarrow & \oplus G & \longrightarrow & M'' & \longrightarrow & G & \longrightarrow & 0 ,
\end{array}$$

using that $\operatorname{proj.dim} G \leq 1$. But $\operatorname{Ext}^1_A(G,G) = 0$ shows that the last sequence splits, thus $\varphi = \varphi'\pi$ for some $\varphi' : G \longrightarrow M$. However, by construction of I, we see that φ' maps into I, thus $\varphi = 0$. Since $C \in \mathcal{F}$, and $\operatorname{Hom}(M,\mathcal{F}) = 0$, it follows that $C = 0$, thus ε is surjective.

<u>Special case 1</u>: If e is an idempotent in A, let ${}_AG = Ae$. Then $\mathcal{F}({}_AG) = A/AeA\text{-mod}$, and $\mathcal{G}({}_AG)$ is the set of all A-modules generated by Ae, thus the set of all A-modules ${}_AM$ with top $M \in \langle \text{top } Ae \rangle$. [One only has to check that

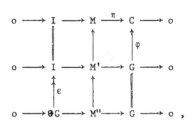

$\mathrm{Hom}_A(Ae,M) = 0$ is equivalent to M being an A/AeA-module. However, $\mathrm{Hom}_A(Ae,M) = 0$ if and only if $AeA \cdot M = 0$].

Special case 2: The case we are mainly interested in in this section is that of ${}_AG$ being a tilting module.

(4) Given a tilting module ${}_AT$ with $B = \mathrm{End}({}_AT)$, we specify two torsion pairs, one in A-mod, the second in B-mod, as follows:

$$F({}_AT) := \{{}_AX \mid \mathrm{Hom}_A(T,X) = 0\} = \{{}_AX \mid {}_AX \text{ is cogenerated by } \tau_A T\},$$

$$G({}_AT) := \{{}_AX \mid \mathrm{Ext}^1_A(T,X) = 0\} = \{{}_AX \mid {}_AX \text{ is generated by } {}_AT\} .$$

Then $(F({}_AT), G({}_AT))$ is a torsion pair in A-mod.

$$Y({}_AT) := \{{}_BN \mid \mathrm{Tor}^B_1(T_B, {}_BN) = 0\} = \{{}_BN \mid {}_BN \text{ is cogenerated by } D(T_B)\}.$$

$$X({}_AT) := \{{}_BN \mid T_B \otimes {}_BN = 0\} = \{{}_BN \mid {}_BN \text{ is generated by } \tau^- D(T_B)\} .$$

Then $(Y({}_AT), X({}_AT))$ is a torsion pair in B-mod.

Given a tilting module ${}_AT$ with $B = \mathrm{End}({}_AT)$, the two functors $\Sigma_T := \mathrm{Hom}_A({}_AT_B,-)$ and $\Sigma'_T := \mathrm{Ext}^1_A({}_AT_B,-)$ play an important role.

Theorem of Brenner-Butler: The functor Σ_T defines an equivalence from $G({}_AT)$ onto $Y({}_AT)$; the functor Σ'_T defines an equivalence from $F({}_AT)$ onto $X({}_AT)$.

The respective inverse functors are given by ${}_AT_B \otimes -$ from $Y({}_AT)$ to $G({}_AT)$, and by $\mathrm{Tor}^B_1({}_AT_B,-)$ from $X({}_AT)$ to $F({}_AT)$. Note that these equivalences $G({}_AT) \approx Y({}_AT)$ and $F({}_AT) \approx X({}_AT)$ are equivalences of exact categories, since the restrictions of the four functors Σ_T, Σ'_T, ${}_AT_B \otimes -$, and $\mathrm{Tor}^B_1({}_AT_B,-)$ to the respective subcategories $G({}_AT)$, $F({}_AT)$, $Y({}_AT)$, and $X({}_AT)$, send short exact sequences to short exact ones.

One of the consequences we will use is the fact that $Y({}_AT)$ is the image of A-mod under the functor Σ, and $X({}_AT)$ is the image of A-mod under the functor Σ'. [Namely, given any A-module ${}_AM$, we have $\Sigma M = \Sigma t_{G({}_AT)}M$, and $\Sigma' M = \Sigma'(M/t_{G({}_AT)}M)$, thus $\Sigma M \in Y({}_AT)$, and $\Sigma' M \in X({}_AT)$, and the theorem of Brenner-Butler asserts that any module in $Y({}_AT)$ is in the image of Σ, any module in $X({}_AT)$ is in the image of Σ'.]

(5) Using the first definition of $G({}_AT)$, it is obvious that the injective cogenerator $D(A_A)$ belongs to $G({}_AT)$. Its image under Σ is

$$\Sigma D({}_AA) = \mathrm{Hom}_A({}_AT_B, D(A_A)) \approx \mathrm{Hom}(A_A, D({}_AT_B)) \approx D(T_B).$$

Denoting as usual by $P_A(a)$ the projective cover, by $Q_A(a)$ the injective envelope of a simple A-module $E_A(a)$, we similarly have $Q_A(a) \in G({}_AT)$, and there is the following recipe for obtaining the τ_B^--translate of $\Sigma Q_A(a)$.

<u>Connecting lemma</u>: $\tau_B^-\Sigma Q_A(a) = \Sigma' P_A(a)$. In particular, $\Sigma Q_A(a)$ is injective if and only if $P_A(a) \in \langle {}_AT \rangle$, and in this case, $\nu \Sigma P_A(a) = \Sigma Q_A(a)$.

<u>Proof</u>. According to (γ''), there is an exact sequence

$$(*) \qquad 0 \longrightarrow P(a) \longrightarrow T' \xrightarrow{\pi} T'' \longrightarrow 0 ,$$

with $T', T'' \in \langle {}_AT \rangle$. Applying $\mathrm{Hom}(-, {}_AT)$, we obtain the exact sequence

$$0 \longrightarrow \mathrm{Hom}(T'', T) \xrightarrow{\mathrm{Hom}(\pi, T)} \mathrm{Hom}(T', T) \longrightarrow \mathrm{Hom}(P(a), T) \longrightarrow 0 ,$$

and this is a projective presentation of the right B-module $\mathrm{Hom}(P(a),T)$. Applying the duality functor D, we obtain the injective presentation

$$0 \longrightarrow D\,\mathrm{Hom}(P(a),T) \longrightarrow D\,\mathrm{Hom}(T',T) \xrightarrow{D\,\mathrm{Hom}(\pi,T)} D\,\mathrm{Hom}(T'',T) \longrightarrow 0,$$

and $D\,\mathrm{Hom}_A(P(a),T) \approx \mathrm{Hom}_A(T,Q(a)) = \Sigma Q(a)$. Under the Nakayama functor ν^-, we obtain from $D\,\mathrm{Hom}_A(\pi,T)$ just $\nu^- D\,\mathrm{Hom}_A(\pi,T) = \mathrm{Hom}(T,\pi)$, and there is the exact sequence derived from $(*)$

$$\mathrm{Hom}_A(T,T') \xrightarrow{\mathrm{Hom}(T,\pi)} \mathrm{Hom}_A(T,T'') \longrightarrow \mathrm{Ext}^1_A(T,P\ (a)) \longrightarrow 0,$$

which shows that $\mathrm{Cok}\,\mathrm{Hom}(T,\pi) = \Sigma' P(a)$. This gives the first part of the lemma.

Of course, $\Sigma Q_A(a)$ is injective if and only if $\tau_B^-\Sigma Q_A(a) = 0$, thus if and only if $\mathrm{Ext}^1_A(T,P_A(a)) = 0$. In this case,

$$\nu_B \Sigma P_A(a) = D\,\mathrm{Hom}_B(\Sigma P_A(a), \Sigma_A T) \approx D\,\mathrm{Hom}_A(P_A(a), {}_AT)$$

$$\approx \mathrm{Hom}_A({}_AT, Q_A(a)) = \Sigma Q_A(a).$$

(6) Let us specify some bounds for the projective dimension of the modules in $\mathcal{Y}({}_AT)$ and in $\mathcal{X}({}_AT)$, and for the vanishing of $\mathrm{Ext}^n(\mathcal{Y}({}_AT), \mathcal{X}({}_AT))$.

(a) If ${}_AM \in G({}_AT)$, then $\mathrm{proj.dim.}_B \Sigma M \le \mathrm{proj.dim.}_A M$.

(b) If ${}_AN \in F({}_AT)$, then $\mathrm{proj.dim.}_B \Sigma' M \le 1 + \max(1, \mathrm{proj.dim.}_A N)$.

(c) For ${}_AM \in G({}_AT)$, with $\mathrm{proj.dim.}_A M \le 1$, $\Sigma \tau_A M = \tau_B \Sigma M$.

(d) If ${}_AM \in G({}_AT)$ has $\mathrm{proj.dim.}_A M \le n$, then $\mathrm{Ext}^n_B(\Sigma M, \mathcal{X}({}_AT)) = 0$.

<u>Proof</u>. Let us consider ${}_AM \in \mathcal{G}({}_AT)$, and assume proj.dim.${}_AM = n$. First, let $n = 0$. The module ${}_AM$ being both projective and in $\mathcal{G}({}_AT)$, has to belong to $<{}_AT>$, thus ΣM is projective, the first assertion of (a).

Next, let proj.dim.${}_AM = n \geq 1$. There exists an exact sequence

$$(*) \qquad 0 \longrightarrow K \longrightarrow T' \longrightarrow M \longrightarrow 0$$

with $T' \in <{}_AT>$, and $K \in \mathcal{G}({}_AT)$, and such that, in case $n = 1$, even $K \in <{}_AT>$. [Namely, let $\varepsilon : T' = \oplus T \longrightarrow M$ be a universal map from a direct sum of copies of T, and let $K = \operatorname{Ker} \varepsilon$. Applying $\operatorname{Hom}(T,-)$ to the exact sequence (*), we obtain an exact sequence

$$\operatorname{Hom}_A(T,\oplus T) \longrightarrow \operatorname{Hom}_A(T,M) \longrightarrow \operatorname{Ext}^1_A(T,K) \longrightarrow 0,$$

with the first map being surjective, due to the universality of ε, thus $\operatorname{Ext}^1_A(T,K) = 0$. Now assume $n = 1$, and apply $\operatorname{Hom}_A(-,K)$ and $\operatorname{Hom}_A(-,T)$ to (*). We obtain

$$0 = \operatorname{Ext}^1(\oplus T,K) \longrightarrow \operatorname{Ext}^1(K,K) \longrightarrow 0,$$

$$0 = \operatorname{Ext}^1(\oplus T,T) \longrightarrow \operatorname{Ext}^1(K,T) \longrightarrow 0,$$

thus $\operatorname{Ext}^1(T \oplus K, T \oplus K) = 0$. Since also proj.dim.$K \leq 1$, we see that K belongs to $<{}_AT>$.] In case $n \geq 2$, using that proj.dim.${}_AT \leq 1$, the sequence (*) shows that proj.dim.$K \leq n-1$. Applying Σ to (*), we obtain an exact sequence with middle term being projective. In case $n = 1$, also ΣK is projective, thus proj.dim.${}_B\Sigma M \leq 1$. In case $n \geq 2$, we use induction:

$$\text{proj.dim.}_B\Sigma M \leq 1 + \text{proj.dim.}_B\Sigma K \leq 1 + \text{proj.dim.}_AK \leq n.$$

Next, let us show (c). Since proj.dim.${}_AM \leq 1$, we have $\tau_AM = D\operatorname{Ext}^1_A({}_AM,{}_AA)$. since also proj.dim.${}_B\Sigma M \leq 1$, we have $\tau_B\Sigma M = D\operatorname{Ext}^1_B(\Sigma_AM,{}_BB) = D\operatorname{Ext}^1_B(\Sigma_AM,\Sigma_AT) \approx D\operatorname{Ext}^1_A({}_AM,{}_AT)$. Also note that $\operatorname{Ext}^1_A({}_AM,{}_AT) \approx \operatorname{Ext}^1_A({}_AM,{}_AA) \otimes {}_AT$. [Namely, starting with a projective resolution

$$0 \longrightarrow P_1 \longrightarrow P_o \longrightarrow {}_AM \longrightarrow 0$$

of M, applying $\operatorname{Hom}_A(-,{}_AA)$, we obtain

$$\operatorname{Hom}(P_o,{}_AA) \longrightarrow \operatorname{Hom}(P_1,{}_AA) \longrightarrow \operatorname{Ext}^1(M,{}_AA) \longrightarrow 0,$$

thus, tensoring with ${}_AT$, we obtain the upper exact sequence of

$$\begin{array}{ccccccc}
\mathrm{Hom}(P_o,{}_AA)\otimes{}_AT & \longrightarrow & \mathrm{Hom}(P_1,{}_AA)\otimes{}_AT & \longrightarrow & \mathrm{Ext}^1_A(M,{}_AA)\otimes{}_AT & \longrightarrow & 0\\
\downarrow{\scriptstyle\gamma_o} & & \downarrow{\scriptstyle\gamma_1} & & & & \\
\mathrm{Hom}(P_o,{}_AT) & \longrightarrow & \mathrm{Hom}(P_1,{}_AT) & \longrightarrow & \mathrm{Ext}^1_A(M,T) & \longrightarrow & 0
\end{array}$$

the lower being obtained by applying $\mathrm{Hom}(-,{}_AT)$ to the resolution. Also, there are the isomorphisms γ_o, γ_1 which give a commutative square, with $p_i[(\varphi_i \otimes t)\gamma_i] = (p_i\varphi_i)t$ for $p_i \in P_i$, $\varphi_i \in \mathrm{Hom}(P_i,{}_AA), t \in T]$. Thus

$$\begin{aligned}
\Sigma\tau_A M &= \mathrm{Hom}_A({}_AT_B, D\,\mathrm{Ext}^1_A(M,{}_AA))\\
&\approx \mathrm{Hom}_A(\mathrm{Ext}^1_A(M,{}_AA), D({}_AT_B))\\
&\approx D(\mathrm{Ext}^1_A(M,{}_AA)\otimes{}_AT_B)\\
&\approx D\,\mathrm{Ext}^1_A(M,{}_AT_B) = \tau_B\Sigma M,
\end{aligned}$$

where the first isomorphism is application of D, the second is the adjunction isomorphism.

(d) Let ${}_AM \in G({}_AT)$, $\mathrm{proj.dim.}_AM \leq 1$, and $X \in X({}_AT)$. Then

$$\mathrm{Ext}^1_B(\Sigma M,X) \approx D\,\mathrm{Hom}(X,\tau\Sigma M) = D\,\mathrm{Hom}(X,\Sigma\tau M) = 0,$$

with the first isomorphism being due to $\mathrm{proj.dim.}\Sigma M \leq 1$. Now, let ${}_AM \in G({}_AT)$, $\mathrm{proj.dim.}_AM = n \geq 2$, and $X \in X({}_AT)$. Consider the exact sequence (*) with $T' \in \langle{}_AT\rangle$, and apply first Σ, then $\mathrm{Hom}_B(-,X)$ to it. We obtain an isomorphism

$$\mathrm{Ext}^{n-1}_B(\Sigma K,X) \longrightarrow \mathrm{Ext}^n_B(\Sigma M,X),$$

due to the fact that $\Sigma T'$ is projective. However, $\mathrm{proj.dim.}_AK \leq n-1$, thus by induction $\mathrm{Ext}^{n-1}_B(\Sigma K,X) = 0$, and therefore $\mathrm{Ext}^n_B(\Sigma M,X) = 0$.

(b) Let $N \in F({}_AT)$. Let $E_1,\ldots,E_s$ be a k-basis of $\mathrm{Ext}^1_A(T,N)$ and take the sequence

(**) $$(E_i)_i:\quad 0 \longrightarrow N \longrightarrow \widetilde{N} \longrightarrow \bigoplus_{i=1}^{s} T \longrightarrow 0.$$

Applying $\mathrm{Hom}(T,-)$, we obtain

$$0 \longrightarrow \mathrm{Hom}(T,\widetilde{N}) \longrightarrow \mathrm{Hom}(T,\bigoplus_{i=1}^{s} T) \xrightarrow{\delta} \mathrm{Ext}^1(T,N) \longrightarrow \mathrm{Ext}^1(T,\widetilde{N}) \longrightarrow 0,$$

the first zero is due to the fact that $N \in F({}_AT)$. Also, the connecting morphism δ is surjective, due to the construction of (**), thus $\mathrm{Ext}^1(T,\widetilde{N}) = 0$. Thus, we deal with the exact sequence

$$0 \longrightarrow \mathrm{Hom}(T,\tilde{N}) \longrightarrow \bigoplus_{i=1}^{s} {}_B B \longrightarrow \mathrm{Ext}^1(T,N) \longrightarrow 0.$$

Since proj.dim.$_A T \leq 1$, the sequence (**) shows that proj.dim.$_A\tilde{N} \leq$ max(1,proj.dim.$_A N$). Since $\mathrm{Ext}^1(T,\tilde{N}) = 0$, we know that $\tilde{N} \in G({}_AT)$, thus

$$\text{proj.dim.}_B\mathrm{Hom}(T,\tilde{N}) \leq \text{proj.dim.}_A\tilde{N},$$

according to (a). Thus

$$\begin{aligned}\text{proj.dim}_B\mathrm{Ext}^1(T,N) &\leq 1 + \text{proj.dim.}\mathrm{Hom}(T,\tilde{N}) \\ &\leq 1 + \max(1,\text{proj.dim.}_AN).\end{aligned}$$

(6*) There are corresponding (even better!) results by Bongartz for the injective dimension of the modules in $Y({}_AT)$ and in $X({}_AT)$.

(a) If ${}_AM \in G({}_AT)$, then inj.dim.$_B\Sigma M \leq 1 +$ inj.dim.$_AM$

(b) If ${}_AN \in F({}_AT)$, then inj.dim.$_B\Sigma' N \leq$ inj.dim.$_AN$

(c) If ${}_AN \in F({}_AT)$ has inj.dim.$_AN = n$, then $\mathrm{Ext}^n(Y({}_AT),\Sigma' N) = 0$.

Corollary. If ${}_AT$ is a tilting module, and $B = \mathrm{End}({}_AT)$, then

$$\text{gl.dim.}A-1 \leq \text{gl.dim.}B \leq \text{gl.dim.}A+1.$$

Proof. The assertions (6) (a), (b) show that for gl.dim.A = n, we have proj.dim.$_BX \leq n+1$ for any module $_BX$ in $X({}_AT) \vee Y({}_AT)$, thus also for any module in $X({}_AT) \urcorner\!\llcorner Y({}_AT) = B$-mod, thus gl.dim.B $\leq$ gl.dim.A+1. But if ${}_AT$ is a tilting module, also $_{B^{op}}T$ is a tilting module, and $\mathrm{End}(_{B^{op}}T) = A^{op}$, thus

$$\text{gl.dim.}A = \text{gl.dim.}A^{op} \leq \text{gl.dim.}B^{op}+1 = \text{gl.dim.}B+1.$$

(7) Given a tilting module ${}_AT$ with $B = \mathrm{End}({}_AT)$, there is a linear transformation $\sigma_T : K_0(A) \longrightarrow K_0(B)$ which measures the change of dimension vectors under the functors Σ_T and Σ_T', namely such that

$$(\underline{\dim}\, M)\sigma_T = \underline{\dim}\, \Sigma_T M - \underline{\dim}\, \Sigma_T' M$$

for any A-module M. We note that σ_T is uniquely determined by this formula, and is, in fact, an isomorphism of groups.

Let us now assume that the Cartan matrix C_A is invertible, and let $T(1),\ldots,T(n)$ be the indecomposable direct summands of ${}_AT$, with $n = \mathrm{rank}\, K_0(A)$. Let $t(i) = \underline{\dim}\, T(i)$, and t the n×n-matrix with $e(i)t = t(i)$.

(a) $\sigma_T = C_A^{-1} t^T$.

<u>Proof.</u> Note that $\Sigma_T T(j) =: P_B(j)$, $1 \leq j \leq n$, are the indecomposable projective B-modules. Since the indecomposable injective A-modules $Q_A(i)$ belong to $\mathcal{G}({}_A T)$, we have

$$\begin{aligned}(q_A(i)\sigma_T)_j &= (\underline{\dim}\, \Sigma_T Q(i) \\ &= \dim \operatorname{Hom}_B(P_B(j), \Sigma_T Q(i)) \\ &= \dim \operatorname{Hom}_A(T(j), Q(i)) \\ &= t(j)_i = (e(i)t^T)_j ,\end{aligned}$$

thus $e(i)C_A\sigma_T = q_A(i)\sigma_T = e(i)t^T$, for all i, therefore $C_A\sigma_T = t^T$.

We can rewrite this as follows:

(a') $(x\sigma_T)_i = \langle t(i), x\rangle$.

<u>Proof.</u> $\langle t(i),x\rangle = t(i)C^{-T}x^T = e(i)tC_A^{-T}x^T = xC_A^{-1}t^T e(i)^T = (x\sigma)_i$.

(b) $C_B = tC_A^{-T}t^T$.

<u>Proof.</u> $e(i)C_B^T = p_B(i) = \underline{\dim}\, \Sigma_T T(i) = (\underline{\dim}\, T(i))\sigma_T = e(i)tC_A^{-1}t^T$, thus $C_B^T = tC_A^{-1}t^T$.

(c) $C_A^{-T} = \sigma_T C_B^{-T}\sigma_T^T$.

<u>Proof.</u> $\sigma_T C_B^{-T}\sigma_T^T = C_A^{-1}t^T \cdot t^{-T}C_A t^{-1} \cdot tC_A^{-T} = C_A^{-T}$.

(c') For all $x,y \in K_o(A)$, we have $\langle x,y\rangle_A = \langle x\sigma_T, y\sigma_T\rangle_B$, and therefore also $\chi_A(x) = \chi_B(x\sigma_T)$.

Applying this to the dimension vectors of A-modules we obtain

(c") <u>If</u> ${}_A M$ <u>belongs to</u> $\mathcal{G}({}_A T)$, <u>then</u> $\chi_A(\underline{\dim}\, M) = \chi_B(\underline{\dim}\, \Sigma M)$. <u>If</u> ${}_A N$ <u>belongs to</u> $\mathcal{F}({}_A T)$, <u>then</u> $\chi_A(\underline{\dim}\, M) = \chi_B(\underline{\dim}\, \Sigma' N)$.

<u>Proof.</u> If ${}_A M \in \mathcal{G}({}_A T)$, then $\underline{\dim}\, \Sigma M = (\underline{\dim}\, M)\sigma$, thus $\chi_B(\underline{\dim}\, \Sigma M) = \chi_B((\underline{\dim}\, M)\sigma) = \chi_A(\underline{\dim}\, M)$. Similarly, for ${}_A N \in \mathcal{F}({}_A T)$, we have $\underline{\dim}\, \Sigma' N = -(\underline{\dim}\, N)\sigma$, thus $\chi_B(\underline{\dim}\, \Sigma N) = \chi_B(-(\underline{\dim}\, N)\sigma) = \chi_A(\underline{\dim}\, N)$.

(d) $\Phi_A\sigma_T = \sigma_T\Phi_B$.

Proof. $\Phi_A\sigma_T = -C_A^{-T}C_A \cdot C_A^{-1}t^T = -C_A^{-T}t^T$

$= -C_A^{-1}t^T \cdot t^{-T}C_At^{-1} \cdot tC_A^{-T}t^T = \sigma_T\Phi_B$.

(e) If $T(1),\dots,T(n)$ all are directing, then

$$F({}_AT) = \langle {}_AM \text{ indecomposable} \mid \langle t(i),\underline{\dim} M\rangle \le 0 \text{ for all } i\rangle,$$

$$G({}_AT) = \langle {}_AM \text{ indecomposable} \mid \langle t(i),\underline{\dim} M\rangle \ge 0 \text{ for all } i\rangle.$$

Proof. Let ${}_AM$ be indecomposable. We cannot have both $\mathrm{Hom}_A(T(i),M) \neq 0$, $\mathrm{Ext}_A^1(T(i),M) \neq 0$, since otherwise we obtain a cycle $T(i) \preceq M \preceq \tau T(i) \prec T(i)$. Therefore, if $\langle t(i),\underline{\dim} M\rangle \le 0$, then $\mathrm{Hom}_A(T(i),M) = 0$, and if $\langle t(i),\underline{\dim} M\rangle \ge 0$, then $\mathrm{Ext}_A^1(T(i),M) = 0$, using that

$$\langle t(i),\underline{\dim} M\rangle = \dim \mathrm{Hom}_A(T(i),M) - \dim \mathrm{Ext}_A^1(T(i),M).$$

Of course, we also have the reverse implications. This finishes the proof.

We may reformulate this as follows:

(e') If $T(1),\dots,T(n)$ are directing, then

$$F({}_AT) = \langle {}_AM \text{ indecomposable} \mid (\underline{\dim} M)\sigma_T < 0\rangle,$$

$$G({}_AT) = \langle {}_AM \text{ indecomposable} \mid (\underline{\dim} M)\sigma_T > 0\rangle,$$

Proof. According to (a'), the components of $x\sigma_T$ are just given by $\langle t(i),x\rangle$; and, for ${}_AM \neq 0$, also $\underline{\dim}\, M \neq 0$, thus $(\underline{\dim}\, M)\sigma_T \neq 0$, since σ_T is invertible.

(8) Remark Let us mention that some subspace categories $\check{U}(K,|\cdot|)$ are of the form $G({}_AT)$, with ${}_AT$ a tilting module, so that we can use tilting theory in order to study these categories, We recall from 2.5 that given a finite Krull-Schmidt category K and an additive functor $|\cdot| : K \to k\text{-mod}$, there is the injective realization of $(K,|\cdot|)$ given by $K = A_o\text{-inj}$, $|\cdot| = \mathrm{Hom}(R,-)$, where A_o is some finite-dimensional algebra, and R an A_o-module. Now assume, A_o is hereditary, and let T_o be an injective cogenerator, thus $K = \langle T_o\rangle$. We form the one-point extension $A = A_o[R]$, and denote by ω the extension vertex. Let $P(\omega)$ be the corresponding indecomposable projective A-module, with radical inclusion $\nu: R \to P(\omega)$ Let $\mu: R \to Q$ be an injective envelope in A_o-mod, and form the push out

$$\begin{array}{ccc} R & \xrightarrow{\ \nu\ } & P(\omega) \\ {\scriptstyle\mu}\downarrow & & \downarrow{\scriptstyle\mu'} \\ Q & \xrightarrow{\ \nu'\ } & G(\omega) \end{array}$$

Lemma. The A-module ${}_AT = T_o \oplus G(\omega)$ is a tilting module, and $\check{U}(K,|\cdot|) \approx G({}_AT)$.

Proof. Since A_o is hereditary, proj.dim.$_{A_o}T_o \leq 1$, thus also proj.dim.$_AT_o \leq 1$. The cokernel of ν is $E(\omega)$, thus we also have the exact sequence

$$(*) \qquad 0 \longrightarrow Q \longrightarrow G(\omega) \longrightarrow E(\omega) \longrightarrow 0,$$

note that $E(\omega)$ is simple injective. Similarly, we have an exact sequence

$$(**) \qquad 0 \longrightarrow P(\omega) \longrightarrow G(\omega) \longrightarrow Q/R \longrightarrow 0,$$

with Q/R being (as a factor module of Q) an injective A_o-module. It follows that the projective dimension of Q/R both as an A_o-module as well as an A-module is ≤ 1. Since $P(\omega)$ is A-projective, we conclude from $(**)$ that proj.dim.$_AG(\omega) \leq 1$. Also, applying $\mathrm{Ext}^1_A(T_o,-)$ to $(*)$, or $\mathrm{Ext}^1_A(-,T_o)$ to $(**)$, we obtain the following two sequences:

$$0 = \mathrm{Ext}^1_A(T_o,Q) \longrightarrow \mathrm{Ext}^1_A(T_o,G(\omega)) \longrightarrow \mathrm{Ext}^1_A(T_o,E(\omega)) = 0,$$

$$0 = \mathrm{Ext}^1_A(Q/R,T_o) \longrightarrow \mathrm{Ext}^1_A(G(\omega),T_o) \longrightarrow \mathrm{Ext}^1_A(P(\omega),T_o) = 0.$$

In both cases, the first term is zero, since this Ext^1_A-group is in fact an $\mathrm{Ext}^1_{A_o}$-group. Thus, both middle terms are zero. In particular, with $\mathrm{Ext}^1_A(G(\omega),T_o)=0$ also $\mathrm{Ext}^1(G(\omega),Q) = 0$, since $Q \in \langle T_o \rangle$. Applying $\mathrm{Ext}^1(G(\omega)),-)$ to $(*)$, we get

$$0 = \mathrm{Ext}^1(G(\omega),Q) \longrightarrow \mathrm{Ext}^1(G(\omega),G(\omega)) \longrightarrow \mathrm{Ext}^1(G(\omega),E(\omega)) = 0,$$

altogether we see that $\mathrm{Ext}^1(T_o \oplus G(\omega),T_o \oplus G(\omega)) = 0$. It follows that $T_o \oplus G(\omega)$ satisfies all the conditions (α), (β), (γ) of a tilting module.

In 2.5.2', we have identified $\check{U}(K,|\cdot|)$ with the module class in A-mod consisting of all A-modules with restrictions to A_o being A_o-injective. We claim that this module class is just $G({}_AT)$. Clearly, if ${}_AM$ is generated by ${}_AT$, then its restriction to A_o is generated by $T_o \oplus Q$, but since A_o is hereditary, any module generated by $T_o \oplus Q$ is A_o-injective. Conversely, let ${}_AM$ be an A-module, and assume its maximal A_o-submodule M_o is A_o-injective. Since M/M_o is a direct sum of copies of $E(\omega)$, the module $T_o \oplus P(\omega)$ generates M. Given a homomorphism

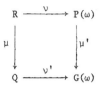

$\varphi : P(\omega) \longrightarrow M$, we claim that we can factor φ through $G(\omega)$. Namely, the restriction of φ to R maps into M_o, and, since M_o is injective, we can extend $\varphi | R$ along $\mu : R \longrightarrow Q$, thus we obtain $\varphi' : Q \longrightarrow M$ with $\nu\varphi = \mu\varphi'$. Since, by definition $G(\omega)$ is the pushout of ν and μ, it follows that φ factors through μ'. This shows that also $T = T_o \oplus G(\omega)$ generates ${}_A M$.

4.2 Tilted algebras

An algebra of the form $\mathrm{End}({}_AT)$, with ${}_AT$ a tilting module, and A hereditary, is said to be a tilted algebra.

(1) Let A be hereditary, ${}_AT$ a tilting-module with $B = \mathrm{End}({}_AT)$. Then $B\text{-mod} = \mathcal{X}({}_AT) \vee \mathcal{Y}({}_AT)$. As a consequence, $\mathcal{Y}({}_AT)$ is closed under predecessors, and $\mathcal{X}({}_AT)$ under successors. In particular, $\mathcal{Y}({}_AT)$ is closed under τ_B, and $\mathcal{X}({}_AT)$ under τ_B^-. Also, any indecomposable injective B-module which does not belong to $\mathcal{X}({}_AT)$ is of the form $\Sigma Q_A(a)$, with $P_A(a) \in \langle {}_AT \rangle$.

Proof. The first assertion is a direct consequence of 4.1.6.d . If M_1, M_2 are indecomposable B-modules, with $\mathrm{Hom}(M_1,M_2) \neq 0$, and M_2 belongs to $\mathcal{Y}({}_AT)$, then also $M_1 \in \mathcal{Y}({}_AT)$, since $\mathrm{Hom}(\mathcal{X}({}_AT),\mathcal{Y}({}_AT)) = 0$. Finally, if $G \in \mathcal{G}({}_AT)$ and ΣG is an indecomposable injective B-module, let Q be an injective envelope of G. The exact sequence

$$(*) \qquad 0 \longrightarrow G \longrightarrow Q \longrightarrow Q/G \longrightarrow 0$$

is mapped under Σ to an exact sequence, since $G \in \mathcal{G}({}_AT)$, and the image sequence splits, thus $(*)$ splits. But this is possible only for $G = Q$. Thus $G = Q_A(a)$, and then use the connecting lemma.

As a consequence, given an Auslander-Reiten sequence of B-mod either all terms are in $\mathcal{Y}({}_AT)$, or all terms are in $\mathcal{X}({}_AT)$, or finally, the left hand term is in $\mathcal{Y}({}_AT)$, the right hand term is in $\mathcal{X}({}_AT)$. There are only a few Auslander-Reiten sequences of the last form, and they can be described explicitly. They will be called the connecting sequences.

(2) If $\quad 0 \longrightarrow N_1 \xrightarrow{f} N_2 \xrightarrow{g} N_3 \longrightarrow 0 \quad$ is an Auslander-Reiten sequence in B-mod with $N_1 \in \mathcal{Y}({}_AT)$, $N_3 \in \mathcal{X}({}_AT)$, then there is some vertex a with $P(a) \notin \langle {}_AT \rangle$ such that $N_1 = \Sigma I(a)$, and $N_3 = \Sigma' P(a)$.

Proof. Let $N_1 = \Sigma X$ for some $X \in \mathcal{G}({}_AT)$, and let Q be the injective envelope of X. The exact sequence

$$(*) \qquad 0 \longrightarrow X \longrightarrow Q \longrightarrow Q/X \longrightarrow 0$$

goes to the following exact sequence in $\mathcal{Y}({}_AT)$.

$$(**) \qquad 0 \longrightarrow \Sigma X \longrightarrow \Sigma Q \longrightarrow \Sigma Q/X \longrightarrow 0.$$

If this sequence does not split, then it induces the Auslander-Reiten sequence starting with $\Sigma X = N_1$. However, $\mathrm{Hom}(N_3,\Sigma Q/X) = 0$, since $N_3 \in \mathcal{X}({}_AT), \Sigma Q/X \in \mathcal{Y}({}_AT)$.

Thus the sequence (**) splits, and therefore also (*), but this is possible only in case $Q/X = 0$, thus X is injective, say $X = I(a)$ for some a. Since N_1 is not injective, $P(a) \notin \langle {}_AT \rangle$. Also, the connecting lemma shows that $N_3 = \tau_B^- N_1 = \Sigma' P(a)$.

In order to determine the middle term of a connecting sequence, we consider the source map for an arbitrary module $\Sigma Q_A(a)$.

(2') The source map for $\Sigma Q_A(a)$ is of the form

$$\Sigma Q_A(a) \longrightarrow \Sigma(Q_A(a)/E_A(a)) \oplus \Sigma' \text{ rad } P_A(a).$$

As a consequence, the connecting sequences are of the form

$$0 \longrightarrow \Sigma Q_A(a) \longrightarrow \Sigma(Q_A(a)/E_A(a)) \oplus \Sigma' \text{ rad } P_A(a) \longrightarrow \Sigma' P_A(a) \longrightarrow 0,$$

where a is any vertex of $\Delta(A)$ with $P_A(a) \notin \langle {}_AT \rangle$.

In order to characterize tilted algebras, let us introduce the concept of a slice S, [deviating from the use of this word in previous publications]. Let B be an algebra, and S a module class in B-mod. Then S will be called a <u>slice</u> provided the following conditions are satisfied:

(α) S is sincere (it contains a (not necessarily indecomposable) sincere module).

(β) S is path closed (if $S_0 \prec M \prec S_1$, and $S_0, S_1 \in S$, then also $M \in S$).

(γ) If M is indecomposable and not projective, then at most one of M, τM belongs to S, and

(δ) If M, S are indecomposable, $f : M \longrightarrow S$ irreducible, and $S \in S$, then either $M \in S$ or M is not injective and $\tau^- M \in S$.

(3) <u>Let A be hereditary, and ${}_AT$ a tilting module with $\operatorname{End}({}_AT) = B$, then $\{\Sigma_A Q \mid {}_AQ$ injective$\}$ is a slice in B-mod. Conversely, any slice in any module category occurs in this way.</u>

<u>Proof</u> (α): The module $\Sigma D(A_A)$ is sincere: given an indecomposable direct summand T' of T, we have $\operatorname{Hom}_A(T', D(A_A)) \neq 0$, thus $\operatorname{Hom}_B(\Sigma T', \Sigma D(A_A)) \neq 0$.

(β) Assume there is given a path $(Y_0, Y_1, \ldots, Y_m)$ in B-mod, with $Y_0 = \Sigma Q_A(a)$, $Y_m = \Sigma Q_A(b)$ where $Q_A(a)$, $Q_A(b)$ are indecomposable injective A-modules. Since $(\mathcal{Y}({}_AT), \mathcal{X}({}_AT))$ is a split torsion pair, all Y_i must belong to $\mathcal{Y}({}_AT)$, since Y_m belongs to $\mathcal{Y}({}_AT)$. Thus $Y_i = \Sigma G_i$ for some $G_i \in \mathcal{G}({}_AT)$, and there is the path $(Q_A(a) = G_0, G_1, \ldots, G_m)$ in $\mathcal{G}({}_AT)$. Since G_0 is injective, and A is hereditary, all G_i are injective.

(γ) Assume M is indecomposable and not projective, and both M, τM belong to $\mathcal{S}$. Let $\tau M = \Sigma Q_A(a)$, then $\Sigma Q_A(a)$ is not injective, thus $M = \tau^{-}\Sigma Q_A(a) = \Sigma' P_A(a)$ belongs to $\mathcal{X}({}_AT)$, a contradiction.

(δ) Let M be indecomposable, $f : M \longrightarrow S$ irreducible and $S \in \mathcal{S}$. With S also M belongs to $\mathcal{Y}({}_AT)$. First, if M is an injective B-module, then $M = \Sigma Q_A(a)$ for some a. Thus, assume M is not injective. Then there exists an irreducible map $S \longrightarrow \tau^{-}M$. If $\tau^{-}M$ is in $\mathcal{Y}({}_AT)$, say $\tau^{-}M = \Sigma G$ for some $G \in \mathcal{G}({}_AT)$, and $S = \Sigma Q$ with Q injective in A-mod, then $\mathrm{Hom}_A(Q,G) \neq 0$ shows that also G is injective, thus $\Sigma G \in \mathcal{S}$. If, on the other hand, $\tau^{-}M$ is in $\mathcal{X}({}_AT)$, then the Auslander-Reiten sequence starting with M is a connecting sequence, thus $M = \Sigma Q_A(a)$ for some $Q_A(a)$.

Now, conversely, let B be an arbitrary finite-dimensional algebra, and $\mathcal{S}$ a slice in B-mod. First, proj.dim.$S \leq 1$, for any $S \in \mathcal{S}$. We can assume S to be indecomposable. Take an indecomposable injective module Q, and assume $\mathrm{Hom}(Q,\tau S) \neq 0$. Since $\mathcal{S}$ is sincere, there is an indecomposable $S' \in \mathcal{S}$ with $\mathrm{Hom}(S',Q) \neq 0$, thus

$$S' \preceq Q \preceq \tau S \prec S$$

implies that with S' and S also τS belongs to $\mathcal{S}$, contrary to condition (γ). Thus proj.dim.$S \leq 1$, according to 2.4.1 . Dually, we also have inj.dim. $S \leq 1$.

Similarly, given two indecomposable modules $S, S' \in \mathcal{S}$, then

$$\mathrm{Ext}^1(S,S') \approx D\,\overline{\mathrm{Hom}}(S',\tau S)$$

has to be zero, since otherwise

$$S' \preceq \tau S \prec S$$

contradicts again (β) and (γ).

By 4.1.1, $\mathcal{S}$ contains only finitely many isomorphism classes of indecomposable modules, thus take a module S in $\mathcal{S}$ with $\mathcal{S} = \langle S\rangle$. According to 4.1 1 , there exists M such that $S \oplus M$ is a tilting module, and such that any non-projective indecomposable direct summand M' of M satisfies $\mathrm{Hom}(M',S) \neq 0$. Actually, since S is sincere, also the projective indecomposable direct summands M' of M have this property. Now, let $0 \neq \varphi_0 : M' \longrightarrow S_0$, where $M' \in \langle M\rangle$, $S_0 \in \mathcal{S}$, both being indecomposable. If M' does not belong to $\mathcal{S}$, we obtain inductively maps $\varphi_i : M' \longrightarrow S_i$, and irreducible maps $f_i : S_i \longrightarrow S_{i-1}$, with S_i indecomposable and in $\mathcal{S}$, such that $\varphi_i f_i \ldots f_1 \neq 0$. [Namely, since $M' \notin \mathcal{S}$, the map φ_i is not an isomorphism, thus we can factor it through the sink map $f'_{i+1} : S'_{i+1} \longrightarrow S_i$, say $\varphi_i = \varphi'_{i+1} f'_{i+1}$. Decomposing $S'_{i+1} = \bigoplus_j S'_{i+1,j}$ with $S'_{i+1,j}$

indecomposable, we can write $\varphi_i = \sum_j \varphi'_{i+1,j} f'_{i+1,j}$, with $\varphi'_{i+1,j} : M' \longrightarrow S'_{i+1,j}$. There is some j with $\varphi'_{i+1,j} \cdot f'_{i+1,j} \cdot f_i \dots f_1 \neq 0$, let $\varphi_{i+1} = \varphi'_{i+1,j}$, $f_{i+1} = f'_{i+1,j}$, and $S_{i+1} = S'_{i+1,j}$ for this j. We claim that S_{i+1} belongs to $\mathcal{S}$. Assume not. Since there is the irreducible map $f_{i+1} : S_{i+1} \longrightarrow S_i$, it follows from ($\delta$) that S_{i+1} is not injective, and that $\tau^- S_{i+1}$ belongs to $\mathcal{S}$. Thus, proj.dim. $\tau^- S_{i+1} \leq 1$, and therefore

$$\mathrm{Ext}^1(\tau^- S_{i+1}, M') = D\,\mathrm{Hom}(M', S_{i+1}) \neq 0,$$

contrary to the fact that $\tau^- S_{i+1}$ is a direct summand of S, and $\mathrm{Ext}^1(S,M') = 0$.] But this is impossible, since all f_i are in rad End(S), and this ideal is nilpotent of some fixed degree. This shows that $M' \in \mathcal{S}$, and therefore S itself is a tilting module. Actually, since we also have inj.dim. $S \leq 1$, S also is a cotilting module.

Let $A = \mathrm{End}({}_B S)$. We want to show that A is hereditary. Let Q be an indecomposable injective A-module, and we want to show that for any irreducible map $p : Q \longrightarrow Q'$ in A-mod, the module Q' is injective again. Since ${}_B S$ is a cotilting module, there is the tilting module ${}_A DS$, and $B = \mathrm{End}({}_A DS)$, and we may consider $\mathcal{Y} = \mathcal{Y}({}_A DS)$, $\mathcal{X} = \mathcal{X}({}_A DS)$. Now, the image of Q under Σ is some indecomposable module S_o in $\mathcal{S}$. Let M be an indecomposable B-module with an irreducible map $S_o \longrightarrow M$. Then M either belongs to $\mathcal{S}$ or to $\mathcal{X}$. [Namely, if M is projective, then there is S_1 indecomposable in $\mathcal{S}$ with $\mathrm{Hom}(M,S_1) \neq 0$, since $\mathcal{S}$ is sincere, thus $S_o \preceq M \preceq S_1$ shows that $M \in \mathcal{S}$. If M is not projective, there is an irreducible map $\tau M \longrightarrow S_o$, and, according to ($\delta$), either τM or M belongs to $\mathcal{S}$. If τM is in $\mathcal{S}$, then τM is the image under Σ_{DS} of an indecomposable injective A-module, and therefore $M = \tau^- \tau M$ belongs to $\mathcal{X}$ due to the connecting lemma.] Thus, the source map for S_o is of the form $f = (f', f'') : S_o \longrightarrow S' \oplus X$, where $S' \in \mathcal{S}$, $X \in \mathcal{X}$, and $f' : S_o \longrightarrow S'$, $f'' : S_o \longrightarrow X$. Now, consider the map $\Sigma p : \Sigma Q = S_o \longrightarrow \Sigma Q'$, it is a (relative) irreducible map in $\mathcal{Y}$, since p is irreducible in $\mathcal{G}({}_A DS)$. We can factor Σp through $f = (f', f'')$, say $\Sigma p = f'h' + f''h''$, however $h'' = 0$, since $X \in \mathcal{X}$ and $\Sigma Q' \in \mathcal{Y}$. Thus $\Sigma p = f'h'$ is a factorization of Σp inside $\mathcal{Y}$, and f' is not split mono, since it is irreducible or zero, thus h' has to be split epi. This shows that $\Sigma Q'$ is a direct summand of S', and therefore in $\mathcal{S}$. Note that Q' belongs to $\mathcal{G}({}_A DS)$, since an irreducible map $p : Q \longrightarrow Q'$ with Q injective, has to be an epimorphism. Thus $\Sigma Q' \in \mathcal{S}$ implies that Q' itself is injective. Altogether, we have shown that for an indecomposable injective A-module Q, also $Q/\mathrm{soc}\, Q$ is injective, thus A is hereditary. This finishes the proof.

We have shown above that any slice S in B-mod is of the form $S = \langle {}_BS \rangle$ for some B-module ${}_BS$, and such a module ${}_BS$ will be called a slice module. Note that ${}_BS$ is a slice module if and only if ${}_BS$ is a cotilting module with $\mathrm{End}({}_BS)$ being hereditary, or, equivalently, if and only if ${}_BS$ is a tilting module with $\mathrm{End}({}_BS)$ being hereditary. [The equivalence of the last two conditions comes from the fact that for A hereditary, ${}_AT$ is a tilting module if and only if it is a cotilting module, this fact being applied to DS.] Note that if S is a slice in B-mod with slice module ${}_BS$, and $A = \mathrm{End}({}_BS)$, then the quiver $\Delta(A)$ of A is just the opposite $\Delta(S)^*$ of the quiver $\Delta(S)$ of S. Namely,

$$\Delta(A) \approx \Delta(A\text{-proj})^* \approx \Delta(A\text{-inj})^* \approx \Delta(S)^*$$

(for the first isomorphism, we refer to 2.4, the second is induced by ν, the third by $\mathrm{Hom}_A({}_ADS,-)$).

(4) Let B be an algebra, and C a component of B-mod. Assume C contains a slice module S with $\mathrm{End}({}_BS) = A$, and let Δ be the quiver of A. Then ΓC can be embedded into $\mathbb{Z}\Delta^*$.

Proof. Let $S = \bigoplus_{i=1}^{n} S_i$ with S_i indecomposable, and, as we may suppose, pairwise indecomposable. Let $s_i = [S_i]$, and consider $s_1, \ldots, s_n$ as the vertices of Δ^*. Given an indecomposable module of the form $\tau^z S_i$, with $z \in \mathbb{Z}$, we want to send $[\tau^z S_i]$ to the pair $(z, s_i) \in \mathbb{Z}\Delta^*$. Note that this is welldefined, since no S_i can be τ-periodic, due to the conditions (β) and (γ) of a tilting module. We claim that any indecomposable module belonging to C is of the form $\tau^z S_i$ for some pair (s, s_i). Namely, assume there is given an irreducible map $\tau^z S_i \longrightarrow M$, with M indecomposable. First, let $z \leq 0$. If $\tau^t M$ is non-zero and projective, for some $0 \leq t \leq -z$, then $\mathrm{Hom}(\tau^t M, S_j) \neq 0$ for some summand S_j of ${}_BS$, due to (α) and therefore $\tau^t M$ is summand of ${}_BS$, due to (β). Otherwise, we obtain an irreducible map $\tau^{-z+1}M \longrightarrow S_i$, and thus either $\tau^{-z+1}M$ or $\tau^{-z}M$ is a summand of S. Similarly, if $z > 0$, and some $\tau^{-t}M$ is non-zero and injective, where $0 \leq t < z$, then $\mathrm{Hom}(S_j, \tau^{-t}M) \neq 0$, due to ($\alpha$), and $\mathrm{Hom}(\tau^{-t}M, \tau^{z-t-1}S_j) \neq 0$, (with $z-t-1 \geq 0$), thus $\tau^{-t}M$ is a summand of S, due to (β). Or, if $z > 0$, and $\tau^{-z}M$ is non-zero, then either $\tau^{-z+1}M$ or $\tau^{-z}M$ is a summand of S, due to (δ). This finishes the proof.

(4') Corollary. A tilted representation finite algebra is directed.

Proof. There is just one component, and it is embedded into some $\mathbb{Z}\Delta^*$.

(4") Corollary. If B is a tilted, representation finite algebra, then χ_B is weakly positive, and dim furnishes a bijection between the indecomposable B-modules and the positive roots of χ_B.

Proof. Since gl.dim. $B \leq 2$, the result follows from (4') and 2.4.9.

(4''') Corollary. Let $\mathcal{P}$ be a preprojective component in B-mod containing a slice $\mathcal{S}$. Let $\Delta(\mathcal{S})$ be the quiver of $\mathcal{S}$, and $\mathcal{O}(\mathcal{P}) = \mathcal{O}(\Gamma\mathcal{P})$ the orbit quiver of $\mathcal{P}$. We can identify the underlying graphs $\overline{\Delta(\mathcal{S})}$ and $\overline{\mathcal{O}(\mathcal{P})}$, associating to an indecomposable module S in $\mathcal{S}$ (considered as a vertex of $\Delta(\mathcal{S})$) its τ-orbit (considered as a vertex of $\mathcal{O}(\mathcal{P})$).

(5) Given a slice module ${}_BS$, say with $A = \mathrm{End}({}_BS)$, let $s = \underline{\dim}\, {}_BS$. Using s, we can characterize various module classes in B-mod numerically, as follows:

$$F({}_BS) = \langle {}_BM \text{ indecomposable} \mid \langle s, \underline{\dim}\, M\rangle < 0 \rangle,$$

$$G({}_BS) = \langle {}_BM \text{ indecomposable} \mid \langle s, \underline{\dim}\, M\rangle > 0 \rangle,$$

$$\mathcal{Y}({}_ADS) = \langle {}_BM \text{ indecomposable} \mid \langle \underline{\dim}\, M, s\rangle > 0 \rangle,$$

$$\mathcal{X}({}_ADS) = \langle {}_BM \text{ indecomposable} \mid \langle \underline{\dim}\, M, s\rangle < 0 \rangle.$$

Proof. Let ${}_BM$ be indecomposable. Since proj.dim. ${}_BS \leq 1$, we have

$$\langle s, \underline{\dim}\, M\rangle = \dim \mathrm{Hom}_B(S,M) - \dim \mathrm{Ext}^1_B(S,M).$$

If M belongs to $F({}_BS)$, then M is generated by S, thus $\mathrm{Hom}_B(S,M) \neq 0$, and $\mathrm{Ext}^1_B(S,M) = 0$, thus $\langle s, \underline{\dim}\, M\rangle > 0$. Similarly, if $M \in G({}_BS)$, then M is cogenerated by τS, thus $0 \neq \mathrm{Hom}(M,\tau S) = D\,\mathrm{Ext}^1(S,M)$, and $\mathrm{Hom}_B(S,M) = 0$, therefore $\langle s, \underline{\dim}\, M\rangle < 0$. Conversely, assume $\langle s, \underline{\dim}\, M\rangle > 0$, thus $\mathrm{Hom}_B(S_i,M) \neq 0$ for some indecomposable summand S_i of S. We must have $\mathrm{Ext}^1_B(S_j,M) = 0$ for any indecomposable summand S_j of S, since otherwise we obtain a path $S_i \preceq M \preceq \tau S_j \prec S_j$, contradicting (β) and (γ), thus $M \in G({}_BS)$. Similarly, if $\langle s, \underline{\dim}\, M\rangle < 0$, say $\mathrm{Ext}^1_B(S_i,M) \neq 0$ for some indecomposable summand S_i of M, then $\mathrm{Hom}(S,M) = 0$. For the second two equalities, we note that by definition $\mathcal{Y}({}_ADS)$ are the B-modules cogenerated by ${}_BS$, that $\mathcal{X}({}_ADS)$ are the B-modules generated by $\tau^- {}_BS$ and B-mod $= \mathcal{Y}({}_ADS) \vee \mathcal{X}({}_ADS)$. Also note that

$$\langle \underline{\dim}\, M, s\rangle = \dim \mathrm{Hom}_B(M,S) - \dim \mathrm{Ext}^1_B(M,S),$$

since inj.dim. ${}_BS \leq 1$. As above, we see that $\mathrm{Hom}(M,S) \neq 0$ implies $\mathrm{Ext}^1(M,S) = 0$. Thus, if M is in $\mathcal{Y}({}_ADS)$, then $\mathrm{Hom}(M,S) \neq 0$, thus $\langle \underline{\dim}\, M, s\rangle > 0$. Conversely, $\langle \underline{\dim}\, M, s\rangle > 0$ implies $\mathrm{Hom}_B(M,S) \neq 0$. But then M cannot belong to $\mathcal{X}({}_ADS)$, since otherwise we obtain a cycle of the form $S_i \prec \tau^- S_i \preceq M \preceq S_j$, with $S_i, S_j \in \langle {}_BS\rangle$. Similarly, if M is in $\mathcal{X}({}_ADS)$, then $0 \neq \mathrm{Hom}_B(\tau^- S,M) = D\,\mathrm{Ext}^1_B(M,S)$, thus $\langle \underline{\dim}\, M, s\rangle < 0$. Conversely, if $\langle \underline{\dim}\, M, s\rangle < 0$, then we have seen above that M does not belong to $\mathcal{Y}({}_ADS)$, thus it has to belong to $\mathcal{X}({}_ADS)$. This finishes the proof.

Let us add that we have

$$G({}_BS) \cap \mathcal{Y}({}_ADS) = \langle {}_BS \rangle .$$

[Clearly, ${}_BS$ belongs to both $G({}_BS)$ and $\mathcal{Y}({}_ADS)$. Conversely, if an indecomposable module is both generated and cogenerated by ${}_BS$, then it must belong to $\langle {}_BS\rangle$, since $\langle {}_BS\rangle$ is path closed]. Thus we have

$$\langle {}_BS\rangle = \langle {}_BM \text{ indecomposable} \mid \langle s,\underline{\dim}\, M\rangle > 0,\ \langle \underline{\dim}\, M,s\rangle > 0 \rangle .$$

In order to construct slices, let us consider components containing an indecomposable sincere representation. Recall that a module class $\mathcal{C}$ is said to be standard provided $\mathcal{C} \approx k(\Gamma\mathcal{C})$. Also, a module class will be called convex, provided given indecomposable modules $C_1 \preceq X \preceq C_2$ with $C_1, C_2 \in \mathcal{C}$, also $X \in \mathcal{C}$. Finally, a module class $\mathcal{C}$ will be said to be directed provided it does not contain cycles. Note that a component which is preprojective or preinjective is standard, convex and directed (2.4.11). In general, given a translation quiver Γ, a path $(x_0|\alpha_1,\ldots,\alpha_t|x_t)$ in Γ is said to be a sectional path provided $\tau x_{i+1} \neq x_{i-1}$, for all $1 < i < t$, where $\alpha_i : x_{i-1} \to x_i$. Given an indecomposable module M in a component $\mathcal{C}$, let $\mathcal{S}(\to M)$ be the module class given by all indecomposable modules X in $\mathcal{C}$ with $X \preceq M$, and such that any path from $[X]$ to $[M]$ in $\Gamma\mathcal{C}$ is sectional. Dually, let $\mathcal{S}(M\to)$ be the module class given by all indecomposable modules Y in $\mathcal{C}$ with $M \preceq Y$, and such that any path from $[M]$ to $[Y]$ in $\Gamma\mathcal{C}$ is sectional.

(6) Let M be an indecomposable sincere module in a standard, convex, and directed component. Then both $\mathcal{S}(\to M)$ and $\mathcal{S}(M\to)$ are slices. More generally, if X is indecomposable and in $\mathcal{S}(\to M)$, then $\mathcal{S}(X\to)$ is a slice; if Y is indecomposable and in $\mathcal{S}(M\to)$, then $\mathcal{S}(\to Y)$ is a slice.

Proof. Let X be indecomposable and in $\mathcal{S}(\to M)$. We verify the conditions (α), (β), (γ), (δ). Since $M \in \mathcal{S}(X\to)$, and M is sincere, condition (α) is satisfied. Assume $S_0 \preceq N \preceq S_1$, with $S_0, S_1 \in \mathcal{S}(X\to)$. Now $X \preceq S_0 \preceq N$ shows that $X \preceq N$. Since $\mathcal{C}$ is convex, N belongs to $\mathcal{C}$. Since $N \preceq S_1$, and $\mathcal{C}$ is a standard component, there is a path w from $[N]$ to $[S_1]$. Since the composition of any path v from $[X]$ to $[N]$ with w is sectional, also v is sectional, thus we see that N belongs to $\mathcal{S}(X\to)$. This shows (β). In order to establish (γ), assume for the contrary that both N and τN belong to $\mathcal{S}(X\to)$. Thus $X \preceq N$, but then $X \preceq N \prec \tau N$ gives a non-sectional path from $[X]$ to $[\tau N]$. Finally, let $S \in \mathcal{S}(X\to)$, and N indecomposable with an irreducible map $N \to S$. Assume that N is not in $\mathcal{S}(X\to)$. Then $X \not\preceq N$, according to (β). In particular, N cannot be injective, since otherwise $\operatorname{Hom}(M,N) \neq 0$, due to the fact that M is sincere, and then $X \preceq M \preceq N$. Assume also $\tau^- N$ does not belong to $\mathcal{S}(X\to)$. Since $X \preceq S \preceq \tau^- N$, the second condition cannot be satisfied, thus there is a non-sectional path

$$[X] = x_0 \xrightarrow{\alpha_1} x_1 \xrightarrow{\alpha_2} \cdots \xrightarrow{\alpha_t} x_t = [\tau^- N]$$

in $\Gamma(C)$. Choose s maximal with $\tau x_{s+1} = x_{s-1}$. Note that x_t is not a projective vertex. Assume one of the vertices x_i, with $s+2 \leq i < t$, is projective, say $x_i = [P]$ for some indecomposable projective module P. Then $\mathrm{Hom}(P,M) \neq 0$, since M is sincere, and we obtain a non-sectional path from $[X]$ to $[M]$ starting with $\alpha_1,\ldots,\alpha_{i-1}$, a contradiction to $X \in S(\to M)$. However, this means that we can apply τ to the path $(x_{s+1}|\alpha_{s+2},\ldots,\alpha_t|x_t)$ and obtain a path from $\tau x_{s+1} = x_{s-1}$ to $\tau x_t = [N]$, thus $X \preceq N$, a contradiction. This shows that $\tau^- N$ belongs to $S(X\to)$, and finishes the proof that $S(X\to)$ is a slice.

In particular, we may consider the special case $X = M$. Note that M belongs to $S(\to M)$, since C is supposed to be directed. It follows that $S(M\to)$ is a slice. By the dual considerations, $S(\to Y)$ is a slice for any $Y \in S(M\to)$, and, in particular, $S(\to M)$ itself is a slice.

(6') Corollary. If the algebra B has a preprojective component containing an indecomposable sincere module, then B is a tilted algebra.

Proof. This is a direct consequence of (6) and (3).

(6") Assume the algebra B has a preprojective component containing an indecomposable sincere module. Then there exists an indecomposable projective module P such that $S(P\to)$ is a slice, and such that P is not a proper predecessor of any indecomposable projective module.

Proof. First, let us note that given a slice S not containing an indecomposable projective module, then also $\tau S = \{\tau S \mid S \in S\}$ is a slice. This may be proved either directly, using just the definition of a slice, or, easier, using the characterization given in (3).

Let C be a preprojective component in B-mod, and M an indecomposable sincere module. Thus $S(\to M)$ is a slice. Since M belongs to a preprojective component, M has only finitely many predecessors, therefore there is some $t \geq 0$ such that $\tau^{t-1}S(\to M)$ does not contain any indecomposable projective module, whereas $\tau^t S(\to M)$ does. Let P be an indecomposable projective module in $\tau^t S(\to M)$ such that P is not a proper predecessor of any other indecomposable projective module in $\tau^t S(\to M)$. Actually, P is not a proper predecessor of any other indecomposable projective module in B-mod. [Namely, assume $P \preceq P'$, with P' indecomposable projective, then $\mathrm{Hom}(P',M) \neq 0$, since M is sincere. Thus $P' = \tau^s S$ for some S in $S(\to M)$, and some $s \geq 0$. By our assumption on t, we have $s \geq t$, and $P \preceq P'$ now implies $s = t$.] Let $P = \tau^t X$, with X in $S(\to M)$. According to (6), $S(X\to)$ is a slice. Note that $\tau^i S(X\to)$, $0 \leq i < t$ does not contain an indecomposable projective module, since otherwise we would obtain an indecomposable projective module having P as a proper predecessor. Thus, also $\tau^t S(X\to) = S(P\to)$ is a slice.

In section 3.3, we have introduced the notation of a wing module, and have noted that a sincere separating wing module W always is separating, and that any indecomposable module not belonging to the interior of a wing of W is generated or cogenerated by W. The proof however had to be delayed until now. It is a consequence of the following result:

(7) Let W be a sincere, directing wing B-module. Then both $S(\to W)$ and $S(W\to)$ are slices in B-mod.

Proof. An indecomposable module X belongs to $S(\to W)$ if and only if [X] belongs to some wing θ of W, and is a projective vertex of θ (with the notation introduced in 3.3, and 3.4, the indecomposable modules in $S(\to W)$ are those of the form $W_{1j}^{(s)}$, where $1 \leq j \leq n_s$, $1 \leq s \leq t$). For, wings are by definition mesh-complete translation subquivers, and no proper predecessor of W can be injective, since W is both sincere and directing. The properties (α), (γ) and (δ) of a slice are obviously satisfied for $S(\to W)$, it only remains to consider property (β). Thus, let $S_0 \prec M \prec S_1$ with $S_0, S_1 \in S(\to W)$. Let S_0 belong to the wing $\theta^{(s)}$, say $S_0 = W_{1j}^{(s)}$, for some $1 \leq j \leq n_s$. There is a sequence of indecomposable modules $S_0 = M_0, M_1, \ldots, M_m = M$ such that $\mathrm{rad}(M_{i-1}, M_i) \neq 0$ for $1 \leq i \leq m$. Using induction on m, it is sufficient to consider the case $m = 1$, thus assume there is $0 \neq f \in \mathrm{rad}(S_0, M)$. According to 3.4.1, we know that [M] cannot belong to the interior of the wing $\theta^{(s)}$. [Note that the proof of 3.4.1 only uses that W is a directing wing module, and not lemma 3.3]. Assume $M \not\approx W_{1u}^{(s)}$, for all $j \leq u < n_s$. We can use the source maps for the various $W_{iv}^{(s)}$, $1 \leq i \leq v$, $j \leq v < n_s$, in order to write f as a sum of maps each of which factors through one of the modules $W_{in_s}^{(s)}$, $1 \leq i \leq n_s$. In particular, $\mathrm{Hom}(W_{in_s}^{(s)}, M) \neq 0$, for some i, therefore $W \preceq W_{in_s}^{(s)} \preceq M$. [Actually, using the full strength of 3.4.1, we can factor f through a directed sum of copies of W.] Since also $M \preceq S_1 \preceq W$, and W is directing, we conclude $M \approx W = W_{1n_s}^{(s)}$. Altogether we see that M is one of the modules $W_{1u}^{(s)}$, $j \leq u \leq n_s$, thus $M \in S(\to W)$. This finishes the proof.

(7') Corollary. Let W be a sincere directing wing B-module. Then W is separating, and any indecomposable B-module not belonging to the interior of a wing of W is either generated or cogenerated by W.

Proof. We consider the slice $S(\to W)$. Of course, all the modules in $S(\to W)$ are generated by W. Let $(n_1, \ldots, n_t)$ be the type of W as a wing module. Let A be the path algebra of the quiver Δ with underlying graph $\bar{\Delta} = \mathbb{T}_{n_1, \ldots, n_t}$ and having the center a of the star as unique sink. Note that Δ^* is the quiver of $S(W\to)$. According to (3), there is a tilting A-module ${}_AT$ such that $\mathrm{End}({}_AT) = B$, and $\{\Sigma_A Q \mid {}_AQ \text{ injective}\} = S(W\to)$, where $\Sigma = \Sigma_T = \mathrm{Hom}({}_AT,-)$. Of course, we have $\Sigma Q(a) = W$. The indecomposable B-modules belonging to the interior of the wings of W are just the images under Σ of the non-injective indecomposable A-modules ${}_AU$ satisfying $\mathrm{Hom}_A(U,Q(a)) = 0$. Now, let M be an indecomposable A-module with $\mathrm{Hom}_A(M,Q(a)) \neq 0$. It is easy to see (and well-known) that this implies that M actually is cogenerated by $Q(a)$, thus there is an exact sequence

$$0 \to M \to \oplus Q(a) \to M' \to 0.$$

If we apply Σ, we obtain an exact sequence

$$(*) \qquad 0 \to \Sigma M \to \oplus W \to \Sigma M' \to 0,$$

where $\Sigma M'$ belongs to $S(W\to)$. In particular, ΣM is cogenerated by W. This shows that the images under Σ of the indecomposable A-modules either belong to $S(W\to)$ (and thus are generated by W), or to the interior of the wings of W, or are cogenerated by W. Next, consider the images of A-mod under $\Sigma' = \mathrm{Ext}^1({}_AT,-)$. Given an A-module ${}_AN$, let ${}_AP \longrightarrow {}_AN$ be a projective cover, then also $\Sigma'{}_AP \longrightarrow \Sigma'{}_AN$ is surjective. Thus, it is sufficient to show: any $\Sigma' P_A(x)$ is generated by W, in order to conclude that all modules in $X({}_AT)$ are generated by W. We will use induction on the distance of x to the center of Δ. Note that $\Sigma' P_A(x) = \tau^- \Sigma Q_A(x)$, according to the connecting lemma $\Sigma' P(a) = 0$ or else the sink map for $\Sigma' P(a)$ is surjective and of the from

$$\Sigma(Q_A(a)/E_A(a)) \longrightarrow \Sigma' P(a),$$

according to (2'). Since $Q_A(a)/E_A(a)$ is injective, we know that $\Sigma(Q_A(a)/E_A(a))$ is generated by W. If $x \neq a$, let $x \longrightarrow x'$ be the unique arrow of Δ starting at x. Then either $\Sigma' P(x) = 0$, or else the sink map for $\Sigma' P(x)$ is surjective and of the form

$$\Sigma(Q_A(a)/E_A(a)) \oplus \Sigma' P(x') \longrightarrow \Sigma' P(x),$$

again according to (2') and using that $\mathrm{rad}\, P(x) = P(x')$. By induction, $\Sigma' P(x')$ is generated by W. Since also $\Sigma(Q_A(a)/E_A(a))$ is generated by W, we see that $\Sigma' P(x)$ is generated by W.

In order to show that W is separating, consider indecomposable modules X, Y with $X \prec W \prec Y$, and a map $f : X \longrightarrow Y$. We must have $X = \Sigma M$ for some indecomposable A-module M, and $\mathrm{Hom}_A(M,Q(a)) \neq 0$. Consider the exact sequence induced from (*) by f, say

$$\begin{array}{ccccccccc} 0 & \longrightarrow & \Sigma M & \longrightarrow & \oplus W & \longrightarrow & \Sigma M' & \longrightarrow & 0 \\ & & \downarrow{\scriptstyle f} & & \downarrow & & \| & & \\ 0 & \longrightarrow & Y & \longrightarrow & Z & \longrightarrow & \Sigma M' & \longrightarrow & 0 \end{array}$$

Now, $\mathrm{Ext}^1_B(\Sigma M', Y) = D\,\mathrm{Hom}(Y, \tau\Sigma M')$, and $\tau\Sigma M'$ is in $\mathcal{Y}({}_AT)$, but not in $\mathcal{S}(M\rightarrow)$, thus $\mathrm{Hom}(Y, \tau\Sigma M') = 0$ [the only indecomposable modules $Y' \in \mathcal{Y}({}_AT)$ with $W \preceq Y'$ are those in $\mathcal{S}(M\rightarrow)$]. The splitting of the lower sequence implies that f can be factored through $\oplus W$. This finishes the proof that W is separating.

(8) Various kinds of tilted algebras will have to be studied in further sections. In particular, we will be interested in algebras of the form $\mathrm{End}({}_AT)$, where ${}_AT$ is a tilting module over a tame hereditary algebra A. In this case, there are three essentially different cases to be considered:

(i) ${}_AT$ contains a non-zero preprojective, and a non-zero preinjective direct summand.

(ii) ${}_AT$ contains no non-zero preinjective direct summand, but a non-zero regular direct summand.

(ii*) ${}_AT$ contains no non-zero preprojective direct summand, but a non-zero regular direct summand.

(iii) ${}_AT$ is preprojective

(iii*) ${}_AT$ is preinjective.

Let $B = \mathrm{End}({}_AT)$. As we will show below, B *is representation finite precisely in case (i)*. The case (ii) will be considered in great detail in section 4.9. Of course, the algebras B obtained in case (ii*) are just the opposite algebras to those obtained in (ii). Finally, the algebras B obtained in case (iii) will be called the *tame concealed* algebras, they coincide with those obtained in case (iii*) and will be treated in the next section.

In order to consider the representation type of $B = \mathrm{End}({}_AT)$, where ${}_AT$ is a tilting module with A being tame hereditary, let us denote by $\mathcal{P}$, $\mathcal{T}$, $\mathcal{Q}$ the module classes in A-mod given by the preprojective, or regular, or preinjective A-modules, respectively. It is obvious that $\mathcal{Q}$ is contained in $\mathcal{G}({}_AT)$, in the cases (ii) and (iii), and $\mathcal{P}$ is contained in $\mathcal{F}({}_AT)$ in the cases (ii*) and (iii*), thus, in these cases, B is representation-infinite. Now, assume ${}_AT$ has an indecomposable preprojective summand T_1, and an indecomposable preinjective summand T_2. We claim that in this case both $\mathcal{F}({}_AT)$ and $\mathcal{G}({}_AT)$ are finite. Namely, let $T_1 = \tau^{-r}P(a)$, and $T_2 = \tau^{s}Q(b)$, with $P(a)$ indecomposable projective, $Q(b)$ indecomposable injective, and $r, s \in \mathbb{N}_0$. If ${}_AM$ is indecomposable and in $\mathcal{F}({}_AT)$, then

$$0 = \mathrm{Hom}_A(\tau^{-r}P(a), M) \approx \mathrm{Hom}_A(P(a), \tau^{r} M),$$

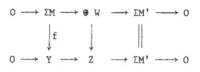

thus either ${}_AM$ is one of the finitely many indecomposable A-modules ${}_AN$ satisfying $\tau^rN = 0$, or else τ^rM is one of the finitely many indecomposable A(a)-modules where A(a) is obtained from A by deleting the vertex a (note that any such algebra A(a) is representation finite). Thus, these are only finitely many possibilities for ${}_AM$. Similarly, if ${}_AM$ is indecomposable and in $\mathcal{G}({}_AT)$, then

$$0 = \mathrm{Ext}^1_A(\tau^sQ(b),M) \approx D\,\mathrm{Hom}_A(M,\tau^{s+1}Q(b)) \approx D\,\mathrm{Hom}_A(\tau^{-s-1}M,Q(b)),$$

thus either ${}_AM$ is one of the finitely many indecomposable A-modules ${}_AN$ satisfying $\tau^{-s-1}N = 0$, or else $\tau^{-s-1}N$ is one of the finitely many indecomposable A(b)-modules. Thus, both $\mathcal{F}({}_AT)$ and $\mathcal{G}({}_AT)$ are finite, therefore $\mathcal{X}({}_AT)$ and $\mathcal{Y}({}_AT)$, are finite, thus B-mod $= \mathcal{Y}({}_AT) \vee \mathcal{X}({}_AT)$ is finite.

Finally, we want to make one remark concerning case (ii). Thus, let ${}_AT = T_0 \oplus T_1$ be a tilting module, with A being tame hereditary, T_0 preprojective, and $0 \neq T_1$ regular. We claim that $\mathrm{End}(T_0)$ itself is representation-infinite. Namely, consider any indecomposable regular module ${}_AM$ of τ-period 1. Then $\mathrm{Hom}(T_1,M) = 0$ and $\mathrm{Ext}^1_A(T_1,M) = 0$, since indecomposable direct summands of T_1 belong to tubes of rank ≥ 2, whereas M belongs to a tube of rank 1. Of course, also $\mathrm{Ext}^1_A(T_0,M) = 0$. Thus $\mathrm{Ext}^1_A({}_AT,{}_AM) = 0$, therefore ${}_AM$ belongs to $\mathcal{G}({}_AT)$, and is generated by ${}_AT$. However, since $\mathrm{Hom}(T_1,M) = 0$, we see that $\Sigma\,{}_AM$ is annihilated by the idempotent e_1 of $B = \mathrm{End}({}_AT)$ corresponding to the projection of ${}_AT = T_0 \oplus T_1$ onto T_1. Thus $\Sigma\,{}_AM$ is a B/Be_1B-module, and $B/Be_1B \approx \mathrm{End}(T_0)$.

Similarly, in case (ii*), let ${}_AT = T_1 \oplus T_2$, A tame hereditary, $0 \neq T_1$ regular, and T_2 preinjective. Then $\mathrm{End}(T_2)$ itself is representation infinite.

(8') Let A be tame hereditary, and ${}_AT$ a tilting module; let $B = \mathrm{End}({}_AT)$. Let χ_B be the quadratic form of B. Then χ_B is positive semi-definite of radical rank 1. Let y be a generator of $\mathrm{rad}\chi_B$.

(i) If y is positive and sincere, then B is a tame concealed algebra.
(ii) If y is positive, but not sincere, then B is representation infinite, but not tame concealed.
(iii) If y is neither positive nor negative, then B is representation finite.

<u>Proof</u>. Let σ_T be the linear transformation corresponding to ${}_AT$. Since σ_T is an isometry, with χ_A also χ_B is positive semi-definite of radical rank 1. Let h be the positive radical generator for χ_A, and H an indecomposable A-module with $\underline{\dim}\, H = h$. Let T(i) be an indecomposable direct summand of ${}_AT$, and $t(i) = \underline{\dim}\, T(i)$. If T(i) is preprojective, then

$$(h\sigma_T)_i = \langle t(i),h\rangle = \dim \operatorname{Hom}(T(i),H) > 0.$$

If $T(i)$ is preinjective, then

$$(h\sigma_T)_i = \langle t(i),h\rangle = -\dim \operatorname{Ext}^1(T(i),H) < 0.$$

Finally, if $T(i)$ is regular, then

$$(h\sigma_T)_i = \langle t(i),h\rangle = 0.$$

Thus, if ${}_AT$ is preprojective, then $h\sigma_T$ is positive and sincere, if ${}_AT$ is preinjective, then $h\sigma_T$ is negative and sincere. Also, if ${}_AT$ has both indecomposable preprojective and indecomposable preinjective direct summands, then $h\sigma_T$ has both positive and negative components. Finally, if ${}_AT$ has an indecomposable regular direct summand, then $h\sigma_T$ is not sincere. Since y is a multiple of $h\sigma_T$, the assertion now follows directly from (8).

4.3 Concealed algebras

Let A be a basic connected, hereditary, representation-infinite algebra, thus A is the path algebra of some connected quiver Δ with underlying graph different from $\mathbb{A}_n$, $\mathbb{D}_n$, $\mathbb{E}_6$, $\mathbb{E}_7$, $\mathbb{E}_8$. According to 2.4.13, there is a preprojective component P in A-mod containing all projective modules, and no indecomposable injective module, and there is a preinjective component Q containing all injective modules, and no indecomposable projective module. Let R be the module class given by the indecomposable A-modules not belonging to P or Q.

An algebra of the form $B = \mathrm{End}({}_AT)$, where A is connected, hereditary, representation-infinite, and ${}_AT$ is a preprojective tilting module, is said to be a <u>concealed</u> algebra. In case A is, in addition, tame, then B is said to be a <u>tame, concealed</u> algebra.

(1) Let A be connected, hereditary, representation-infinite. An algebra B is of the form $B = \mathrm{End}({}_AT)$ with ${}_AT$ a preprojective tilting module if and only if it is of the form $B = \mathrm{End}({}_AT')$, with ${}_AT'$ a preinjective tilting module.

<u>Proof.</u> We recall from 2.4.13 that P, Q are standard components with $\Gamma(P) = \mathbb{N}_0\Delta$, $\Gamma(Q) = (-\mathbb{N}_0)\Delta$, where A is the path algebra of Δ, thus $P \approx k(\mathbb{N}_0\Delta)$, $Q \approx k((-\mathbb{N}_0)\Delta)$. Thus, given a preprojective tilting module ${}_AT$, say, with $\tau^{t+1}T = 0$ for some $t \in \mathbb{N}_0$, then ${}_AT$ belongs to the module class P^t given by all modules ${}_AM$ with $\tau^{t+1}M = 0$, and $P^t \approx k([0,t]\Delta)$. If we denote by Q^t the module class given by all modules ${}_AM$ with $\tau^{-t-1}M = 0$, then $Q^t \approx k([-t,0]\Delta)$, thus the categories P^t and Q^t are isomorphic, with an isomorphism $\eta : P^t \longrightarrow Q^t$ satisfying $\eta\tau_AM \approx \tau_A\eta M$ for M indecomposable, and not projective. Now, let ${}_AT' = \eta T$, this is a module in Q^t and $\mathrm{End}({}_A\eta T) \approx \mathrm{End}({}_AT')$. In order to see that ηT is a tilting module, let T_i, T_j be indecomposable direct summands of T. Then

$$\mathrm{Ext}^1_A(\eta T_i,\eta T_j) \approx D\,\mathrm{Hom}_A(\eta T_j,\tau\eta T_i) \approx D\,\mathrm{Hom}_A(\eta T_j,\eta\tau T_i)$$

$$\approx D\,\mathrm{Hom}_A(T_j,\tau T_i) \approx \mathrm{Ext}^1_A(T_i,T_j) = 0,$$

in case T_i is not projective, so that $\tau\eta T_i = \eta\tau T_i$. On the other hand, if T_i is projective, then $\eta T_i = \tau^t Q$ for some indecomposable injective module Q, thus $\mathrm{Hom}_A(M,\tau^{t+1}Q) = 0$ for all $M \in Q^t$, and therefore also in this case

$$\mathrm{Ext}^1_A(\eta T_i,\eta T_j) \approx D\,\mathrm{Hom}_A(\eta T_j,\tau\eta T_i) = D\,\mathrm{Hom}_A(\eta T_j,\tau^{t+1}Q) = 0.$$

This shows that ${}_AT'$ is a preinjective tilting module. Of course, in this way, we construct all preinjective tilting modules in Q^t. Since t may be taken arbitrarily large, this finishes the proof.

<u>Remark.</u> We may rephrase the proof above as follows: Let A be the path algebra

of the connected quiver Δ, and assume A is representation-infinite. We consider the mesh category $k(\mathbb{Z}\Delta)$. A subset $\{x_1,\dots,x_n\}$ of pairwise different vertices of $\mathbb{Z}\Delta$ will be called a tilting set in $\mathbb{Z}\Delta$, provided n is the number of vertices of Δ, and we have $\operatorname{Hom}_{k(\mathbb{Z}\Delta)}(x_i,\tau x_j) = 0$ for all x_i,x_j. Then, the endomorphism rings of multiplicity-free preprojective tilting A-modules are just the rings $\operatorname{End}_{k(\mathbb{Z}\Delta)}(\bigoplus_{i=1}^{n} x_i)$, with $\{x_1,\dots,x_n\}$ a tilting set in $\mathbb{Z}\Delta$.

(2) Let A be a connected, hereditary, representation-infinite algebra, and ${}_AT$ a preprojective tilting module. Let $B = \operatorname{End}({}_AT)$. Now $F({}_AT)$ is finite, and contained in P, and $G({}_AT)$ is cofinite, and consists of a cofinite subcategory of P, and of all of R and all of Q. [Namely, any indecomposable module not belonging to $G({}_AT)$ is a predecessor of some τT_i, where T_i is an indecomposable summand of T, and there are only finitely many such predecessors, all belonging to P.] We obtain the module category B-mod by taking $Y({}_AT)$ ($\approx G({}_AT)$) and $X({}_AT)$ ($\approx F({}_AT)$), and pasting them together using the connecting sequences. Note that in $Y({}_AT)$, we have three different parts, namely the images under Σ of $P \cap G({}_AT)$, R, and Q, and since all injective A-modules belong to Q, the connecting sequences start in ΣQ. In particular, $\Sigma(P \cap G({}_AT))$ and ΣR both are closed under irreducible maps. Also, since $(Y({}_AT), X({}_AT))$ is a torsion pair, considering the four module classes

$$\Sigma(P \cap G({}_AT)),\ \Sigma R,\ \Sigma Q,\ X({}_AT),$$

there are non-zero maps only from any of these classes to itself and to the module classes to the right. We may visualize the categories A-mod and B-mod as follows:

A-mod

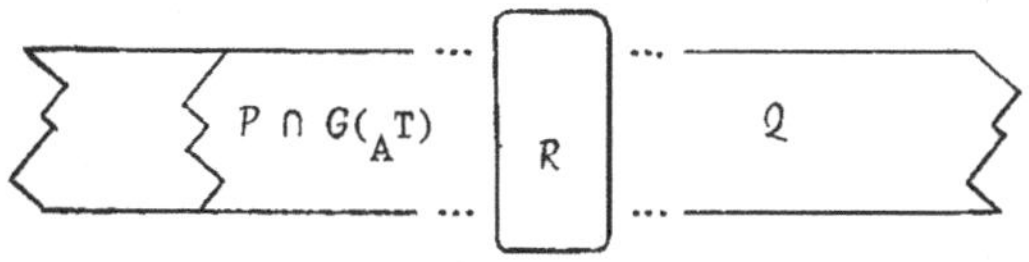

B-mod

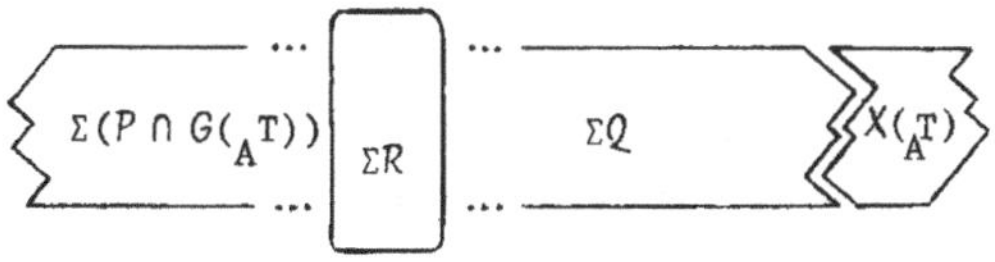

Note that $\Sigma(P \cap G({}_AT))$ is a preprojective component, and the modules in $\Sigma(P \cap G({}_AT))$ will be called the preprojective B-modules. Actually, $\Gamma\Sigma(P \cap G({}_AT))$ contains a full translation subquiver which is isomorphic to ΓP. [Namely, $\Gamma P = \mathbb{N}_0\Delta$, if A is the path algebra of Δ, and if the isomorphism classes of the indecomposable summands $T(i)$ of T are contained in $[0,z]\Delta$, then $[z,\infty)\Delta$ is contained in $\Gamma(P \cap G({}_AT))$, since we have $\operatorname{Ext}^1(T(i),M) = D\operatorname{Hom}(M,\tau T(i)) = 0$ for $[M] \in [z,\infty)\Delta$

and $T(i) \in [0,z]\Delta$, and, of course, $[z,\infty)\Delta$ is isomorphic to $\mathbb{N}_0\Delta$.] Also, $\Sigma\mathcal{Q} \vee X({}_AT)$ is a preinjective component (see 4.2.4), with $\Gamma(\Sigma\mathcal{Q} \vee X({}_AT))$ containing the full translation subquiver $\Gamma(\Sigma\mathcal{Q})$ which is isomorphic to $\Gamma\mathcal{P}$; the modules in $\Sigma\mathcal{Q} \vee X({}_AT)$ will be called preinjective.

(2') Let A be a connected, hereditary representation-infinite algebra, ${}_AT$ and ${}_AT'$ preprojective tilting modules, and $B = \mathrm{End}({}_AT)$, $B' = \mathrm{End}({}_AT')$. Then there exists a preprojective tilting B-module ${}_BU$ with $B' = \mathrm{End}({}_BU)$.

Proof. We only note that $\Gamma(\mathcal{P} \cap \mathcal{G}({}_AT))$ contains a full translation subquiver which is isomorphic to $\Gamma\mathcal{P}$.

(3) Let us now consider the special case of tame concealed algebras. If $B = \mathrm{End}({}_AT)$, where A is tame hereditary, and ${}_AT$ a preprojective tilting module, then the tubular type of B is, by definition, the tubular type of A, thus it is of the form $(n_1,\ldots,n_r)$, with $\mathbb{T}_{n_1,\ldots,n_r}$ being a Dynkin graph.

Theorem. Let B be a tame concealed algebra of tubular type $(n_1,\ldots,n_r)$. Then B has a preprojective component $\mathcal{P}$ with orbit quiver of type $\tilde{\mathbb{T}}_{n_1,\ldots,n_r}$, a preinjective component $\mathcal{Q}$ with orbit quiver of type $\tilde{\mathbb{T}}_{n_1,\ldots,n_r}$, and a stable separating tubular $\mathbb{P}_1k$-family $\mathcal{T}$ of type $(n_1,\ldots,n_r)$, separating $\mathcal{P}$ from $\mathcal{Q}$, and in this way, we obtain all indecomposable B-modules. Also, B-mod is controlled by χ_B. Let us denote by w the minimal positive radical vector of χ_B in $K_0(B)$. Then $\mathcal{P}$, $\mathcal{T}$, $\mathcal{Q}$ are the module classes given by the indecomposable B-modules ${}_BM$ satisfying $\langle w, \underline{\dim} M\rangle < 0$, $= 0$, or > 0, respectively.

As we have mentioned above, the modules in $\mathcal{P}$ are said to be preprojective, those in $\mathcal{Q}$ preinjective. The modules belonging to $\mathcal{T}$ are said to be regular, those lying at the mouth of a tube in $\mathcal{T}$ are said to be simple regular. Also note that in case B is of tubular type $(n_1,\ldots,n_r)$, then the number of vertices of B is given by $2 + \sum_{i=1}^{r}(n_i-1)$.

Proof. Let A be tame hereditary, let ${}_AT$ be a preprojective tilting module with $B = \mathrm{End}({}_AT)$. The first assertions follow directly from (2) using the structure theorem 3.6.5 for A-mod.

Now, denote by h the minimal positive radical vector for χ_A. Let $T(1),\ldots,T(n)$ be the indecomposable direct summands of ${}_AT$, and $t(1),\ldots,t(n)$ their dimension vectors. Since any $T(i)$ is preprojective, we have

$$\langle t(i),h\rangle = -\langle h,t(i)\rangle > 0,$$

thus $h\sigma_T$ has all its coefficients positive (4.1.7.a'). It follows that $h\sigma_T$ is sincere, and that it is the minimal positive radical generator for χ_B, thus $h\sigma_T = w$.

Denote by P_h, T_h, Q_h the module classes of the preprojective, regular, or preinjective A-modules, respectively. Since we have $\langle w,x\sigma\rangle = \langle h,x\rangle$, it follows that $P = \Sigma(P_h \cap G({}_AT)) \subseteq P_w$, $T = \Sigma T_h \subseteq T_w$, $\Sigma Q_h \subseteq Q_w$, and also $\Sigma'(P_h \cap F({}_AT)) \subseteq Q_w$ (note that for an indecomposable module $M \in P_h \cap F({}_AT)$, we have $\langle h, \underline{\dim} M\rangle < 0$, and $\underline{\dim}\,\Sigma' M = -(\underline{\dim}\, M)\sigma$, thus $\langle w, \underline{\dim}\,\Sigma' M\rangle = \langle w, -(\underline{\dim}\, M)\sigma\rangle = -\langle h, \underline{\dim} M\rangle > 0$). The last two inclusions together give $Q = \Sigma Q_h \vee \Sigma'(P_h \cap F({}_AT)) \subseteq Q_w$, and since B-mod $= P \vee T \vee Q$, it follows that $P = P_w$, $T = T_w$, $Q = Q_w$.

In order to see that B-mod is controlled by χ_B, we first note that $\chi_B(\underline{\dim}_B M) = 0$ or 1 for any indecomposable B-module ${}_BM$, using that A-mod is controlled by χ_A and 4.1.7.c''. Also, given a positive multiple x of w, then $x\sigma_T^{-1}$ is a positive multiple of h, thus there are infinitely many indecomposable A-modules ${}_AN$ with $\underline{\dim}_A N = x\sigma_T^{-1}$, and under Σ, we obtain B-modules ΣN with $\underline{\dim}\,\Sigma N = x$. If x is a positive root in $K_0(B)$, then $\chi_A(x\sigma_T^{-1}) = 1$, thus $y = x\sigma_T^{-1}$ is either positive or negative. In case y is positive, there is a unique indecomposable A-module ${}_AN$ with $\underline{\dim}\, N = y$, and ${}_AN$ belongs to $G({}_AT)$, according to 4.1.7.e'. Thus, ΣN is the unique B-module ${}_BM$ with $\underline{\dim}_B M = x$. On the other hand, if y is negative, there is a unique indecomposable A-module ${}_AN$ with $\underline{\dim}\, N = -y$, and ${}_AN$ belongs to $F({}_AT)$, according to 4.1.7.e'. Thus $\Sigma'N$ is the unique B-module ${}_BM$ with $\underline{\dim}_B M = x$. This finishes the proof.

Let us remark that the above proof shows that for a concealed algebra B, any positive root is connected. [Namely, if x is a positive root, then we obtain an indecomposable B-module ${}_BM$ with $\underline{\dim}_B M = x$, namely ${}_BM = \Sigma N$ where ${}_AN$ is indecomposable, and $\underline{\dim}_A N = x\sigma_T^{-1}$, in case $x\sigma_T^{-1} > 0$, and ${}_BM = \Sigma' N$, where ${}_AN$ is indecomposable, and $\underline{\dim}_A N = -x\sigma_T^{-1}$, in case $x\sigma_T^{-1} < 0$. However, dimension vectors of indecomposable modules always are connected.]

(4) The concealed algebras can be classified completely. This has been done by Happel-Vossieck, and we reproduce their list in an appendix. Note that the list gives only the possible frames (see 4.4). For $\tilde{\mathbb{A}}_n$, any orientation, but the cyclic one, is allowed. In case $\tilde{\mathbb{D}}_n$, all (unoriented) edges may be endowed with an arbitrary orientation. In the cases $\tilde{\mathbb{E}}_6$, $\tilde{\mathbb{E}}_7$, $\tilde{\mathbb{E}}_8$, again we may choose any orientation for the (unoriented) edges; however, in addition, we may replace any one of these arms by an arbitrary branch, as defined in the next section 4.4.

(5) We recall that given the tubular type $(n_1,\dots,n_r)$ of a tame concealed algebra B, then $\mathbb{T}_{n_1,\dots,n_r}$ is a Dynkin diagram. Let us note that for any possible tubular type, there is precisely one tame concealed algebra, which is, in addition, canonical, namely the unique canonical algebra with quiver $\Delta(n_1,\dots,n_r)$ [the uniqueness of such an algebra follows from $r \leq 3$].

Proof. We have to show that the canonical algebras with quiver $\Delta(n_1,\dots,n_r)$, where $\mathbf{T}_{n_1,\dots,n_r}$ is a Dynkin diagram, are tame concealed. For $r = 2$, nothing has to be shown, since the corresponding algebras are even tame hereditary. For the remaining cases we want to indicate corresponding tilting sets in $\mathbb{Z}\widetilde{\mathbf{T}}_{n,2,2}$, $\mathbb{Z}\widetilde{\mathbf{T}}_{3,3,2}$, $\mathbb{Z}\widetilde{\mathbf{T}}_{4,3,2}$, and $\mathbb{Z}\widetilde{\mathbf{T}}_{5,3,2}$. Actually, in case $\mathbb{Z}\widetilde{\mathbf{T}}_{n,2,2}$, we exhibit only the case $n = 7$.

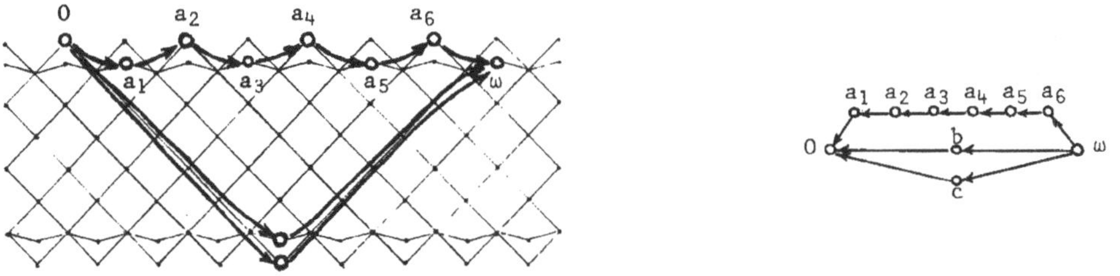

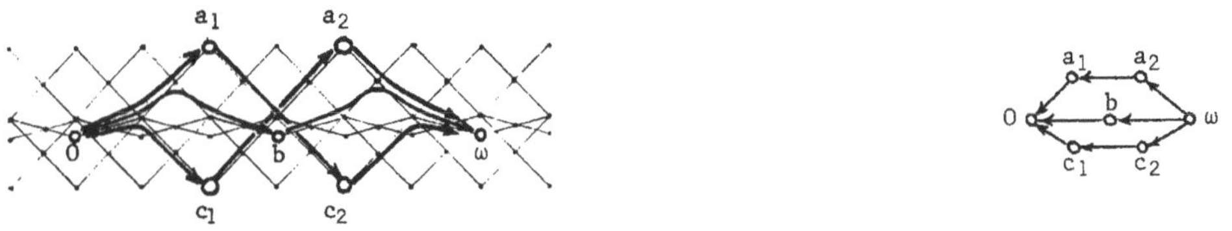

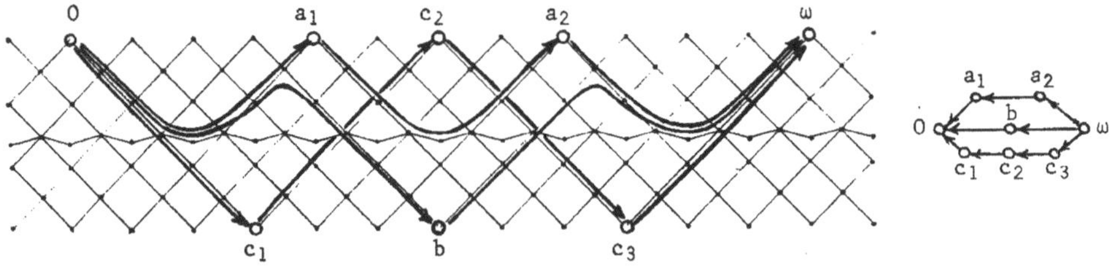

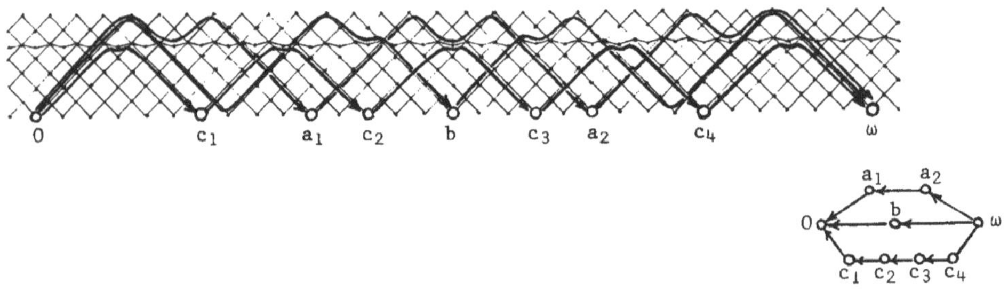

We call an algebra B *minimal representation-infinite* provided B is representation infinite, but B/BeB is representation finite for any idempotent $0 \neq e$ of B.

(6) *Let A be an algebra with a preprojective component containing all projective modules. If e is an idempotent of A, then A/AeA has preprojective components such that their union contains all indecomposable projective A/AeA-modules. In particular, if A is representation-infinite, then there exists an idempotent e' of A such that A/Ae'A is minimal representation-infinite and has a preprojective component.*

The assertion (6) is a special case of a general result on torsion pairs. We recall that given a Krull-Schmidt category K, we have defined inductively ${}_dK$ as follows: ${}_{-1}K = \langle 0 \rangle$, and an indecomposable object $X \in K$ belongs to ${}_dK$ if and only if any indecomposable object $Y \in K$ with $\mathrm{rad}(Y,X) \neq 0$ belongs to ${}_{d-1}K$. Also, let ${}_\infty K$ be the union of all ${}_dK$ with $d \in \mathbb{N}_0$.

(6') *Let A be an algebra and (F,G) a torsion pair in A-mod. Let X be an indecomposable A-module in ${}_d$(A-mod). Then X/t_GX belongs to ${}_dF$.*

Proof, by induction on d. We write t instead of t_G. We may assume that X is indecomposable. If $d = 0$, then X is a simple projective A-module. Thus, X/tX is either zero, or else $= X$, and then $0 \longrightarrow X$ is its sink map also in F, thus $X \in {}_0F$.

Now, assume $d \geq 1$. Let X' be an indecomposable direct summand of X/tX, say $X/tX = X' \oplus X''$, and let Y' be an indecomposable module in F with $\mathrm{rad}(Y',X') \neq 0$, say, let $0 \neq \varphi' : Y' \longrightarrow X'$ be non-invertible. We obtain an induced exaxt sequence

$$\begin{array}{ccccccccc} 0 & \longrightarrow & tX & \longrightarrow & X & \longrightarrow & X/tX & \longrightarrow & 0 \\ & & \| & & \uparrow{\scriptstyle\tilde{\varphi}} & & \uparrow{\scriptstyle[\varphi',0]} & & \\ 0 & \longrightarrow & tX & \longrightarrow & \tilde{Y} & \longrightarrow & Y' & \longrightarrow & 0. \end{array}$$

If $\tilde{Y}$ decomposes, say $\bigoplus_i \tilde{Y}_i$, then also $\tilde{Y}/t\tilde{Y} = \bigoplus_i \tilde{Y}_i/t\tilde{Y}_i$. However, $t\tilde{Y}$ is the image of tX under the induced map, thus $\tilde{Y}/t\tilde{Y} \approx Y'$ is indecomposable. Consequently, there is one index i with $\tilde{Y}_i/t\tilde{Y}_i \approx Y'$, and all other $\tilde{Y}_j$ are torsion. Thus, write $\tilde{Y} = Y \oplus Y''$, with Y'' being torsion, and let $\varphi : Y \longrightarrow X$ be the restriction of $\tilde{\varphi}$ to Y. Now modulo torsion, we obtain from φ just the map φ', thus φ is neither zero, nor invertible. It follows that Y belongs to ${}_{d-1}$(A-mod), thus, by induction, $Y' = Y/tY$ belongs to ${}_{d-1}F$, and therefore $X/tX \in {}_dF$.

Proof of (6): We apply (6') to the case of F = A/AeA-mod, and $G = G(Ae)$, with e an idempotent of A. Note that the projective A/AeA-modules are of the form $P/t_G(P)$, with P a projective A-module. We note that an algebra A has preprojective components such that their union contain all indecomposable projective modules if and only if the projective modules belong to ${}_\infty$(A-mod). Assume that A has a preprojective component, containing all projective modules, then any indecomposable projective A-module belongs to ${}_\infty$(A-mod), thus it follows that any indecomposable projective A/AeA-module belongs to ${}_\infty$(A/AeA-mod). This shows the first part of (6). Now assume that A is connected, representation infinite, and has a preprojective component $\mathcal{P}$. According to 2.4.10, $\mathcal{P}$ must be infinite. If an indecomposable projective A-module Ae, with e an idempotent of A, does not belong to $\mathcal{P}$, then e annihilates all modules in $\mathcal{P}$, thus $\mathcal{P}$ is even a component of (A/AeA)-mod. Thus, without loss of generality, we may assume that $\mathcal{P}$ contains all projective modules. Now, take an idempotent e' of A such that $A/Ae'A$ is minimal representation infinite. According to (7'), $A/Ae'A$ has preprojective components such that their union contains all indecomposable projective $A/Ae'A$-modules. One of these components must be infinite, since otherwise A would be representation finite, and actually it must contain all projective modules, since otherwise $A/Ae'A$ would not be minimal.

(7) Proposition (Happel-Vossieck). An algebra B is minimal representation infinite with a preprojective component if and only if either B is the path algebra of a quiver $\circ \overset{\leftarrow}{\underset{\leftarrow}{\vdots}} \circ$ with two vertices and at least two arrows (all pointing in one direction) or else B is tame concealed.

Proof: Let B be a minimal representation infinite algebra with a preprojective component $\mathcal{P}$. As above, we see that $\mathcal{P}$ is infinite and must contain all projective modules. Since there are only finitely many non-sincere indecomposable modules, $\mathcal{P}$ contains a sincere indecomposable module. According to 4.2.6', B is a tilted algebra, say $B = \operatorname{End}({}_AT)$, with A hereditary, and ${}_AT$ a tilting module. Let $n = \operatorname{rank} K_0(B) \geq 3$. We claim that the quadratic form χ_B is critical. Namely, assume χ_B is weakly positive. Then there are only finitely many positive roots for χ_B; the indecomposable modules in $\mathcal{P}$ are directing, thus we obtain infinitely many positive roots in the form $\underline{\dim}_B M$, with ${}_BM$ indecomposable in $\mathcal{P}$, see 2.4.8. On the other hand, if e is a proper idempotent of B, then B/BeB has preprojective components such that their union contains all indecomposable projective modules, and is representation finite, thus any indecomposable B/BeB-module Y belongs to a preprojective component, and therefore is directing in (B/BeB)-mod; in particular, $\operatorname{Ext}^1_B(Y,Y) = \operatorname{Ext}^1_{B/BeB}(Y,Y) = 0$. In order to show that the restriction of χ_B to any proper coordinate subspace is weakly positive, let $x \in K_0(B)$ be positive, and non-sincere, and choose a B-module X with $\underline{\dim}\, X = x$, and $\dim \operatorname{End}({}_BX)$ smallest possible. Let $X = \oplus X_i$, with all X_i indecomposable. According to 2.3.1, we have $\operatorname{Ext}^1_B(X_i,X_j) = 0$ for all $i \neq j$, and, since any X_i is non-sincere, we also have $\operatorname{Ext}^1_B(X_i,X_i) = 0$, thus $\operatorname{Ext}^1_B(X,X) = 0$. Since B is a tilted algebra, $\operatorname{gl.dim} B \leq 2$, thus

$$\chi_B(x) = \chi_B(\underline{\dim}\, X) = \dim \mathrm{End}(X) + \dim \mathrm{Ext}^2(X,X) > 0.$$

This shows that χ_B is critical. Using theorem 2 of Ovsienko (1.0), we conclude that χ_B is positive semi-definite with radical rank 1, thus the same is true for χ_A, and therefore A is tame hereditary. However, we have seen (4.2.6) that, in case ${}_AT$ has both non-zero preprojective and non-zero preinjective direct summands, B is of finite representation type and that in case ${}_AT$ has non-zero regular summands (and either no non-zero preinjective or no non-zero preprojective summand), B is not minimally representation-infinite. Thus ${}_AT$ is either preprojective, or preinjective, and therefore B is tame concealed. - It remains to consider the case n = 2. However, since B has a preprojective component, and n = 2, it is obvious that the quiver of B cannot have oriented cycles, thus it is given by two vertices and m arrows all pointing in one direction. Since B is representation infinite, $m \geq 2$, and since there are no paths of length ≥ 2, there cannot be any relation. This finishes the proof of proposition 4.3.7.

Let us give a criterion for an algebra B to be tame concealed which will be very useful in our subsequent discussions.

(8) Let B be a finite-dimensional algebra of finite global dimension with χ_B positive semi-definite such that the radical $\mathrm{rad}\chi_B$ is one-dimensional, with a sincere generator. Assume there is a component P in B-mod, such that any indecomposable module in P is directing and such that $\mathrm{Hom}(D(B_B),P) = 0$. Then B is tame concealed (and P is the preprojective component).

Proof. We want to show that any τ-orbit in P contains an indecomposable projective module, so that P is a preprojective component. Now, B cannot be representation finite, since otherwise P would contain all indecomposable injective modules. According to (6), some algebra of the form B/BeB is tame concealed, with e an idempotent of B. However, for any non-zero idempotent e', and B' = B/Be'B, the quadratic form $\chi_{B'}$ is positive definite, so B' cannot be tame concealed. This shows that e = 0, and therefore B itself is tame concealed.

Given any algebra B and any B-module M, define its predecessor-support algebra $B_p(M)$ to be the restriction of B to the full subquiver of all vertices a with $P(a) \preceq M$. Of course, M itself is a $B_p(M)$-module, even the minimal projective resolution of M consists only of $B_p(M)$-modules. In particular, $\mathrm{proj.dim}_B M = \mathrm{proj.dim}_{B_p(M)} M$. Also, if M is indecomposable, and $M' \longrightarrow M$ is its sink map, then M' is a $B_p(M)$-module, therefore $\tau_B(M) = \tau_{B_p(M)} M$, since both are given by the kernel $M' \longrightarrow M$. On the other hand, consider again M indecomposable, and assume $\mathrm{Hom}_B(M, {}_BB_p(M)) \neq 0$. If M is directing, then M actually is projective. [Namely, if $\mathrm{Hom}_B(M,P(a)) \neq 0$ for some vertex a of $B_p(M)$, then both

$P(a) \preceq M$, $M \preceq P(a)$ together give $P(a) = M$.]. As a consequence, for any directing module M with $\text{proj.dim}_B M = 1$, we have

$$\underline{\dim}\, \tau M = (\underline{\dim}\, M)\Phi_{B_p(M)}$$

[Namely, we use 2.4.4 for M considered as a $B_p(M)$-module]. Also, if M is a directing indecomposable module with $B_p(M) = B$ then $\text{inj.dim}_B M \leq 1$. [Namely, otherwise $\text{Hom}(\tau^- M, {}_B B) \neq 0$, according to 2.4.1*, say $\tau^- M \preceq P(a)$ for some indecomposable projective module $P(a)$, however $P(a) \preceq M \prec \tau^- M \preceq P(a)$ gives a cycle.] Thus, for a directing indecomposable module ${}_B M$ with $B_p(M) = B$ and $\text{Hom}(D(B_B), M) = 0$, we have

$$\underline{\dim}\, \tau^- M = (\underline{\dim}\, M)\Phi_B^{-1},$$

according to 2.4.4*.

Thus, let us consider for a moment the operation of Φ. Since Φ_B leaves invariant the quadratic form χ_B, it operates also on the factorspace $K_o(B)/\text{rad}\chi_B$. Since the induced quadratic form $\bar{\chi}_B$ on $K_o(B)/\text{rad}\chi_B$ is positive definite, by assumption, and since the action of Φ_B on $K_o(B)/\text{rad}\chi_B$ leaves invariant $\bar{\chi}_B$, it follows that Φ_B has finite order on $K_o(B)/\text{rad}\chi_B$, say order d. Let w be a generator of $\text{rad}\chi_B$, with at least one of its components being positive. For any $x \in K_o(B)$, we have $x\Phi_B^d - x \in \text{rad}\chi_B$, thus we may define a (linear) map $\partial : K_o(B) \longrightarrow \mathbb{Z}$ by

$$x\Phi_B^d - x = \partial(x)\cdot w\,.$$

Since w is Φ_B-invariant, according to 2.4.d, it follows that

$$\partial(x\Phi_B) = \partial(x).$$

Consider now an indecomposable module ${}_B M$ from P. Since $\text{Hom}(D(B_B), P) = 0$, and P is closed under τ, we know that $\text{proj.dim.}M \leq 1$, according to 2.4.1. Thus, either M is projective, or else $\underline{\dim}\, \tau M = (\underline{\dim}\, M)\Phi_{B_p(M)}$. Also, in case $B_p(M) = B$, we always have $\underline{\dim}\, \tau^- M = (\underline{\dim}\, M)\Phi_B^{-1}$, and the module $\tau^- M$ satisfies again $B_p(\tau^- M) = B$. We claim that in case $B_p(M) = B$, we must have $\partial(\underline{\dim}\, M) < 0$. Namely, for any $n \in \mathbb{N}_1$,

$$\underline{\dim}\, \tau^{-dn} M = (\underline{\dim}\, M)\Phi_B^{-dn} = \underline{\dim}\, M - n\partial(\underline{\dim}\, M)w$$

is positive. However, this is possible only in case $\partial(\underline{\dim}\, M) \leq 0$, since w has a positive component. On the other hand, if $\partial(\underline{\dim}\, M) = 0$, then

$$\underline{\dim}\, \tau^{-d} M = \underline{\dim}\, M$$

implies that $\tau^{-d} M \approx M$ using that M is directing (see 2.4.8), but this would give a cycle containing M, impossible. Thus $\partial(\underline{\dim}\, M) < 0$.

Also, we claim that for some $n \in \mathbb{N}_0$, the module $\tau^n M$ is projective, or satisfies $B_p(\tau^n M) \neq B$. Namely, otherwise

$$\underline{\dim}\, \tau^{dn} M = (\underline{\dim}\, M)\Phi_B^{dn} = \underline{\dim}\, M + n\partial(\underline{\dim}\, M)w,$$

but this is impossible, the left side being positive for all n, whereas the right side cannot stay positive for large n. Thus, considering any τ-orbit in $\mathcal{P}$, we know that it contains some projective module, or some module N with $B_p(N) \neq B$. Assume some τ-orbit does not contain a projective module, and choose some N in this τ-orbit with $B' = B_p(N)$ being minimal. It follows that $B_p(\tau^n N) = B'$ for all $n \in \mathbb{N}_1$. Since the radical generator w is assumed to be sincere, and $B' \neq B$, the restriction $\chi_B | K_0(B')$ is positive definite, however this is just $\chi_{B'}$ (since B'-mod is closed under minimal projective resolutions in B-mod). It follows that $\Phi_{B'}$ has finite order d', and therefore

$$\underline{\dim}\, \tau^{d'} N = (\underline{\dim}\, N)\Phi_{B'}^{d'} = \underline{\dim}\, N .$$

Using again that N is directing, we get a contradiction. This shows that any τ-orbit contains a projective module, and therefore $\mathcal{P}$ is a preprojective component. This finishes the proof of (8).

4.4 Branches

The following quiver with relations is called the complete branch in b

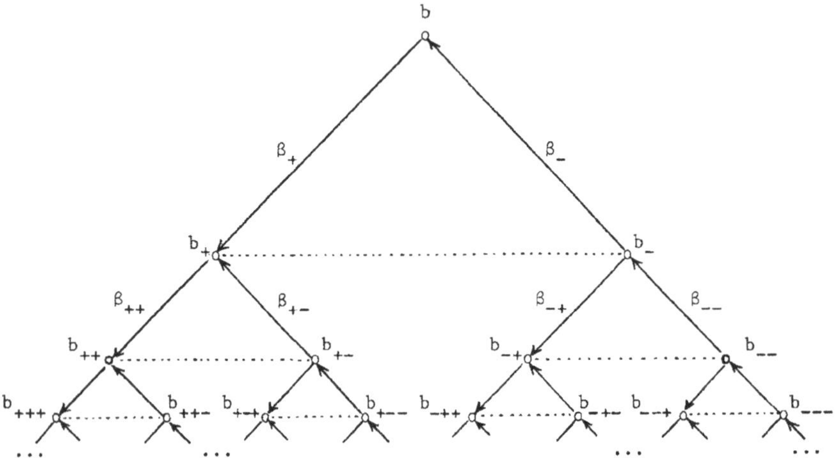

its vertices are of the form $b_{i_1\dots i_n}$ indexed by all possible sequences $i_1,\dots,i_n$, with $i_1,\dots,i_n \in \{+,-\}$, the vertex for the empty sequence (when $n = 0$) being denoted just by b; there are arrows

$$\beta_{i_1\dots i_n -} : b_{i_1\dots i_n -} \longrightarrow b_{i_1\dots i_n} ,$$

and

$$\beta_{i_1\dots i_n +} : b_{i_1\dots i_n} \longrightarrow b_{i_1\dots i_n +} ,$$

and there are the relations $\beta_{i_1\dots i_n -}\beta_{i_1\dots i_n +}$, for all $i_1,\dots,i_n$. A vertex $b_{i_1\dots i_n}$ is said to depend on a vertex $b_{j_1\dots j_m}$ provided $n \geq m$, and $j_1 = i_1,\dots,j_m = i_m$. (Of course, the vertices depending on a fixed vertex b' form again a complete branch.) A finite full connected subquiver Δ with the induced relations, containing the vertex b and consisting of n vertices (or also the corresponding algebra B) is called a branch in b of length n, we also call the empty quiver a branch of length 0. The length of the branch B will be denoted by $|B|$. Note that the underlying quiver Δ of a branch, together with the specified vertex b, uniquely determines the relations. The (unique) branch of length n containing the vertex $b_{i_1\dots i_n}$ with $i_1 = \dots = i_n = +$ is called the factorspace branch of length $n+1$, the (again unique) branch of length n containing the vertex

$b_{i_1 \ldots i_n}$ with $i_1 = \ldots = i_n = -$ is called the subspace branch of length $n+1$. (Note that both the factorspace branch and the subspace branch are linearly ordered quivers of type $\mathbf{A}_n$, but they are distinguished by the chosen endpoint b.)

Given a branch B in a, then any vertex b' of the quiver of B determines a branch $B(b')$ in b', namely the restriction to the full subquiver of all vertices in B depending on b'. (Actually, we may define $B(b')$ in this way for any b' in the complete branch in b, for $b' \notin B$ we will obtain the empty branch. We let $B^+ = B(b_+)$, and $B^- = B(b_-)$.) We define a function ℓ_B on the set of vertices of B by $\ell_B(b') = |B(b')|$, the length of the branch $B(b')$; we call ℓ_B the branch length function. Let us give an example of a branch together with its branch length function:

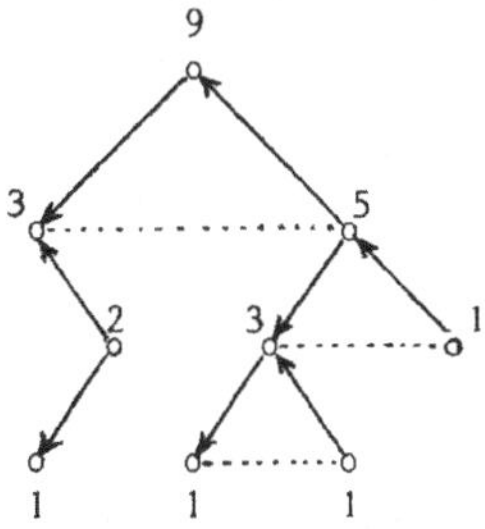

Given an algebra C with quiver Λ and relations σ_i, and a vertex b of Λ, then the restriction B of C to some full subquiver Λ' of Λ will be said to be a branch of C in b of length n, provided, first, B is a branch in b of length n; second, there is a full subquiver Δ such that $\Lambda = \Lambda' \cup \Delta$, $\Lambda' \cap \Delta = \{b\}$, and such that, moreover, any relation σ_i has its support either in Λ' or in Δ. If we denote by $\{\rho_i\}$ the set of all relations σ_j having support in Δ, and let $A = A(\Delta, \{\rho_i\})$, then C is said to be obtained from A by adding the branch B in b.

Considering the category C-mod, where C is an algebra with a branch B, it may be convenient to replace B by a different branch of the same length, and our aim is to show that in this way, the category C-mod is changed only slightly. Actually, in writing down specific examples of algebras, we usually will not specify the form of the branches. The convention will be as follows:
Let $(\Delta, \{\rho_j\})$ be a quiver with relations, and $b_1, \ldots, b_t$ (not necessarily distinct) vertices of Δ; also let $n_1, \ldots, n_t$ be natural numbers ≥ 2. Let $\bar{\Delta}^{(i)}$ be a copy of the (unoriented) graph $\mathbf{A}_{n_i}$, and $\bar{b}_i$ one of its endpoints. We form the disjoint union of Δ and the various $\bar{\Delta}^{(i)}$, $1 \leq i \leq t$, and identify b_i with $\bar{b}_i$, for all $1 \leq i \leq t$. We obtain in this way (taking into account also the relations $\{\rho_j\}$) what we will call a frame, the subgraphs $\bar{\Delta}^{(i)}$ being its branches. An

algebra C, or the corresponding quiver with relations, will be said to have this frame provided C is obtained from $A(\Delta,\{\rho_i\})$ by adding, for any $1 \le i \le t$, a branch B_i in b_i of length n_i. For example, the Happel-Vossieck list of all tame concealed algebras of type $\widetilde{\mathbb{E}}_6, \widetilde{\mathbb{E}}_7, \widetilde{\mathbb{E}}_8$ is given in A.2 by writing down all possible frames. [Also, it will be necessary at some stage, to cut off branches from a frame. This will mean that we delete an edge, say b'—b'' in order to obtain two connected components with one of them, say the one containing b', being a graph of type $\mathbb{A}_m$ having b' as its endpoint. In this case, b' is said to cut off a branch of length m. In general, given a frame constructed as above, then any vertex of $\bar{\Delta}^{(i)}$ different from $\bar{b}_i$ cuts off a branch.]

Let A be a finite dimensional algebra. We need some general result concerning wings. We recall that given an indecomposable module W, a wing for $[W]$ in $\Gamma(A)$ of length n is a mesh-complete subquiver θ of $\Gamma(A)$ which is of the form $\theta(n)$ (= the Auslander-Reiten quiver of the linearly ordered quiver of type $\mathbb{A}_n$), and the corresponding module class $\mathcal{W}$ given by all indecomposable modules M with $[M] \in \theta$ will be said to be a wing (of length n) of the module W. For the indecomposable modules in $\mathcal{W}$, we will use the notation W_{ij}, $1 \le j \le n$, as introduced in 3.3, with $W_{1n} = W$.

(1) Let W be an indecomposable module, and let $\mathcal{W}$ be a wing of length n for W, consisting of modules of projective dimension ≤ 1. Let $_AT$ be a partial tilting module. Let $W \in \langle {}_AT \rangle$ and $W_{1j} \notin \langle {}_AT \rangle$ for $s \le j < n$ and s being minimal with this property, and that $\operatorname{Hom}(W_{s+1,n}, W_{s,n-1}) = 0$. Then $_AT \oplus W_{s+1,n}$ is a partial tilting module.

Proof. We choose maps $\mu_{ij} : W_{ij} \longrightarrow W_{i,j+1}$ and $\varepsilon_{ij} : W_{ij} \longrightarrow W_{i+1,j}$ as in 3.4. In addition, let $W_{i,i-1} = 0$ for all $1 \le i \le n$, with corresponding zero maps $\mu_{i,i-1} : W_{i,i-1} \longrightarrow W_{ii}$ and $\varepsilon_{ii} : W_{ii} \longrightarrow W_{i+1,i}$. Actually, usually we will denote the maps μ_{ij}, and also their compositions, just by μ, the maps ε_{ij} and their compositions, just by ε.

Assume that $W_{1j} \notin \langle {}_AT \rangle$ for $s \le j < n$, and that s is minimal with this property. First, let us note that none of the modules W_{ij}, with $1 \le i \le s$, $s \le j < n$ belongs to $\langle {}_AT \rangle$. Thus, consider some module W_{ij}, with $1 < i \le s$, $s \le j < n$. Since $1 < s$, we see that the module $W_{1,s-1}$ has to belong to $\langle {}_AT \rangle$. However, there is the exact sequence

$$0 \longrightarrow W_{1,s-1} \xrightarrow{[\varepsilon\mu]} W_{i,s-1} \oplus W_{1j} \xrightarrow{\begin{bmatrix}\mu\\ \varepsilon\end{bmatrix}} W_{ij} \longrightarrow 0 ,$$

and this sequence does not split. Thus $\operatorname{Ext}^1_A(W_{ij}, W_{1,s-1}) \neq 0$, therefore $\operatorname{Ext}^1_A(W_{ij}, T) \neq 0$. This shows that W_{ij} cannot belong to $\langle {}_AT \rangle$.

Using that no direct summand of ${}_A T$ is isomorphic to any W_{ij} with $1 \leq i \leq s$, $s \leq j < n$, we are able to show that any map $f : {}_A T \longrightarrow W_{s,n-1}$ factors through τW_{1n} (where $\tau P = 0$ for P projective). Namely, using the sink map of the various W_{ij} with $1 < i \leq s$, $s \leq j < n$, and the relation $\mu_{is-1}\varepsilon_{is} \cdots \varepsilon_{s-1,s} = 0$ for $1 < i < s$, we see that f factors through $W_{1,n-1}$. Since we assume that proj.dim. $W_{1,j+1} \leq 1$ for all $1 \leq j \leq n-1$, we know from 2.4.1 that there is no map from an indecomposable injective to $W_{1,j}$ with $1 \leq j \leq n-1$. In particular, there is no irreducible map from an indecomposable injective module to any $W_{1,j}$, thus, the sink map for $W_{1,j}$ is of the form $[\mu *] : W_{1,j-1} \oplus \tau W_{1,j+1} \longrightarrow W_{1,j}$. Factoring our map further through the sink maps of $W_{1,j}$ with $s \leq j < n$, and using now $\mu_{1,s-1}\varepsilon_{1s} \cdots \varepsilon_{s-1,s} = 0$, we see that f actually factors through τW_{1n}. However, $\mathrm{Hom}_A({}_A T, \tau W_{1n}) \approx D\,\mathrm{Ext}^1_A(W_{1n}, T) = 0$, and therefore $\mathrm{Hom}_A({}_A T, W_{s,n-1}) = 0$. Thus

$$\mathrm{Ext}^1_A(W_{s+1,n}, T) \approx D\,\mathrm{Hom}_A({}_A T, \tau W_{s+1,n}) = D\,\mathrm{Hom}_A({}_A T, W_{s,n-1}) = 0.$$

Also, by assumption,

$$\mathrm{Ext}^1_A(W_{s+1,n}, W_{s+1,n}) \approx D\,\mathrm{Hom}_A(W_{s+1,n}, W_{s,n-1}) = 0.$$

Finally, the epimorphism $W_{1n} \longrightarrow W_{s+1,n}$ gives rise to an epimorphism

$$\mathrm{Ext}^1_A(T, W_{1n}) \longrightarrow \mathrm{Ext}^1_A(T, W_{s+1,n}),$$

thus, since the first term is zero, also the second term is zero. This shows that $\mathrm{Ext}^1_A(T \oplus W_{s+1,n}, T \oplus W_{s+1,n}) = 0$, and therefore $T \oplus W_{s+1,n}$ is a partial tilting module.

(2) As an application, let us classify all tilting A-modules, where A is given by the linearly ordered quiver of type $\mathbb{A}_n$. Note that there is a unique indecomposable projective-injective A-module, and, as an indecomposable projective-injective module, it is a direct summand of any tilting module. Note that the vertices of the quiver of an endomorphism ring $\mathrm{End}({}_A M)$ correspond to the indecomposable summands of ${}_A M$.

Proposition. Let A be given by the linearly ordered quiver of type $\mathbb{A}_n$, and ${}_A T$ a (multiplicity-free) tilting A-module. Then $B = \mathrm{End}({}_A T)$ is a branch of length n in a, where a corresponds to the unique projective-injective A-module. Any branch in a of length n occurs in this way.

Proof. Recall that we have denoted by $\Theta(n)$ the Auslander-Reiten quiver of A, thus we will denote the indecomposable A-modules by W_{ij}, $1 \leq i \leq j \leq n$; of course, A-mod is just a wing for the projective-injective vertex W_{1n}. Let ${}_A T$ be a (multiplicity free) tilting module. First, assume that $W_{1,n-1}$ is a direct summand of ${}_AT$. Since any module W_{in} with $2 \leq i \leq n$ satisfies $\mathrm{Ext}^1(W_{in},W_{1,n-1}) \neq 0$, we see that all indecomposable direct summands of ${}_AT$ different from W_{1n} belong to the wing $\langle W_{ij} \mid 1 \leq i \leq j \leq n-1\rangle$ of $W_{1,n-1}$, and we use induction. We obtain in this way all branches of a of length n which do not contain the vertex a_-. Dually, in case ${}_AT$ has W_{2n} as direct summand, we obtain all branches of a of length n which do not contain the vertex a_+. Now assume ${}_AT$ has neither $W_{1,n-1}$ nor W_{2n} as direct summand. Choose s minimal such that no W_{1j}, $s \leq j < n$, is a direct summand of ${}_AT$. Then $s \geq 2$, since $s = 1$ would imply that W_{2n} is a direct summand of ${}_AT$, according to (1). By assumption, also $s \leq n-1$. By the choice of s, we know that $W_{1,s-1}$ is a direct summand of ${}_AT$, the corresponding vertex of B will be denoted by a_+. According to (1), also $W_{s+1,n}$ is a direct summand of ${}_AT$, and the corresponding vertex of B will be denoted by a_-. Note that no module of the form W_{ij} with $1 \leq i \leq s$, $s \leq j \leq n$, except W_{1n}, can be a direct summand of ${}_AT$ (see the first part of the proof of (1), and also its dual). Thus, the remaining indecomposable summands belong to the wing $\mathcal{W}' = \langle W_{ij} \mid 1 \leq i \leq j \leq s-1\rangle$ of $W_{1,s-1}$, and to the wing $\mathcal{W}'' = \langle W_{ij} \mid s+1 \leq i \leq j \leq n\rangle$ of $W_{s+1,n}$. Now $\mathcal{W}'$ is a wing of length $s-1$, thus, it can contain at most $s-1$ isomorphism classes of indecomposable summands of ${}_AT$, and $\mathcal{W}''$ is a wing of length $n-s$, thus it can contain at most $n-s$ isomorphism classes of indecomposable summands. In addition to the summands in $\mathcal{W}'$ and $\mathcal{W}''$, there is the single isomorphism class of W_{1n}, and, since $1 + (s-1) + (n-s) = n$, we conclude by induction that $\mathcal{W}'$ gives rise to a branch in a_+ of length $s-1$, and $\mathcal{W}''$ gives rise to a branch in a_- of length $n-s$. Note that there are non-zero maps $W_{1,s-1} \longrightarrow W_{1n}, W_{1n} \longrightarrow W_{s+1,n}$, but their composition is zero. In the quiver of B, we obtain arrows $a_+ \longleftarrow a \longleftarrow a_-$, and a relation going from a_- to a_+. Altogether we obtain a branch in a of length n containing both a_+ and a_-, and, again by induction, we actually obtain all such branches.

(3) We have denoted by $\Theta(n)$ the Auslander-Reiten quiver $\Gamma(A)$, where A is given by the linearly ordered quiver of type $\mathbb{A}_n$. A subset Ξ of $\Theta(n)_0$ consisting of the isomorphism classes of the indecomposable summands of a tilting module ${}_AT$, will be called a <u>tilting set of</u> $\Theta(n)$. The tilting sets of $\Theta(n)$ may be characterized inductively as follows: $\Xi = \{x_1,\ldots,x_n\}$ is a tilting set if and only if the following conditions are satisfied:

(i) The projective-injective vertex of $\Theta(n)$ belongs to Ξ.

(ii) If $x_i \in \Xi$ is not the projective-injective vertex, and $W^{(i)}$ is the (unique) wing of x_i in $\Theta(n)$, then $\Xi \cap W^{(i)}$ is a tilting set in $W^{(i)}$.

(iii) If $w_{1j} \notin \Xi$ for $s \leq j < n$, and s is minimal with this property, then $w_{s+1,n} \in \Xi$.

Given a branch B of length n, we obtain the corresponding tilting set $\Xi(B)$ in $\theta(n)$ as follows: recall that we have denoted $B^+ = B(b_+)$ (= the branch consisting of all vertices in B which are dependent on b_+), and $B^- = B(b_-)$. Then, inductively

$$\Xi(B) = \{w_{1n}\} \cup \Xi(B^+) \cup \tau^{-1-|B^+|}\Xi(B^-) ,$$

where, for $m \leq n$, we consider $\theta(m)$ as a full subquiver of $\theta(n)$. For example, for the branch B of length 9 considered above, we obtain the following tilting set $\Xi(B)$ (its elements are marked by an encircled star ⊛).

B

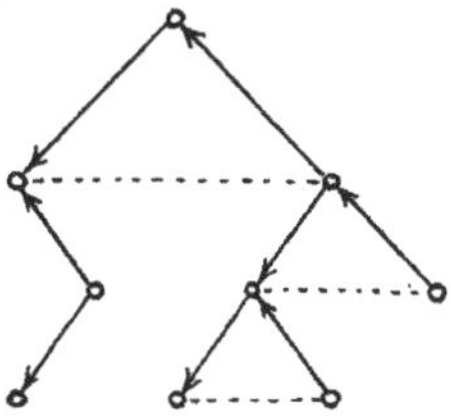

$\Xi(B)$

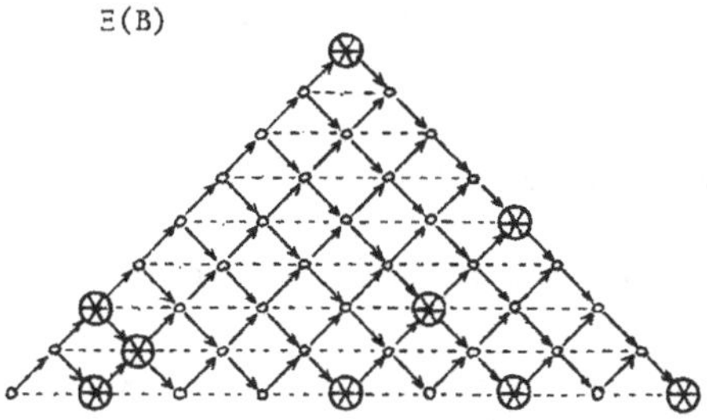

Also, given a tilting set Ξ in $\theta(n)$, let us denote by $G(\Xi)$ the full subcategory of $k(\theta(n))$ given by all vertices y with $\mathrm{Hom}_{k(\theta(n))}(y,\tau x) = 0$ for all $x \in \Xi$. Of course, considering Ξ as being the set of indecomposable summands of the tilting module ${}_AT$, we see that $G(\Xi) = G({}_AT)$, using the description $G({}_AT) = \{{}_AX \mid \mathrm{Ext}^1({}_AT,{}_AX) = 0\}$. On the other hand, since $G({}_AT)$ can also be described as the set of all modules ${}_AX$ generated by ${}_AT$, we see that w_{ij} belongs to $G(\Xi)$ if and only if there is some $w_{i'j} \in \Xi$, with $i' \leq i$. In particular, $G(\Xi)$ always contains all the vertices w_{in}, $1 \leq i \leq n$. In our example $\Xi(B)$ above, we again mark the elements of $\Xi(B)$ by an encircled star ⊛, the remaining elements of $G(\Xi(B))$ by a single star $*$, and also give $\Gamma G(\Xi(B))$:

$G(\Xi(B))$

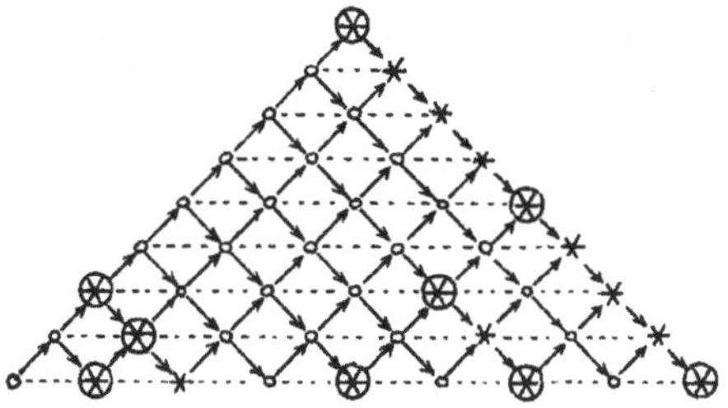

$\Gamma G(\Xi(B))$

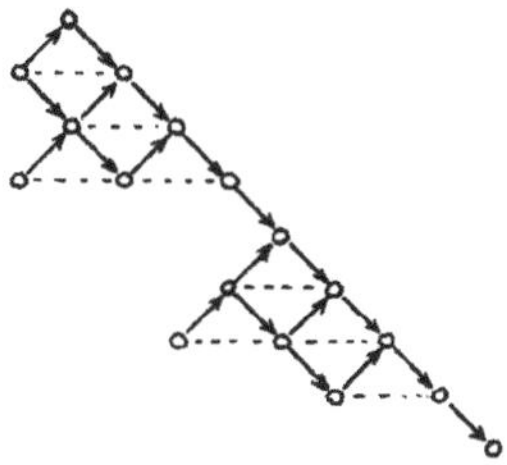

Let us denote by $M_i(B)$ the indecomposable B-module $M_i(B) = \mathrm{Hom}({}_A T, W_{in})$, where W_{in} is indecomposable in $\mathcal{G}(\Xi)$, with $[W_{in}] = w_{in}$. Note that $\mathrm{Hom}_B(P_B(b), M_i(B)) \neq 0$, since $P_B(b) = \mathrm{Hom}({}_A T, W_{1n})$, and $\mathrm{Hom}_B(P_B(b), M_i(B)) = \mathrm{Hom}_A(W_{1n}, W_{1i})$. The support of $M_i(B)$ is a connected subquiver of B containing b, and without any relation on it, thus the support of $M_i(B)$ is uniquely determined by some vertex b' of B (being given by the unique walk joining b and b').

(4) Let A be a finite dimensional algebra, and b a vertex of A. Let C be obtained from A by adding the factorspace branch B of length n in b. Let ${}_C T$ be a (multiplicity-free) tilting module, with $P_C(a) \in \langle {}_C T \rangle$, for all vertices a of A. Then $C' = \mathrm{End}({}_C T)$ is obtained from A by adding some branch of length n in b, and any algebra obtained from A by adding some branch of length n in b is of this form C'. Also, the torsion theory $(\mathcal{Y}({}_C T), \mathcal{X}({}_C T))$ splits, and the indecomposable C-modules not belonging to $\mathcal{G}({}_C T)$, are B_o-modules, where B_o is obtained from B by deleting b.

<u>Proof</u>. Let the quiver of B be given as follows:

$$\underset{b_1}{\circ} \xleftarrow{\beta_1} \underset{b_2}{\circ} \xleftarrow{\beta_2} \underset{b_3}{\circ} \quad \cdots \quad \underset{b_{n-1}}{\circ} \xleftarrow{\beta_{n-1}} \underset{b_n = b}{\circ} ,$$

let $W_{1j} = P_C(b_j)$ note that for $1 \leq j < n$, we have $P_C(b_j) = P_B(b_j) = P_{B_o}(b_j)$, however the canonical map $P_C(b) \longrightarrow P_B(b)$ usually is a proper epimorphism). Also, for $1 < i \leq j \leq n$ let W_{ij} be the cokernel of the map $P(b_{i-1}) \longrightarrow P(b_j)$ which is the composition of maps of the form $P(\beta_s^*)$. Note that in this way, wo obtain a wing $\mathcal{W} = \langle W_{ij} \mid 1 \leq i \leq j \leq n \rangle$ of length n for $W_{1n} = P_C(b)$. [One proves by induction on j, with $1 \leq j < n-1$, and i, with $1 \leq i \leq j$, that the Auslander-Reiten-sequence starting at W_{ij} is of the required form.] By construction, all modules in $\mathcal{W}$ have projective dimension ≤ 1.

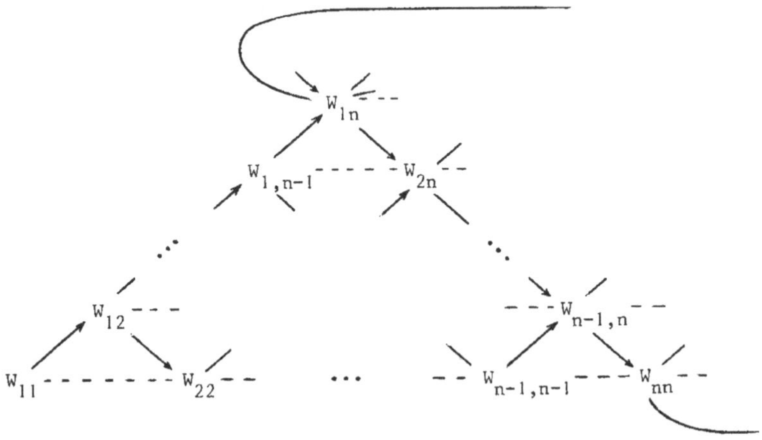

Let ${}_C T' = \oplus P_C(a)$, where a runs through the vertices of A. If ${}_C T$ is any tilting module having ${}_C T'$ as direct summand, then we conclude from (1) that the isomorphism classes of the indecomposable summands of ${}_C T$ which belong to $\mathcal{W}$ form a tilting set X of $\Gamma(\mathcal{W})$, and conversely, that any tilting set of $\Gamma(\mathcal{W})$ arises in this way. It follows that $C' = \mathrm{End}({}_C T)$ is obtained from $A = \mathrm{End}({}_C T')$ by adding a branch B' of length n in b, namely the branch corresponding to the tilting set X.

Given an indecomposable module ${}_C M$ not belonging to $\mathcal{G}({}_C T)$, then $0 \neq \mathrm{Ext}^1_C(T,M) \approx D\,\mathrm{Hom}_C(M,\tau T)$, but the indecomposable summands of τT are of the form W_{ij} with $j < n$, and the modules of this form are closed under predecessors, and all of them are B_o-modules, thus ${}_C M$ is a B_o-module. Also, an injective resolution of W_{ij} with $j < n$ is given by

$$0 \longrightarrow W_{ij} \longrightarrow Q_C(b_i) \longrightarrow Q_C(b_{j+1}) \longrightarrow 0,$$

therefore inj.dim. $W_{ij} \leq 1$. In particular, inj.dim.${}_C N \leq 1$ for any module ${}_C N$ in $\mathcal{F}({}_C T)$, and therefore $(\mathcal{Y}({}_C T), \mathcal{X}({}_C T))$ splits, according to 4.1.6.d.

(5) Let A be a finite dimensional algebra, and b a vertex of A. Let C be obtained from A by adding a branch B of length n in b. Let B^s be the subspace branch of length n, say with quiver $b = b_1 \longleftarrow b_2 \longleftarrow \dots \longleftarrow b_n$ and let C^s be obtained from A by adding B^s in b. There exists a tilting module ${}_C T$ with $\mathrm{End}({}_C T) = C^s$, and the following properties:

(a) The indecomposable C^s-modules not belonging to $\mathcal{Y}({}_C T)$ are B^s_o-modules, where B^s_o is obtained from B^s by deleting b.

(b) Γ(C-mod) is obtained from the disjoint union of $\Gamma\mathcal{Y}({}_C T)$ and $\Gamma\mathcal{G}(\Xi(B))$ by identifying the full subquiver of $\Gamma\mathcal{G}({}_C T)$ given by the vertices $[P(b_i)]$, $1 \leq i \leq n$, with the full subquiver of $\Gamma\mathcal{G}({}_C T)$ given by the vertices w_{in}, $1 \leq i \leq n$, always identifying $T(b_i)$ with w_{in}, for $1 \leq i \leq n$.

Proof. Let B^f be the factorspace branch of length n, and let C^f be obtained from A by adding B^f in b. There are tilting C^f-modules T', T^s such that the projective C^f-modules $P(a)$, with a vertex of a of A, are all contained both in $\langle T' \rangle$ and in $\langle T^s \rangle$, and such that $\mathrm{End}(T') = C$, $\mathrm{End}(T^s) = C^s$. We have $\mathcal{G}(T^s) \subseteq \mathcal{G}(T')$, in particular, $T^s \in \mathcal{G}(T')$, and the C-module $T = \mathrm{Hom}(T',T^s)$ is in fact a tilting C-module. Of course, $\mathrm{End}(T) = \mathrm{End}(T^s) = C^s$. Also, $\mathcal{G}(T^s)$ is mapped under $\mathrm{Hom}(T',-)$ into $\mathcal{G}(T)$, as a consequence, $\mathcal{Y}(T^s) \subseteq \mathcal{Y}(T)$. Thus, the indecomposable C^s-modules not belonging to $\mathcal{Y}(T)$ also do not belong to $\mathcal{Y}(T^s)$, thus they belong to $\mathcal{X}(T^s)$, since $(\mathcal{Y}(T^s),\mathcal{X}(T^s))$ is a split torsion pair, and therefore are B^s_o-modules. For the last statement, we note that the C-modules M not belonging to $G({}_C T)$ are B-modules with $M_b = 0$. And, an indecomposable B-module does not belong to $G({}_C T)$ if and only if $\langle \ell_B, \underline{\dim}\, M\rangle < 0$, where ℓ_B is the branch length function.

Let us add the following remarks concerning the branch length function.

(6) Let A be given by the linearly ordered quiver of type $\mathbb{A}_n$, and ${}_A T = \bigoplus_{i=1}^{n} T(i)$ a tilting module with indecomposable summands $T(i)$. Let $B = \mathrm{End}({}_A T)$. Then the branch length function ℓ_B takes the following values: $\ell_B(i)$ is the length of the module $T(i)$, thus $\ell_B(i) = \langle \underline{\dim}\, T(i), \underline{\dim}\, D(A_A)\rangle$.

Proof. Consider the case where B is a branch in a containing both a_+, and a_-, (The other cases being similar.) We have seen in the proof of (2) that a_+ will correspond to some module $W_{1,s-1}$, and a_- to some module $W_{s+1,n}$, and that there is a wing W' which gives rise to a branch in a_+, of length $s-1$, and a wing W'' which gives rise to a branch in a_-, say of length $n-1$. Now, the length of $W_{1,s-1}$ as an A-module is $s-1$, and this is just the branch length of a_+, and the length of $W_{s+1,n}$ is $n-s$, and is just the branch length of a_-. Using induction, we see that the branch length of any i coincides with the length of $T(i)$. Of course, we may calculate the length of a module ${}_A M$ over a basic algebra A by evaluating $\underline{\dim}\, \mathrm{Hom}({}_A M, D(A_A))$, and this is just $\langle \underline{\dim}\, {}_A M, D(A_A)\rangle$.

(7) Let B be a branch in b of length n. There exists a unique tilting module ${}_B T$ having $P_B(b)$ as direct summand and such that $\mathrm{End}({}_B T)$ is a subspace branch in the vertex corresponding to the summand $P_B(b)$. For this tilting module ${}_B T$, the torsion pair $(F({}_B T), G({}_B T))$ splits, and

$$F({}_B T) = \langle {}_B N \text{ indecomposable} \mid \langle \ell_B, \underline{\dim}\, {}_B N\rangle < 0\rangle,$$

$$G({}_B T) = \langle {}_B N \text{ indecomposable} \mid \langle \ell_B, \underline{\dim}\, {}_B N\rangle > 0\rangle.$$

Proof. We write $B = \mathrm{End}({}_A T')$, with A the factor-space branch of length n, and ${}_A T'$ a tilting module. Let Σ : A-mod $\longrightarrow$ B-mod be given by $\Sigma = \mathrm{Hom}_A({}_A T', -)$, and let ${}_B T = \Sigma D(A_A)$. Note that $P_B(b) = \Sigma P_A(b)$ is a direct summand of ${}_B T$, since $P_A(b)$ is projective-injective. Since ${}_B T$ is a slice module, it is a tilting-module, and also $(F({}_B T), G({}_B T))$ splits. We have calculated in (6) that $\ell_B = \underline{\dim}\, \Sigma D(A_A) = \underline{\dim}\, {}_B T$, thus the characterization of $F({}_B T)$ and $G({}_B T)$ in terms of ℓ_B follows from 4.2.4.

(8) Let B be a branch of C. Let $h \in \operatorname{rad} \chi_C$. Then $h \mid K_0(B)$ is a scalar multiple of the branch length function ℓ_B.

<u>Proof</u>. Let B be a branch in b, and let Δ be its quiver. For $y \in B$, let $N(y)$ be the set of neighbors in Δ which are dependent on y; of course, $N(y)$ consists of at most two vertices. By the definition of ℓ_B, we have

$$\ell_B(y) = 1 + \sum_{x \in N(y)} \ell_B(x).$$

Now let h be in $\operatorname{rad} \chi_C$, thus $D_y\chi_C(h) = 0$ for all vertices y (see 1.1). Now assume y is dominated by z, and let us calculate $D_y\chi_C(h)$. Considering the bigraph of χ_A, there are solid edges from y to z and to any $x \in N(y)$, and there is a dotted edge from y to $y' \in N(z)$, with $y' \neq y$, provided y' exists. Thus

$$0 = D_y\chi_C(h) = 2h(y) - h(z) - \sum_{x \in N(y)} h(x) + \sum_{\substack{u \in N(z) \\ u \neq y}} h(u) ,$$

thus
$$h(y) - \sum_{x \in N(y)} h(x) = h(z) - \sum_{u \in N(z)} h(u).$$

This shows that this type of difference $h(y) - \sum_{x \in N(y)} h(x)$ takes the same value for all $y \in \Delta$, say the value c. (We use that Δ is a connected quiver.) By induction on $\ell_B(y)$, we show that $h(y) = c \cdot \ell_B(y)$. In case $\ell_B(y) = 1$, we have $N(y) = \emptyset$, thus $h(y) = c = c \cdot \ell_B(y)$. Now assume for the $x \in N(y)$, we know already that $h(x) = c \cdot \ell_B(x)$. Then

$$h(y) = c + \sum_{x \in N(y)} h(x) = c + \sum_{x \in N(y)} c\, \ell_B(x)$$

$$= c(1 + \sum_{x \in N(y)} \ell_B(x)) = c\, \ell_B(y).$$

This finishes the proof.

(8') <u>Corollary</u>. Let A be a finite dimensional algebra, let b be a source of the quiver of A. Let C be obtained from A by adding the branch B in b. Let $h \in \operatorname{rad} \chi_C$, and suppose $h \mid K_0(B)$ is non-zero. Then, given any indecomposable C-module M with support in B, we have $\langle h, \underline{\dim}_C M\rangle \neq 0$.

<u>Proof</u>. According to the previous result, $h \mid K_0(B)$ is equal to some $c\ell_B$, with $c \neq 0$. On the other hand, since b is a source of A, we have for any $x \in K_0(C)$ with support in $K_0(B)$ that $\langle h,x\rangle_C = \langle h \mid K_0(B),x\rangle_B$. Thus $\langle h,\underline{\dim}_C M\rangle_C = \langle c\ell_B,\underline{\dim}_B M\rangle_B = c\langle \ell_B,\underline{\dim}_B M\rangle_B$, and $\langle \ell_B,\underline{\dim}_B M\rangle \neq 0$ for any indecomposable B-module, according to (7).

(9) Remark. Consider the category $\check{U}(S)$ of representations of a finite partially ordered S which is the disjoint union of chains, say let

$$S = \coprod_{s=1}^{t} I(m_s) ,$$

with $m_s \in \mathbb{N}_1$. Note that $A_0 = A(S)$ is hereditary, it is the path algebra of the disjoint union of linearly oriented quivers of type $\mathbb{A}_{m_s}$, $1 \leq s \leq t$. Let R be the canonical A(S)-module (it is the minimal A(S)-module which is both injective and faithful). The algebra $A = A_0[R]$ is the path algebra of the star $\mathbb{T}_{m_1+1,\dots,m_t+1}$ with factorspace orientation, the center of the star being the extension vertex (A is obtained from the one-dimension algebra k given by the extension vertex, by adding t factorspace branches of lengths $m_1+1,\dots,m_t+1$). We denote the center of the star by ω , the neighbors of the center by $b_+^{(s)}$, $1 \leq s \leq t$. Let us say that the support of an A-module is away from the center provided it does contain neither ω , nor any $b_+^{(s)}$, $1 \leq s \leq t$. Note that the A-modules away from the center form a subcategory which is the category of representations of the disjoint union of linearly oriented quivers of type $\mathbb{A}_{m_s-1}$, $1 \leq s \leq t$.

Proposition. $\check{U}(S) \approx M$, where M is the module class obtained from A-mod by deleting the indecomposable A-modules with support away from the center. Also, M is the class of torsion modules of a split torsion pair in A-mod.

Proof. Since A_0 is hereditary, we can apply 4.1.8. Thus, let T_0 be the minimal injective cogenerator in A_0-mod . Since R is an injective A_0-module, the module $G(\omega)$ constructed there is just $P(\omega)$. Thus, the A-module $_AT = T_0 \oplus P(\omega)$ is a tilting module, and $\check{U}(S) \approx G(_AT)$. We claim that $_AT$ is a slice module and that $F(_AT)$ is the module class of all A-modules having support away from the center of A . Namely, any branch of $\mathbb{T}_{m_1+1,\dots,m_t+1}$ gives rise to a wing of $P(\omega)$. Let us consider the branch with index s , and denote as usual the indecomposable modules in the corresponding wing $W^{(s)}$ by $W_{ij}^{(s)}$, $1 \leq i \leq j \leq m_s+1$. First of all, $W_{1,m_s+1} = P(\omega)$ is a direct summand of $_AT$. Also, all the modules $W_{i,m_s}^{(s)}$, $1 \leq i \leq m_s$ are direct summands of $_AT$, since these are just the indecomposable injective A_0-modules belonging to $W^{(s)}$. It follows that $_AT$ is a slice module, and that $F(_AT)$ is given by the indecomposable modules $W_{ij}^{(s)}$, with $j < m_s$. This finishes the proof.

As an example , let us consider the case $(m_1,m_2,m_3) = (4,3,1)$, thus S is of the form

We sketch the Auslander-Reiten quiver of A-mod , and indicate by * the position of

the indecomposable summands of ${}_{A}T$. The indecomposable A-modules belonging to $F({}_{A}T)$ are those in the shaded areas:

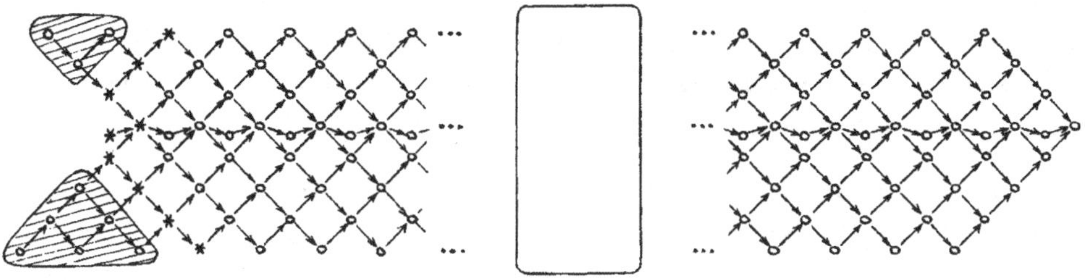

The Auslander-Reiten quiver of $\check{U}(S)$ is therefore of the form

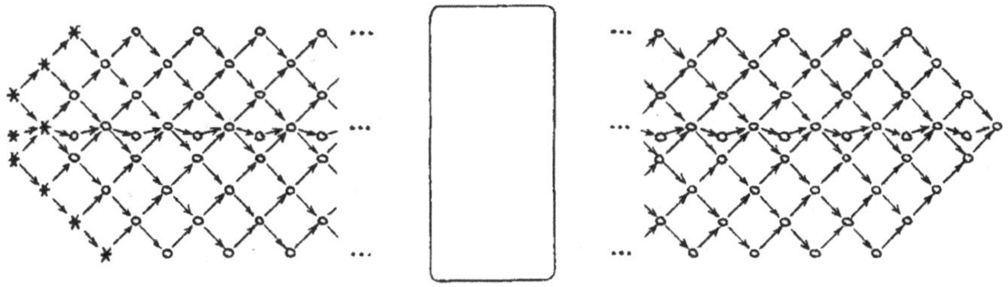

4.5 Ray modules

We are going to consider special one-point extensions of algebras and want to study the change of components. We will be interested in extensions of concealed algebras which will be (up to changing branches) built by forming successively one-point extensions using ray modules, as defined below. Therefore, we now concentrate on such one-point extensions.

Let Γ be a translation quiver without multiple arrows. We recall that a finite or infinite path, say

$$x_1 \longrightarrow x_2 \longrightarrow \dots \longrightarrow x_{n-1} \longrightarrow x_n$$

is said to be sectional provided $\tau x_{i+1} \neq x_{i-1}$, for all possible i (here, all $2 \leq i \leq n-1$). A vertex v in Γ is said to be a ray vertex provided there exists an (infinite) sectional path

$$v = v[1] \xrightarrow{\nu_1} v[2] \longrightarrow \dots \longrightarrow v[i] \xrightarrow{\nu_i} v[i+1] \longrightarrow \dots$$

with pairwise different vertices $v[i]$, $i \in \mathbb{N}_1$, such that for any i, the path $(v[1] \mid \nu_1,\dots,\nu_i \mid v[i+1])$ is the only sectional path of length i starting at v. In dealing with a ray vertex v, we usually keep the notation $v[i]$ for the end-point of the unique sectional path of length $i-1$ starting at v, and ν_i for the arrow $v[i] \longrightarrow v[i+1]$. To give an example of a ray vertex, consider a stable tube $\mathbb{Z}\mathbb{A}_\infty/n$; any vertex belonging to the mouth of $\mathbb{Z}\mathbb{A}_\infty/n$ is a ray vertex.

Given a branch B, and a ray vertex v of the translation quiver Γ, let us define a translation quiver $\Gamma[v,B]$ as follows: First, consider the case of a subspace branch; say, let us denote by B^s the subspace branch of the same length as B. The vertices of $\Gamma[v,B]$ are those of Γ different from $v[i]$, for any i, together with all triples (v,i,j), where $i \in \mathbb{N}_1$, and $0 \leq j \leq |B^s|$ (of course, it would be enough to use the pairs (i,j) keeping in mind the ray vertex v, however, since often we will consider different ray vertices at the same time, it seems more convenient to work with these triples). The arrows of $\Gamma[v,B]$ are, first of all, those arrows of Γ not involving any $v[i]$; in addition there are the arrows

$$(v,i,j) \longrightarrow (v,i+1,j) \quad \text{for all } i \in \mathbb{N}_1,\ 0 \leq j \leq |B^s|,$$

$$(v,i,j) \longrightarrow (v,i,j+1) \quad \text{for all } i \in \mathbb{N}_1,\ 0 \leq j < |B^s|,$$

and, for any arrow $x \longrightarrow v[i]$ in Γ, with $i = 1$, or both $i > 1$ and $x \neq v[i-1]$, there is an arrow $x \longrightarrow (v,i,0)$; similarly, for any arrow $v[i] \longrightarrow y$ in Γ, with $y \neq v[i+1]$, there is an arrow $(v,i,|B^s|) \longrightarrow y$. We let

$\tau_{\Gamma[v,B^s]}(v,i,j) = (v,i-1,j-1)$, provided $i > 1$, $j \geq 1$, and in case $v[i]$ is not projective in Γ, let $\tau_{\Gamma[v,B^s]}(v,i,0) = \tau_\Gamma v[i]$ provided $\tau_\Gamma v[i]$ is not of the form $v[j]$ for any j, and $\tau_{\Gamma[v,B^s]}(v,i,0) = (v,j,|B^s|)$ provided $\tau_\Gamma v[i] = v[j]$. Finally, let z be in Γ, not projective, and not of the form $v[i]$ for any i. Then, let $\tau_{\Gamma[v,B^s]}(z) = \tau_\Gamma z$ provided $\tau_\Gamma z$ is not of the form $v[j]$ for any j, and $\tau_{\Gamma[v,B^s]}(z) = (v,j,|B^s|)$ in case $\tau_\Gamma z = v[j]$. In this way, wo obtain $\Gamma[v,B^s]$. In order to form $\Gamma[v,B]$, we consider the disjoint union of $\Gamma[v,B^s]$ and $\Gamma G(\Xi(B))$, and identify the full subquiver of $\Gamma[v,B^s]$ given by the vertices $(v,1,j)$ with the full subquiver of $\Gamma G(\Xi(B))$ given by the vertices $w_{j,|B|}$, $1 \leq j \leq |B|$, always identifying $(v,1,j)$ with $w_{j,|B|}$. We will say that $\Gamma[v,B]$ is <u>obtained</u> from Γ <u>by ray insertions</u>. For example, in case B is the following branch in b

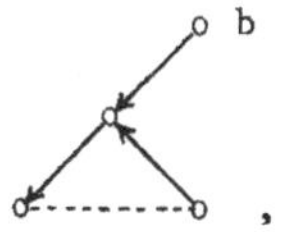

then the translation quiver $\Gamma[v,B]$ obtained from some Γ by ray insertions has the following shape (where we write vij instead of (v,i,j):

Γ

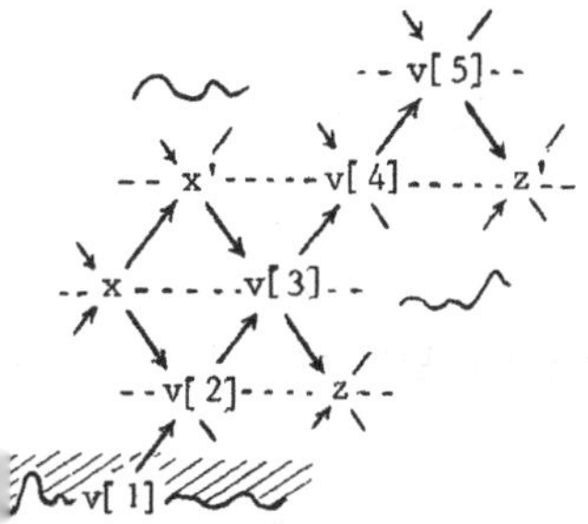

$\Gamma[v,B]$

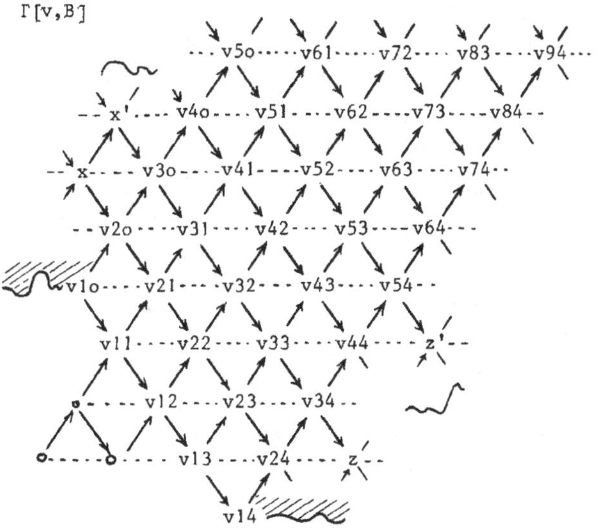

[We may think of $\Gamma[v,B]$ being obtained from Γ by cutting Γ along the ray, obtaining in this way two copies of the ray, the new vertices corresponding to $v[i]$ being denoted by $(v,i,0)$, and $(v,i,|B|)$, respectively. Any arrow

$\nu_i : v[i] \longrightarrow v[i+1]$ gives rise to two arrows $(v,i,0) \longrightarrow (v,i+1,0)$ and $v(i,|B|) \longrightarrow (v,i+1,|B|)$. The arrows different from ν_i ending in $v[i]$ should now end in $(v,i,0)$, those starting in $v[i]$ should start in $(v,i,|B|)$. In the gap between the two new rays we insert $|B| - 1$ additional rays $(v,i,1) \longrightarrow (v,i,2) \longrightarrow \ldots$, with $0 < i < |B|$, and connecting all the new rays by using arrows $(v,i,j) \longrightarrow (v,i,j+1)$, and corresponding extensions. The actual form of B is only used for adding at most $\frac{1}{2}|B|(|B| - 1)$ vertices, all of which being iterated τ-translates of vertices of the form $(v,1,j)$, with $2 \leq j \leq |B|$.]

Let C be a standard component of A_0-mod with trivial modulation. An indecomposable module V in C will be called a <u>ray module</u> provided [V] is a ray vertex in ΓC. If V is a ray module in C, then we denote by $V[i]$ any indecomposable module with isomorphism class $[V][i]$.

Also, given an algebra A_0, an A_0-module M, and a branch B, let $A_0[M,B]$ be obtained from the one-point extension $A_0[M]$ by adding the branch B in the vertex ω of $A_0[M]$. In case B is the empty branch, then $A_0[M,B]$ should mean just A_0.

(1) <u>Proposition. Let</u> C <u>be a standard component of</u> A_0<u>-mod,</u> V <u>a ray module in</u> C<u>, and</u> B <u>a non-empty branch. Let</u> $A = A_0[V,B]$.

<u>For any</u> $i \in \mathbb{N}_1$, <u>and any</u> $1 \leq j \leq |B|$, <u>there is a unique indecomposable A-module</u> $V[i]_j$ <u>having</u> $V[i]$ <u>as a submodule and such that</u> $V[i]_j / V[i] = M_j(B)$. <u>Let</u> C' <u>be the module class given by the modules from</u> C, <u>the modules</u> $V[i]_j$, $i \in \mathbb{N}_1$, $1 \leq j \leq |B|$, <u>and the indecomposable B-modules</u> X <u>with</u> $\langle \ell_B, \underline{\dim}\, X\rangle < 0$. <u>Then</u> C' <u>is a standard-component of</u> A, <u>and</u> $\Gamma C' = (\Gamma C)[v,B]$, <u>where</u> $v = [V]$. <u>If</u> M <u>is an indecomposable A-module not belonging to</u> C', <u>then no indecomposable summand of its restriction to</u> A_0 <u>belongs to</u> C, <u>and any indecomposable summand</u> X <u>of</u> $M/(M|A_0)$ <u>satisfies</u> $\langle \ell_B, \underline{\dim}\, X\rangle > 0$.

<u>Proof.</u> First, let us outline why it is sufficient to consider the case of B being the branch of length 1, thus to deal with the one-point extension $A_0[V]$ of A_0 by the ray module V.

We recall that any $A_0[V]$-module can be written in the form $(M_0, M_\omega, \gamma_M)$, with M_0 an A_0-module, M_ω a k-space, and $\gamma_M : V \otimes M_\omega \longrightarrow M_0$ A_0-linear, and $M_0 = M|A_0$ is called the restriction of M to A_0. Also, given an A_0-module M_0, we identify M_0 with $(M_0,0,o)$, and we define $\overline{M_0} = (M_0, \mathrm{Hom}(V,M_0), \varepsilon_M)$, with $\varepsilon_{M_0} : V \otimes \mathrm{Hom}(V,M_0) \longrightarrow M_0$ the evaluation map.

Note that our proof of the proposition in the case of B being of length 1 will show, in particular, that in the component C' of $A_0[V]$-mod, the module $\overline{V}$

is a ray module, thus, we can use induction, and obtain in this way the proposition for $A_o[V,B^s]$, with B^s being any subspace branch. Given any branch B, we let B^s be the subspace branch of the same length; also, let $C = A_o[V,B]$, $C^s = A_o[V,B^s]$. Now the component C' of C^s-mod given by the proposition does not contain any C^s-module with support on B^s_o, thus C' is contained in $\mathcal{Y}({}_C T)$, where ${}_C T$ is the tilting module given by 4.4.5, thus we obtain the corresponding component of C-mod by identifying the vertices $[P(b_i)]$ of $\Gamma C'$ with the vertices w_{in} in $\Gamma G(\Xi(B))$. It is easy to calculate that the C-module $V[i]_j$ corresponding to the vertex (v,i,j) has the module $V[i]$ (which corresponds to $(v,i,0)$) as submodule and that $V[i]_j/V[i] = M_j(B)$. Let F denote the module class given by all indecomposable B-modules X with $\langle \ell_B, \underline{\dim}\, X\rangle < 0$. Then the indecomposable modules from F correspond to the vertices w_{ij} of $\Gamma G(\Xi(B))$ with $j \neq |B|$, in particular, if X is indecomposable and in F, and $X' \longrightarrow X$ is its sink map, then X' is in F. Since F is finite, it follows that for any indecomposable module M not belonging to C', we have $\operatorname{Hom}(M,F) = 0$, using sink maps in F. In particular, for any indecomposable module M outside C', no non-zero summand of $M/(M \mid A_o)$ can belong to F.

Thus, let us consider now the special case $A = A_o[V]$ of a one-point extension, with V a ray module in C.

Given the translation quiver Γ, and v a ray vertex, we let $\Gamma[v] = \Gamma[v,*]$, with $*$ the branch of length 1. However, we want to change slightly the notation, we denote in $\Gamma[v]$ the vertices $(v,i,0)$ by $v[i]$, the vertices $(v,i,1)$ by $\overline{v[i]}$. The arrows $v[i] \longrightarrow v[i+1]$ in Γ or in $\Gamma[v]$ will be denoted by ν_i, the arrows $\overline{v[i]} \rightarrow \overline{v[i+1]}$ in $\Gamma[v]$ by $\overline{\nu_i}$, finally, the arrows $v[i] \longrightarrow \overline{v[i]}$ in $\Gamma[v]$ will be denoted by ι_i.

Since C is assumed to be standard, the maps in C are given by linear combinations of paths in ΓC, and we denote the map in $k(\Gamma C)$ given by the path $(X \mid \alpha_1,\dots,\alpha_n \mid Y)$ by $\alpha_1\dots\alpha_n$.

First, let us show that a map $\alpha_1\dots\alpha_n$ given by any non-sectional path starting at V is zero. If not, choose such a path with n being minimal. Note that $n \geq 2$. Since $\alpha_1\dots\alpha_{n-1} \neq 0$, the corresponding path must be sectional and we have $\alpha_i = \nu_i$, for $1 \leq i \leq n-1$. Write $\alpha_n : V[n] \longrightarrow X$ for some indecomposable X. Since we started with a non-sectional path, $\tau X = V[n-1]$. Let β_j, $1 \leq j \leq s$ be the arrows ending in X and different from α_n. Then $\alpha_{n-1}\alpha_n + \sum_j \sigma(\beta_j)\beta_j = 0$, according to the mesh relations. However, by the minimality assumption, $\alpha_1\dots\alpha_{n-2}\sigma(\beta_j) = 0$ for all j, thus also $\alpha_1\dots\alpha_{n-1}\alpha_n = 0$, a contradiction.

As a consequence, $\dim \mathrm{Hom}(V,V[i]) = 1$, and $\mathrm{Hom}(V,\nu_i)$ is bijective, for all $i \in \mathbb{N}_1$. Namely, since there is just one sectional path from V to $V[i]$, we know that $\mathrm{Hom}(V,V[i])$ is generated by $\nu_1\ldots\nu_{i-1}$, thus we only have to show that $\mathrm{Hom}(V,\nu_i)$ is injective, for all i. Assume not, and take i minimal. Note that $V[i+1]$ cannot be projective, since otherwise ν_i would be mono, and therefore also $\mathrm{Hom}(V,\nu_i)$. Let $\begin{bmatrix}\nu_i\\ \varphi\end{bmatrix} : V[i] \oplus Y \longrightarrow V[i+1]$ be the sink map for $V[i+1]$, and factor $[\nu_1\ldots\nu_{i-1},0]$ through its kernel, thus through $\tau V[i+1]$. However, this means that we write $\nu_1\ldots\nu_{i-1}$ as a linear combination of paths from V to $V[i]$ which go via $\tau V[i+1]$, no one of which can be sectional (since there is just one sectional path from V to $V[i]$, and $V[i-1] \not\to \tau V[i+1]$.). By our previous result, $\nu_1\ldots\nu_{i-1} = 0$, but this contradicts the minimality of i (in case $i = 1$, we obtain $1_V = 0$, also impossible).

Let us derive some consequences for the A-modules. First of all $\overline{V[i]}$ is never an A_0-module. Second, any arrow $\varphi : V[i] \longrightarrow X$, with $\varphi \neq \nu_i$, gives rise to a map $(\varphi,0) : \overline{V[i]} \longrightarrow X$. Namely, we have to verify that the diagram

$$\begin{array}{ccc} V\otimes \mathrm{Hom}(V,V[i]) & \longrightarrow & 0 \\ \Big\downarrow{\scriptstyle \varepsilon_{V[i]}} & & \Big\downarrow \\ V[i] & \xrightarrow{\ \varphi\ } & X \end{array}$$

commutes, however, starting with an element $r \otimes \nu_1\ldots\nu_{i-1} \in V \otimes \mathrm{Hom}(V,V[i])$, we obtain under $\varepsilon_{V[i]}\varphi$ just the composition $\nu_1\ldots\nu_{i-1}\varphi$ applied to r, and $\nu_1\ldots\nu_{i-1}\varphi = 0$, as we have shown above.

Actually, let us determine the source map for $\overline{V[i]}$, using 2.5.5. Let $\varphi_j : V[i] \longrightarrow X_j$ be all arrows starting in $V[i]$ and different from ν_i, say $1 \leq j \leq t$. Then $[\nu_i,\varphi_1,\ldots,\varphi_t] : V[i] \longrightarrow V[i+1] \oplus X_1 \oplus \ldots \oplus X_t$ is the source map in A_0-mod. In A-mod, the source map for $\overline{V[i]}$ is given by

$$\begin{array}{ccc} V\otimes \mathrm{Hom}(V,V[i]) & \longrightarrow & V\otimes \mathrm{Hom}(V,V[i]) \\ \Big\downarrow{\scriptstyle \varepsilon_{V[i]}} & & \Big\downarrow{\scriptstyle \varepsilon_{V[i]}[\nu_i,\varphi_1,\ldots,\varphi_t]}. \\ V[i] & \xrightarrow{[\nu_i,\varphi_1,\ldots,\varphi_t]} & V[i+1]\oplus X_1 \oplus \ldots \oplus X_t, \end{array}$$

however, since $\mathrm{Hom}(V,\nu_i)$ is bijective, we obtain an isomorphism

$$(1,\mathrm{Hom}(V,\nu_i)) : (V[i+1],\mathrm{Hom}(V,V[i]),\varepsilon_{V[i]}\nu_i) \longrightarrow \overline{V[i+1]},$$

thus the source map above is isomorphic to

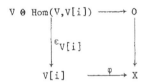

$$[\bar{\nu}_i,(\varphi_1,0),\ldots,(\varphi_t,0)] : \overline{V[i]} \longrightarrow V[i+1] \oplus X_1 \oplus \ldots \oplus X_t$$

with $\bar{\nu}_i = (\nu_i, \mathrm{Hom}(V,\nu_i))$. Also note that for the indecomposable modules M in C not of the form $V[i]$ for any i, the source map in C also is a source map in A-mod, since $\mathrm{Hom}(V,M) = 0$.

In this way, we have determined all arrows $[M] \to [N]$ in A-mod, with $M = \overline{V[i]}$ for some i, or with $M \in C$, and M not of the form $V[i]$, for any i. Namely, in case $M = \overline{V[i]}$, we have either $N = \overline{V[i+1]}$, with irreducible map $\bar{\nu}_i : \overline{V[i]} \longrightarrow \overline{V[i+1]}$, or else $N = X$, with an irreducible map $(\varphi,0) : \overline{V[i]} \longrightarrow X$ such that $\varphi : V[i] \longrightarrow X$ is irreducible in A_o-mod. Finally, for $M \in C$, and not of the form $V[i]$ for any i, we have $N \in C$, with an irreducible map $M \longrightarrow N$ in A_o-mod.

Similarly, let us determine some sink maps in A-mod. Consider first the sink map for $V[i]$, with $i \geq 2$. In A_o-mod, the sink map is of the form $\begin{bmatrix}\nu_{i-1}\\ \psi\end{bmatrix} : V[i-1] \oplus Y \longrightarrow V[i]$, where $\psi = \begin{bmatrix}\psi_1\\ \vdots\\ \psi_t\end{bmatrix} : Y = Y_1 \oplus \ldots \oplus Y_t \longrightarrow V[i]$ is given by all the arrows $\psi_j : Y_j \longrightarrow V[i]$ ending in $V[i]$ and different from ν_{i-1}. Applying $\mathrm{Hom}(V,-)$ to this map, and using that $\mathrm{Hom}(V,\nu_{i-1})$ is bijective, we see that the kernel of $\mathrm{Hom}(V,\begin{bmatrix}\nu_{i-1}\\ \psi\end{bmatrix})$ is just given by $\mathrm{Hom}(V,Y)$. Thus, the sink map for $V[i]$ in A-mod is given by

$$[\nu_{i-1},(\psi,0)] \;:\; V[i-1] \oplus \bar{Y} \longrightarrow V[i].$$

In particular, we see that the maps ν_i remain irreducible in $A_o[V]$-mod.

Now, let $\iota_i = (-1)^{i-1}(1,0) : V[i] \longrightarrow \overline{V[i]}$, so we obtain the exact sequence

$$0 \longrightarrow V[i] \xrightarrow{[\nu_i,\iota_i]} V[i+1] \oplus \overline{V[i]} \xrightarrow{\begin{bmatrix}\iota_{i+1}\\ \bar{\nu}_i\end{bmatrix}} \overline{V[i+1]} \longrightarrow 0 .$$

By induction on i, we claim that these are Auslander-Reiten sequences, or equivalently, that $[\nu_i,\iota_i]$ is a source map. Note that we know already that the maps ν_i and $\overline{\nu_i}$ are all irreducible in A-mod. Also, the existence of these non-split exact sequences shows that no $\overline{V[j]}$, with $j \geq 2$, is projective. Consider first the case $i = 1$. Since there is the irreducible map $\overline{\nu_1} : \overline{V[1]} \longrightarrow \overline{V[2]}$, and $\overline{V[2]}$ is not projective, we know that there has to be an irreducible map $\tau_A\overline{V[2]} \longrightarrow \overline{V[1]}$. However, $\overline{V[1]}$ is projective, and its radical is indecomposable, namely $V[1]$, thus $\tau_A\overline{V[2]} = V[1]$. Also, the inclusion map of $V[1]$ into $\overline{V[1]}$ is just the map ι_1, thus ι_1 is irreducible. Since both maps ν_1,ι_1 are irreducible, also $[\nu_1,\iota_1]$ is irreducible, and since the cokernel of $[\nu_1,\iota_1]$ is just

$\tau_A^{-}V[1]$, it follows that the sequence above for $i = 1$ is an Auslander-Reiten sequence. Now assume, we have shown that the sequence is an Auslander-Reiten sequence for some i. It follows (considering the given sink map of $\overline{V[i+1]}$) that ι_{i+1} is irreducible. Thus again, we know that $[\nu_{i+1}, \iota_{i+1}]$ is irreducible. Also, the existence of the irreducible map $\overline{\nu_{i+1}} : \overline{V[i+1]} \longrightarrow \overline{V[i+2]}$ shows that there has to be an irreducible map $\tau_A\overline{V[i+2]} \longrightarrow \overline{V[i+1]}$. Now, the only indecomposable A-modules having an irreducible map to $\overline{V[i+1]}$ are $V[i+1]$ and $\overline{V[i]}$. However, it is impossible that $\tau_A\overline{V[i+2]} = \overline{V[i]}$, since $\tau_A^{-}\overline{V[i]}$ has to be an A_o-module (according to 2.5.6)), thus $\tau_A\overline{V[i+2]} = \overline{V[i+1]}$. Taking into account that $[\nu_{i+1}, \iota_{i+1}]$ is irreducible, we conclude that the given sequence for i replaced by $i+1$ is an Auslander-Reiten sequence.

It follows that the indecomposable modules from C together with the modules $\overline{V[i]}$ form a component of A-mod. Namely, given any such module M, and let $M \longrightarrow N$ be its source map in A-mod, then N decomposes into a direct sum of such modules. Similarly, taking its sink map $N' \longrightarrow M$, then N' is a direct sum of such modules. Namely, in case M is projective, then either $M = \overline{V[1]}$ and $N' = V[1]$, or else M is an indecomposable projective A_o-module, and in C, thus its radical N' already has belonged to C. Or, in case M is not projective, then either M belongs to C, thus $\tau_A M = \overline{\tau_{A_o} M}$ (2.5.6) is in C or of the form $\overline{V[j]}$, or else $M = \overline{V[i]}$, and then $i \geq 2$ and $\tau_A\overline{V[i]} = V[i-1]$ is in C, and always we have the source map $\tau_A M \longrightarrow N'$.

The determination of all the source maps shows that the component obtained is of the form $(\Gamma C)[v]$, and also, it is easy to see that this component is standard.

It remains to be shown that any indecomposable A-module M with non-zero restriction to A_o belongs to C. Thus, let $M = (M_o, M_\omega, \gamma_M)$, and assume M_o has a direct summand X in C. Since M is indecomposable, and $M_o \neq o$, we must have $\mathrm{Hom}(V,X) \neq 0$, thus $X = V[i]$ for some i. Let $p : M_\omega \rightarrow V[i]$ be a projection map. say $mp = 1_{V[i]}$, for some $m : V[i] \longrightarrow M_o$. Let $\tilde{p} : M_\omega \longrightarrow \mathrm{Hom}(V,[i])$ be the adjoint of $\gamma_M p : V \otimes M_\omega \longrightarrow V[i]$, thus we have a map $(p,\tilde{p}) : M \longrightarrow \overline{V[i]}$. Note that $\tilde{p} \neq 0$, since otherwise $(p,0)$ maps in fact to $V[i]$, and therefore is split epi with coretraction $(m,0)$. Thus, we have a map $f = (f_o, f_\omega) : M \longrightarrow \overline{V[i]}$ with $f_\omega \neq 0$, and we choose such a map with i minimal. If $i = 1$, then f does not map into the radical of the projective A-module $\overline{V[1]}$, thus f is surjective, and therefore split epi, thus $M = \overline{V[1]}$. If $i > 1$, and M is not isomorphic to $\overline{V[i]}$ then we factor f through the sink map $\begin{bmatrix} \iota_i \\ \overline{\nu_{i-1}} \end{bmatrix}$ of $\overline{V[i]}$, thus $f = f'\iota_i + f''\overline{\nu}_{i-1}$ for some $f' : M \longrightarrow V[i]$, $f'' : M \longrightarrow \overline{V[i-1]}$. Since $f_\omega \neq 0$, we must have $f''_\omega \neq 0$, however, this contradicts the minimality of i. Thus $M = \overline{V[i]}$. This finishes the proof.

4.6 Tubes

We recall that a tube is a translation quiver Γ with $|\Gamma| = S^1 \times \mathbb{R}_0^+$ and such that Γ contains a cyclic path. The tubes we will be interested in will always be given very explicitly. However, in order to formulate certain results, it seems convenient to have some general concepts about tubes available.

(1) Let Γ be a tube, with Γ_1 its set of arrows. There exists a unique function $d : \Gamma_1 \longrightarrow \{\pm 1\}$ with the following properties:

(a) If α, β are different arrows with the same starting point or the same end point, then $d(\alpha) \neq d(\beta)$.

(b) If $\alpha : x \longrightarrow y$ is in Γ_1 and y is not projective, then $d(\sigma\alpha) \neq d(\alpha)$.

(c) For any infinite path

$$x_1 \xrightarrow{\alpha_1} x_2 \longrightarrow \ldots \longrightarrow x_i \xrightarrow{\alpha_i} x_{i+1} \longrightarrow \ldots$$

with pairwise different arrows, there exists some i with $d(\alpha_i) = 1$.

(c*) For any infinite path

$$\ldots\ y_{i+1} \xrightarrow{\beta_i} y_i \longrightarrow \ldots \longrightarrow y_2 \xrightarrow{\beta_1} y_1$$

with pairwise different arrows, there exists some i with $d(\beta_i) = -1$.

The arrows α with $d(\alpha) = 1$ are said to point to infinity, the arrows with $d(\alpha) = -1$ are said to point to the mouth.

An infinite path

$$(*) \qquad x = x[1] \longrightarrow x[2] \longrightarrow \ldots \longrightarrow x[i] \longrightarrow x[i+1] \longrightarrow \ldots$$

in a tube, with pairwise different vertices and all arrows pointing to infinity, is said to be a ray. Note that rays always are sectional paths. In dealing with rays, we usually will stick to the notation $x[i]$ for the starting point of the i-th arrow. A ray which is not properly contained in any other ray is said to be maximal.

(2) Any ray is contained in a unique maximal ray. If two rays have a vertex in common, then one of them is contained in the other.

The dual notion is that of a coray, a coray being an infinite path

$$(**) \qquad \ldots \longrightarrow x_{i+1} \longrightarrow x_i \longrightarrow \ldots \longrightarrow x_2 \longrightarrow x_1$$

with pairwise different vertices and all arrows pointing to the mouth. A coray not properly contained in any other coray is said to be maximal. Any coray is contained in a unique maximal coray, and given two corays which have a vertex in common, then one of them is contained in the other.

Let us note the following: if $v_0 \xrightarrow{\alpha} v_1 \xrightarrow{\beta} v_2$ are two arrows in a tube, not pointing in the same direction (that means, $d(\alpha) \neq d(\beta)$), and if this path is sectional, then v_0 is injective, v_2 is projective [namely, if for example v_2 is not projective, then both α and $\sigma(\beta)$ have the same direction and the same end-point, thus $\alpha = \sigma(\beta)$, therefore $v_0 = \tau v_2$; and dually, the same holds, in case we assume that v_0 is not injective]. As a consequence, if we assume that Γ is a tube with either no injective, or no projective vertex, then any path (*) with pairwise different vertices is a ray, any path (**) with pairwise different vertices is a coray.

(3) Let Γ be a tube, and v a ray vertex in Γ, with sectional path

$$v = v[1] \xrightarrow{\nu_1} v[2] \longrightarrow \ldots \longrightarrow v[i] \xrightarrow{\nu_i} v[i+1] \longrightarrow \ldots$$

starting at v. Then this is a ray.

Proof. Assume there exists some $i \geq 2$ with ν_{i-1} pointing to the mouth, ν_i pointing to infinity. Then $v[i-1]$ has to be injective, $v[i+1]$ projective. Also, $\nu_i : v[i] \longrightarrow v[i+1]$ is the only arrow starting at $v[i]$, since for any arrow $\beta : v[i] \longrightarrow z$, the path $\nu_1 \ldots \nu_{i-1}\beta$ is sectional, due to the fact that $v[i-1]$ is injective. Since $v[i+1]$ is projective, we conclude that for any arrow $x \longrightarrow v[i]$, the vertex x is injective. However, then the vertex x has a connected neighborhood which becomes non-connected, when removing x. Since this is impossible in $S^1 \times \mathbb{R}_0^+$, we conclude that for any arrow ν_i pointing to infinity, with $i \geq 2$, also ν_{i-1} points to infinity. Now according to 4.5.1, there are arbitrarily large i with ν_i pointing to infinity, thus all ν_i point to infinity.

(4) If Γ is a tube, v a ray vertex in Γ, and B a branch, then $\Gamma[v,B]$ is again a tube. More general, given $v_1,\ldots,v_t$ ray vertices in Γ which generate pairwise disjoint rays, and $B_1,\ldots,B_t$ (not necessarily different, and possibly empty) branches, we define $\Gamma[v_i,B_i]_{i=1}^t$ inductively by
$\Gamma[v_i,B_i]_{i=1}^t = (\Gamma[v_i,B_i]_{i=1}^{t-1})[v_t,B_t]$.

We denote by $\Omega(n) = \mathbb{Z}\vec{\mathbb{A}}_\infty/n$, the stable tube of rank n. Recall that $\vec{\mathbb{A}}_\infty$ is the quiver with $\mathbb{N}_1$ as set of vertices, and with arrows $i \longrightarrow i+1$, for all $i \in \mathbb{N}_1$. Thus, the vertices of $\Omega(n)$ are of the form $(\bar{z},i)$, with $\bar{z} \in \mathbb{Z}/n$, $i \in \mathbb{N}_1$, there are arrows $(\bar{z},i) \longrightarrow (\bar{z},i+1)$ and $(\bar{z},i+1) \longrightarrow (\overline{z+1},i)$, and $\tau(\bar{z},i) = (\overline{z-1},i)$, for all $\bar{z} \in \mathbb{Z}/n$, and all $i \in \mathbb{N}_1$. The element $(\bar{z},i)$ will be said to be of length i, those of length 1 are said to lie on the mouth of $\Omega(n)$. Note that the arrows $(\bar{z},i) \longrightarrow (\bar{z},i+1)$ point to infinity, those of the form $(\bar{z},i+1) \longrightarrow (\overline{z+1},i)$ point to the mouth. We denote by $e_0,\dots,e_{n-1}$ the elements of the mouth of $\Omega(n)$, with $\tau e_z = e_{z-1}$, for all z, say we let $e_z = (\bar{z},1)$ for all z. If these are given the subspace branches B_z, $0 \leq z \leq e_{n-1}$, then we denote the vertices of $\Omega(n)[e_z,B_z^S]_z$ by triples $(\bar{z},i,j)$, with $\bar{z} \in \mathbb{Z}/n\mathbb{Z}$, $i \in \mathbb{N}_1$, and $0 \leq j \leq |B_z^S|$. For example, in case $n = 6$, with B_z^S being, for $i = 0,\dots,5$ of lengths 3,0,1,4,0,0 respectively, we obtain the following tube (we have to identify the dashed vertical lines, vertices with the same name being identified. Instead of $(\bar{z},i,j)$, we write zij):

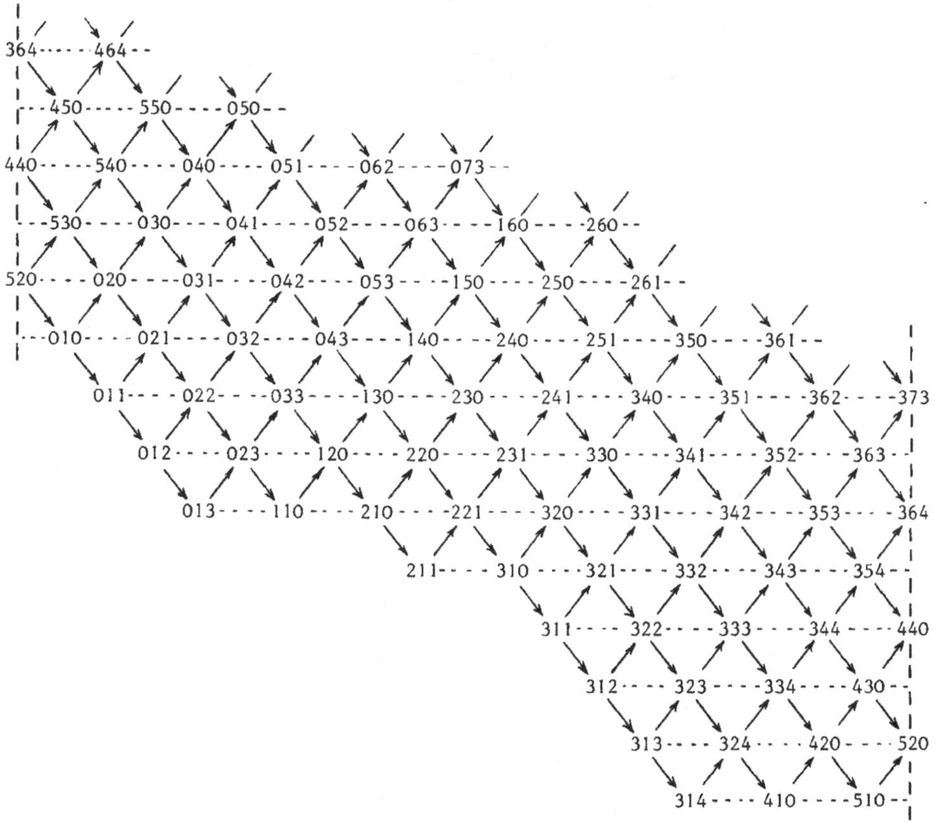

The general case of $\Omega(n)[e_z,B_z]_z$ with B_z being branches is obtained from $\Omega(n)[e_z,B_z^S]_z$ by identifying the full subquiver of $\Omega(n)[e_z,B_z^S]_z$ given by the vertices $(\bar{z},1,j)$ with $1 \leq j \leq |B_z^S|$ with the full subquiver of $\Gamma G(\Xi(B_z))$, or better of an indexed copy $\{z\} \times \Gamma G(\Xi(B_z))$ of it, given by the vertices $(z, w_{j,|B_z|})$, $1 \leq j \leq |B_z|$, always identifying $(\bar{z},1,j)$ with $(z, w_{j,|B_z|})$. For example, choosing as $B_0, \ldots, B_5$ the following branches

(+) $B_0 =$ 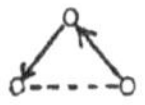 , $B_1 = \emptyset$, $B_2 = \circ$, $B_3 =$ 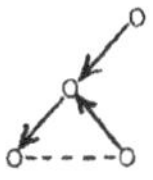 , $B_4 = \emptyset$, $B_5 = \emptyset$,

the translation quiver $\Omega(6)[e_z, B_z]_z$ has the following shape:

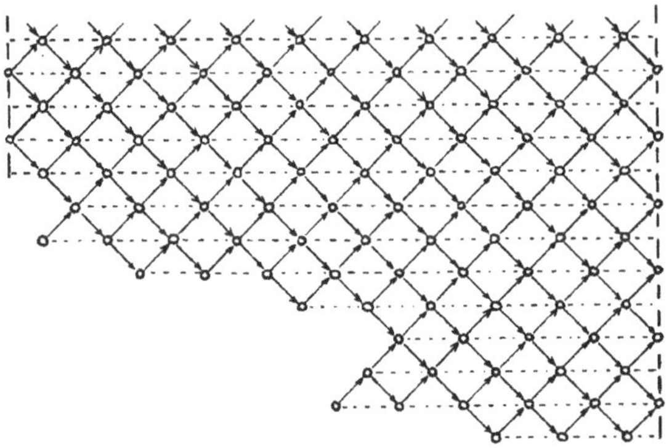

Note that the rays in $\Omega(n)[e_z, B_z^s]_z$ are given by vertices with fixed first and third coordinate, and that every vertex even of $\Omega(n)[e_z, B_z]_z$ belongs to a ray . (We may use this observation in order to introduce coordinates $(\bar{z},i,j)$ also for the vertices outside $\Omega(n)[e_z, B_z^s]_z$, using numbers $i \leq 0$). We may consider $\Omega(n)[e_z, B_z]_z$ as being obtained from $\Omega(n)$ by <u>ray insertion</u> (inserting those rays with last coordinate $j > 0$.

Let us consider a translation quiver $\Gamma = \Omega(n)[e_z, B_z]_z$ obtained from $\Omega(n)$ by ray insertion, and let $k(\Gamma)$ be its mesh category. The vertices $x = (\bar{z},i,0)$ in Γ are characterized by the property that $\mathrm{Hom}_{k(\Gamma)}(p,x) = 0$ for any projective vertex p of Γ. In fact, the full subcategory given by these objects is just the mesh category $k(\Omega(n))$.

<u>Remark.</u> A description of the mesh category $k(\Omega(n)[e_z, B_z]_z)$ as a category of modules over some (infinite-dimensional) algebra will be given at the end of this section.

As we have mentioned, we call $\Omega(n)[e_z,B_z]_z$ an extension of $\Omega(n)$ by ray insertion, and $m = n + \sum_{z=0}^{n-1} |B_z|$ will be said to be the extension rank. Our next aim is to recover $\Omega(n)[e_z,B_z]_z$ from $\Omega(m)$, where m is the extension rank.

(5) First, let us note that for any vertex $x \in \Omega(m)$ of length $\ell < m$, there exists a wing $W(x)$ of length ℓ in $\Omega(m)$ such that x is the projective-injective vertex of $W(x)$. For example, the wing $W((1,\ell))$ in $\Omega(m)$, where $\ell < m$, consists of all vertices (i,j) with $i+j \leq \ell+1$. If we want to identify $W((1,\ell))$ with $\theta(\ell)$, we have to identify (i,j) with $w_{i,i+j-1}$.

A subset Ξ of the vertex set of $\Omega(m)$ is said to be a closed set provided two conditions are satisfied: first of all,

(α) $\mathrm{Hom}_{k(\Omega(m))}(x,\tau y) = 0$ for all $x,y \in \Xi$; thus, in particular, the length of any $x \in \Xi$ has to be $< m$, and therefore $W(x)$ is defined for any $x \in \Xi$, and second:

(β) $W(x) \cap \Xi$ is a tilting set in $W(x)$, for any $x \in \Xi$.

Let $(B_0,\ldots,B_{n-1})$ be a sequence of (not necessarily different, and possibly empty) branches, and let $m = n + \sum_{z=0}^{n-1} |B_z|$. We associate with $(B_0,\ldots,B_{n-1})$ a closed set $\Xi(B_0,..,B_{n-1})$ inside $\theta(m-1)$, and therefore inside $\Omega(m)$:

$$\Xi(B_0,\ldots,B_{n-1}) = \bigcup_{z=0}^{n-1} \tau^{-r_z}\,\Xi(B_z), \text{ where } r_z = \sum_{u=0}^{z-1} (|B_u|+1).$$

As an example, we show the closed set $\Xi(B_0,\ldots,B_5)$, for the sequence (+) of branches considered above. The elements of $\Xi(B_0,\ldots,B_5)$ are marked by ⊛, also, we have marked by ◎ the vertices $v_z = (r_z,|B_z|+1)$, since they are rather useful, and we have labelled the component z of any vertex $(z,1)$ on the mouth:

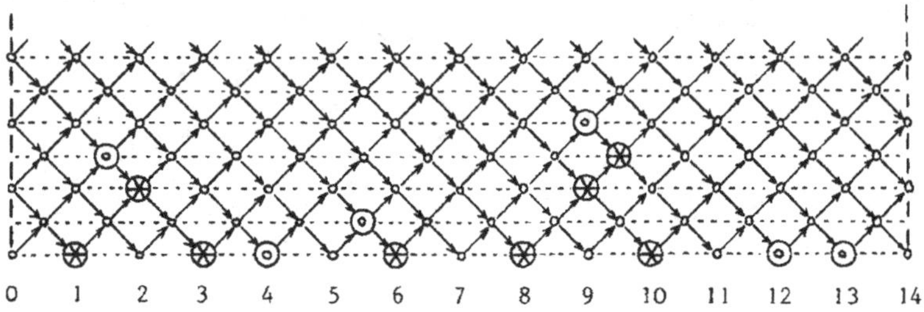

Conversely, any closed set Ξ in $\Omega(m)$ can be obtained (up to a cyclic permutation of $\Omega(m)$) in this way. Let us outline how to obtain from Ξ the sequence $(B_0,\ldots,B_{n-1})$ as well as the vertices $v_0,\ldots,v_{n-1}$, with $n = m - |\Xi|$. In case Ξ

is empty, let $B_z = \emptyset$ and $v_z = (z,1)$, for all z. Otherwise, choose some $x_0 \in \Xi$ of maximal length, say of length ℓ_0, and choose the coordinates in $\Omega(m)$ so that $x_0 = (1,\ell_0)$. Let B_0 be the branch corresponding to the tilting set $W(x_0) \cap \Xi$ in $W(x_0)$, and let $v_o = (0,\ell_o+1)$. Now assume, $(B_0,\dots,B_z)$ are defined, let

$r_{z+1} = \sum_{u=o}^{z} (|B_u| + 1)$. Consider the ray of all vertices with first component $r_{z+1}+1$. If this ray should not contain an element from Ξ, then let $B_{z+1} = \emptyset$, and $v_{z+1} = (r_{z+1},1)$. If, on the other hand, this ray contains the vertex $x_{z+1} = (r_{z+1}+1,\ell_{z+1}) \in \Xi$, and ℓ_{z+1} is maximal with this property, then let B_{z+1} be the branch corresponding to the tilting set $W(x_{z+1}) \cap \Xi$ in $W(x_{z+1})$, and $v_{z+1} = (r_{z+1},\ell_{z+1}+1)$. (The vertices v_z, $0 \le z \le n-1$ may be called the <u>branch indicators</u>, and for B_z being a non-empty branch, we may say that B_z is <u>a branch in</u> the vertex $x_z = (r_z+1,|B_z|)$. Note that there is an arrow $v_z \longrightarrow x_z$ pointing to the mouth.)

(6) Given a closed set Ξ in $\Omega(m)$, let us define $G(\Xi)$ to be the full subcategory of $k(\Omega(m))$ given by all vertices y in $\Omega(m)$ satisfying $\mathrm{Hom}_{k(\Omega(m))}(y,\tau x) = 0$ for all $x \in \Xi$. (Of course, this definition is parallel to that of $G(\Xi)$, where Ξ is a tilting set in $\theta(n)$. However, for Ξ a tilting set in $\theta(n)$, we also could describe $G(\Xi)$ by a generation condition; a similar statement is no longer true for Ξ being a closed set in a stable tube). Note that for any $(z,i) \in G(\Xi)$, with $i \ge 2$, also $(z+1,i-1)$ belongs to $G(\Xi)$. Also, if $(z,i) \in G(\Xi)$, and not in Ξ, then $(z-1,i+1)$ belongs to $G(\Xi)$. We want to indicate $G(\Xi(B_0,\dots,B_5))$ for the case (+); the elements belonging to $\Xi(B_0,\dots,B_5)$ being marked by an encircled star ⊛ , those belonging to $G(\Xi(B_0,\dots,B_5))$, but not to $\Xi(B_0,\dots,B_5)$, by a single star $*$:

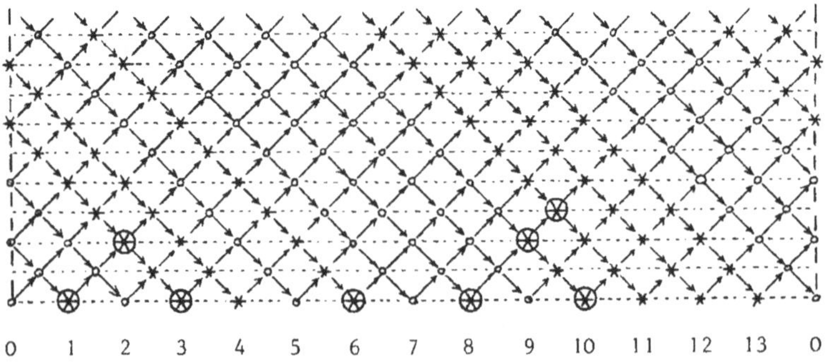

<u>Proposition.</u> <u>Let</u> $B_0,\dots,B_{n-1}$ <u>be branches, let</u> $m = n + \sum_{z=o}^{n-1} B_z$. <u>Then</u> $\Gamma G(\Xi(B_0,\dots,B_{n-1})) \approx \Omega(n)[e_z,B_z]_z$.

<u>Proof.</u> Let us give a second description of $\Omega(n)[e_z,B_z]_z$, the "coray" des-

cription. We claim that $\Omega(n)[e_z,B_z]_z$ has precisely n maximal corays, they are ending in the vertices $(z,1,|B_z|)$, and that a vertex x belongs to a maximal coray if and only if x belongs to the subquiver $\Omega' = \Omega(n)[e_z,B^s_z]_z$.

Let us define Ω'' as being the translation quiver with vertices $\{\bar{z},i\}$, $\bar{z} \in \mathbf{Z}/n$, $i \in \mathbb{N}_1$, with arrows $\{\bar{z},i+1\} \longrightarrow \{\bar{z},i\}$, and $\{\bar{z},i\} \longrightarrow \{\overline{z+1},i+1+|B_{z+1}|\}$, for all $\{\bar{z},i\}$, and with translation $\tau\{\bar{z},i\} = \{\overline{z-1},i-|B_z|\}$ for $\{\bar{z},i\}$ with $i > |B_z|$, whereas the vertices $\{\bar{z},i\}$ with $i \leq |B_z|$ are supposed to be projective. In Ω'', the maximal corays are given by the vertices with a fixed first coordinate. Again, considering the example $n = 6$, and $|B_z| = 3,0,1,4,0,$ or 0, for $z = 0,1,\dots,5$ respectively, Ω'' has the following form

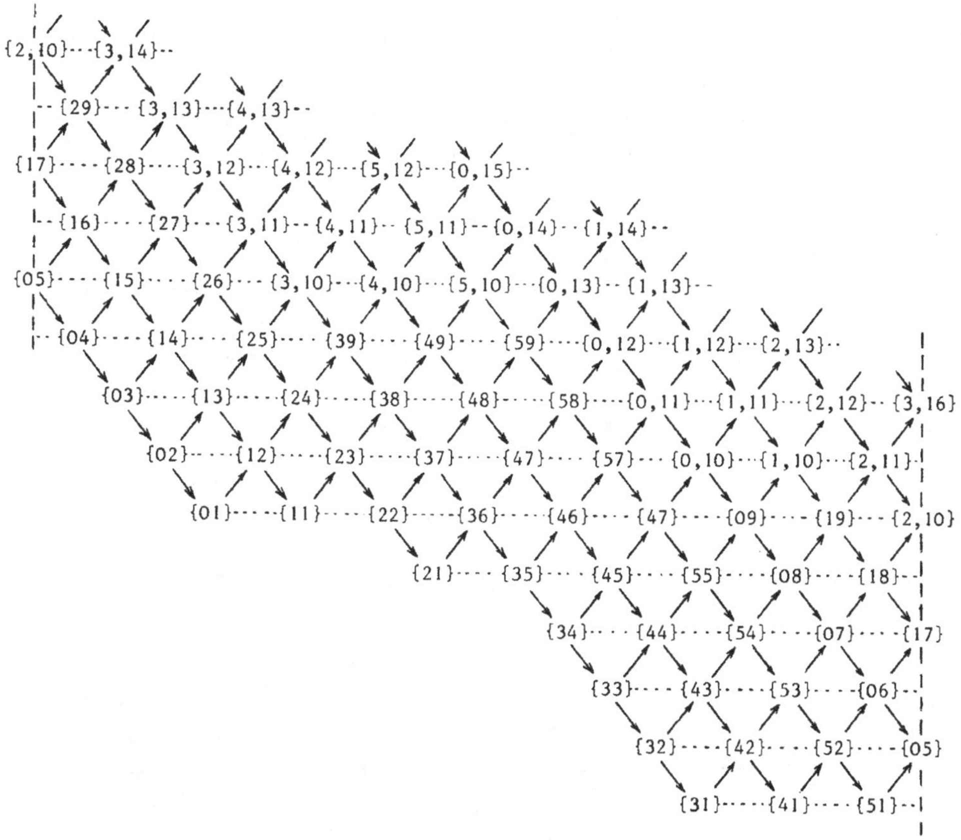

In general, we can identify Ω' and Ω'', by identifying $(\bar{z},i,j) \in \Omega'$ with

$$\{\overline{z+i-1},\ i-j + \sum_{u=z}^{z+i-1} |B_u|\} \in \Omega''.$$

In order to prove that this defines a bijection of the vertices, one may use

induction on i. In this way, one identifies the rays of Ω' with rays in Ω'', and it then remains to be checked that this identification is compatible with the remaining arrows, and with τ.

In a similar, but easier, way, we may describe the vertices $(\bar{z},i)$ of $\Omega(m)$ also in the coray notation: Here, the vertex $(\bar{z},i)$ corresponds to $\{\overline{z+1-1},i\}$.

For the proof of the proposition, let us restrict to the case of all B_z being subspace branches, the general case follows from this without difficulties.

Let us define an injective mapping $(\Omega'')_0 \longrightarrow (\Omega(m))_0$, sending $\{\bar{z},i\}$ with $0 \leq z < n$ to $\{\overline{s(z)},i\}$, where $s(z) = \sum_{u=o}^{z} |B_u|$. This map sends corays to corays, and the image is the set of all vertices belonging to the corays in $\Omega(m)$ ending in $\{\overline{s(z)},1\}$. However, this is just the set of objects in $G := G(\Xi(B_o^s,\ldots,B_{n-1}^s))$. In order to calculate the arrows in G, we note that G is closed under epimorphisms in $k(\Omega(m))$, thus given an irreducible map $\alpha : X \longrightarrow Y$ in G, with X,Y indecomposable, then either this is an epimorphism, and then α is already irreducible in $k(\Omega(m))$ (these are the arrows belonging to corays) or else α is a monomorphism, and then α, as a map of $k(\Omega(m))$, is a composition of irreducible maps all of which are monomorphisms (any $\{\overline{s(z)},i\}$ is the starting point of such a map, giving rise to an arrow ending in $\{\overline{s(z+1)},i + |B_{z+1}| + 1\}$, in case $0 \leq z < n-1$, and in $\{\overline{s(0)},i + |B_0| + 1\}$, in case $z = n-1$). In this way, one shows that the Auslander-Reiten quiver of $G(\Xi(B_o^s,\ldots,B_{n-1}^s))$ actually is equal to Ω''.

(7) Let us just mention also the dual notion. Starting with a sequence $(B_o,\ldots,B_{n-1})$ of branches, we call ${}_z[B_z,e_z]\Omega(n) = (\Omega(n)[e_z,B_z^*]_z)^*$ the coextension obtained from the stable tube $\Omega(n)$ by <u>coray insertions</u>, and $m = n + \sum_{u=o}^{n-1} |B_z|$ the <u>coextension rank</u> of ${}_z[B_z,e_z]\Omega(n)$.

<u>Remark.</u> Let us mention that we can realize $\Omega(n)[e_z,B_z]_z$ as the Auslander-Reiten quiver of some module class also as follows: Let A be the (infinite dimensional!) path algebra of the quiver $\vec{\tilde{\mathbb{A}}}_{n-1}$ of type $\tilde{\mathbb{A}}_{n-1}$ with cyclic orientation. According to 3.7, it has an Auslander-Reiten quiver $\Gamma(A)$, and $\Gamma(A)$ is an $\mathbb{A}^1(k)$-family of stable tubes, with one component being of rank n, all other being homogeneous. In case $n \geq 2$, let us denote by $T(0)$ the component of A-mod of rank n, in case $n = 1$, let $T(0)$ be the component containing the simple one-dimensional A-module $E(0)$ corresponding to the unique vertex 0. Thus, always $T(0)$ is the full subcategory of all A-modules annihilated by some power of the ideal $(k\vec{\tilde{\mathbb{A}}}_{n-1})^+$ generated by the arrows of $\vec{\tilde{\mathbb{A}}}_{n-1}$. Identifying $\Gamma T(0)$ with $\Omega(n)$, we may consider e_z as the isomorphism class of the simple A-module $E(z)$ corres-

ponding to the vertex z of $\tilde{\mathbb{A}}_{n-1}$, with $0 \le z \le n-1$. Let $A' = A[E(z),B_z]_z$ be obtained from A by forming the various one-point extensions $A[E(z_1)]\dots[E(z_t)]$ for all z_i with B_{z_i} non-empty, say obtaining in this way new vertices $\omega_1,\dots,\omega_t$, and adding the branch B_{z_i} in ω_i. For example, in the case $n = 6$ with the sequence (+) of branches considered above, the corresponding algebra A' is given by the following quiver with relations:

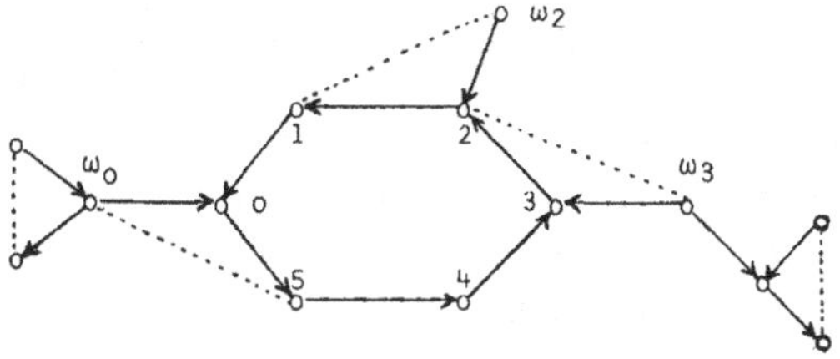

Let us denote by $\mathcal{T}'(0)$ the module class given by all indecomposable A'-modules with restriction to A being non-zero and in $\mathcal{T}(0)$, together with all indecomposable A'-modules M having the support in some branch B_z, such that $\langle \ell_{B_z}, \underline{\dim}\, M\rangle < 0$. Then $\mathcal{T}'(0)$ is a standard component of A'-mod, and $\Gamma\mathcal{T}'(0) = \Omega(n)[e_z,B_z]_z$. (For the proof, we refer to the next section.)

Note that the remaining components $\mathcal{T}(\rho)$ of A-mod, $\rho \in \mathbb{A}^1(k)$, $\rho \neq 0$, remain components of A'-mod. There are only finitely many additional indecomposable A'-modules, namely those indecomposable A'-modules M having the support in some branch B_z, and satisfying $\langle \ell_{B_z}, \underline{\dim}\, M\rangle > 0$. Note that not all of these modules can have sink maps! (Of course, A' is not a finite-dimensional algebra).

4.7 Tubular extensions

Let A_o be an algebra, $E_1,\dots,E_t$ A_o-modules, and $K_1,\dots,K_t$ (usually non-empty) branches. Let $A_o[E_i,K_i]_{i=1}^t$ be inductively defined: Recall that $A_o[E_1,K_1]$ is obtained from the one-point extension $A_o[E_1]$ with extension vertex ω_1 by adding the branch K_1, in ω_1, and let $A_o[E_i,K_i]_{i=1}^t = (A_o[E_i,K_i]_{i=1}^{t-1})[E_t,K_t]$. Now assume there is a standard tubular family $\mathcal{T}$ in A_o-mod, separating $\mathcal{P}$ from $\mathcal{Q}$. The algebra $A = A_o[E_i,K_i]_{i=1}^t$ is called a <u>tubular extension of</u> A_o <u>using modules from</u> $\mathcal{T}$ provided the modules $E_1,\dots,E_t$ are pairwise orthogonal ray modules from $\mathcal{T}$. Note that if $\mathcal{T}$ is a family of stable tubes, then an indecomposable module in $\mathcal{T}$ is a ray module if and only if it belongs to the mouth of one of the tubes, and any two different ray modules in $\mathcal{T}$ are orthogonal. (Occasionally, it will be convenient to allow some of the branches K_i to be empty; we recall that for the empty branch K, we have defined $A_o[E,K] = A_o$).

<u>Example 1</u>. Let A be a canonical algebra with source 0 and sink ω, let $\iota : K_o(A) \longrightarrow \mathbf{Z}$ be the linear form given by $-X_o + X_\omega$, and $\mathcal{T}$ the module class given by all indecomposable A-modules M with $(\underline{\dim}\, M)\iota = 0$. A tubular extension $A[E_i,K_i]_{i=1}^t$ using modules from $\mathcal{T}$ will be called a <u>canonical tubular extension</u> of the canonical algebra A.

<u>Example 2</u>. Let A_o be a tame concealed algebra. There is a unique separating tubular family, so we do not have to specify $\mathcal{T}$.

Let $A = A_o[E_i,K_i]_{i=1}^t$ be a tubular extension of A_o, using modules from a stable I-family $\mathcal{T}$, say with tubes $\mathcal{T}(\rho)$, $\rho \in I$. Let r_ρ be the rank of $\mathcal{T}(\rho)$. The extension type of A over A_o will be given by the the function $n : I \longrightarrow \mathbf{N}_1$, with

$$n_\rho = r_\rho + \sum_{E_i \in \mathcal{T}(\rho)} |K_i|.$$

Of course, as in 3.1, we usually will choose some finite subset I' of I such that $n_\rho = 1$ for $\rho \in I \setminus I'$, say $I' = \{\rho_1,\dots,\rho_t\}$ with pairwise different ρ_s, denote n_{ρ_i} just by n_i, and call $(n_1,\dots,n_t)$ the extension type of A. Note that in case A has extension type $(n_1,\dots,n_t)$, the rank of $K_o(A)$ is given by $2-t+\sum_{i=1}^t n_i$.

(1) <u>Theorem. Let</u> A_o <u>be an algebra with a separating tubular family</u> $\mathcal{T}$, <u>separating</u> $\mathcal{P}$ <u>from</u> $\mathcal{Q}$. <u>Let</u> $A = A_o[E_i,K_i]_{i=1}^t$ <u>be a tubular extension using modules from</u> $\mathcal{T}$. <u>We define module classes</u> $\mathcal{P}_o$, $\mathcal{T}_o$, $\mathcal{Q}_o$ <u>in</u> A-mod <u>as follows: let</u> $\mathcal{P}_o = \mathcal{P}$, <u>let</u> $\mathcal{T}_o$ <u>be the module class given by all indecomposable</u> A-<u>modules</u> M <u>with either</u> $M|A_o$ <u>non-zero and in</u> $\mathcal{T}$, <u>or else the support of</u> M <u>being contained in some</u> K_i,

and $\langle \ell_{K_i}, \underline{\dim} M\rangle < 0$. Let $\mathcal{Q}_0$ be the module class given by all indecomposable A-modules M with either $M|A_0$ non-zero and in $\mathcal{Q}$, or else the support of M being contained in some K_i, and $\langle \ell_{K_i}, \underline{\dim} M\rangle > 0$.

Then $A\text{-mod} = \mathcal{P}_0 \vee \mathcal{T}_0 \vee \mathcal{Q}_0$, with $\mathcal{T}_0$ being a separating tubular family separating $\mathcal{P}_0$ from $\mathcal{Q}_0$. Also, $\Gamma\mathcal{T}_0 = (\Gamma\mathcal{T})[e_i,K_i]_{i=1}^t$, with $e_i = [E_i]$.

It may seem strange that we have denoted module classes in A_0-mod by $\mathcal{P}$, $\mathcal{T}$, $\mathcal{Q}$, those in A-mod by $\mathcal{P}_0$, $\mathcal{T}_0$, $\mathcal{Q}_0$. However, the reason for denoting module classes in A-mod by $\mathcal{P}_0$, $\mathcal{T}_0$, $\mathcal{Q}_0$ is to stress that these module classes are defined by considering the restriction of A-modules to A_0. Also, for the tubular algebras defined in chapter 5, this notation $\mathcal{P}_0$, $\mathcal{T}_0$, $\mathcal{Q}_0$ will turn out to fit well into the general notation of module classes $\mathcal{P}_\gamma$, $\mathcal{T}_\gamma$, $\mathcal{Q}_\gamma$, with $\gamma \in \mathbb{Q}_0$.

Proof of the theorem. First, let us consider the case $t = 1$. Given an A-module M, let $M_0 = M|A_0$ be its restriction to A_0. Let $\mathcal{T}(\rho)$ be the tube of A_0-mod containing E_1, and let $\mathcal{T}'$ be the module class of A_0-mod given by the remaining tubes in $\mathcal{T}$. According to 4.5.1 , there is a component of A-mod, denoted by $\mathcal{T}_0(\rho)$, which contains $\mathcal{T}(\rho)$. This component $\mathcal{T}_0(\rho)$ is standard and, $\Gamma\mathcal{T}_0(\rho) = (\Gamma\mathcal{T}(\rho))[e_1,K_1]$. For any indecomposable A-module $M \notin \mathcal{T}_0(\rho)$, its restriction M_0 belongs to $\mathcal{P} \vee \mathcal{T}' \vee \mathcal{Q}$. Now, if $M_0 \neq M$, then actually $M_0 \in \mathcal{Q}$, since $\mathrm{Hom}(E_1, \mathcal{P} \vee \mathcal{T}') = 0$. Of course, if $M_0 = 0$, then M has its support on K_1, and $\langle \ell_{K_1}, \underline{\dim} M\rangle$ is either positive or negative. Altogether, this shows that $A\text{-mod} = \mathcal{P}_0 \vee \mathcal{T}_0 \vee \mathcal{Q}_0$, and $\mathcal{T}_0 = \mathcal{T}_0(\rho) \vee \mathcal{T}'$. In particular, $\mathcal{T}_0$ is a standard tubular family, and $\Gamma\mathcal{T}_0 = (\Gamma\mathcal{T})[e_1,K_1]$.

In order to see that $\mathcal{T}_0$ separates $\mathcal{P}_0$ from $\mathcal{Q}_0$, let us first consider the modules M with support in K_1. We have $K_1\text{-mod} = \mathcal{F} \vee \mathcal{G}$, where $\mathcal{F}$ is given by the indecomposable K_1-modules M with $\langle \ell_{K_1}, \underline{\dim} M\rangle < 0$, and $\mathcal{G}$ by those with $\langle \ell_{K_1}, \underline{\dim} M\rangle > 0$. By definition, $\mathcal{F} \subseteq \mathcal{T}_0$, and $\mathcal{G} \subseteq \mathcal{Q}_0$. Since $\mathcal{P}_0 \subseteq A_0$-mod, we have $\mathrm{Hom}(K_1\text{-mod}, \mathcal{P}_0) = 0$. Since $\mathcal{T}_0 \vee \mathcal{Q}_0 \subseteq (\mathcal{T} \vee \mathcal{Q})^{\urcorner} \llcorner K_1\text{-mod}$, we have $\mathrm{Hom}(\mathcal{T}_0 \vee \mathcal{Q}_0, \mathcal{P}_0) = 0$. Now assume there are given indecomposable modules $N \in \mathcal{Q}_0$, $M \in \mathcal{T}_0$, and a map $f : N \longrightarrow M$. The restriction $f_0 : N_0 \longrightarrow M_0$ of f to A_0 must be zero, since $N_0 \in \mathcal{Q}$, $M_0 \in \mathcal{T}$. Thus $f = 0$ in case $M = M_0$. Also $f = 0$ in case $M \in \mathcal{F}$, since $N/N_0 \in \mathcal{G}$. Thus, it remains to consider the case of $M_0 \neq 0$, and $M/M_0 = M_j(K_1)$ for some $1 \leq j \leq |K_1|$. Since f vanishes in N_0, we may replace N by some indecomposable summand of N/N_0, thus we may assume $N \in \mathcal{G}$. Using the structure of $\Gamma\mathcal{T}_0$, we see that we can factor f through an A_0-module [namely, $[M] = (v,i,j)$ for some $i \in \mathbb{N}_1$, with $v = [E_i]$ and we use induction on $i+j$. Now the sink map for M is of the form $M' \oplus M'' \longrightarrow M$ with always $[M'] = (v,i,j-1)$, and with $[M''] = (v,i-1,j)$ in case $i > 1$, $M'' \in \mathcal{F}$ in case $i = 1$. Note that $\mathrm{Hom}(N,\mathcal{F}) = 0$

and that the modules M''' with $[M'''] = (v,i,0)$ are $A_o$modules]. Thus $f = 0$. This shows that $\mathrm{Hom}(\mathcal{Q}_o, \mathcal{T}_o) = 0$. In order to see that any map f from $\mathcal{P}_o$ to $\mathcal{Q}_o$ factors through any tube of $\mathcal{T}_o$, we only note that $\mathcal{P}_o = \mathcal{P} \subseteq A_o$-mod, thus the image of f lies in $\mathcal{Q}$, thus we may factor f even through a module belonging to a fixed tube of $\mathcal{T}$. This finishes the proof for $t = 1$.

The general case follows by induction. We only have to note that the ray vertices $e_2,\dots,e_t$ of $\Gamma\mathcal{T}$ remain ray vertices in $(\Gamma\mathcal{T})[e_1,K_1]$, thus $E_2,\dots,E_t$ are ray modules in $A_o[E_1,K_1]$.

Given an extension $A = A_o[E_i,K_i]_{i=1}^t$, where A_o is an algebra with invertible Cartan-matrix, also the Cartan-matrix of A will be invertible, and we can calculate the quadratic form χ_A as follows. Let us denote by ω_i the vertex which belongs both to $A_o[E_i]$ and to K_i. Given an element $x \in K_o(A)$, denote by x_o its restriction to $K_o(A_o)$, by x_i its restriction to $K_o(K_i)$, and by x_a its coefficient at the vertex a. Then, for $x \in K_o(A)$

$$\chi_A(x) = \chi_{A_o}(x_o) + \sum_i \chi_{K_i}(x_i) - \sum_i x_{\omega_i} \langle \underline{\dim}\, E_i, x_o \rangle$$

This follows from the fact that $\langle e(a), e(b)\rangle = \langle e(b), e(a)\rangle = 0$ for a a vertex of A_o, and b a vertex of K_i, with $b \neq \omega_i$. Similarly, as a consequence of 2.5.12 we obtain:

(2) Let $A = A_o[E_i,K_i]_{i=1}^t$, and let w be some vector in $K_o(A_o)$. Then w, considered as an element of $K_o(A)$, belongs to $\mathrm{rad}\chi_A$ if and only if $w \in \mathrm{rad}\chi_{A_o}$ and $\langle w, \underline{\dim}\, E_i\rangle = 0$ for all $1 \leq i \leq t$.

In particular, if $A = A_o[E_i,K_i]_{i=1}^t$ is a tubular extension of a tame concealed algebra, then the vectors of $K_o(A_o)$ which are radical vectos for χ_{A_o} are also, considered as elements of $K_o(A)$, radical vectors for χ_A.

(3) Let A_o be an algebra, let $\iota : K_o(A_o) \longrightarrow \mathbf{Z}$ be a linear form such that $\mathcal{T}$ is a sincere, stable tubular family, separating $\mathcal{P}$ from $\mathcal{Q}$, where $\mathcal{P}, \mathcal{T}, \mathcal{Q}$ are the module classes given by the indecomposable A_o-modules M_o with $(\underline{\dim}\, M_o)\iota < 0$, $= 0$, or > 0, respectively. Let $A = A_o[E_i,K_i]_{i=1}^t$ be a tubular extension using modules from $\mathcal{T}$, and define $\mathcal{P}_o, \mathcal{T}_o, \mathcal{Q}_o$ as in (1). Extend ι to $K_o(A)$ by $e(a)\iota = 0$ for any vertex a outside A_o.

Let X be an indecomposable module in $\mathcal{T}_o$ which is not an A_o-module. Then $\underline{\dim}\, X$ is a positive root, $(\underline{\dim}\, X)\iota = 0$, and if Y is any indecomposable A-module with $\underline{\dim}\, X = \underline{\dim}\, Y$, then $X \approx Y$.

<u>Proof</u>. For any A-module M, let M_o denote its restriction to A_o. Since X_o is in $\mathcal{T}$, we have $(\underline{\dim}\, X)\iota = (\underline{\dim}\, X_o)\iota = 0$. Note that the support of X/X_o lies in some K_i, say K_1. If X itself is a K_1-module, then clearly $\underline{\dim}\, X$ is a positive root, and $\underline{\dim}\, X$ uniquely determines the isomorphism class of X. Thus, we may suppose that X_o is an indecomposable A_o-module in $\mathcal{T}$, say, belonging to the tube $\mathcal{T}(\rho)$, and $X/X_o = M_j(K_1)$ for some $1 \le j \le |K_1|$.
We obtain

$$\chi_A(\underline{\dim}\, X) = \chi_{A_o}(\underline{\dim}\, X_o) + \chi_{K_1}(\underline{\dim}\, M_j(K_1)) - \langle \underline{\dim}\, E_1, \underline{\dim}\, X_o\rangle,$$

using that $(\underline{\dim}\, X)_{\omega_1} = (\underline{\dim}\, M_j(K_1))_{\omega_1} = 1$. Now, X_o belongs to the ray starting at E_1, thus $\dim \operatorname{Hom}(E_1, X_o) = 1$. Let ℓ be the length of X_o in the abelian category $\mathcal{T}(\rho)$, and n the rank of $\mathcal{T}(\rho)$. If $\ell \equiv 0 \pmod n$, then $\chi_{A_o}(\underline{\dim}\, X_o) = 0$, and $\langle \underline{\dim}\, E_1, \underline{\dim}\, X_o\rangle = 0$. If $\ell \not\equiv 0 \pmod n$, then $\chi_{A_o}(\underline{\dim}\, X_o) = 1$, and $\langle \underline{\dim}\, E_1, \underline{\dim}\, X_o\rangle = 1$ (see 3.1.3'). Since always $\chi_{K_1}(\underline{\dim}\, M_j(K_1)) = 1$, we conclude that $\chi_A(\underline{\dim}\, X) = 1$.

Now, let Y be indecomposable, and $\underline{\dim}\, X = \underline{\dim}\, Y$. Note that proj.dim.$X \le 1$, since all indecomposable injective modules belong to $\mathcal{Q}_o$, and $\operatorname{Hom}(\mathcal{Q}_o, \tau X) = 0$. Thus

$$1 = \chi(\underline{\dim}\, X) = \langle \underline{\dim}\, X, \underline{\dim}\, Y\rangle = \dim \operatorname{Hom}(X,Y) - \dim \operatorname{Ext}^1(X,Y)$$

shows that $\operatorname{Hom}(X,Y) \neq 0$, thus $Y \in \mathcal{T}_o \vee \mathcal{Q}_o$. However, $(\underline{\dim}\, Y_o)\iota = (\underline{\dim}\, Y)\iota = (\underline{\dim}\, X)\iota = 0$ shows that all indecomposable summands of Y_o, and therefore Y_o itself, have to belong to $\mathcal{T}$. Thus Y belongs to $\mathcal{T}_o$, and therefore Y_o is indecomposable. Since X_o and Y_o both belong to the ray starting at E_1, and since all the maps

$$E_1 = E_1[1] \longrightarrow E_1[2] \longrightarrow \dots \longrightarrow E_1[i] \longrightarrow E_1[i+1] \longrightarrow \dots$$

in the ray are monomorphisms, we conclude from $\underline{\dim}\, X_o = \underline{\dim}\, Y_o$ that $X_o \approx Y_o$. Thus $X \approx Y$, since there is only one indecomposable A-module X with given restriction X_o and given factor X/X_o.

(4) Let A_o be an algebra, let $\mathcal{T}$ be a tubular family, separating $\mathcal{P}$ from $\mathcal{Q}$. Let $A = A_o[E_i, K_i]_{i=1}^t$ be a tubular extension of A_o using modules from $\mathcal{T}$. Let T_o be a tilting A_o-module belonging to $\mathcal{P}$, with $\operatorname{End}(T_o) = B_o$ and with functor $\Sigma_o = \operatorname{Hom}({}_{A_o}(T_o)_{B_o}, -) : A_o\text{-mod} \longrightarrow B_o\text{-mod}$ and let T_p be the direct sum of all indecomposable projective A-modules P(a), with a outside A_o. Then, ${}_AT = T_o \oplus T_p$ is a tilting A-module, and $\operatorname{End}({}_AT) = B_o[\Sigma_o(E_i), K_i]_{i=1}^t$.

<u>Proof.</u> Since $\text{proj.dim}_{A_0} T_0 \leq 1$ implies that $\text{proj.dim}_A T \leq 1$, and since the number of isomorphism classes of indecomposable summands of T_0 equals the rank of $K_0(A)$, we only have to show $\text{Ext}^1_A(T,T) = 0$. However, the only interesting term is $\text{Ext}^1_A(T_0,T_p) = D\,\text{Hom}(T_p,\tau T_0) = 0$, since $T_p \in \mathcal{T}_0$ and $\tau T_0 \in \mathcal{P}_0$. Thus, ${}_AT$ is a tilting module. It remains to show that $\text{End}({}_AT) = B_0[\Sigma_0(E_i),K_i]_{i=1}^t$. Using induction, it is sufficient to consider the case $t = 1$. Also, we may restrict to the case of K_1 being a branch of length 1, since in general the branch K_1 attached in the extension vertex of $A_0[E_1]$ will carry over to the same branch K_1 being attached to the extension vertex of $B_0[\Sigma(E_1)]$. Thus, let ω be the extension vertex of $A = A_0[E]$, with corresponding projective module $P(\omega)$, and let ${}_AT = T_o \oplus P(\omega)$ and $\Sigma_0 = \text{Hom}_{A_0}(T_0,-)$. Then $\text{End}_A(T_0 \oplus P(\omega))$ is in matrix notation given by

$$\begin{bmatrix} \text{End}(T_0) & \text{Hom}(T_0,P(\omega)) \\ 0 & \text{End}(P(\omega)) \end{bmatrix} = \begin{bmatrix} B_0 & \Sigma_0(E) \\ 0 & k \end{bmatrix} = B_0[\Sigma_0(E)] ,$$

using that all homomorphisms $T_0 \longrightarrow P(\omega)$ map into the radical of $P(\omega)$, thus into E. This finishes the proof.

(5) Let us write down at least some of the dual concepts: Given an algebra A_0, with a separating tubular family $\mathcal{T}$, then also the family $\mathcal{T}^*$ consisting of the dual modules in A_0^{op}-mod is a separating tubular family, and we call $A = {}_{i=1}^{t}[K_i,E_i]A_0$ <u>a tubular coextension of</u> A_0 <u>using modules from</u> $\mathcal{T}$, provided $A^{op} = A_0^{op}[DE_i,K_i^{op}]_{i=1}^t$ is a tubular extension of A_0^{op} using modules from $\mathcal{T}^*$. In case $\mathcal{T}$ is a stable tubular family, the <u>coextension type of</u> A <u>over</u> A_0 is, by definition, the extension type of A^{op} over A_0^{op}. (For a general discussion of coextensions of algebras, see p. 257 below.)

4.8 Examples: Canonical tubular extensions of canonical algebras.

We recall that a canonical tubular extension C of the canonical algebra C_o is an algebra of the form $C = C_o[S_i,K_i]_i$, where the S_i are pairwise orthogonal ray modules belonging to the tubular family $\mathcal{T}$ constructed in 3.7, and the K_i are branches. Note that $\mathcal{T}$ is given by all indecomposable C_o-modules M with $(\underline{\dim}\, M)_o = (\underline{\dim}\, M)_\omega$. We are going to give a description of these algebras C as one-point coextensions.

An algebra B will be called a bush algebra with center ω and branches $B^{(s)}$, $1 \le s \le t$, provided B is obtained from the one-dimensional algebra with vertex ω by adding the branches $B^{(s)}$, $1 \le s \le t$, in ω. In case $|B^{(s)}| = n_s$, we will say that the branching type of B is $(n_1,\ldots,n_t)$, or also $\mathbb{T}_{n_1,\ldots,n_t}$. [Special cases of bush algebras of branching type $(n_1,\ldots,n_t)$ are given by endowing $\mathbb{T}_{n_1,\ldots,n_t}$ with an arbitrary orientation.]

Given a bush algebra B, with center ω and branches $B^{(s)}$, $1 \le s \le t$, a coordinate module ${}_BM = (M_a,\alpha)$ is given as follows: M_ω is a two-dimensional vectorspace, and the restriction of ${}_BM$ to the branch $B^{(s)}$ is the restriction of a representation of the complete branch to $B^{(s)}$ which has the following form:

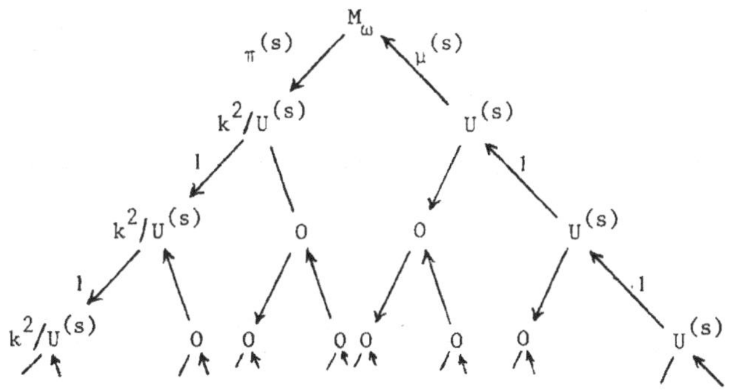

where $\mu^{(s)} : U^{(s)} \longrightarrow M_\omega$ is the inclusion of a one-dimensional subspace, with projection map $\pi^{(s)} : M_\omega \longrightarrow M_\omega/U^{(s)}$, with the various $U^{(s)}$, $1 \le s \le t$ being pairwise different subspaces of M_ω.

(1) An algebra C is a canonical tubular extension of a canonical algebra C_o if and only if C is a coextension of a bush algebra C_∞ by a coordinate module. The extension type of C over C_o is equal to the branching type of the bush algebra C_∞.

Proof. Let C_o be a canonical algebra, say given by the quiver $\Delta_o = \Delta(m_1,\dots,m_t)$, with $m_1 \geq m_2 \geq \dots \geq m_t$, and a generic ideal J of relations. In order to introduce a notation for the vertices and arrows of $\Delta(m_1,\dots,m_t)$, we may consider $\Delta(m_1,\dots,m_t)$ as being obtained from the disjoint union of quivers

$$\overset{a_o^{(s)}}{\circ} \xleftarrow{\;\alpha_1^{(s)}\;} \overset{a_1^{(s)}}{\circ} \xleftarrow{\;\alpha_2^{(s)}\;} \overset{a_2^{(s)}}{\circ} \quad \cdots \quad \overset{a_{m_s-1}^{(s)}}{\circ} \xleftarrow{\;\alpha_{m_s}^{(s)}\;} \overset{a_{m_s}^{(s)}}{\circ}$$

with $1 \leq s \leq t$, by identifying all sinks $a_o^{(s)}$ in order to form a single vertex 0, and all sources $a_{m_s}^{(s)}$ in order to form a vertex ω. The composition $\alpha_{m_s}^{(s)} \dots \alpha_1^{(s)}$ will be denoted by $\alpha^{(s)}$. We denote by $F_i^{(s)}$ the cokernel of the map $P(\alpha_i^{(s)*}) : P(a_{i-1}^{(s)}) \longrightarrow P(a_i^{(s)})$ of C_o-modules, and note that $F_i^{(s)} = E(a_i^{(s)})$ for all $1 \leq i \leq m_s-1$, $1 \leq s \leq t$. Let $C = C_o[S_j,K_j]_j$ be a canonical tubular extension of C_o, and let Δ be the quiver for C. By increasing, if necessary, t, we may assume that all S_j are of the form $F_i^{(s)}$ for some i,s. [Namely, if some S_j is not yet of the form $F_i^{(s)}$, then we note that S_j, considered as a representation of $\Delta(m_1,\dots,m_t)$, is given by vector spaces all of which are one-dimensional, and maps all of which are isomorphisms. Now replace $\Delta(m_1,\dots,m_t)$ by $\Delta(m_1,\dots,m_t,1)$ and J by $J' = J + \langle\alpha^{(1)} + \lambda\alpha^{(2)} + \alpha^{(t+1)}\rangle$, where $\alpha^{(1)} + \lambda\alpha^{(2)}$ vanishes on S_j. One easily checks that J' is generic, again. Any representation of $\Delta(m_1,\dots,m_t)$ satisfying J can be considered as a representation of $\Delta(m_1,\dots,m_t,1)$, by assigning to the additional arrow $\alpha_1^{(t+1)}$ the map $-\alpha^{(1)} -\lambda\alpha^{(2)}$, and then the relations J' are satisfied. Note that in this way, $S_j = F_1^{(t+1)}$].

Let us show that the algebra C_∞ obtained from C by deleting the vertex 0 is a bush algebra and that $Q(0)/E(0)$ is a coordinate module. If $F_i^{(s)} = S_j$, for some j, let $B_i^{(s)} = K_i$, it is a branch in the vertex $\omega_i^{(s)}$; otherwise, let $B_i^{(s)}$ be the empty branch. The centre of C_∞ will be the vertex ω. Let us show that any s with $n_s \geq 2$, or with some $B_i^{(s)} \neq \emptyset$ gives rise to a branch $B^{(s)}$ in ω of length $n_s + \sum_i |B_i^{(s)}|$. Namely, let $B^{(s)}$ be the restriction of C to the full subquiver given by the vertices $a_i^{(s)}$ and those in $B_i^{(s)}$, where $1 \leq i \leq m_s$. We define an embedding $\varphi^{(s)}$ of $B^{(s)}$ into the complete branch in b, by sending $a_i^{(s)}$ to $b_{+\dots+}$, with m_s-i indices $+$, and, such that $\varphi^{(s)}$ restricted to $B_i^{(s)}$ is the unique embedding of the branch $B_i^{(s)}$ in $\omega_i^{(s)}$ into the complete branch such that $\varphi^{(s)}(\omega_i^{(s)}) = b_{+\dots+-}$ with m_s-i indices $+$, and one index $-$. Note that, in case $B_i^{(s)}$ exists, there is a unique arrow in Δ starting in $\omega_i^{(s)}$ and ending in Δ_o, namely $\omega_i^{(s)} \longrightarrow a_i^{(s)}$, and there is a (up to scalar multiples) unique relation starting in $\omega_i^{(s)}$ and ending in Δ_o, namely the composition of the arrow $\gamma_i^{(s)}$ with the arrow $\alpha_i^{(s)}$. Altogether, this shows that $B^{(s)}$ is a branch in ω.

Now, let us consider $Q(0)/E(0)$. We denote by $Q_0(0)$ the indecomposable injective C_o-module with socle $E(0)$. Since C_o is a canonical algebra, we know that $Q_o(0)/E(0)$ is a coordinate module for the algebra obtained from C_o by deleting the vertex 0. Now, $Q_o(0)$ is the restriction of $Q(0)$ to C_o, and we want to recover $Q(0)$ from $Q_o(0)$. We consider $Q(0) = (Q_a,\alpha)$ as a representation of Δ. First, we deal with those branches $B_i^{(s)}$, with $1 \le i < m_s$, which are non-empty. Since for $1 \le i < m_s$, $\alpha_i^{(s)} : Q_{a_i^{(s)}} \longrightarrow Q_{a_{i-1}^{(s)}}$ is an isomorphism, and we have the relation $\gamma_i^{(s)}\alpha_i^{(s)} = 0$, it follows that $Q_{\omega_i^{(s)}} = 0$, and therefore $Q_b = 0$ for any vertex $b \in B_i^{(s)}$. It remains to consider the non-empty branches $B_{m_s}^{(s)}$. Note that $\dim Q_\omega = 2$, and that $\alpha_{m_s}^{(s)} : Q_\omega \longrightarrow Q_{a_{m-1}^{(s)}}$ is surjective, with a one-dimensional kernel, say denote this kernel by $U^{(s)}$. Then the relation $\gamma_{m_s}^{(s)}\alpha_{m_s}^{(s)} = 0$ shows that up to isomorphism, we have $Q_{\omega_{m_s}^{(s)}} = U^{(s)}$, with $\gamma_{m_s}^{(s)}$ the inclusion map. Also, for the remaining vertices b of $B_{m_s}^{(s)}$ having a path to $\omega_{m_s}^{(s)}$, we can choose $Q_b = U^{(s)}$, and for the corresponding arrows the identity map, whereas for all other vertices b' of $B_{m_s}^{(s)}$, we must have $Q_{b'} = 0$, since there is no path from such a b' to 0. Since the various subspaces $U^{(s)}$ of Q_ω, $1 \le s \le t$, are derived from the coordinate module $Q_o(0)/E(0)$, these subspaces are pairwise different. Altogether we see that $Q(0)/E(0)$ is a coordinate module for C_∞.

Conversely, assume C is the coextension $[M]B_\infty$ of a bush algebra B_∞ with centre ω by a coordinate module M. Let B'_∞ be the restriction of B_∞ to the full subquiver given by all vertices b of B_∞ which can be reached from ω by a path, and let M' be the restriction of M to B'_∞. Since M is a coordinate B_∞-module, M' is a coordinate B'_∞-module, thus the coextension $B_o = [M']B'_\infty$ of B'_∞ by M' is a canonical algebra, and one shows similar to the proof above that C is a canonical tubular extension of B_o. This finishes the proof.

It will be convenient to have some characterizations of bush algebras as tilted algebras available.

(2) Let A be given by the star $\mathbb{T}_{n_1,\dots,n_t}$ with factor-space orientation and with centre ω. The bush algebras of branching type $(n_1,\dots,n_t)$ are the endomorphism rings $B = \mathrm{End}(_AT)$ where $_AT$ is a multiplicity-free tilting module with $P(\omega) \in \langle _AT\rangle$. A B-module is a coordinate module if and only if it is of the form $\mathrm{Hom}_A(_AT_B, {}_AM)$ for some coordinate module $_AM$.

Proof. The first assertion follows from 4.4.4. Now, let ${}_AT$ be a tilting module, with $P(\omega) \in \langle {}_AT \rangle$, and let ${}_AM$ be a coordinate module for A. We want to show that $\mathrm{Hom}_A({}_AT_B, {}_AM)$ is coordinate module for B. Let

$$b_1^{(s)} \longleftarrow b_2^{(s)} \longleftarrow \dots \longleftarrow b_n^{(s)} = \omega$$

be some branch in the quiver of A. This branch gives rise, in $\Gamma(A)$, to a wing $\mathcal{W}^{(s)}$ of length n_s containing besides the indecomposable projective modules $P(b_i^{(s)})$, $1 \le i \le n$, also the quotients $P(b_j^{(s)})/P(b_i^{(s)})$ with $1 \le i < j \le n_s$. According to our investigation of branches, there are n_s isomorphism classes of indecomposable direct summands of ${}_AT$ belonging to $\mathcal{W}^{(s)}$, and one of these is $P(\omega)$. If we want to consider $\mathrm{Hom}_A({}_AT_B, {}_AM)$ as a representation of the quiver of B, we have to calculate the vector-spaces $\mathrm{Hom}_A(T(i), M)$, where $T(i)$ are the indecomposable direct summands of ${}_AT$. For $T(i) = P(\omega)$, we obtain $\mathrm{Hom}_A(P(\omega), M) = M_\omega$, a two-dimensional vector-space. For any summand of ${}_AT$ of the form $P(b_i^{(s)})$, $1 \le i < n$, we obtain $\mathrm{Hom}(P(b_i^{(s)}), M) = M_{b_i^{(s)}}$, and this space is of the form $M_\omega/U^{(s)}$ for some one-dimensional subspace $U^{(s)}$ of M_ω. For a summand of ${}_AT$ of the form $P(\omega)/P(b_i^{(s)})$, $1 \le i < n$, we obtain as $\mathrm{Hom}(P(\omega)/P(b_i^{(s)}), M)$ the kernel of the given map $M_\omega \longrightarrow M_{b_i^{(s)}}$, thus just $U^{(s)}$. Finally, for a summand of ${}_AT$ of the form $P(b_j^{(s)})/P(b_i^{(s)})$ with $1 \le i < j < n_s$, we obtain $\mathrm{Hom}(P(b_j^{(s)})/P(b_i^{(s)}), M) = 0$. It follows that we obtain just a coordinate module for B, and conversely, also all coordinate modules for B.

(3) Let B_o be a bush algebra of branching type $\mathbb{T}$, with $\mathbb{T}$ being a Dynkin graph. Let M be a coordinate module for B_o. Then $B = B_o[M]$ has a preprojective component $\mathcal{P}$ containing a slice module ${}_BS$ with $\mathrm{End}({}_BS)$ hereditary, of tubular type $\mathbb{T}$.

Proof. Let A be the hereditary algebra given by the quiver $\mathbb{T}$ with factor space orientation, and let z be the centre of $\mathbb{T}$. Let ${}_AT$ be a tilting module with $P(z) \in \langle {}_AT \rangle$, let $B_o = \mathrm{End}({}_AT)$, and $\Sigma = \mathrm{Hom}({}_AT, -) : A\text{-mod} \longrightarrow B_o\text{-mod}$.

Let $\mathbb{T} = \mathbb{T}_{n_1, \dots, n_t}$ and consider first the case $t \le 2$. In this case, $M = \Sigma Q(a) \oplus \Sigma Q(b)$, where a and b are the two endpoints of $\mathbb{T}$ (or, in case $\mathbb{T}$ consists of a unique vertex a, then $a = b$). Note that a slice module in B-mod is given by ${}_BS = \Sigma D(A_A) \oplus P(\omega)$, where ω is the extension vertex of $B_o[M]$, and the quiver of $\mathrm{End}({}_BS)$ is given by $\tilde{\mathbb{T}}_{n_1, n_2}$, with one path going from $P(\omega)$ via $\Sigma Q(a)$ to $\Sigma Q(z)$, the other from $P(\omega)$ via $\Sigma Q(b)$ to $\Sigma Q(z)$.

Next, consider the possible cases with $t > 2$, namely, $\mathbb{T}_{n22}, \mathbb{T}_{332}, \mathbb{T}_{432}, \mathbb{T}_{532}$. Note that in these cases M will be indecomposable. Let ${}_AT'$ be the tilting module with $P(z) \in \langle {}_AT' \rangle$ and such that $\mathrm{End}({}_AT')$ is the hereditary algebra given by the quiver $\mathbb{T}$ with subspace orientation. Then $G({}_AT')$ is mapped under Σ to a full subcategory of B_0-mod, in such a way that $\Gamma G({}_AT')$ corresponds to a full convex subquiver of Γ(B-mod). Note that the unique coordinate A-module M' belongs to $G({}_AT')$, since it is sincere, and that its image under Σ is the (unique) coordinate B_0-module M. Let us indicate the position of M' in $G({}_AT')$ for the various cases (marked by ◎). It follows, that the one-point extension $B_0[M]$ has a slice given by a subquiver of the form $\tilde{\mathbb{D}}_{n-2}, \tilde{\mathbb{E}}_6, \tilde{\mathbb{E}}_7$, or $\tilde{\mathbb{E}}_8$, respectively.

$\mathbb{D}_8 = \mathbb{T}_{622}$

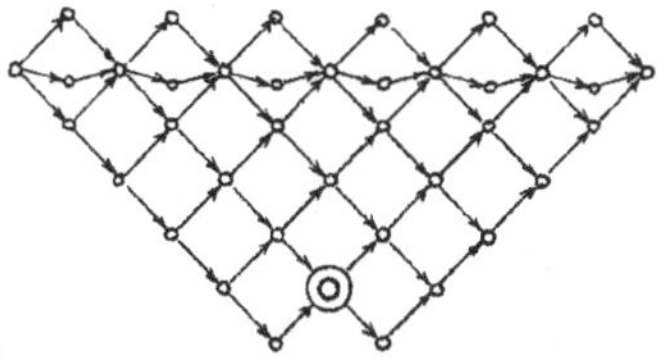

$\mathbb{E}_6 = \mathbb{T}_{332}$

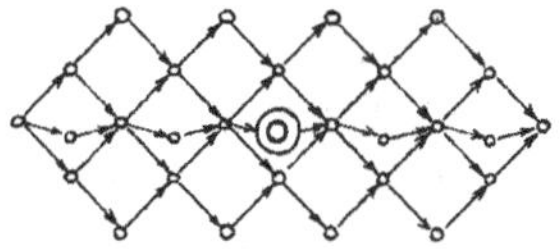

$\mathbb{E}_7 = \mathbb{T}_{432}$

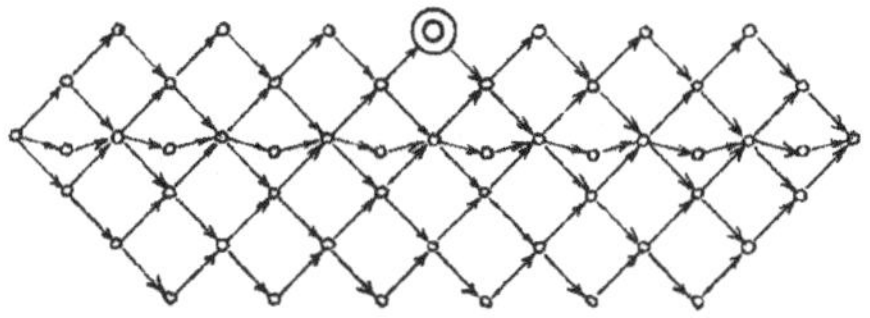

$\mathbb{E}_8 = \mathbb{T}_{532}$

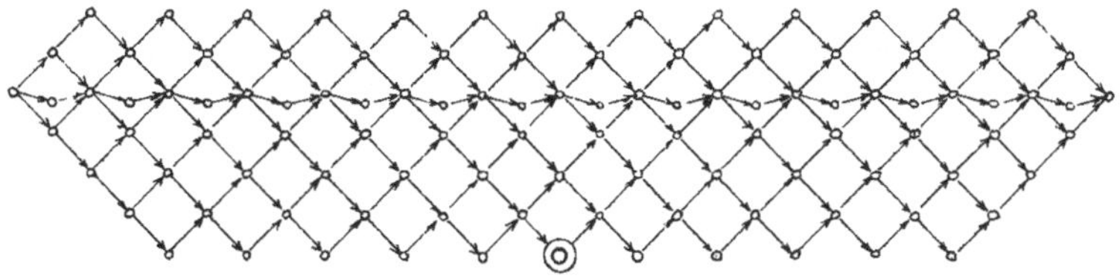

(4) Let B be a bush algebra of branching type $\tilde{\mathbb{D}}_4, \tilde{\mathbb{E}}_6, \tilde{\mathbb{E}}_7$, or $\tilde{\mathbb{E}}_8$, let M be a coordinate module for B. Then B is a tame concealed algebra, and M is simple regular of period 1, 2, 3, or 5, respectively.

Proof. The algebra A given by $\tilde{\mathbb{D}}_4, \tilde{\mathbb{E}}_6, \tilde{\mathbb{E}}_7$ or $\tilde{\mathbb{E}}_8$ with factorspace orientation, is hereditary, and if ${}_A T$ is a tilting module having $P(\omega)$ as direct summand, where ω is the centre of the star, then ${}_A T$ is preprojective, thus $B = \mathrm{End}({}_A T)$ is a tame concealed algebra.

In dealing with coordinate modules, it is sufficient to deal with the case $B = A$. Thus, let M be a coordinate A-module. Consider first the case $\tilde{\mathbb{D}}_4$. Then, $\underline{\dim}\, M$ is a radical vector for χ_A, and, since M is indecomposable, we see that M is regular. Assume, M is not simple regular. A simple regular submodule of M has (up to permuting the vertices) a dimension vector of the form $\begin{array}{c}1\\1\\0^{1}\\0\end{array}$, however this means that there is a one-dimensional subspace of M_ω which lies in the kernel (and therefore is the kernel) of two of the maps $M_\omega \longrightarrow M_a$, (a being a vertex different from ω). This contradicts the fact that M is a coordinate module.

In cases $\tilde{\mathbb{E}}_6, \tilde{\mathbb{E}}_7$, and $\tilde{\mathbb{E}}_8$, we write down a complete Φ_A-orbit of $\underline{\dim}\, M$. Since also in these cases, M is indecomposable, it follows from $\underline{\dim}\, M$ being Φ_A-periodic that M is regular. Since in all cases, the sum $\Sigma(\underline{\dim}\, M)\Phi^i$ over a complete Φ_A-orbit is just the minimal positive radical vector for χ_A, it follows that M is simple regular, and of the asserted period. In the following table, Φ_A operates as shift to the left, and the image of the first vector under Φ_A being the last:

$$\tilde{\mathbb{E}}_6: \quad \begin{array}{c}11\\112\\11\end{array} \qquad \begin{array}{c}01\\011\\01\end{array}$$

$$\tilde{\mathbb{E}}_7: \quad \begin{array}{c}111\\1112\\1\end{array} \qquad \begin{array}{c}011\\0111\\0\end{array} \qquad \begin{array}{c}001\\0011\\1\end{array}$$

$$\tilde{\mathbb{E}}_8: \quad \begin{array}{r}11111\\112\\1\;\;\end{array} \qquad \begin{array}{r}01111\\011\\0\;\;\end{array} \qquad \begin{array}{r}00111\\001\\1\;\;\end{array} \qquad \begin{array}{r}00011\\111\\0\;\;\end{array} \qquad \begin{array}{r}00001\\011\\1\;\;\end{array}$$

4.9 Domestic tubular extensions of tame concealed algebras

Let B be a tubular extension of a tame concealed algebra B_o of extension type $(m_1,\dots,m_t)$. In this section, we study the case of $\mathbb{T}_{m_1,\dots,m_t}$ being a Dynkin graph, and call such an extension a domestic tubular extension.

(1) Theorem. Let $\mathbb{T}_{m_1,\dots,m_t}$ be a Dynkin graph, and A tame concealed of tubular type $(m_1,\dots,m_t)$. If ${}_AT$ is a tilting module without non-zero preinjective direct summands, say ${}_AT = T_o \oplus T_1$ with T_o preprojective, T_1 regular, then $B_o = \mathrm{End}({}_AT_o)$ is a tame concealed algebra, and $B = \mathrm{End}({}_AT)$ is a tubular extension of B_o of extension type $(m_1,\dots,m_t)$. Conversely, given any tame concealed algebra B'_o and a tubular extension B' of B'_o of extension type $(m_1,\dots,m_t)$, there exists a tilting module ${}_AT' = T'_o \oplus T'_1$ with T'_o preprojective, T'_1 regular, $B'_o = \mathrm{End}({}_AT'_o)$, and $B' = \mathrm{End}({}_AT')$.

Proof. Let us denote by $\mathcal{P}$, $\mathcal{T}$, and $\mathcal{Q}$ the module classes of all preprojective, regular, and preinjective A-modules, respectively. Let h be the positive radical generator in $K_o(A)$, and let $\mathcal{T}(\rho_o)$ be some tube of rank 1, say containing an indecomposable module H with $\underline{\dim}\, H = h$.

First, assume that A is hereditary. Let ${}_AT = T_o \oplus T_1$ be a tilting module with $T_o \in \mathcal{P}$, $T_1 \in \mathcal{T}$. Note that this implies that $\mathcal{Q} \subseteq \mathcal{G}({}_AT)$, and also $\mathcal{T}(\rho_o) \subseteq \mathcal{G}({}_AT)$ [namely, $\mathrm{Ext}^1(T,M) = D\,\underline{\mathrm{Hom}}(\tau^- M,T) = 0$, for any $M \in \mathcal{T}(\rho_o) \vee \mathcal{Q}$, since with M also $\tau^- M$ belongs to $\mathcal{T}(\rho_o) \vee \mathcal{Q}$, and no non-zero direct summand of T_1 can belong to $\mathcal{T}(\rho_o)$].

Let $B_o = \mathrm{End}({}_AT_o)$, $B = \mathrm{End}({}_AT)$, and $\Sigma = \mathrm{Hom}_A({}_AT_B,-) : A\text{-mod} \longrightarrow B\text{-mod}$, with corresponding linear transformation $\sigma = \sigma_T$. Let $\mathcal{P}' = \Sigma(\mathcal{P} \cap \mathcal{G}({}_AT))$, and note that the B-modules in $\mathcal{P}'$ are actually B_o-modules [namely, for $X \in \mathcal{P}$, we have $\mathrm{Hom}(T_1,X) = 0$]. Similarly, also ΣH is in fact a B_o-module. Note that $\underline{\dim}\, \Sigma H = h\sigma$, thus the radical generator $h\sigma$ of χ_B actually lies in $K_o(B)$, and it is a sincere vector in $K_o(B)$. [Namely, for any indecomposable summand $T(i)$ of T_o, we have $\mathrm{Hom}(T(i),H) \neq 0$, since the injective envelope Q of $T(i)$ lies in $\mathcal{Q}$, and we can factor the inclusion map $T(i) \longrightarrow Q$ through $\mathcal{T}(\rho)$.] It follows that χ_{B_o} is positive semi-definite of radical rank one, with a positive sincere radical generator.

Let M,N be indecomposable B-modules, and $M \in \mathcal{P}'$. Then, $N \preceq M$ implies $N \in \mathcal{P}'$. [Namely, first of all we must have $N \in \mathcal{Y}({}_AT)$, since $(\mathcal{Y}({}_AT),\mathcal{X}({}_AT))$, is a split torsion pair, and then $N \in \mathcal{P}'$, since $\mathcal{P}$ is closed under predecessor in A-mod.] In particular, we also note that any indecomposable module $M \in \mathcal{P}'$ is directing [since those in $\mathcal{P}$ are directing]. Also, if there exists an irreducible map

$M \longrightarrow N$, then $N \in \mathcal{P}'$. [Namely, let $M = \Sigma X$ for some $X \in \mathcal{P} \cap \mathcal{G}(_A T)$. Since X is not injective, it follows that $N \in \mathcal{Y}(_A T)$, say $N = \Sigma Y$ for some indecomposable module $Y \in \mathcal{G}(_A T)$, and we deal with a map $f : X \longrightarrow Y$ such that Σf is irreducible in B-mod. Let $g : Y' \longrightarrow Y$ be the sink map for Y in A-mod, and $f = f'g$. If $Y \in \mathcal{Q}$, then also $Y' \in \mathcal{Q} \subseteq \mathcal{G}(_A T)$, thus $\Sigma f = (\Sigma f')(\Sigma g)$ is a proper factorization, impossible. Assume $Y \in \mathcal{T}$, thus also $Y' \in \mathcal{T}$ and let Y'' be the torsion submodule of Y, say with inclusion $\mu : Y'' \longrightarrow Y'$, with respect to $(\mathcal{F}(_A T), \mathcal{G}(_A T))$. Note that Y'/Y'' is in $\mathcal{F}(_A T)$, and, as a factor module of Y', in $\mathcal{T} \vee \mathcal{Q}$. However, since $\mathcal{Q} \subseteq \mathcal{G}(_A T)$, we see that Y'/Y'' can have no non-zero summand in $\mathcal{Q}$, thus $Y'/Y'' \in \mathcal{T}$ and therefore also $Y'' \in \mathcal{T}$. Thus $Y'' \in \mathcal{T} \cap \mathcal{G}(_A T)$. Since the image of f' is contained in Y'', we can write $f' = f''\mu$, and $\Sigma f = (\Sigma f'')(\Sigma(\mu g))$ is again a proper factorization of Σf. Altogether we see that Y has to belong to $\mathcal{P}$.] Finally, given an indecomposable injective B_o-module Q_o, then $\mathrm{Hom}_B(\Sigma H, Q_o) \neq 0$, since ΣH is a sincere B_o-module, and since $\mathcal{P}'$ is closed under predecessors, and ΣH does not belong to $\mathcal{P}'$, we conclude that $\mathrm{Hom}(Q_o, \mathcal{P}') = 0$, thus $\mathrm{Hom}(D(_{B_o}B_o), \mathcal{P}') = 0$.

We see that we can apply 4.3.8 [to any component of B_o-mod contained in $\mathcal{P}'$] and conclude that B_o is a tame concealed algebra and that $\mathcal{P}'$ is the set of all preprojective B_o-modules.

Next, we want to see that B is a tubular extension of B_o, of extension type $(m_1,\dots,m_t)$. First of all, we note that for any tube $\mathcal{T}(\rho)$, the image $\Sigma(\mathcal{T}(\rho) \cap \mathcal{G}(_A T))$ will be closed under irreducible maps. [Let M,N be indecomposable B-modules, and $M = \Sigma X$ for some $X \in \mathcal{T}(\rho) \cap \mathcal{G}(_A T)$. If there is an irreducible map $N \longrightarrow M$, then N belongs to $\mathcal{Y}(_A T)$, say $N = \Sigma Y$, with $Y \in \mathcal{P} \vee \mathcal{T}(\rho)$, however, we have shown above that $Y \in \mathcal{P}$ is impossible. If there is an irreducible map $M \longrightarrow N$, then again $N \in \mathcal{Y}(_A T)$, since M is not injective, say again $N = \Sigma Y$, now with $Y \in \mathcal{T}(\rho) \vee \mathcal{Q}$. However, considering the sink map $Y' \longrightarrow Y$ for Y, we see as above that $M \longrightarrow N$ factors through $\Sigma Y'$, therefore we must have $Y \in \mathcal{T}(\rho)$.] Recall that for $\mathcal{T}(\rho)$ a homogeneous tube, no indecomposable summand of $_A T$ belongs to $\mathcal{T}(\rho)$, thus $\mathcal{T}(\rho) \subseteq \mathcal{G}(_A T)$, and we denote by $\mathcal{T}(\rho_1),\dots,\mathcal{T}(\rho_t)$ the (pairwise different) non-homogeneous tubes of A-mod, with $\mathcal{T}(\rho_s)$ being of rank m_s. For any s, let Ξ_s be the set of isomorphism classes of indecomposable summands of $_A T$ belonging to $\mathcal{T}(\rho_s)$. We claim that always $\Xi_s \subseteq \Gamma\mathcal{T}(\rho_s)$ is a closed set (in the sense of 4.6.5). Of course, the first condition $\mathrm{Hom}(x, \tau y) = 0$ in the category $k(\Gamma\mathcal{T}(\rho_s))$, for all $x,y \in \Xi_s$, is just a restatement of the property $\mathrm{Ext}^1(T', T'') = 0$ for direct summands T', T'' of T belonging to $\mathcal{T}(\rho_s)$, since $\mathcal{T}(\rho_s)$ is a standard component containing only modules of projective dimension 1. The second condition of $W(x) \cap \Xi_s$ being a tilting set in $W(x)$ for any $x \in \Xi_s$ has been shown in 4.4.1. Now, the module class $\mathcal{T}(\rho_s) \cap \mathcal{G}(_A T)$ is the set of all modules M in $\mathcal{T}(\rho_s)$ satisfying $\mathrm{Ext}^1_A(_A T, M) = 0$, or equivalently, satisfying $\mathrm{Ext}^1_A(T', M) = 0$ for any direct summand T' of ${}_AT$ belonging to $T(\rho_s)$. Thus, $T(\rho_s) \cap G({}_AT)$ is nothing else but $G(\Xi_s)$.

According to 4.6.5, we can identify $\Gamma T(\rho_s)$ with $\Omega(m_s)$ in such a way that $\Xi_s = \Xi(K_0^{(s)},\dots,K_{n_s-1}^{(s)})$ for some sequence $K_0^{(s)},\dots,K_{n_s-1}^{(s)}$ of branches, where $n_s = m_s - |\Xi_s|$; and then 4.6.6 gives a description of $\Gamma G(\Xi_s)$. Thus, we obtain

$$\Gamma(T(\rho_s) \cap G({}_AT)) = \Gamma G(\Xi_s) = \Gamma G(\Xi(K_0^{(s)},\dots,K_{n_s-1}^{(s)}))$$

$$\approx \Omega(n_s)[e_z^{(s)},K_z^{(s)}]_{z=0}^{n_{s-1}},$$

where the vertices $e_0^{(s)},\dots,e_{n_s-1}^{(s)}$ are the vertices of $\Omega(n_s)$ lying on the mouth, with $\tau e_z^{(s)} = e_{z-1}^{(s)}$, and they correspond in $\Omega(m_s)$ to the branch indicators, say $e_z^{(s)} \in \Omega(n_s)$ corresponding to $v_z^{(s)} \in \Omega(m_s)$. Let $V_z^{(s)}$ be an indecomposable A-module with $[V_z^{(s)}] = v_z^{(s)}$. Since $v_z^{(s)}$ belongs to $\Gamma G(\Xi_s)$, we know that $V_z^{(s)} \in T(\rho_s) \cap G({}_AT)$; and let $E_z^{(s)} = \Sigma V_z^{(s)}$. As we have mentioned at the end of 4.6.4, the set of vertices x in $\Gamma_s = \Omega(n_s[e_z^{(s)},K_z^{(s)}]_z$, with $\mathrm{Hom}_{k(\Gamma_s)}(p,x) = 0$ for all projective vertices p of Γ_s, is $\Omega(n_s)_0$ (considered as a subset of $(\Gamma_s)_0$), and in fact, the full subcategory of $k(\Gamma_s)$, given by these vertices is $k(\Omega(n_s))$. Note that the vertices $x \in \Gamma_s$ with $\mathrm{Hom}_{k(\Gamma_s)}(p,x) = 0$ for all projective vertices p of Γ_s correspond to the indecomposable modules M in $T(\rho_s) \cap G({}_AT)$ satisying $\mathrm{Hom}(T',M) = 0$ for any indecomposable summand T' of ${}_AT$ in $T(\rho_s)$, or, equivalently, satisfying $\mathrm{Hom}(T_1,M) = 0$, and these are precisely those indecomposable modules M in $T(\rho_s) \cap G({}_AT)$ with ΣM being a B_0-module. Thus, we see that $\Sigma(T(\rho_s) \cap G({}_AT)) \cap B_0$-mod is of the form $k(\Omega(n_s))$. Actually this is a component of B_0-mod. [Namely, let us denote $\Sigma(T(\rho_s) \cap G({}_AT)) \cap B_0$-mod by T'. Since $T' \approx k(\Omega(n_s))$, any indecomposable module in T' is contained in a cycle, thus it is a regular B_0-module; also, any two indecomposable modules in T' are connected by a path, thus T' is contained in a single regular component of B_0-mod, call it T''. Now T'' is a stable tube, thus it is sufficient to show that given an irreducible map $M \longrightarrow N$ in B_0-mod. with M,N indecomposable and $N \in T'$, we also have $M \in T'$. However, since $\mathrm{Hom}(M,N) \neq 0$, we must have $M \in Y({}_AT)$, and, since P' is closed under irreducible maps, it follows that $M = \Sigma X$ for some $X \in T(\rho_s) \cap G({}_AT)$, thus $M \in T'$. This shows that T' is a component of B_0-mod.] It follows that the modules $E_z^{(s)}$, $0 \le z \le n_s-1$, are the simple regular B_0-modules in some tube of B_0-mod of rank n_s.

Also, any non-empty branch $K_z^{(s)}$ is a branch in some vertex $x_z^{(s)}$ $(=(r_z^{(s)} + 1,|K_z^{(s)}|)$ in the notation of 4.6.5), with $x_z^{(s)}$ belonging to Ξ_s.

Let $T_z^{(s)}$ be an indecomposable module with $[T_z^{(s)}] = x_z^{(s)}$, and $P_z^{(s)} = \Sigma T_z^{(s)}$ the corresponding projective B-module. There is an arrow $v_z^{(s)} \longrightarrow x_z^{(s)}$ corresponding to an irreducible map $E_s^{(z)} \longrightarrow P_z^{(s)}$, thus $E_s^{(z)}$ is a direct summand of the radical of the indecomposable projective B-module $P_z^{(s)}$. An additional radical summand of $P_z^{(s)}$ will be due to the addition of the branch $K_z^{(s)}$ in the extension vertex of $B_0[E_z^{(s)}]$. Altogether we see that B is obtained from B_0 by making the various one-point extensions using the modules $E_z^{(s)}$ (in case $B_z^{(s)} \neq \emptyset$) and adding the branch $B_z^{(s)}$. It follows that $B = B_0[E_z^{(s)}, K_z^{(s)}]_{s,z}$. Thus B is a tubular extension of B_0, and the extension type is given by $(m_1,\ldots,m_t)$, since $m_s = n_s + \sum_{s=0}^{n_s-1} |K_z^{(s)}|$.

Let us consider now the general case of A tame concealed, and not necessarily hereditary. There exists a preprojective tilting module ${}_{A'}S$, with A' being hereditary, and $\mathrm{End}({}_{A'}S) = A$. Let $\mathcal{P}'$, $\mathcal{T}'$, $\mathcal{Q}'$ be the preprojective, regular, and preinjective A'-modules, respectively. Then $\Psi = \mathrm{Hom}_{A'}({}_{A'}S_A,-) : A'\text{-mod} \longrightarrow A\text{-mod}$ maps $\mathcal{P}' \cap \mathcal{G}({}_{A'}S)$ onto $\mathcal{P}$, and $\mathcal{T}'$ onto $\mathcal{T}$, and $\Psi' = {}_{A'}S_A \otimes -$ is the reverse functor. Thus, if ${}_AT = T_0 \oplus T_1$ is a tilting A-module, with $T_0 \in \mathcal{P}$, $T_1 \in \mathcal{T}$, let $T_0' = \Psi'(T_0)$, $T_1' = \Psi'(T_1)$, and $T' = T_0' \oplus T_1'$. Then T' satisfies all conditions of a tilting module, and $\mathrm{End}({}_{A'}T') \approx \mathrm{End}({}_AT)$ with $\mathrm{End}({}_{A'}T_0')$ corresponding to $\mathrm{End}({}_AT)$. Thus, using the previous case, it follows that $\mathrm{End}({}_AT_0)$ is tame concealed, and that $\mathrm{End}({}_AT)$ is a tubular extension of $\mathrm{End}({}_AT_0)$, with extension type equal to the tubular type of both A and A'.

Let us consider now the reverse implication. We fix again a tame concealed algebra A of tubular type $(m_1,\ldots,m_t)$, and consider a tubular extension C of some tame concealed algebra C_0 of extension type $(m_1,\ldots,m_t)$, say $C = C_0[E_i,K_i]_{i=1}^t$, with simple regular C_0-modules E_i, and branches K_i. We want to show that we can realize C as the endomorphism ring of some tilting A-module ${}_AT$ without preininjective direct summands. Note that this implies that for ${}_AT = T_0 \oplus T_1$ with T_0 preprojective, T_1 regular, we must have $\mathrm{End}(T_0) = C_0$. Namely, the considerations above show that also $\mathrm{End}(T_0)$ is tame concealed, and C is a tubular extension of $\mathrm{End}(T_0)$. According to 4.7.2, there is an element of $\mathrm{rad}\chi_C$ which is a sincere element of $K_0(C_0)$, and one which is a sincere element of $K_0(\mathrm{End}(T_0))$. Since $\mathrm{rad}\chi_C$ is one-dimensional, it follows that the support of these two elements coincide, and both C_0 and $\mathrm{End}(T_0)$ are the restrictions of C to the support of any radical generator, thus $\mathrm{End}(T_0) = C_0$.

Let us denote by $(n_1,\ldots,n_t)$ the tubular type of C_0, and let B_0 the canonical tame concealed algebra of the same tubular type (see 4.3.5). Since B_0 and C_0 are tame concealed algebras of the same tubular type, there exists a preprojective tilting B_0-module ${}_{B_0}S_0$ with $\mathrm{End}({}_{B_0}S_0) = C_0$. Note that the functor $\gamma_0 = \mathrm{Hom}({}_{B_0}S_{0\,C_0},-) : B_0\text{-mod} \longrightarrow C_0\text{-mod}$ gives an equivalence from $\mathcal{G}({}_{B_0}S_0)$ onto $\mathcal{Y}({}_{B_0}S_0)$, with inverse $\gamma'_0 = {}_{B_0}S_{0\,C_0} \otimes -$, and that all the preprojective and all the regular C_0-modules are contained in $\mathcal{Y}({}_{B_0}S_0)$. In particular, let $F_i = \gamma'_0(E_i)$, $1 \leq i \leq t$, these are pairwise non-isomorphic simple regular B_0-modules, and let $B = B_0[F_i,K_i]_{i=1}^t$. This is a tubular extension of B_0, of extension type $(m_1,\ldots,m_t)$. Let P be the direct sum of all indecomposable projective B-modules $P(b)$, with b a vertex outside B_0. Then ${}_BS = S_0 \oplus P$ is a tilting B-module with $\mathrm{End}({}_BS) = C$, according to 4.7.4.

Since B is a (canonical) tubular extension of the canonical algebra B_0, it is a coextension of a bush algebra B_∞ by a coordinate module, and the branching type of B_∞ is the extension type of B over B_0, thus $(m_1,\ldots,m_t)$, see 4.8.1. According to the dual of 4.8.3, there is in B-mod a preinjective component $\mathcal{B}$ with a slice module with endomorphism ring A' being hereditary of tubular type $(m_1,\ldots,m_t)$. According to 4.2.3, there is a tilting A'-module ${}_{A'}U$ with $\mathrm{End}({}_{A'}U) = B$, and a functor $\beta = \mathrm{Hom}_{A'}({}_{A'}U_B,-) : A'\text{-mod} \longrightarrow B\text{-mod}$, such that $\beta(D(A'_{A'}))$ is a slice module in $\mathcal{B}$. Note that the torsion pair $(\mathcal{Y}({}_{A'}U),\mathcal{X}({}_{A'}U))$ splits and that $\mathcal{X}({}_{A'}U)$ contains only modules from $\mathcal{B}$. In particular, the module ${}_BS$ belongs to $\mathcal{Y}({}_{A'}U)$ [namely, S_0 belongs to the preprojective component of B-mod, and the indecomposable summands of P belong to tubes]. Let $\beta' = {}_{A'}U_B \otimes -$. Note that $\beta'({}_BS)$ is a direct sum of preprojective and regular A'-modules, since the preinjective A'-modules are mapped under β to $\mathcal{B}$.

Finally, the given algebra A is tame concealed of the same tubular type as A', thus there exists a preprojective tilting A-module ${}_AV$ with $\mathrm{End}({}_AV) = A'$, and we denote $\alpha = \mathrm{Hom}({}_AV_{A'},-) : A\text{-mod} \longrightarrow A'\text{-mod}$, and $\alpha' = {}_AV_{A'} \otimes -$. The preprojective and the regular A'-modules lie in $\mathcal{Y}({}_AV)$, thus are images under α, and we can use the inverse functor α'. In particular, this applies to $\beta'S$, and we obtain $\alpha'\beta'S$, an A-module with preprojective and regular summands, thus having projective dimension ≤ 1. Also, since ${}_BS$ is a tilting module, we also have $\mathrm{Ext}^1_{A'}(\beta'S,\beta'S) = 0$, and $\mathrm{Ext}^1_A(\alpha'\beta'S,\alpha'\beta'S) = 0$, and the number of isomorphism classes of indecomposable summands of S', $\beta'S$, and $\alpha'\beta'S$ are the same. This shows that $\alpha'\beta'S$ is a tilting A-module. Finally, $\mathrm{End}_A(\alpha'\beta'S) \approx \mathrm{End}_{A'}(\beta'S) \approx \mathrm{End}({}_BS) \approx C$. This shows that we obtain C as endomorphism ring of a tilting A-module, and finishes the proof.

(1') Corollary. Let B be a domestic tubular extension of a tame concealed algebra. Then χ_B is positive semi-definite with radical rank 1.

We obtain the following structure theorem for B-mod, where B is a domestic tubular extension of some tame concealed algebra.

(2) Theorem. Let B be a domestic tubular extension of the tame concealed algebra B_0, of extension type $(m_1,\ldots,m_t)$, and let B_0 be of tubular type $(n_1,\ldots,n_t)$. Then B has a preprojective component $\mathcal{P}$, of orbit type $\widetilde{\mathbb{T}}_{n_1,\ldots,n_t}$, given by the preprojective B_0-modules; a preinjective component $\mathcal{Q}$ of orbit type $\widetilde{\mathbb{T}}_{m_1,\ldots,m_t}$, and a tubular $\mathbb{P}_1k$-family $\mathcal{T}$ of extension type $(m_1,\ldots,m_t)$ obtained from the tubular $\mathbb{P}_1k$-family of all regular B_0-modules by ray insertion. The tubular family $\mathcal{T}$ separates $\mathcal{P}$ from $\mathcal{Q}$, and, in this way, we obtain all indecomposable A-modules. Also, B-mod is controlled by χ_B.

Proof. Let $B = B_0[E_i,K_i]_{i=1}^t$. Let $\mathcal{P}$ be the set of preprojective B_0-modules. Let $\mathcal{T}$ be the module class given by all indecomposable B-modules M with either $M|B_0$ non-zero and a regular B_0-module, or else the support of M is contained in some K_i, and $\langle \ell_{K_i}, \underline{\dim}\, M\rangle < 0$. Let $\mathcal{Q}$ be the module class given by all indecomposable B-modules M with either $M|B_0$ being a non-zero preinjective B_0-module, or else the support of M is contained in some K_i, and $\langle \ell_{K_i}, \underline{\dim}\, M\rangle > 0$. According to 4.7.1, we have B-mod $= \mathcal{P} \vee \mathcal{T} \vee \mathcal{Q}$, with $\mathcal{T}$ separating $\mathcal{P}$ from $\mathcal{Q}$, and $\mathcal{T}$ being obtained from the tubular family of all regular B_0-modules by ray insertion, and of extension type $(m_1,\ldots,m_t)$.

In order to consider $\mathcal{Q}$, let A be some hereditary algebra of tubular type $(m_1,\ldots,m_t)$, with the preprojective, regular, and preinjective module classes denoted by $\mathcal{P}^A$, $\mathcal{T}^A$, $\mathcal{Q}^A$, respectively. Let $m = 2-t+\sum_{i=1}^t m_i$, it is the rank of $K_0(A)$. According to (1), there is a tilting module ${}_AT = T_o \oplus T_1$, with T_0 preprojective, T_1 regular, $\operatorname{End}({}_AT_0) = B_0$, $\operatorname{End}({}_AT) = B$, and let $\Sigma = \operatorname{Hom}_A({}_AT_B,-)$. We have seen in the proof of (1) that $\Sigma(\mathcal{P}^A \cap \mathcal{G}({}_AT)) = \mathcal{P}$, and that $\Sigma(\mathcal{T}^A \cap \mathcal{G}({}_AT))$ are the components of B-mod containing indecomposable regular B_0-modules, thus $\Sigma(\mathcal{T}^A \cap \mathcal{G}({}_AT)) = \mathcal{T}$. Since $\mathcal{Q}^A \subseteq \mathcal{G}({}_AT)$, we see that Σ defines an equivalence from $\mathcal{Q}^A$ onto $\mathcal{Q} \cap \mathcal{Y}({}_AT)$. Also, $\mathcal{X}({}_AT) \subseteq \mathcal{Q}$, thus $\mathcal{Q} = (\mathcal{Q} \cap \mathcal{Y}({}_AT)) \vee \mathcal{X}({}_AT)$. Now, $\mathcal{F}({}_AT)$ is a finite category. [Namely, ${}_AT$ has a preprojective summand, say $\tau^{-t}P(a)$. If ${}_AM$ is an indecomposable module in $\mathcal{F}({}_AT)$, then $0 = \operatorname{Hom}(\tau^{-t}P(a),M) = \operatorname{Hom}(P(a),\tau^tM)$, thus M is either one of the finitely many indecomposable A-modules X with $\tau^tX = 0$, or else it is one of the finitely many τ^t-translates of the indecomposable A(a)-modules, with A(a) being obtained from A by deleting a.] Therefore, $\mathcal{X}({}_AT)$ is a finite category. Now $\Sigma D(A_A)$ is a slice module in B-mod, thus the component of B-mod containing $\Sigma D(A_A)$ can be embedded into $\mathbb{Z}\Delta^*$, where Δ is the quiver of A. Now $\Sigma\mathcal{Q}^A$ just furnishes the subquiver $\mathbb{N}_o^-\Delta^*$ of $\mathbb{Z}\Delta^*$. Since there are only finitely many indecomposable modules remaining (those of $\mathcal{X}({}_AT)$), and there cannot be a finite component in B-mod, it follows that $\mathcal{Q}$ is a single component, embeddable into $\mathbb{Z}\Delta^*$, and in fact, into some $(\infty,z)\Delta^*$, with $z \in \mathbb{N}$. This shows that $\mathcal{Q}$ is a preinjective component and that its orbit type is that of $\mathcal{Q}^A$, thus $\widetilde{\mathbb{T}}_{m_1,\ldots,m_t}$.

Next, we show that B-mod is controlled by χ_B. We use induction on rank $K_o(B)$. Of course, if ${}_BM$ is an indecomposable B-module, then ${}_BM$ is the image under Σ of an indecomposable module in $\mathcal{G}({}_AT)$, or under $\mathrm{Ext}^1_A({}_AT,-)$ of an indecomposable module in $\mathcal{F}({}_AT)$, thus $\chi_B(\underline{\dim}\,{}_BM) = 0$ or 1 according to 4.1.c".
Now, conversely, let x be a connected positive vector in $K_o(B)$, with $\chi_B(x) = 0$ or 1. If $\chi_B(x) = 0$, then x is a positive radical vector of $K_o(B_o)$, and according to 4.3.3, applied to B_o, there are infinitely many indecomposable B_o-modules X with $\dim X = x$. Thus, it remains to consider the case of a root. First, assume x is not sincere, say $x_a = 0$ for some vertex a. Let $B(a)$ be the algebra obtained from B by deleting a. If a is outside B_o, then $B(a)$ is the product of a domestic tubular extension B' and at most two branches K_1', K_2'. Since x is connected, it is either a root in $K_o(B')$ or in one $K_o(K_i')$. If $x \in K_o(B')$, we find by induction an indecomposable B'-module X with $\underline{\dim}\, X = x$, if $x \in K_o(K_i')$, there also exists an indecomposable K_i'-module X with $\underline{\dim}\, X = x$, see 4.2.4". Now, assume a is a vertex of B_o. Since the radical generator is sincere in $K_o(B_o)$, there are only finitely many roots x with $x_a = 0$. On the other hand, with B also $B(a)$ has a union of preinjective components $\mathcal{C}_i$ containing all the indecomposable injective modules (4.3.7'), these components $\mathcal{C}_i$ must be finite [since for X indecomposable in $\mathcal{C}_i$, $\underline{\dim}\, X$ determines X, according to 2.4.8, and $\underline{\dim}\, X$ is a root for χ_B, as we already know], thus $B(a)$ is a directed algebra. In particular, $\mathrm{Ext}^1(Y,Y) = 0$ for Y any indecomposable $B(a)$-module. Now, taking any $B(a)$-module X with $\underline{\dim}\, X = x$ and $\dim \mathrm{End}(X)$ smallest possible, it follows from $\mathrm{gl.dim}\, B \leq 2$, and $\chi_B(x) = 1$ that X is indecomposable (using 2.3.1).

It remains to consider the case of x being sincere. Let $\sigma : K_o(A) \longrightarrow K_o(B)$ be the linear transformation corresponding to Σ. Let $T = \oplus T(i)$, with $T(i)$ indecomposable, and $t(i) = \underline{\dim}\, T(i)$. Let $y = x\sigma^{-1}$. Since y is a root for χ_A, it is either positive or negative. First, let $y > o$. Note that

$$\langle t(i),y\rangle = \langle p_B(i),y\sigma\rangle = x_i > 0.$$

There exists an indecomposable A-module Y with $\underline{\dim}\, Y = y$, and we claim that $Y \in \mathcal{G}({}_AT)$. Namely, assume for some $T(i)$, we have $\mathrm{Ext}^1_A(T(i),Y) \neq 0$. Since

$$0 < \langle t(i),y\rangle = \dim \mathrm{Hom}_A(T(i),Y) - \dim \mathrm{Ext}^1_A(T(i),Y),$$

it follows that $\dim \mathrm{Hom}_A(T(i),Y) \geq 2$. Also, $0 \neq \mathrm{Ext}^1_A(T(i),Y) = D\,\mathrm{Hom}(Y,\tau T(i))$ shows that there is a cycle $Y \preceq \tau T(i) \prec T(i) \preceq Y$ containing both Y and $T(i)$, thus Y and $T(i)$ belong to some tube $\mathcal{T}(\rho)$, say a tube of rank r. Since the $\mathcal{T}(\rho)$-length of $T(i)$ has to be $< r$, due to the fact that $\mathrm{Ext}^1(T(i),T(i)) = 0$, we have $\dim \mathrm{Hom}(T(i),Z) \leq 1$ for any indecomposable module Z in $\mathcal{T}(\rho)$. This contradiction shows that $Y \in \mathcal{G}({}_AT)$, and then ΣY is an indecomposable B-module with $\underline{\dim}\,\Sigma Y = x$. Similarly, if $y < 0$, then there exists an indecomposable A-module Y' with $\underline{\dim}\,Y' = -y$. If we would have $\mathrm{Hom}_A(T(i),Y') \neq 0$ for some $T(i)$, then $\langle t(i),-y\rangle < 0$ implies $\dim \mathrm{Ext}^1_A(T(i),Y') \geq 2$. However, again there is a cyclic path $Y' \preceq \tau T(i) \prec T(i) \preceq Y'$ containing Y', $\tau T(i)$, so they belong to some tube, and then we must have $\dim \mathrm{Ext}^1_A(T(i),Y') \leq 1$, since $T(i)$ is a partial tilting module. This shows that $Y' \in \mathcal{F}({}_AT)$, and then $X' = \mathrm{Ext}^1_A({}_AT,Y')$ is an indecomposable B-module with $\underline{\dim}\,X' = x$. This finishes the proof.

<u>Remark.</u> We also may derive directly from the structure of B-mod that B is controlled by χ_B. (See the corresponding proof of 5.2.6).

(3) It seems to us rather exciting that we can derive from (1) some very interesting <u>full embeddings</u> $k(\mathbb{Z}\Delta') \subset k(\mathbb{Z}\Delta)$, <u>where</u> Δ, <u>and</u> Δ' <u>both are quivers of Euclidean type.</u>

Let Δ, Δ' be quivers (without oriented cycles) of Euclidean type, let A be the path algebra of Δ, and A' the path algebra of Δ'. Thus, the preprojective component $\mathcal{P}$ of A is equivalent to $k(\mathbb{N}_0\Delta)$, the preprojective component $\mathcal{P}'$ of A' is equivalent to $k(\mathbb{N}_0\Delta')$. The type of Δ can be written in the form $\tilde{\mathbb{T}}_{m_1,\ldots,m_t}$, with $\mathbb{T}_{m_1,\ldots,m_t}$ being a Dynkin graph, similarly, the type of Δ' can be written in the form $\tilde{\mathbb{T}}_{m'_1,\ldots,m'_t}$, with $\mathbb{T}_{m'_1,\ldots,m'_t}$ a Dynkin graph (note that we can use the same t, using, if necessary, some $m_s = 1$ or some $m'_s = 1$; also note that we always will assume $m_1 \geq m_2 \geq \ldots \geq m_t$, and $m'_1 \geq m'_2 \geq \ldots \geq m'_t$). Recall that the tubular type of A is $(m_1,\ldots,m_t)$, that of A' is $(m'_1,\ldots,m'_t)$. Let us now assume that $m'_s \leq m_s$ for all $1 \leq s \leq t$. In this case, we can construct a tubular extension B of A' of extension type $(m_1,\ldots,m_t)$, and (1) asserts that $B = \mathrm{End}({}_AT)$ for some tilting module ${}_AT$. Also, the preprojective component of B-mod is just the preprojective component $\mathcal{P}'$ of A'-mod, and $\mathrm{Hom}({}_AT,-)$ furnishes an equivalence from $\mathcal{P} \cap \mathcal{G}({}_AT)$ onto $\mathcal{P}'$. Actually, let ${}_AT = T_0 \oplus T_1$, with T_0 preprojective, T_1 regular, and assume $\tau^{p+1}T_0 = 0$ for some $p \in \mathbb{N}_0$. The indecomposable modules in $\mathcal{P}$ are of the form $\tau^{-z}P$, with $z \in \mathbb{N}_0$, and P indecomposable projective. If $X = \tau^{-z}P$, and $z \geq p$, then X belongs to $\mathcal{G}({}_AT)$ if and only if $\mathrm{Ext}^1_A(T_1,X) = 0$ [namely, $\mathrm{Ext}^1_A(T_0,X) = D\,\mathrm{Hom}(X,\tau T_0) = D\,\mathrm{Hom}(\tau^{-z}P,\tau T_0) =$

$D\,Hom(P,\tau^{z+1}T_o) = 0$.] Since T_1 is τ-periodic, say with τ-period q, it follows that with X also $\tau^{-q}X$ belongs to $G({}_AT)$. Let us identify P with $k(\mathbb{N}_o\Delta)$, and consider $k(\mathbb{N}_o\Delta)$ as a subcategory of $k(\mathbb{Z}\Delta)$. We may consider the full subcategory G of $k(\mathbb{Z}\Delta)$ given by all vertices (z,a), $z \in \mathbb{Z}$, a vertex of Δ, such that for some i, we have $z+iq \geq p$, and such that the module in P identified with $(z+iq,a)$ belongs to $G({}_AT)$. The subcategory G obtained in this way is invarient under τ^q, and it is equivalent to $k(\mathbb{Z}\Delta')$, thus we obtain, in this way, a full embedding of $k(\mathbb{Z}\Delta')$ into $k(\mathbb{Z}\Delta)$.

Let us order the possible tubular types $(m_1,\ldots,m_t)$ using the componentwise ordering. We have constructed a full embedding $k(\mathbb{Z}\widetilde{\mathbb{T}}_{m_1',\ldots,m_t'}) \hookrightarrow k(\mathbb{Z}\widetilde{\mathbb{T}}_{m_1,\ldots,m_t})$, in case $(m_1',\ldots,m_t') \leq (m_1,\ldots,m_t)$. Of particular interest will be the case of $(m_1',\ldots,m_t')$ being maximal under $(m_1,\ldots,m_t)$, thus $m_s' = m_s$ for all s but one, and $m_{s_o}' = m_{s_o} - 1$, for the remaining index s_o. We want to exhibit the corresponding embeddings $k(\mathbb{Z}\widetilde{\mathbb{T}}_{m_1',\ldots,m_t'}) \hookrightarrow k(\mathbb{Z}\widetilde{\mathbb{T}}_{m_1,\ldots,m_t})$. In the appendix A.1, we have listed positive periodic additive functions f_s on $\mathbb{Z}\Delta$, Δ being Euclidean, and the subcategories G to be considered are given by the vertices on which some fixed f_s vanishes. In the following table, we always indicate $G \subseteq k(\mathbb{Z}\Delta)$, as well as the equivalent category $k(\mathbb{Z}\Delta')$. Of course, for Δ being of type $\widetilde{\mathbb{A}}_n$ or $\widetilde{\mathbb{D}}_n$, we just consider typical special cases.

Case: Δ of type $\widetilde{\mathbb{A}}_n$

$k(\mathbb{Z}\widetilde{\mathbb{A}}_{3,3}) \hookrightarrow k(\mathbb{Z}\widetilde{\mathbb{A}}_{4,3})$

$\widetilde{\mathbb{A}}_{3,3}$

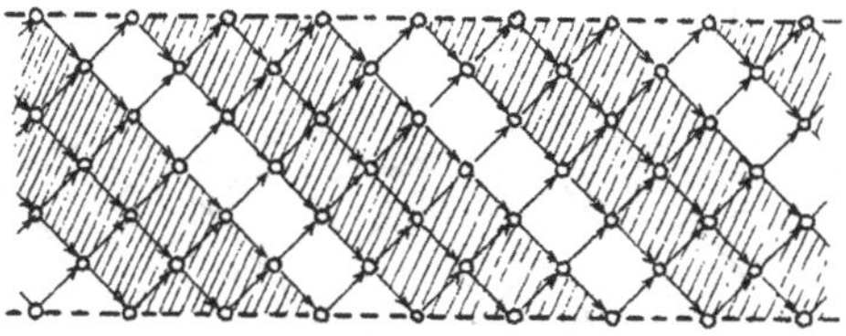

$\widetilde{\mathbb{A}}_{4,3}$

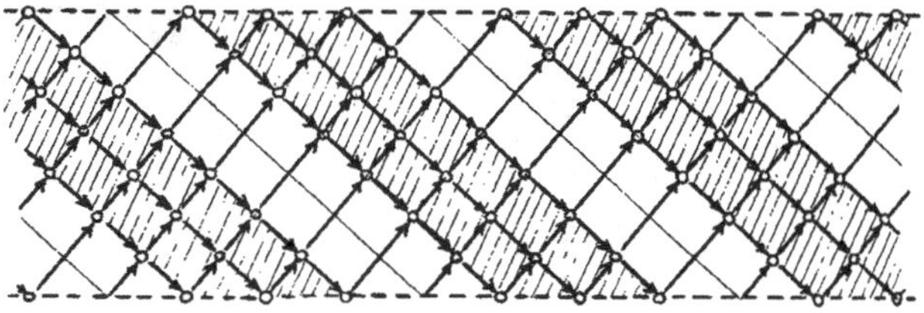

(One has to identify for $\widetilde{\mathbb{A}}_{p,q}$ the upper and the lower rim in order to obtain a cylinder).

Case: Δ of type $\tilde{\mathbb{D}}_n$

$k(\mathbb{Z}\tilde{\mathbb{D}}_8) \hookrightarrow k(\mathbb{Z}\tilde{\mathbb{D}}_9)$

$\tilde{\mathbb{D}}_8$

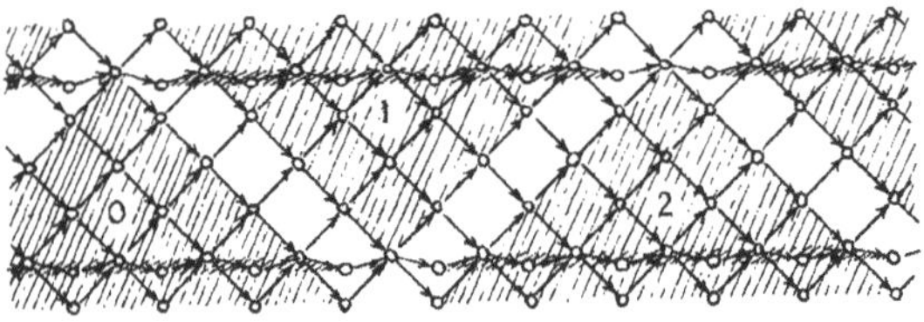

$\tilde{\mathbb{D}}_9$

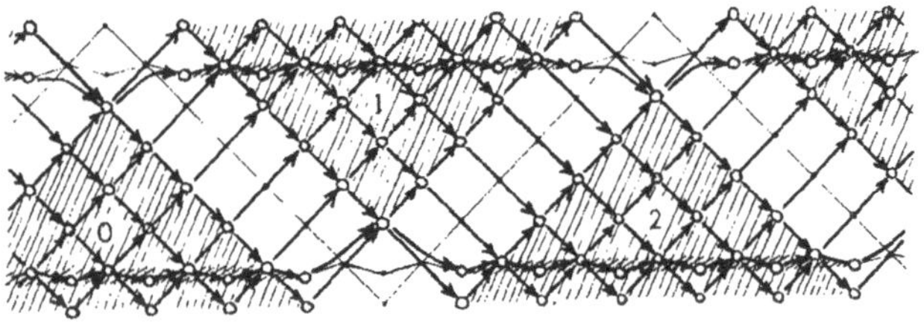

$k(\mathbb{Z}\tilde{\mathbb{A}}_{7,2}) \hookrightarrow k(\mathbb{Z}\tilde{\mathbb{D}}_9)$

$\tilde{\mathbb{A}}_{7,2}$

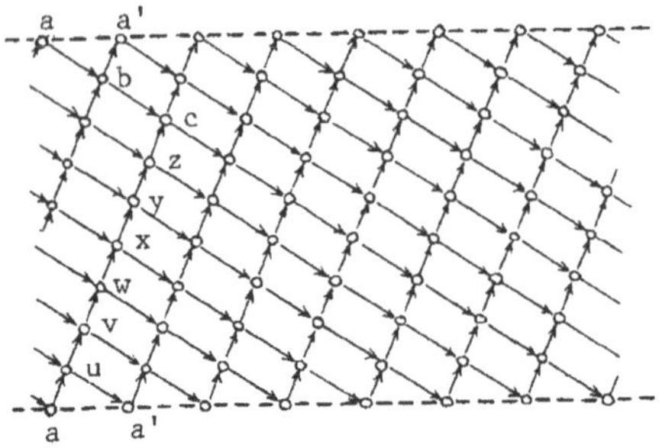

$\tilde{\mathbb{D}}_9$

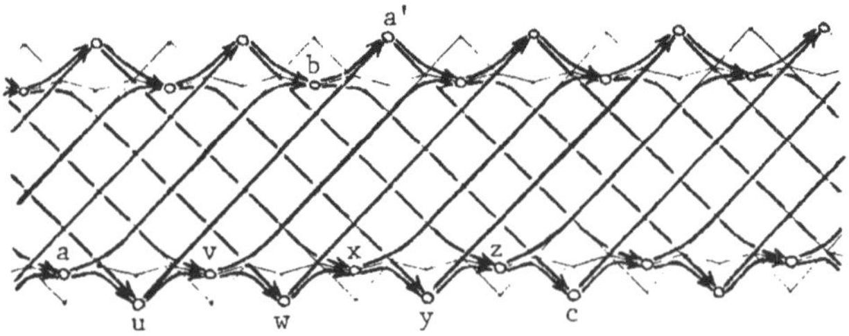

Case: Δ of type $\tilde{\mathbb{E}}_6$

$k(\mathbb{Z}\tilde{\mathbb{D}}_5) \hookrightarrow k(\mathbb{Z}\tilde{\mathbb{E}}_6)$

$\tilde{\mathbb{D}}_5$

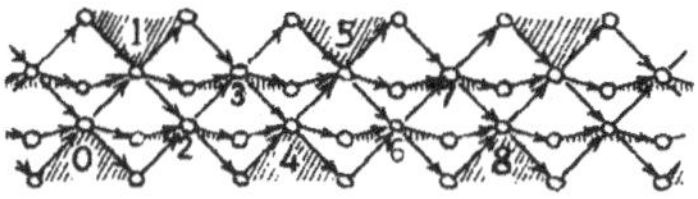

$\tilde{\mathbb{E}}_6$

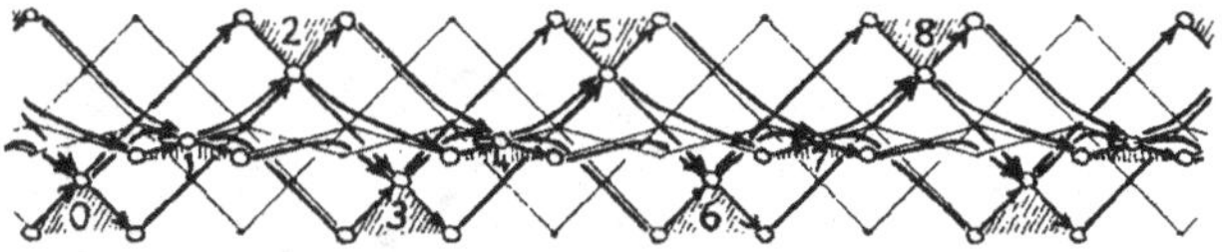

$k(\mathbb{Z}\tilde{\mathbb{A}}_{3,3}) \hookrightarrow k(\mathbb{Z}\tilde{\mathbb{E}}_6)$

$\tilde{\mathbb{A}}_{3,3}$

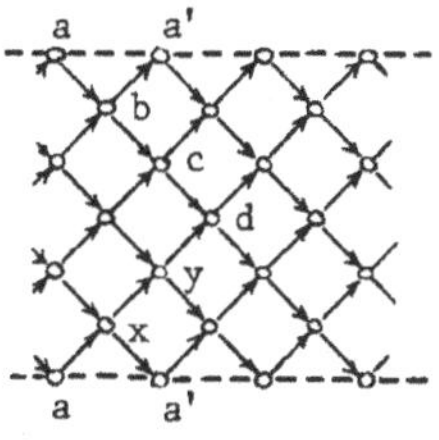

$\tilde{\mathbb{E}}_6$

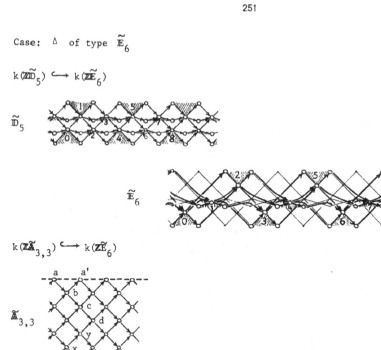

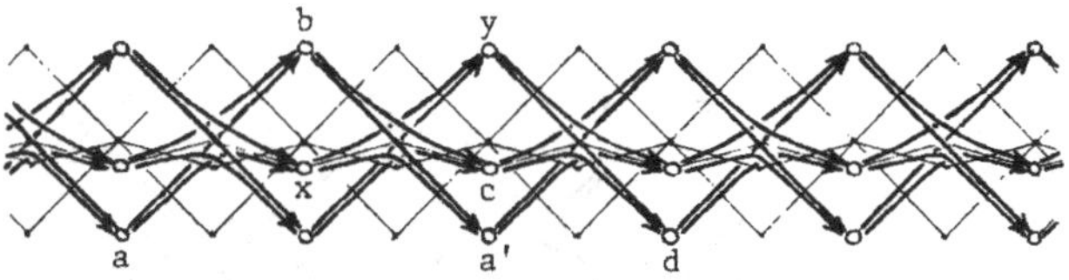

Case: Δ of type $\tilde{\mathbb{E}}_7$

$k(\mathbb{Z}\tilde{\mathbb{E}}_6) \hookrightarrow k(\mathbb{Z}\tilde{\mathbb{E}}_7)$

$\tilde{\mathbb{E}}_6$

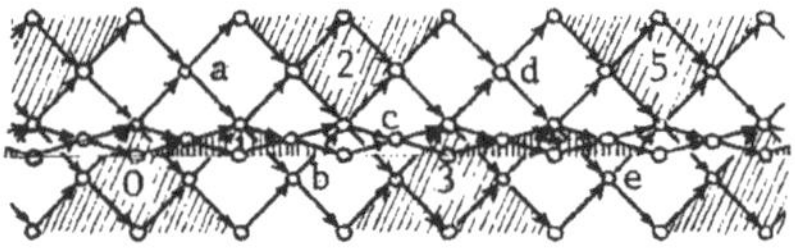

$\tilde{\mathbb{E}}_7$

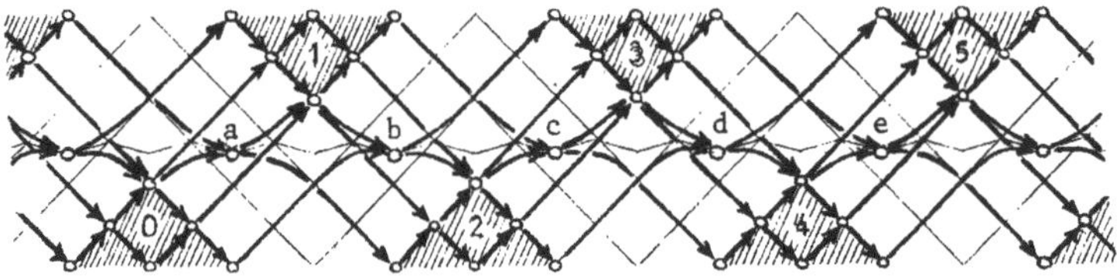

$k(\mathbb{Z}\tilde{\mathbb{D}}_6) \hookrightarrow k(\mathbb{Z}\tilde{\mathbb{E}}_7)$

$\tilde{\mathbb{D}}_6$

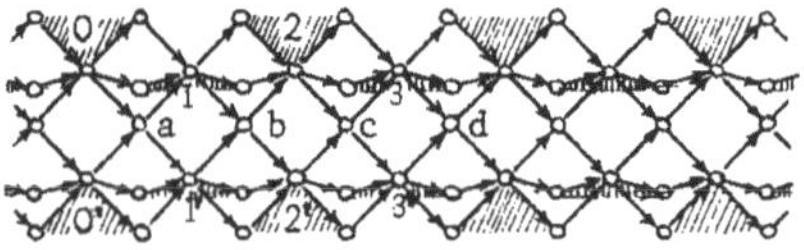

$\tilde{\mathbb{E}}_7$

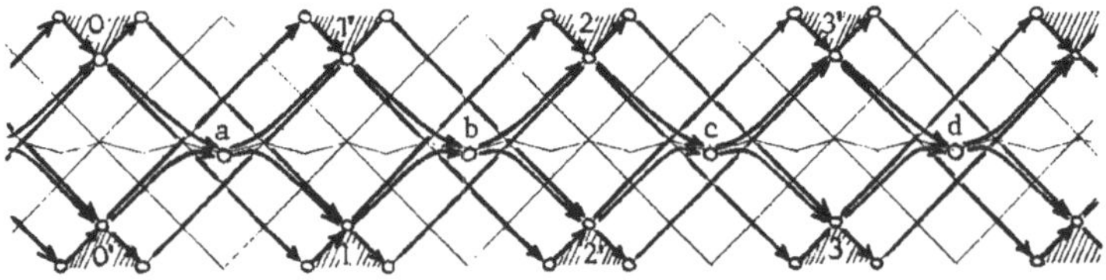

$k(\mathbb{Z}\tilde{\mathbb{A}}_{4,3}) \hookrightarrow k(\mathbb{Z}\tilde{\mathbb{E}}_7)$

$\tilde{\mathbb{A}}_{4,3}$

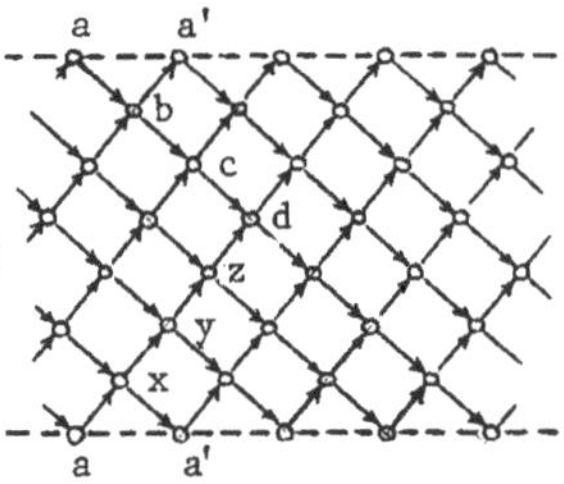

$\tilde{\mathbb{E}}_7$

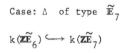

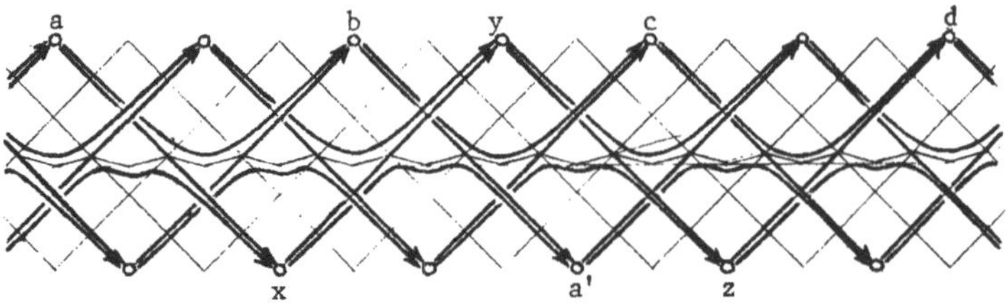

Case: Δ of type $\tilde{\mathbb{E}}_8$

$k(\mathbb{Z}\tilde{\mathbb{E}}_7) \hookrightarrow k(\mathbb{Z}\tilde{\mathbb{E}}_8)$

$\tilde{\mathbb{E}}_7$

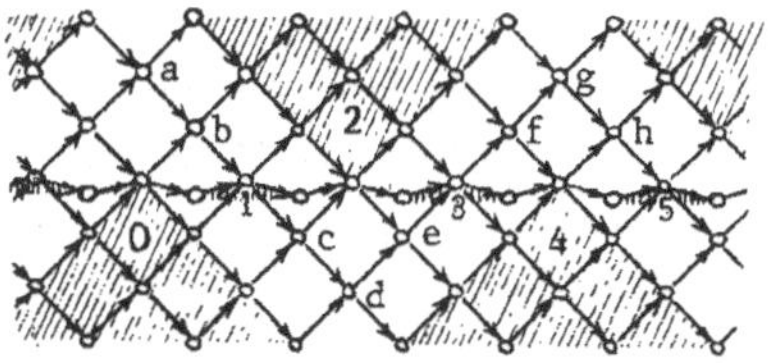

$\tilde{\mathbb{E}}_8$

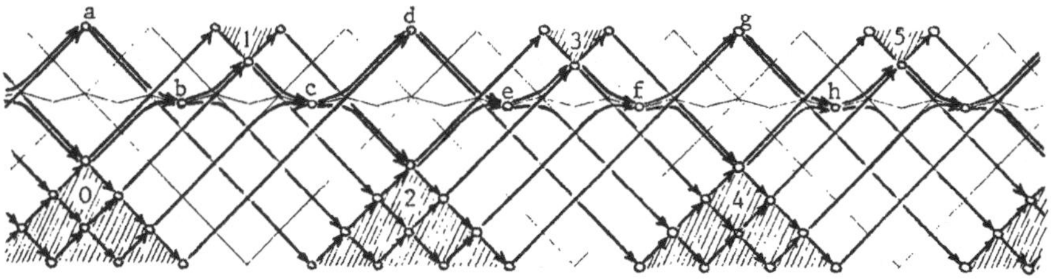

$k(\mathbb{Z}\tilde{\mathbb{D}}_7) \hookrightarrow k(\mathbb{Z}\tilde{\mathbb{E}}_8)$

$\tilde{\mathbb{D}}_7$

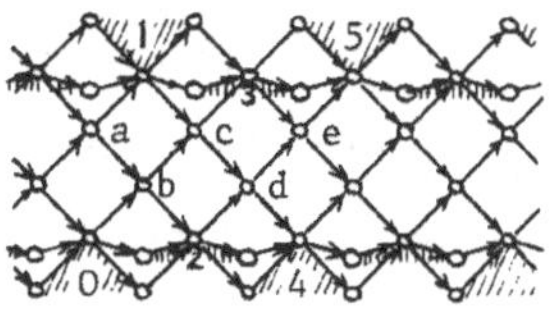

$\tilde{\mathbb{E}}_8$

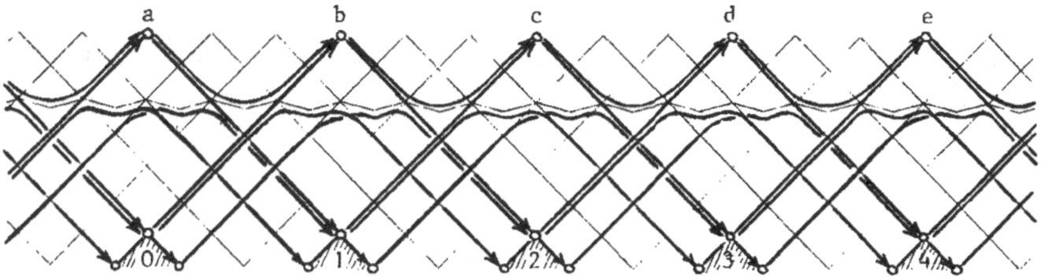

$k(\mathbb{Z}\tilde{\mathbb{A}}_{5,3}) \hookrightarrow k(\mathbb{Z}\tilde{\mathbb{E}}_8)$

$\tilde{\mathbb{A}}_{5,3}$

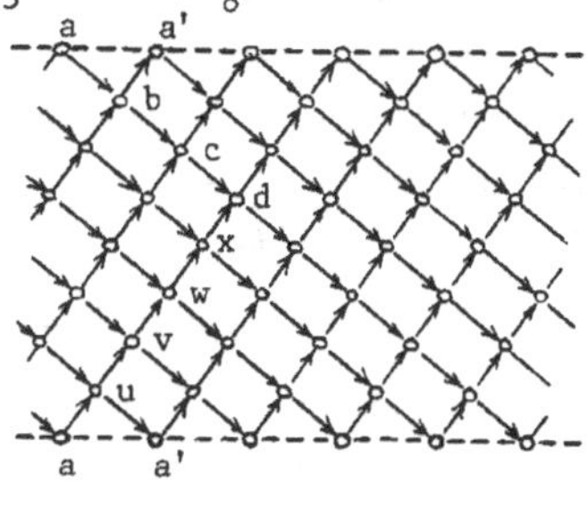

$\tilde{\mathbb{E}}_8$

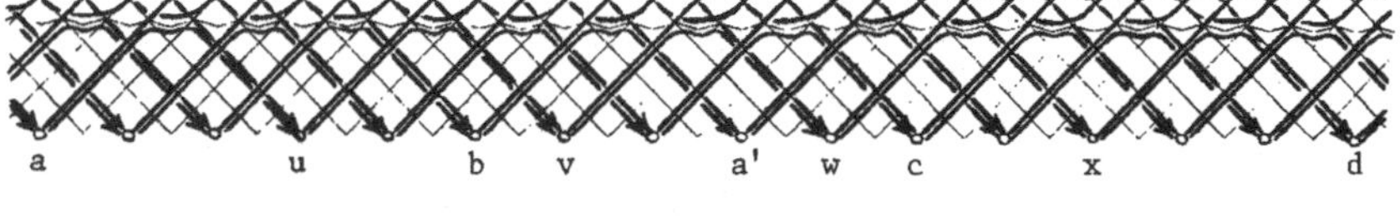

4.10 The critical directed vectorspace categories and their tubular extensions

In 2.6, we have determined all the critical directed vectorspace categories: there are just six such vectorspace categories, denoted by $C(1),\dots,C(6)$. Now, four of them, $C(2)$, $C(3)$, $C(4)$, and $C(6)$ are given by finite partially ordered sets which are disjoint unions of three chains, so in these cases, we can use the considerations in 4.4.9. Of course, we are also familiar with the case $C(1)$, since the subspace category of this vectorspace category is just the category of Kronecker modules considered in 3.2. It remains to consider the case $C(5)$.

Let A be given by the quiver

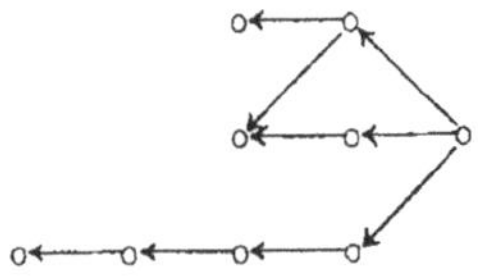

with commutativity relation. Direct calculation shows that A has a preprojective component P with orbit quiver of type $\tilde{\mathbb{T}}_{5,3,2} = \mathbb{T}_{6,3,2}$; we will exhibit part of P below. Since P contains an indecomposable sincere module, A is a tilted algebra, according to 4.2.6'. The radical of the quadratic form χ_A is generated by the vector $\begin{smallmatrix} & 24 \\ & 135 \\ 1234 & \end{smallmatrix}$, and since this vector is positive and sincere, it follows that A is tame concealed.

We can use this algebra A in order to calculate the Auslander-Reiten quiver of $\breve{\mathcal{U}}(S)$, where S is the partially ordered set of type $C(5)$:

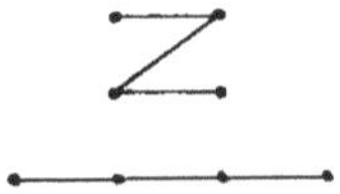

Let $A_0 = A(S)$ be the incidence algebra of S, and R the canonical A_0-module; thus $(A_0\text{-inj},\mathrm{Hom}(R,-))$ is an injective realization of the vectorspace category given by S, and we have $A = A_0[R]$. Let T_0 be the minimal injective cogenerator in A_0-mod. Also, let $G(\omega)$ be the pushout of the inclusions $R \longrightarrow P(\omega)$ and $R \longrightarrow Q$, where Q is the injective envelope of R in A_0-mod, thus $\underline{\dim}\, G(\omega) = \begin{smallmatrix} 12 \\ 111 \\ 1111 \end{smallmatrix}$. According to 4.1.8, ${}_AT = T_0 \oplus G(\omega)$ is a tilting module, and $\breve{\mathcal{U}}(S) \approx G({}_AT)$. Let us exhibit now part of the preprojective component P of A, showing at the same time that ${}_AT$ belongs to P. We have encircled the dimension vectors of the indecomposable summands of ${}_AT$.

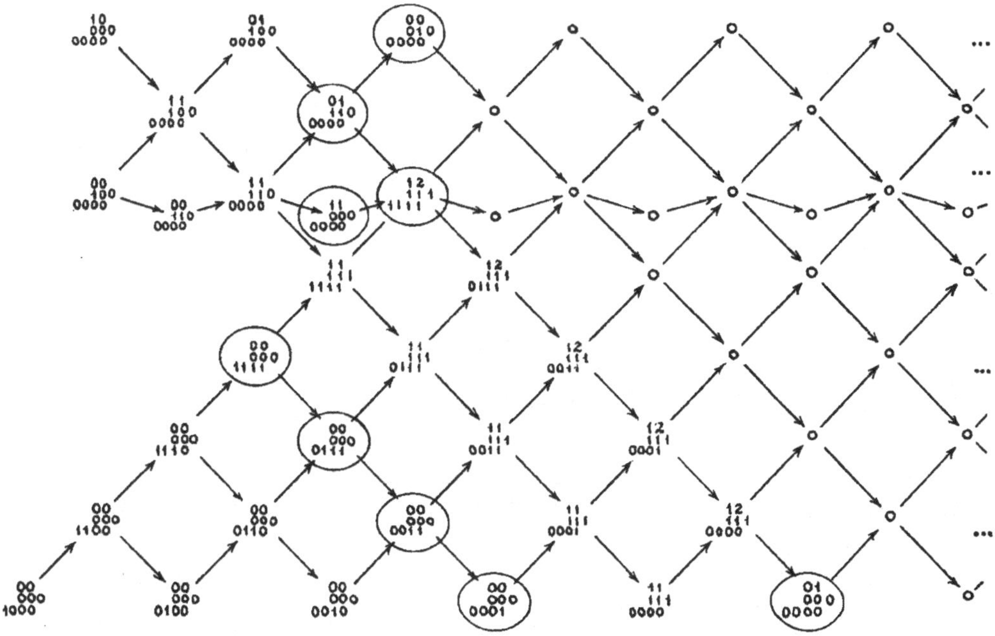

Since $_A T$ is a preprojective tilting module, only finitely many indecomposable A-modules do not belong to $\mathcal{G}(_A T)$, and all of them are in $\mathcal{P}$. We sketch the Auslander-Reiten quiver of A-mod; the indecomposable A-modules not belonging to $\mathcal{G}(_A T)$ are those in the shaded areas (we indicate again the position of the indecomposable summands of $_A T$, now using $*$):

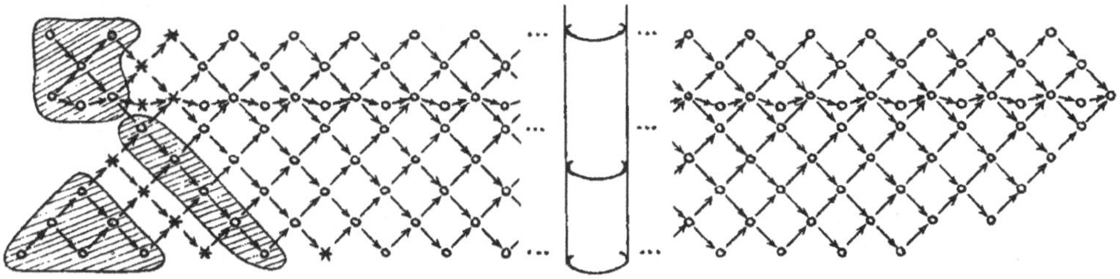

It follows that the Auslander-Reiten quiver of $\mathcal{G}(_A T)$, and therefore of $\check{\mathcal{U}}(S)$, is of the form

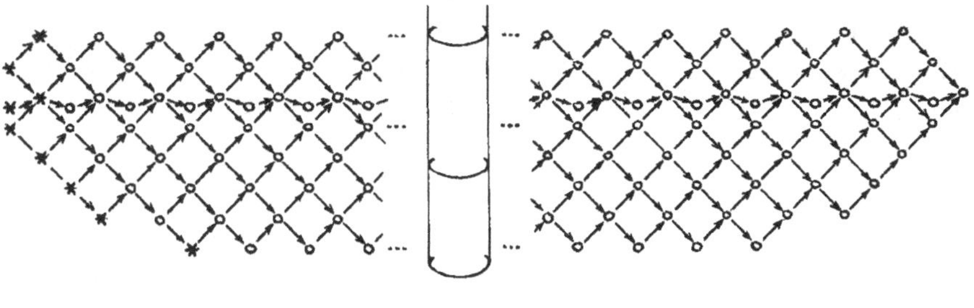

We introduce the tubular type of a critical directed vectorspace category as follows:

case	C(1)	C(2)	C(3)	C(4)	C(5)	C(6)
tubular type	(1)	(2,2,2)	(3,3,2)	(4,3,2)	(5,3,2)	(5,3,2)

In this way, we can formulate the considerations above in a unified way as

(1) Let $(K,|\cdot|)$ be a critical directed vectorspace category, let $(n_1,\ldots,n_r)$ be its tubular type. Then there exists a tame concealed algebra A of tubular type $(n_1,\ldots,n_r)$, and a preprojective tilting module ${}_AT$ such that $\check{\mathcal{U}}(K,|\cdot|) \approx G({}_AT)$.

(2) Theorem. Let $(K,|\cdot|)$ be a critical directed vectorspace category, let $(n_1,\ldots,n_r)$ be its tubular type. Then $\check{\mathcal{U}}(K,|\cdot|)$ has the following components: a preprojective component $\mathcal{P}$ with orbit quiver of type $\tilde{\mathbb{T}}_{n_1,\ldots,n_r}$, a preinjective component $\mathcal{Q}$ with orbit quiver of type $\tilde{\mathbb{T}}_{n_1,\ldots,n_r}$, and a stable separating tubular $\mathbb{P}_1k$-family $\mathcal{T}$ of type $(n_1,\ldots,n_r)$, separating $\mathcal{P}$ from $\mathcal{Q}$. Also, $\check{\mathcal{U}}(K,|\cdot|)$ is controlled by χ_K. Let h be the minimal positive radical vector of χ_K. Then an indecomposable object X of $\check{\mathcal{U}}(K,|\cdot|)$ belongs to $\mathcal{P}$, $\mathcal{T}$, or $\mathcal{Q}$ if and only if $\langle h,\underline{\dim}_K X\rangle < 0$, $= 0$, or > 0, respectively.

Proof. It remains to prove the last assertions. We denote by $\theta : \check{\mathcal{U}}(K,|\cdot|) \to G({}_AT)$ the canonical equivalence and by $\vartheta : G(A_0\text{-inj}) \times \mathbb{Z} \to K_0(A)$ the corresponding linear transformation. Given any indecomposable object V in $\check{\mathcal{U}}(K,|\cdot|)$, we have

$$\chi_K(\underline{\dim}_K V) = \chi_A(\underline{\dim}\,\theta V),$$

according to 2.5.a, and, since θV is an indecomposable A-module, $\chi_A(\underline{\dim}\,\theta V)$ is either $= 0$ or $= 1$. Conversely, let x be a positive root of χ_K. According to 2.5.b,c, we know that $x\vartheta$ is a positive and connected root of χ_A. Therefore, there exists an indecomposable A-module Y with $\underline{\dim}\, Y = x\vartheta$. Note that ${}_AT$ is preprojective, therefore any indecomposable direct summand of ${}_AT$ is directing. According to 4.1.e, we know that $G({}_AT)$ is the module class given by all indecomposable A-modules M with $\langle t(i),\underline{\dim}\, M\rangle \geq 0$ for all i. Thus Y belongs to $G({}_AT)$. Similarly, if x is a positive radical vector of χ_K, then $x\vartheta$ is a positive and connected radical vector of χ_A. Therefore, there exists an infinite family of isomorphism classes of indecomposable A-modules Y with $\underline{\dim}\, Y = x\vartheta$. Since

<t(i),dim Y> ≥ 0 for all i , it follows that all Y belong to $G({}_A T)$. Finally, $h\vartheta$ is the minimal positive radical vextor of χ_A in $K_o(A)$, and $\langle h, \underline{\dim}_K X\rangle = \langle h\vartheta, \underline{\dim}\vartheta X\rangle$. This finishes the proof.

Let $(K, |\cdot|)$ be a critical directed vectorspace category. The objects in $\check{U}(K, |\cdot|)$ belonging to the tubular family T are called regular, the indecomposable ones belonging to the mouth of a tube in T are said to be simple regular. Also, an indecomposable object in $\check{U}(K, |\cdot|)$ belonging to a homogeneous tube (= a tube of rank 1) is said to be homogeneous. Note that the homogeneous simple regular objects all have the same dimension vector, namely the positive radical generator h for χ_K , whereas the non-homogeneous simple regular objects are uniquely determined by their dimension vectors. Of course, in the Kronecker case $C(1)$, all simple regular objects are homogeneous (and h = (1[1)). In the remaining cases, we list the dimension vectors of the simple regular objects. First, we note h , then the remaining dimension vectors, always collecting together full τ-orbits (the action of τ being given by the shift to the left).

C(2)	1 $2[^1_1$ 1	1 $1[^1_0$ 0	0 $1[^0_1$ 1	1 $1[^0_1$ 0	0 $1[^1_0$ 1	0 $1[^1_1$ 0	1 $1[^0_0$ 1			
C(3)	11 3[11 11	10 1[01 00	01 1[00 10	00 1[10 01	01 1[10 00	00 1[01 10	10 1[00 10	10 2[10 10	01 1[01 01	
C(4)	2 4[111 111	1 1[010 000	0 1[001 100	1 1[000 010	0 1[100 001	1 2[100 100	0 1[010 010	1 1[001 001	1 2[101 010	1 2[010 101
C(5)	21 5[12 1111	10 1[01 0000	10 1[00 1000	00 1[01 0100	10 1[00 0010	00 1[10 0001				
		01 1[01 0010	10 2[10 0100	10 2[01 1001	11 3[01 1010	10 2[11 0101				
C(6)	3 6[22 11111	1 2[10 10000	0 1[01 01000	1 1[00 00100	0 1[10 00010	1 1[01 00001				
		1 2[11 00100	1 2[01 10010	1 2[10 01001	2 3[11 01010	1 3[11 10101				

Proof. Let (n_1, n_2, n_3) be the tubular type of $(K, |\cdot|)$ and $(A_o\text{-inj}, \mathrm{Hom}(R,-))$ an injective realization . We may use a preprojective tilting functor from the canonical algebra B of type (n_1, n_2, n_3) to $A_o[R]$. In remark 1 of 3.7, we have written down the shape of the simple regular B-modules, thus we can calculate in this way the simple regular $A_0[R]$-modules, and therefore the simple regular objects in $\check{U}(K,|\cdot|)$.

As an alternative proof, we may delete in $S = C(i)$ the following encircled vertex

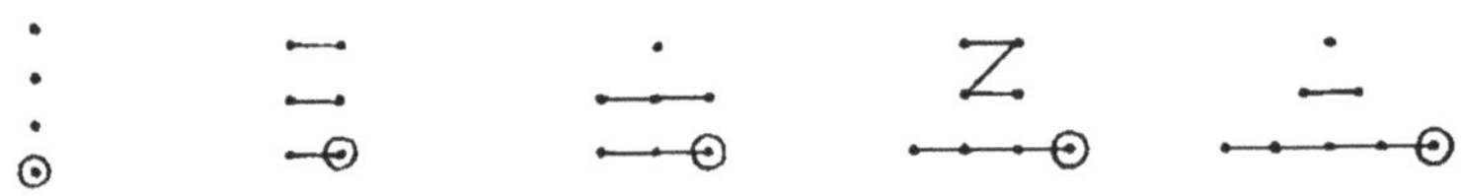

and obtain the partially ordered set $S' = F(i)$. In any of the sequences of dimension vectors $x_1,\ldots,x_n$ (with $n \geq 2$) which are proposed to belong to full τ-orbit of simple regular objects in T , we have arranged the order in such a way that $x_1,\ldots,x_{n-1}$ actually belong to $G(K(S')) \times \mathbb{Z}$. Using $\Gamma(U(S'))$, one easily checks that there are indecomposable objects $X_1,\ldots,X_{n-1}$ in $\check{U}(S')$ such that $\underline{\dim}_{K(S')}X_i = x_i$, $1 \leq i \leq n-1$, and $\tau_{S'}X_i = X_{i-1}$, $2 \leq i \leq n-1$ (we write $\tau_{S'}$ instead of $\tau_{\check{U}(S')}$), and that, moreover, the $\tau_{S'}$-orbit of $[X_i]$ has precisely one neighbor in the orbit quiver of $\check{U}(S')$. We have

(*) $\quad \mathrm{Ext}_{K(S')}(X_i,X_{i-1}) \neq 0$, $2 \leq i \leq n-1$ and $\mathrm{Ext}_{K(S')}(X_1,X_{n-1}) = 0$,

using that $\tau X_i = X_{i-1}$, $2 \leq i \leq n-1$, and that there are no cycles in $U(S')$. Also, since no $[X_i]$ is a projective vertex in $\Gamma(\check{U}(S'))$, and since its $\tau_{S'}$-orbit has precisely one neighbor in the orbit quiver, we see that there is no irreducible K-split monomorphism $Y \to X_i$. We also can check without difficulties that $\langle h,\underline{\dim}_K X_i\rangle = 0$, for $1 \leq i \leq n-1$, note that the coefficients of the linear form $\langle h,-\rangle$ are as follows:

$$2\left[\begin{smallmatrix}-1\\-1\\-1\\-1\end{smallmatrix}\right., \quad 3\left[\begin{smallmatrix}-2 & -1\\-2 & -1\\-2 & -1\end{smallmatrix}\right., \quad 4\left[\begin{smallmatrix} & -2 & \\ -3 & -2 & -1\\ -3 & -2 & -1\end{smallmatrix}\right., \quad 5\left[\begin{smallmatrix} & -3 & -1 & \\ & -4 & -2 & \\ -4 & -3 & -2 & -1\end{smallmatrix}\right., \quad 6\left[\begin{smallmatrix} & & -3 & & \\ & & -4 & -2 & \\ -5 & -4 & -3 & -2 & -1\end{smallmatrix}\right. .$$

As a consequence, the objects X_i , $1 \leq i \leq n-1$, all belong to T . Note that they are even simple regular, since otherwise, there would exist an irreducible K-split monomorphism in $\check{U}(S)$, and therefore in $\check{U}(S')$. Now, $\check{U}(S')$ is a full, exact, and extension closed subcategory of $U(S)$, thus we can replace $\mathrm{Ext}_{K(S')}$ by $\mathrm{Ext}_{K(S)}$, in (*). It follows that $X_1,X_2,\ldots,X_{n-1}$ form part of a τ-orbit of simple regular objects with orbit length $\geq n$. and we define $X_n = \tau_S^{-}X_{n-1}$. In case $C(2)$, any such τ-orbit is of length 2 , thus $n = 2$, and $\underline{\dim}_K X_2 = h - x_1 = x_2$. In the remaining cases, we consider all such τ_S-orbits at the same time. Since we know that the tubular type of $\check{U}(S)$ is (n_1,n_2,n_3) , and since the objects $X_1^{(1)},\ldots,X_{n_1-1}^{(1)}$; $X_1^{(2)},\ldots,X_{n_2-1}^{(2)}$; $X_1^{(3)},\ldots,X_{n_3-1}^{(3)}$ given as above are pairwise non-isomorphic, it follows that the length of the τ_S-orbit of $X_1^{(s)}$ is precisely n_s , and that $\underline{\dim}_K X_{n_s}^{(s)} = h - \sum_{i=1}^{n_s-1} x_i^{(s)} = x_{n_s}^{(s)}$. This finishes the proof.

In order to introduce extensions and coextensions of vectorspace categories, we will have to consider coextensions of algebras; this is the dual concept of an extension of an algebra. We use the following notation: Given an algebra A and an A-module V, the one-point coextension $[V]A$ of A by V is defined by

$$[V]A = (A^{op}[DV])^{op} \approx \begin{bmatrix} k & DV \\ 0 & A \end{bmatrix}$$

(recall that D denotes the duality with respect to the base field k). The vertex of the quiver of $[V]A$ belonging to the simple projective module $\begin{bmatrix} k \\ 0 \end{bmatrix}$ will be denoted by $-\omega$, and called the coextension vertex. The $[V]A$-modules can be written as triples

$$W = (W_{-\omega}, W_o, \delta_W)$$

where $W_{-\omega}$ is a k-vectorspace, W_o an A-module and

$$\delta_W : DV \otimes_A W_o \longrightarrow W_{-\omega}$$

a k-linear map. We also denote W_o by $W|A$ and call it the restriction of W to A. Let $Q(-\omega)$ be the indecomposable injective $[V]A$-module belonging to the coextension vertex. Then

(a) $$Q(-\omega) = (V, k, \varepsilon),$$

with $\varepsilon : DV \otimes_A V \longrightarrow k$ the evaluation map. More generally, let

$$[m,V]A = (A^{op}[DV, K^s(m)])^{op} = \left[\begin{array}{ccc|c} k & & & DV \\ & k & & \vdots \\ 0 & & k & DV \\ \hline & 0 & & A \end{array}\right],$$

(with the block of k's of size m), where $K^s(m)$ is the subspace branch of length m. Given A-modules $V^{(1)}, \ldots, V^{(t)}$, a general coextension of A using these modules is of the form

$$\prod_{i=1}^{t} [m_i, V^{(i)}]A = (A^{op}[DV^{(i)}, K^s(m_i)]_{i=1}^{t})^{op}$$

(the extension is called proper provided $t \geq 1$). Of course, $\prod_{i=1}^{t}[m_i, V^{(i)}]A$ can be obtained from A by a sequence of successive one-point coextensions. The vertices of the quiver of $\prod_{i=1}^{t}[m_i, V^{(i)}]A$ not belonging to A may be denoted by $-\omega_{ij}$, $1 \leq j \leq m_i$, $1 \leq i \leq t$, with arrows

$$\overset{-\omega_{i,j+1}}{\circ} \longleftarrow \overset{-\omega_{ij}}{\circ}$$

Note that given an indecomposable injective $\prod_{i=1}^{t}[m_i, V^{(i)}]A$-module $Q(-\omega_{ij})$, its restriction to A is just $V^{(i)}$.

Consider now the case that A itself is a one-point extension, say $A = A_o[R]$. Given an A-module V, we can write it in the form $V = (V_o, V_\omega, \gamma_V)$, with a k-linear map $\gamma_V : V_\omega \to \mathrm{Hom}_{A_o}(R,V_o)$. Dualizing, we obtain from γ_V the map $D\gamma_V : D\mathrm{Hom}_{A_o}(R,V_o) \to DV$. Also,

$$\mathrm{Hom}_{A_o}(R,V_o) \approx \mathrm{Hom}_{A_o}(R,DDV_o) \approx \mathrm{Hom}_k(DV_o \otimes_{A_o} R, k) = D(DV_o \otimes_{A_o} R)$$

canonically, thus also $\mathrm{Hom}_{A_o}(R,V_o) \approx DV_o \otimes_{A_o} R$ canonically. We denote the composition

$$DV_o \otimes_{A_o} R \xrightarrow{\sim} \mathrm{Hom}_{A_o}(R,V_o) \xrightarrow{D\gamma_V} DV_\omega$$

by $\tilde{\gamma}_V$. We get the following formula:

(b) $$[V](A_o[R]) = ([V_o]A_o)[(DV_\omega,R,\tilde{\gamma}_V)] \ ;$$

in matrix form, this algebra is

$$[V](A_o[R]) = \begin{bmatrix} k & DV_o & DV_\omega \\ 0 & A_o & R \\ 0 & 0 & k \end{bmatrix} ,$$

with multiplication given by the map $\tilde{\gamma}_V : DV_o \otimes_{A_o} R \to DV_\omega$.

Let $(K,|\cdot|)$ be a finite vectorspace category, say given by its injective realization $(K,|\cdot|) = (A_o\text{-inj},\mathrm{Hom}(R,-))$. Given $V \in \check{U}(K,|\cdot|)$, say $V = (V_o,V_\omega,\gamma_V)$, the <u>one-point coextension</u> $[V](K,|\cdot|)$ of $(K,|\cdot|)$ by V is defined to be

$$[V](K,|\cdot|) = ([V_o]A_o\text{-inj},\mathrm{Hom}((DV_\omega,R,\tilde{\gamma}_V),-)) .$$

An indecomposable $[V_o]A_o$-injective module is either an A_o-module, and of course one of the injective A_o-modules, or else the $[V_o]A_o$-injective module belonging to the coextension vertex $-\omega$, and denoted by $Q_o(-\omega)$. Note that the object $\overline{Q_o(-\omega)}$ in $\check{U}([V](K,|\cdot|))$ defined as

$$\overline{Q_o(-\omega)} = (Q_o(-\omega),\mathrm{Hom}((DV_\omega,R,\tilde{\gamma}_V),Q_o(-\omega)),1) ,$$

is just the indecomposable injective $[V](A[R])$-module belonging to the coextension vertex, according to (b).

(c) The source map for $\overline{Q_o(-\omega)}$ in $\check{U}([V](K,|\cdot|))$ is of the form $\overline{Q_o(-\omega)} \to V$.

<u>Proof</u>. The source map for $\overline{Q_o(-\omega)}$ in $[V](A[R])$ is $\overline{Q_o(-\omega)} \to V$, since $\overline{Q_o(-\omega)}$ is the indecomposable injective $[V](A[R])$-module belonging to the coextension vertex, so that we can use (a). Since V belongs to $\check{U}(K,|\cdot|) \subseteq \check{U}([V](K,|\cdot|))$, it follows that this map is a source map also in this subcategory $\check{U}([V](K,|\cdot|))$.

The general case of a <u>coextension</u> of $(K,|\cdot|) = (A_o\text{-inj},\mathrm{Hom}(R,-))$ is a vectorspace category of the form

$$\bigoplus_{i=1}^{t}[m_i,V^{(i)}](K,|\cdot|) = (\bigoplus_{i=1}^{t}[m_i,V_o^{(i)}]A_o\text{-inj},\mathrm{Hom}(R',-)),$$

where R' is defined by the equality

$$\bigoplus_{i=1}^{t}[m_i,V^{(i)}](A_o[R]) = (\bigoplus_{i=1}^{t}[m_i,V_o^{(i)}]A_o)[R'] .$$

Here, $V^{(1)},\dots,V^{(t)}$ are objects in $\check{U}(K,|\cdot|)$, or, equivalently, $A_o[R]$-modules with restriction $V_o^{(i)}$ to A_o being injective. Note that as a consequence of the injectivity, with A_o also $\bigoplus_{i=1}^{t}[m_i,V_o^{(i)}]A_o$ is hereditary. In particular, in this case, we can apply lemma 4.1.8. Given a vectorspace category $(K,|\cdot|) =$ $= (A_o\text{-inj},\mathrm{Hom}(R,-))$, with A_o being hereditary, the tilting module ${}_{A_o[R]}T = T_o \oplus G(\omega)$ defined in 4.1.8 will be said to be the <u>canonical tilting module for</u> $(K,|\cdot|)$.

Consider now the case of $(K,|\cdot|)$ being a critical directed vectorspace category, let $V^{(1)},\dots,V^{(t)}$ be pairwise different simple regular objects of $\check{U}(K,|\cdot|)$. A vectorspace category of the form $\bigoplus_{i=1}^{t}[m_i,V^{(i)}](K,|\cdot|)$ will be called a <u>tubular co-extension</u> of $(K,|\cdot|)$. The <u>coextension type</u> is given by the function $n : \mathbb{P}_1k \to \mathbb{N}_1$, with

$$n_\rho = r_\rho + \sum_{V^{(i)}\in T(\rho)} m_i \quad , \text{ for } \rho\in\mathbb{P}_1k ,$$

where $T = (T(\rho))_\rho$ is the tubular family of $\check{U}(K,|\cdot|)$. Let $(L,|\cdot|') =$ $= \bigoplus_{i=1}^{t}[m_i,V^{(i)}](K,|\cdot|)$ be a tubular coextension of the critical directed vectorspace category $(K,|\cdot|) = (A_o\text{-inj},\mathrm{Hom}(R,-))$, where we assume that A_o is basic. Let $A = A_o[R]$. If we consider the injective realization $(L,|\cdot|') = (B_o\text{-inj},\mathrm{Hom}(R',-))$ with B_o basic, and define $B = B_o[R']$, then $B = \bigoplus_{i=1}^{t}[m_i,V^{(i)}]A$. Note that A is tame concealed, B is a tubular coextension of A, and the coextension type of $(L,|\cdot|')$ is just the coextension type of B over A. Now, $A\text{-mod} = P^A \vee T^A \vee Q^A$, with P^A, T^A, Q^A the module classes of all preprojective, regular, and preinjective A-modules, respectively. According to the dual of 4.7.1, we know that $B\text{-mod} = P^B \vee T^B \vee Q^B$, with T^B again a separating tubular family, separating P^B from Q^B, and $P^A \subseteq P^B$, $T^A \subseteq T^B$, $Q^A = Q^B$.

(3) <u>Lemma</u>. Let $(L,|\cdot|')$ be a tubular coextension of a critical directed vectorspace category with injective realization $(L,|\cdot|') = (B_o\text{-inj},\mathrm{Hom}(R',-))$, and let $B = B_o[R']$. Then, the canonical tilting module for $(L,|\cdot|')$ belongs to P^B.

<u>Proof</u>. Let $(K,|\cdot|) = (A_o\text{-inj},\mathrm{Hom}(R,-))$, and $A = A_o[R]$. We consider the minimal positive radical vector w of χ_A, and the linear form $\langle w,-\rangle$ on $K_o(A)$. We exhibit in the various cases the quiver of A with the vertices replaced by the corresponding coefficients of w, and below the corresponding coefficients of $\langle w,-\rangle$.

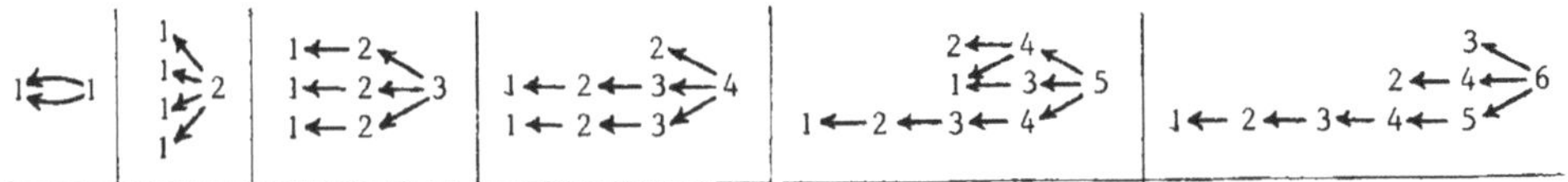

-1 1	-1 -1 2 -1 -1	-1 -1 -1 -1 3 -1 -1	-2 -1 -2 -3 4 -1 -2 -3	-2 -1 -1 -2 5 -1 -1 -1 -1	-3 -2 -2 6 -1 -1 -1 -1 -1

It follows that for any indecomposable A_0-module X, we have $\langle w,\underline{\dim}X\rangle < 0$, therefore X belongs to $\mathcal{P}^A$, according to 4.3.3. If a is any vertex of the quiver of A_0, and $Q_0(a)$ the corresponding injective A_0-module, then $Q_0(a)$ belongs to $\mathcal{P}^A \subseteq \mathcal{P}^B$. Now, if $-\omega_{ij}$ is any vertex of the quiver of B_0, and not belonging to A_0, and $Q_0(-\omega_{ij})$ the corresponding indecomposable injective B_0-module (considered as B-module), then the restriction of $Q_0(-\omega_{ij})$ to A (or, what is the same, to A_0) is just $V^{(i)}|A_0$, thus a non-zero A_0-module, and therefore in $\mathcal{P}^A$. According to the dual of 4.7.1, the module $Q_0(-\omega_{ij})$ belongs to $\mathcal{P}^B$. Finally, consider the direct summand $G(\omega)$ of the canonical tilting module for $(\mathcal{L},|\cdot|')$. By definition, $G(\omega)$ has the indecomposable projective B-module $P_B(\omega)$ as a submodule, and $G(\omega)/P_B(\omega) \approx Q'/R'$, where Q' is an injective envelope of R' in B_0-mod. In particular, $G(\omega)/P_B(\omega)$ is a B_0-module. Now, the restriction of $P_B(\omega)$ to A is just the indecomposable projective A-module $P_A(\omega)$ belonging to ω, thus $P_B(\omega)|A$ is non-zero and in $\mathcal{P}^A$. Also, the restriction of $G(\omega)/P_B(\omega)$ to A is an A_0-module, thus in $\mathcal{P}^A$. As a consequence, $G(\omega)|A$ is non-zero and in $\mathcal{P}^A$, thus the indecomposable B-module $G(\omega)$ again belongs to $\mathcal{P}^B$.

(4) Corollary. Let $(\mathcal{L},|\cdot|')$ be a tubular coextension of a critical directed vectorspace category $(\mathcal{K},|\cdot|)$, with injective realization $(\mathcal{L},|\cdot|') = (B_0\text{-inj},\mathrm{Hom}(R',-))$, and let $B = B_0[R']$. Then both $\mathcal{T}^B$ and $\mathcal{Q}^B$ are contained in $\check{\mathcal{U}}(\mathcal{L},|\cdot|')$, thus

$$\check{\mathcal{U}}(\mathcal{L},|\cdot|) = (\check{\mathcal{U}}(\mathcal{L},|\cdot|) \cap \mathcal{P}^B) \vee \mathcal{T}^B \vee \mathcal{Q}^B ,$$

and $\mathcal{T}^B$ is a separating tubular $\mathbb{P}_1k$-family, separating $(\check{\mathcal{U}}(\mathcal{L},|\cdot|) \cap \mathcal{P}^B)$ from $\mathcal{Q}^B$. Also, $\mathcal{Q}^B$ is the preinjective component of $(\mathcal{K},|\cdot|)$.

Proof. Let ${}_BT$ be the canonical tilting module for $(\mathcal{L},|\cdot|')$. Since ${}_BT \in \mathcal{P}^B$, we have $\mathrm{Ext}^1({}_BT,\mathcal{T}^B \vee \mathcal{Q}^B) = 0$, thus $\mathcal{T}^B \vee \mathcal{Q}^B \subseteq G({}_BT) = \check{\mathcal{U}}(\mathcal{L},|\cdot|')$.

Let $(\mathcal{L},|\cdot|')$ be a tubular coextension of the critical directed vectorspace category $(\mathcal{K},|\cdot|)$, of coextension type $(n_1,\ldots,n_t)$. For case $\mathbb{T}_{n_1,\ldots,n_t}$ is a Dynkin graph, we call such a coextension a domestic tubular coextension.

(5) Theorem. Let $(\mathcal{L},|\cdot|')$ be a domestic tubular coextension of the critical directed vectorspace category $(\mathcal{K},|\cdot|)$. Let $(n_1,\ldots,n_t)$ be the tubular type of

$(K,|\cdot|)$, and $(n_1',\ldots,n_t')$ the coextension type of $(L,|\cdot|')$. Then $\check{\mathcal{U}}(L,|\cdot|')$ has the following components: a preprojective component P of orbit type $\tilde{\mathbb{T}}_{n_1',\ldots,n_t'}$, a preinjective component Q of orbit type $\tilde{\mathbb{T}}_{n_1,\ldots,n_t}$ (with Q being contained in $\check{\mathcal{U}}(K,|\cdot|)$), and a tubular $\mathbb{P}_1k$-family T of coextension type $(n_1',\ldots,n_t')$ (obtained from the tubular family of all regular objects in $\check{\mathcal{U}}(K,|\cdot|)$ by coray insertion). The tubular family T separates P from Q, and in this way, we obtain all of $\check{\mathcal{U}}(L,|\cdot|')$. Also $\check{\mathcal{U}}(L,|\cdot|')$ is controlled by χ_L.

Proof. Let $(K,|\cdot|) = (A_o\text{-inj},\mathrm{Hom}(R,-))$, $(L,|\cdot|') = (B_o\text{-inj},\mathrm{Hom}(R',-))$, and $A = A_o[R]$, $B = B_o[R']$. Since B is a domestic tubular coextension of A of coextension type $(n_1',\ldots,n_t')$, we can use the dual of theorem 4.9.2. We see that P^B is a preprojective component of orbit type $(n_1',\ldots,n_t')$. Since the canonical tilting module ${}_BT$ for $(L,|\cdot|')$ belongs to P^B, it follows that also $P^B \cap G({}_BT)$ is a preprojective component of orbit type $(n_1',\ldots,n_t')$. It remains to show that $\check{\mathcal{U}}(L,|\cdot|')$ is controlled by χ_L. We denote by $\theta : \check{\mathcal{U}}(L,|\cdot|') \to G({}_BT)$ the canonical equivalence, and by $\vartheta : G(B_o\text{-inj}) \times \mathbb{Z} \to K_o(B)$ the corresponding linear transformation. Since for $V \in \check{\mathcal{U}}(K,|\cdot|)$, we have

$$\chi_L(\underline{\dim}_L V) = \chi_B(\underline{\dim}\,\theta V),$$

(see 2.5.a), we know that $\chi_L(\underline{\dim}\, V) = 0$ or 1, for V indecomposable. Conversely, let x be a positive root for χ_L. According to 2.5.b,c, we know that $x\vartheta$ is a positive and connected root of χ_B, thus there exists a unique indecomposable B-module Y with $\underline{\dim}\, Y = x\vartheta$. Also, ${}_BT$ is a preprojective tilting module, therefore any indecomposable direct summand of ${}_BT$ is directing, so we can apply 4.1.e. Since $0 \le \langle t(i), x\vartheta\rangle = \langle t(i), \underline{\dim}\, Y\rangle$ for all i, it follows that Y belongs to $G({}_BT)$. Similarly, given a positive radical vector x of χ_L, one concludes that there are infinitely many indecomposable B-modules Y with $\underline{\dim}\, Y = x\vartheta$, and that all of them belong to $G({}_BT)$.

(6) The domestic tubular coextensions $(L,|\cdot|')$ of the critical directed vectorspace categories can easily be listed. In order to do so, we will use the following convention: The isomorphism classes of the indecomposable objects X of L will be represented by a dot • in case $\dim|X|' = 1$, and by a black square ■ in case $\dim|X|' = 2$. We will draw an edge from left to right, say [X] —— [Y], provided there is an irreducible map $X \longrightarrow Y$ in L. In case two edges start at a square say

,

this shall include the information that the kernels of $|f_1|$ and $|f_2|$ for two non-zero maps $f_1 : X \to Y_1$ and $f_2 : X \to Y_2$ are different. Dually, in case two edges end at a square, say

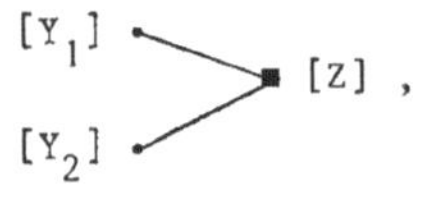

this shall include the information that the images of $|g_1|$ and $|g_2|$ for two non-zero maps $g_1 : Y_1 \to Z$, $g_2 : Y_2 \to Z$, are different. Note that in all cases exhibited below, both L and $|\cdot|'$ are (up to isomorphism) uniquely determined by these rules.

List of all proper domestic coextensions $(L,|\cdot|')$ of a critical directed vectorspace category $C(i)$; if the coextension type is $(n_1,\ldots,n_t)$, the vectorspace category $(L,|\cdot|')$ is listed in the row entitled $\mathbb{T}_{n_1,\ldots,n_t}$.

	C(1)	C(2)	C(3)	C(4)
A_n				
D_n				
E_6				
E_7				
E_8				

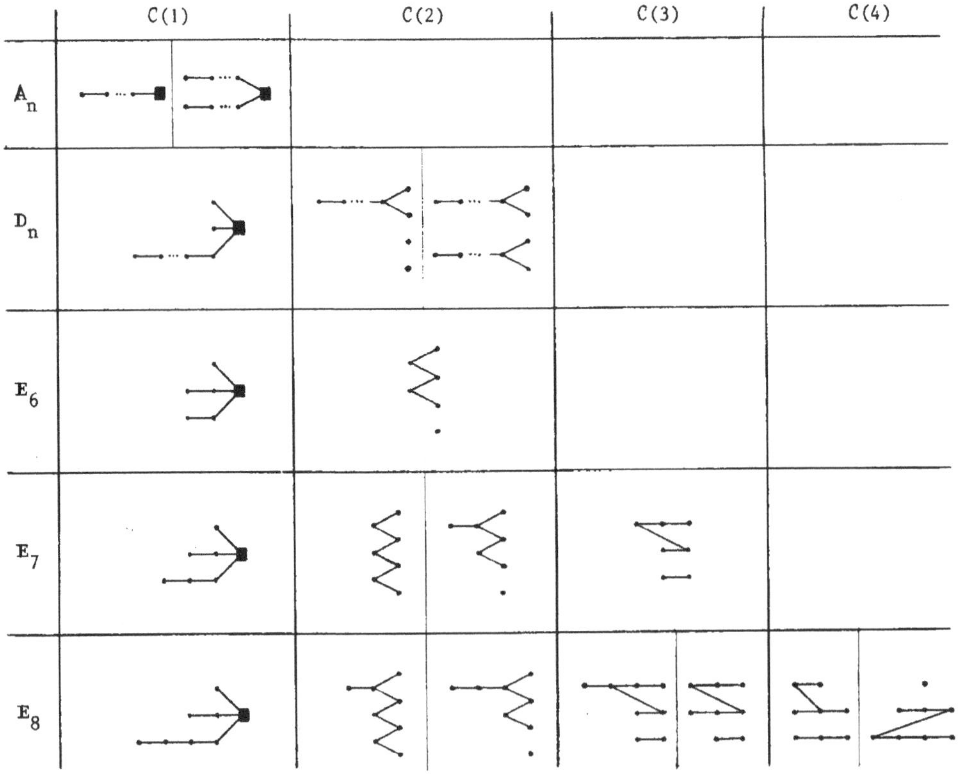

(7) The dual notions are as follows: A vectorspace category $(L,|\cdot|')$ is said to be an extension of extension type $\mathbb{T}$ of the vectorspace category $(K,|\cdot|)$, provided $(L^{op},D|\cdot|)$ is a coextension of $(K^{op},D|\cdot|)$ of coextension type $\mathbb{T}$, and, in this way, we may dualize all the definitions and results considered above. For example, we obtain the list of all proper domestic tubular extensions of critical directed vectorspace categories, by using the left-right-symmetry to the table pre-

sented in (6); for example, the domestic tubular extensions of $C(1)$ are as follows:

And, the dual result of (5) asserts that a domestic tubular extension $(L,|\cdot|')$ of a critical directed vectorspace category $(K,|\cdot|)$, with $(K,|\cdot|)$ being of tubular type $(n_1,\ldots,n_t)$, and such that $(L,|\cdot|')$ has extension type $(n_1',\ldots,n_t')$, has a preprojective component P of orbit type $\widetilde{\mathbb{T}}_{n_1,\ldots,n_t}$, a preinjective component Q of orbit type $\widetilde{\mathbb{T}}_{n_1',\ldots,n_t'}$, and a tubular $\mathbb{P}_1k$-family T of extension type $(n_1',\ldots,n_t')$, separating P from Q, and such that $\check{U}(L,|\cdot|')$ is controlled by χ_L.

References and comments

The notion of a tilting module (as used in these notes) was introduced in a joint paper with Happel [HR1], it is based on earlier investigations of Brenner and Butler [BB] where a slightly more restrictive notion was considered (the "tilting modules" in the sense of [BB] are those tilting modules T which have the additional property that there is a projective direct summand T' of T which generates T). Tilting theory explicitly generalizes the use of special functors by Gelfand-Ponomarev [GP] who worked with "Coxeter functors" for the four subspace quiver, by Bernstein-Gelfand-Ponomarev [BGP], where "reflection functors" and "Coxeter functors" were introduced for any quiver without cyclic paths, and then by Auslander-Platzeck-Reiten [APR] who considered "partial Coxeter functors"; a short outline concerning these latter functors will be given below.

In our presentation of tilting modules and tilting functors we have omitted most of the proofs, and we refer to the original paper [HR1], or, better, to the elegant treatment given by Bongartz [Bo1]. Actually, the sequence of the concepts and results given here is not the one in which proofs are arranged most easily. In [Bo1], Bongartz has shown the equivalence of conditions (γ) and (γ') in the definition of a tilting module (under the assumption of conditions (α), (β)). Since the set of conditions (α), (β), (γ) seems to be most natural, and is the one which usually will be checked in applications, we start with these conditions as defining a tilting module. However, the proof of the equivalence of (γ) and (γ') in [Bo1] is based on the main results of tilting theory which are shown before for modules satisfying the conditions (α), (β), (γ'). The proof of the equivalence uses lemma 4.1.1. We recall the proof of 4.1.1 from [Bo1], at the same time strengthening the assertion. The equivalence of $\mathcal{G}({}_AT)$ and $\mathcal{Y}({}_AT)$ stated in the theorem of Brenner and Butler has been shown in [BB], a proof of the complete result may be found in [HR1] or [Bo1]. As an illustration, we copy the proof of the connecting lemma from [HR1]. 4.1.6* and its corollary are due to Bongartz [Bo1]; we give in 4.1.6 a proof of the corresponding assertion concerning projective dimensions; note that the bound in (6)(b) is not as nice as the corresponding bound in (6*)(a). The linear transformation σ_T induced by a tilting module T has been introduced already in [HR1]. The results 4.2.1 and 4.2.2 are in [HR1], the proofs given here are due to Bongartz [Bo1]; for a proof of 4.2.2', we refer to [HR1].

We have announced above to give a short outline about the functors considered by Auslander-Platzeck-Reiten [APR]. In this paper, one-point coextensions $[R]A$ are considered, where R is an injective A-module. Note that R is an injective A-module if and only if we have $\mathrm{Hom}(\tau^-E(-\omega), [R]A) = 0$. Namely, $R = Q(-\omega)/E(-\omega)$, and R is A-injective, if and only if R is $[R]A$-injective, if and only if $\mathrm{inj.dim.}E(\omega) \leq 1$, if and only if $\mathrm{Hom}(\tau^-E(-\omega), [R]A) = 0$, according to 2.4.1*.] Let ν be the Nakayama functor for A. Then, if R is A-injective, it follows that ν^-R is A-projective, and

$$DR \otimes_A M \approx \mathrm{Hom}_A(\nu^-R, M)$$

for any A-module M. We can define a functor

$$F : [R]A\text{-mod} \longrightarrow A[\nu^-R]\text{-mod}$$

with

$$F(W_{-\omega}, W_0, \delta_W) = (W_0, \mathrm{Ker}\,\delta_W, \iota) \text{ where } \iota : \mathrm{Ker}\,\delta_W \to DR \otimes_A W_0 \approx \mathrm{Hom}_A(\nu^-R, W_0)$$

is the inclusion map. Obviously, $F(E(-\omega)) = 0$, and F induces an equivalence

$$[R]A\text{-mod} \setminus \langle E(-\omega)\rangle \xrightarrow{\sim} A[\nu^-R]\text{-mod} \setminus \langle E(\omega)\rangle .$$

Let us denote by $E(-\omega) \oplus P$ a minimal projective generator of $[R]A$-mod. If $R \neq 0$, then $P \oplus \tau^-E(-\omega)$ is a tilting module, and $F \approx \mathrm{Hom}(P \oplus \tau^-E(-\omega), -)$. These functors F are the "partial Coxeter functors" considered in [APR]. Of course, the "reflection functors" of Bernstein-Gelfand-Ponomarev are special cases of these "partial Coxeter functors".

Sectional paths and related concepts play an important role in many investigations concerning finite dimensional algebras and their Auslander-Reiten quivers. Sectional paths have been introduced by Riedtmann [Rm]. Several definitions were proposed for "slices", "complete slices", "sections", "complete sections" ([Ba2], [BB], [G4], [HR1], [Bo1]), often with the intention to characterize tilted algebras. The definition of a slice given in 4.2 is indebted to these sources, its aim is assertion 4.2.3, and then also 4.2.4. Note that the corollaries 4.2.4',4" and 6' are already in [HR1]. The various possibilities for a tilting module over a tame hereditary algebra and the representation type of the endomorphism rings have been determined in [HR2]; we reproduce this in 4.2.7.

The definition of a concealed algebra follows [HV] where, however, only the tame case was considered. (In [Ri4], a definition for "concealed quiver algebras" was proposed, with the intention to have the structure theory of the module category as outlined in the lemma of section 2.3 of [Ri4] available; we are indebted to Vossieck for pointing out that this lemma is incorrect.) The assertions 4.3.1 and 4.3.2 are from [HR2], the notion of a tilting set is borrowed from Happel [Ha]. The characterization in 4.3.7 of the tame concealed algebras as minimal representation-infinite algebras is due to Happel-Vossieck [HV]. In [Bo4], Bongartz has characterized the tame concealed algebras of tubular type $(n,2,2)$ and $(m,3,2)$, with $n \geq 2$ and $m = 3,4,5$ as those representation-infinite algebras B which have a preprojective component whose orbit graph is a tree and such that B/BeB is representation-finite for any idempotent e corresponding to a sink or a source of the quiver of B. Also, in this paper, Bongartz asked whether any representation-infinite algebra with a preprojective component "contains" a minimal representation-infinite algebra with a preprojective component. This question is answered in the affirmative in 4.3.6, also. In appendix 2, we reproduce the list of the frames of all tame concealed algebras given by Happel-Vossieck in [HV].

The properties of the special tilting functors considered in section 4.4, though rather elementary, are of great importance. For example, the table of all tame concealed algebras would be too long and incomprehensible if one would try to list the isomorphism classes instead of the frames. The tilted algebras of type $\mathbb{A}_n$, and corresponding branches of algebras have been considered by various authors ([BG], [HR2], [M2]). In particular, 4.4.2 and 4.4.3 may be found both in [BG] and in [HR2].

Sections 4.5-4.7 are elaborations of our joint work with d'Este [ER]. With the exception of lemma 4.6.1, we give full proofs for all results; for a proof of 4.6.1, we refer to [ER]. The notion of a ray vertex is here defined for an arbitrary translation quiver and a ray module may belong to an arbitrary, but standard, component. Proposition 4.5.1 is essentially in [ER], its proof however is now based on lemma 2.5.5. The definition of a tube, and of tubular extensions are taken from [ER]. Note that theorem 4.7.1 is copied from [ER], with only minor modifications.

In section 4.9, we study the endomorphism rings B of tilting modules without non-zero preinjective direct summand over a tame concealed algebra. Note that the structure of the category B-mod has already been determined in [HR2].

In the final section 4.10, we derive some consequences concerning subspace categories of vectorspace categories. In particular, we give the full classification of the indecomposable objects in $\breve{\mathcal{U}}(K,|\cdot|)$, where $(K,|\cdot|)$ is any of the six critical directed vectorspace categories $C(i)$, $1 \leq i \leq 6$. Since the case $C(1)$ is just the case of Kronecker modules discussed in 3.2, we can assume that $(K,|\cdot|)$ is given by a partially ordered set S. In these cases, the classification of the indecomposable objects in $\breve{\mathcal{U}}(S)$ is due to Nazarova: the case $C(2)$ was already solved in [N1], the remaining cases in [N3]. The list of the dimension vectors of the simple regular objects is also in [Z2]. The partially ordered sets occurring as domestic coextensions of a critical partially ordered set have been considered by Otraševskaja [Ot], Bünermann [Bü1,2], and also Zavadskij [Z2]. Otraševskaja has determined their indecomposable representations, and Bünermann the full Auslander-Reiten quiver.

5. Tubular algebras

A tubular extension A of a tame concealed algebra of extension type $\mathbb{T}$ = (2,2,2,2), (3,3,3), (4,4,2), or (6,3,2), will be called a tubular algebra, note that the rank of $K_0(A)$ is 6, 8, 9, or 10, respectively. We are going to determine the complete module structure of any tubular algebra. It will turn out, as the name should suggest, that most of the indecomposable modules of a tubular algebra belong to tubes, that there are many such tubes, and that tubes play a fundamental role for the whole categorical structure of the module category: in particular, we will see that any cycle in the module category occurs inside a tube.

The algebra Δ will be said to be cotubular, provided the opposite algebra A^{op} is tubular. Since for a tubular algebra A, the rank of $K_0(A)$ uniquely determines the tubular extension type, we see that the tubular extension type and the tubular coextension type of a tubular and cotubular algebra A coincide, it will be called the type of A. First, we will consider algebras A which are assumed to be both tubular and cotubular. Later, we will see that actually any tubular algebra also is cotubular.

5.1 $K_0(A)$ <u>for a tubular and cotubular algebra</u> A

(1) Let A be a tubular and cotubular algebra. Assume that A is an extension of A_0, and a coextension of A_∞, where A_0 and A_∞ both are tame concealed algebras. Let h_0 be the positive radical generator of $K_0(A_0)$, and h_∞ the positive radical generator of $K_0(A_\infty)$. Then the pair (h_0,h_∞) is uniquely determined by A, therefore also the pair (A_0,A_∞). The radical $\mathrm{rad}\chi_A$ of χ_A has rank 2, the elements h_0, h_∞ generate a subgroup of $\mathrm{rad}\chi_A$ of finite index, $h_0 + h_\infty$ is sincere, and the non-sincere positive radical vectors are the positive multiples of h_0 and h_∞.

Proof. First, assume that A is actually a tubular extension of A_0, and a tubular coextension of A_∞ (there exist such algebras A_0, A_∞, since A is assumed to be tubular and cotubular). Then, according 4.7.2 and its dual, both h_0, h_∞ belong to $\mathrm{rad}\chi_A$.

Let Δ, Δ_0, Δ_∞ be the quiver of A, A_0, and A_∞, respectively. [This should not lead to any confusion; in the sequel, we will not need any notation for the set of vertices of Δ.] Note that Δ_0, Δ_∞ both are proper subquivers of Δ, since the tubular type of a tame concealed algebra is always a Dynkin graph. Note that the support of h_0 and h_∞ is Δ_0, and Δ_∞, respectively. Since Δ is an extension of Δ_0, all arrows connecting vertices from Δ_0 with vertices outside Δ_0 have their endpoints in Δ_0, similarly, all arrows connecting vertices from Δ_∞ with vertices outside Δ_∞ have their starting point inside Δ_∞. Thus, $\Delta_0 \neq \Delta_\infty$. In particular, h_0, h_∞ generate a subgroup of rank 2. Of course, then $\Delta_\infty \nsubseteq \Delta_0$, since $\mathrm{rad}\chi_{A_0}$ is one-dimensional, and χ_{A_0} is the restriction of χ_A to $K_0(A_0)$, and similarly, $\Delta_0 \nsubseteq \Delta_\infty$. Now, A is obtained from A_0 by making one-point extensions and adding branches in the extension-vertices. If we delete an endpoint a of some branch, or, in case the branch is of length 1, the extension vertex a, then the restriction $A(a)$ of A obtained in this way will be a domestic tubular extension of A_0, thus the corresponding restricted quadratic form $(\chi_A)^a = \chi_{A(a)}$ has a radical of rank 1, according to 4.9.1', and containing h_0, thus $(h_\infty)_a \neq 0$. Since h_∞ is connected, it follows that h_∞ does not vanish outside Δ_0, thus $h_0 + h_\infty$ is sincere. Also, it follows that $\mathrm{rad}\chi_A$ has rank 2, thus the subgroup of $\mathrm{rad}\chi_A$ generated by h_0, h_∞ is of finite index. Therefore, given $h \in \mathrm{rad}\chi_A$, there are $\alpha_0,\alpha_\infty \in \mathbb{Q}$ such that $h = \alpha_0 h_0 + \alpha_\infty h_\infty$. Assume h is not a multiple of h_0 or h_∞, thus $\alpha_0 \neq 0$, $\alpha_\infty \neq 0$. If h is positive, then $\alpha_0 > 0$ (considering $\Delta \setminus \Delta_\infty$), and $\alpha_\infty > 0$ (considering $\Delta \setminus \Delta_0$), thus h is sincere. As a consequence, only the positive multiples of h_0, h_∞ are positive non-sincere radical vectors. In particular, h_0, h_∞ are uniquely determined by A.

Now assume, A is an extension of A_0', with A_0' tame concealed, and let h_0' be the positive radical generator of $K_0(A_0')$. Then $\chi_A(h_0') = \chi_{A_0'}(h_0') = 0$, thus h_0' is a positive, non-sincere radical vector, and therefore it coincides with either h_0 or h_∞. But $h_0' \neq h_\infty$, considering again the direction of arrows, thus $h_0' = h_0$, and therefore $A_0' = A_0$. Similarly, if A is a coextension of some tame concealed algebra A_∞', then $A_\infty' = A_\infty$.

Assume that A is a tubular and cotubular algebra. The elements $h_0, h_\infty \in K_0(A)$ given by (1) will be called the <u>canonical radical elements</u> in $K_0(A)$. Let us quote from 4.4.8 a further property of these elements:

(2) Let A be a tubular and cotubular algebra, with canonical radical elements h_0, h_∞. Assume $A = A_0[E_i,K_i]_{i=1}^t$, with A_0 tame concealed, E_i simple regular A_0-modules and K_i branches, $1 \leq i \leq t$, then $h_\infty \mid K_0(K_i)$ is a positive multiple of the branch length function ℓ_{K_i}. Similarly, for $A = {}_{j=1}^{s}[L_j,F_j]A_\infty$, with A_∞ tame concealed, F_j simple regular A_∞-modules, and L_j branches, $1 \leq j \leq s$, then $h_0 \mid K_0(L_j)$ is a positive multiple of the branch length function ℓ_{L_j}.

According to 4.4.8', we have the following consequence:

(3) Let A be a tubular and cotubular algebra, say an extension of A_0 and a coextension of A_∞ where A_0, A_∞ are tame concealed algebras with canonical radical elements h_0, h_∞. Let M be an indecomposable A-module. If $M \mid A_0 = 0$, then $\langle h_\infty, \underline{\dim}\, M\rangle \neq 0$. If $M \mid A_\infty \neq 0$, then $\langle h_0, \underline{\dim}\, M\rangle \neq 0$.

Assume that A is a tubular and cotubular algebra, with canonical radical elements h_0, h_∞. In order to give a description of the module category A-mod, we will invoke linear forms

$$\langle h,-\rangle : K_0(A) \longrightarrow \mathbb{Z}$$

defined by radical elements h of χ_A. First of all, let

$$\iota_0 = \langle h_0,-\rangle, \quad \iota_\infty = \langle h_\infty,-\rangle.$$

Given $\gamma \in \mathbb{Q} \cup \{\infty\}$, we can write it in the form $\gamma = \dfrac{\gamma_\infty}{\gamma_0}$, where $\gamma_\infty \in \mathbb{Z}$, $\gamma_0 \in \mathbb{N}_0$, and such that γ_0, γ_∞ have no common divisor different from 1, and γ_0, γ_∞ are uniquely determined by γ. Then, let

$$h_\gamma = \gamma_0 h_0 + \gamma_\infty h_\infty ,$$

and

$$\iota_\gamma = \langle h_\gamma,-\rangle = \gamma_0 \iota_0 + \gamma_\infty \iota_\infty .$$

As it turns out, the behaviour of the indecomposable A-modules will be rather different according to the signs of $(\underline{\dim}\, M)\iota_0$ and $(\underline{\dim}\, M)\iota_\infty$. In order to deal with those indecomposable modules satisfying $(\underline{\dim}\, M)\iota_0 > 0$ and $(\underline{\dim}\, M)\iota_\infty < 0$, it will be convenient to consider the ratio of $(\underline{\dim}\, M)\iota_0$ and $(\underline{\dim}\, M)\iota_\infty$. Actually, as we will see later, this ratio is defined for all indecomposable A-modules: always at least one of $(\dim)\iota_0$ and $(\underline{\dim}\, M)\iota_\infty$ is non-zero (see 5.2.1). Given any $x \in K_0(A)$, we will say that the <u>index of</u> x <u>is defined</u> provided not both values $x\iota_0$, $x\iota_\infty$ are zero, and, in this case, we call the quotient

$$\operatorname{index}(x) = -\frac{x\iota_0}{x\iota_\infty}$$

the <u>index</u> of x, it is an element of $\mathbb{Q} \cup \{\infty\}$. The index of the dimension vector of a module M will also be called the index of M, written $\operatorname{index}(M)$. Let us note the following simple observation:

(a) Let $x \in K_0(A)$, and assume its index is defined. Let $\gamma \in \mathbb{Q} \cup \{\infty\}$. Then $\operatorname{index}(x) = \gamma$ if and only if $x\iota_\gamma = 0$.

<u>Proof</u>. First assume, $\operatorname{index}(x) = \gamma = \frac{\gamma_\infty}{\gamma_0}$. By definition, $\operatorname{index}(x) = -\frac{x\iota_0}{x\iota_\infty}$, thus $0 = \gamma_0 x\iota_0 + \gamma_\infty x\iota_\infty = x\iota_\gamma$. Also conversely, if $x\iota_\gamma = 0$, where $\gamma = \frac{\gamma_\infty}{\gamma_0}$, and if the index of γ is defined, then $\gamma_0 x\iota_0 + \gamma_\infty x\iota_\infty = 0$ implies $-\frac{x\iota_0}{x\iota_\infty} = \frac{\gamma_\infty}{\gamma_0} = \gamma$.

We will see in the next section that $h_\infty\iota_0 \neq 0$. Using this, there is the following consequence:

(b) For any non-zero element in the radical of χ_A, its index is defined, namely

$$\operatorname{index}(\alpha_0 h_0 + \alpha_\infty h_\infty) = \frac{\alpha_\infty}{\alpha_0} \quad \text{for } \alpha_0, \alpha_\infty \in \mathbb{Q}, (\alpha_0, \alpha_\infty) \neq (0,0).$$

In particular, $\operatorname{index}(h_\gamma) = \gamma$.

<u>Proof</u>. Let $c = h_\infty\iota_0$. Since h_0 is in $\operatorname{rad}\chi_A$, we also have $\langle h_\infty, h_0\rangle = -\langle h_0, h_\infty\rangle = -c$. Of course, $\langle h_0, h_0\rangle = 0 = \langle h_\infty, h_\infty\rangle$. Thus

$$(*) \qquad (\alpha_0 h_0 + \alpha_\infty h_\infty)\iota_0 = \alpha_\infty c, \quad (\alpha_0 h_0 + \alpha_\infty h_\infty)\iota_\infty = -\alpha_0 c.$$

Under the assumption $c \neq 0$, we therefore see that for $\alpha_0 h_0 + \alpha_\infty h_\infty \neq 0$, not both values $(*)$ are non-zero, and

$$\operatorname{index}(\alpha_0 h_0 + \alpha_\infty h_\infty) = -\frac{\alpha_\infty c}{-\alpha_0 c} = \frac{\alpha_\infty}{\alpha_0}.$$

(c) The intersection of $\mathrm{rad}\chi_A$ with the kernel of ι_γ has rank 1 (and contains h_γ).

<u>Proof</u>. According to (a) and (b), given β and $\gamma \in \mathbb{Q} \cup \{\infty\}$, then h_β lies in the kernel of ι_γ if and only if $\beta = \gamma$. Thus $0 \neq \mathrm{rad}\chi_A \cap \ker \iota_\gamma \neq \mathrm{rad}\chi_A$, and note that $\mathrm{rad}\chi_A/(\mathrm{rad}\chi_A \cap \ker \iota_\gamma)$ is torsion free.

<u>Remark</u>. We should note that in general, h_γ does not generate $\mathrm{rad}\chi_A \cap \ker \iota_\gamma$. A typical example is any canonical tubular algebra A, and $\gamma = 1$.

5.2 The structure of the module category of a tubular algebra

Let A be an algebra which is tubular and cotubular, of type $\mathbb{T}$, say an extension of A_0, and a coextension of A_∞, with A_0, A_∞ both being tame concealed. For any $\gamma \in \mathbb{Q}_0^\infty$, we want to define module classes $\mathcal{P}_\gamma$, $\mathcal{T}_\gamma$, $\mathcal{Q}_\gamma$. Actually, for $\gamma = 0$, the module classes have been defined in 4.7.1; by duality, one similarly has these module classes for $\gamma = \infty$. Let us recall these definitions: $\mathcal{P}_0$ is the module class of all preprojective A_0-modules, $\mathcal{T}_0$ is the module class given by all indecomposable A-modules M with either $M|A_0$ a non-zero regular A_0-module, or else $M|A_0$ being zero, and $(\underline{\dim}\, M)\iota_\infty < 0$; finally, $\mathcal{Q}_0$ is the module class given by all indecomposable A-modules M with either $M|A_0$ a non-zero preinjective A_0-module, or else $M|A_0$ being zero, and $(\underline{\dim}\, M)\iota_\infty > 0$. [Note that we use 5.1.2 , in order to express the inequalities in terms of ι_∞; and we remark that for M in $\mathcal{P}_0$, $\mathcal{T}_0$, or $\mathcal{Q}_0$, we have $(\underline{\dim}\, M)\iota_0 < 0$, $= 0$, or ≥ 0, respectively, and a module M from $\mathcal{Q}_0$ satisfies $(\underline{\dim}\, M)\iota_0 = 0$ only in case $M|A_0 = 0$ and $(\underline{\dim}\, M)\iota_\infty > 0$.] Dually, $\mathcal{Q}_\infty$ is the module class of all preinjective A_∞-modules, $\mathcal{T}_\infty$ is the module class given by all indecomposable A-modules M with either $M|A_\infty$ a non-zero regular A_∞-module, or else $M|A_\infty$ being zero, and $(\underline{\dim}\, M)\iota_0 > 0$; Finally, $\mathcal{P}_\infty$ is the module class given by all indecomposable A-modules M with either $M|A_\infty$ a non-zero preprojective A_∞-module, or else $M|A_\infty$ being zero, and $(\underline{\dim}\, M)\iota_0 < 0$. [Again, we have used 5.1.2, and we remark that for M in $\mathcal{P}_\infty$, $\mathcal{T}_\infty$, or $\mathcal{Q}_\infty$, we have $(\underline{\dim}\, M)\iota_\infty \leq 0$, $= 0$, or > 0, respectively, and a module from $\mathcal{P}_\infty$ satisfies $(\underline{\dim}\, M)\iota_\infty = 0$ only in case $M|A_\infty = 0$ and $(\underline{\dim}\, M)\iota_0 < 0$.] Note that $\mathcal{Q}_0 \cap \mathcal{P}_\infty$ is the module class given by all indecomposable A-modules M with both $(\dim M)\iota_0 > 0$ and $(\underline{\dim}\, M)\iota_\infty < 0$. This module class will have to be investigated rather carefully in this chapter.

We also want to quote the structural results which already have been obtained. Of course, $\mathcal{P}_0$ is a preprojective component, $\mathcal{Q}_\infty$ is a preinjective component. According to 4.7.1, we know that $\mathcal{T}_0$ is a separating tubular $\mathbb{P}_1k$-family, separating $\mathcal{P}_0$ from $\mathcal{Q}_0$, being obtained from the tubular family of all regular A_0-modules by ray-insertion, and being of extension type $\mathbb{T}$. By duality, $\mathcal{T}_\infty$ is a separating tubular $\mathbb{P}_1k$-family, separating $\mathcal{P}_\infty$ from $\mathcal{Q}_\infty$, being obtained from the tubular family of all regular A_∞-modules by coray-insertion, and being of extension type $\mathbb{T}$.

(1) The following module classes are pairwise disjoint:

$$\mathcal{P}_0,\ \mathcal{T}_0,\ \mathcal{Q}_0 \cap \mathcal{P}_\infty,\ \mathcal{T}_\infty,\ \mathcal{Q}_\infty,$$

and they give the complete module category A-mod. In particular,

$$\mathcal{Q}_0 = (\mathcal{Q}_0 \cap \mathcal{P}_\infty) \vee \mathcal{T}_\infty \vee \mathcal{Q}_\infty,\quad \mathcal{P}_\infty = \mathcal{P}_0 \vee \mathcal{T}_0 \vee (\mathcal{Q}_0 \cap \mathcal{P}_\infty)\ ,$$

and, for any indecomposable A-module M, at least one of $(\underline{\dim}\, M)\iota_0$, $(\underline{\dim}\, M)\iota_\infty$ is non-zero.

Proof. Note that $\mathcal{P}_0$ is a preprojective component containing no indecomposable injective module, whereas $\mathcal{T}_\infty$ consists of tubes, and $\mathcal{Q}_\infty$ is a preinjective component. Thus $\mathcal{P}_0 \cap \mathcal{T}_\infty = 0$, $\mathcal{P}_0 \cap \mathcal{Q}_\infty = 0$. By duality, also $\mathcal{T}_0 \cap \mathcal{Q}_\infty = 0$. Next, we want to show $\mathcal{T}_0 \cap \mathcal{T}_\infty = 0$. Let $\mathcal{T}_0(\rho)$ be a tube in $\mathcal{T}_0$. It is obtained by ray insertion from a tube consisting of regular A_0-modules. First, consider the case that $\mathcal{T}_0(\rho)$ contains an indecomposable projective module. Then $\mathcal{T}_0(\rho) \cap \mathcal{T}_\infty = 0$, since the tubes in $\mathcal{T}_\infty$ are obtained from stable tubes by coray insertion, so they do not contain indecomposable projective modules. Second, in case $\mathcal{T}_0(\rho)$ contains no projective module, it is a stable tube consisting of A_0-modules. Also, it will contain an indecomposable A_0-module M with $\underline{\dim}\, M = h_0$. On the other hand, the stable tubes belonging to $\mathcal{T}_\infty$ consist of A_∞-modules. Since $h_0 \notin K_0(A_\infty)$, it follows that $\mathcal{T}_0(\rho)$ cannot belong to $\mathcal{T}_\infty$, thus also in this case $\mathcal{T}_0(\rho) \cap \mathcal{T}_\infty = 0$. This shows that $\mathcal{T}_0 \cap \mathcal{T}_\infty = 0$. Since the module classes $\mathcal{P}_0$, $\mathcal{T}_0$, $\mathcal{Q}_0$ are pairwise disjoint, and also the module classes $\mathcal{P}_\infty$, $\mathcal{T}_\infty$, $\mathcal{Q}_\infty$, it follows that $\mathcal{Q}_0 \cap \mathcal{P}_\infty$ is disjoint to all of $\mathcal{P}_0$, $\mathcal{T}_0$, $\mathcal{T}_\infty$, $\mathcal{Q}_\infty$. Of course, since A-mod $= \mathcal{P}_\infty \vee \mathcal{T}_\infty \vee \mathcal{Q}_\infty$, any indecomposable module from $\mathcal{Q}_0$ belongs to one of $\mathcal{Q}_0 \cap \mathcal{P}_\infty$, $\mathcal{T}_\infty$, $\mathcal{Q}_\infty$, thus A-mod $= \mathcal{P}_0 \vee \mathcal{T}_0 \vee \mathcal{Q}_0$ shows that also A-mod $= \mathcal{P}_0 \vee \mathcal{T}_0 \vee (\mathcal{Q}_0 \cap \mathcal{P}_\infty) \vee \mathcal{T}_\infty \vee \mathcal{Q}_\infty$. This gives the first part of the assertion. In particular, we have $\mathcal{Q}_0 = (\mathcal{Q}_0 \cap \mathcal{P}_\infty) \vee \mathcal{T}_\infty \vee \mathcal{Q}_\infty$, since an indecomposable A-module belongs to $\mathcal{Q}_0$ if and only if it does not belong to $\mathcal{P}_0 \vee \mathcal{T}_0$ if and only if it does belong to $(\mathcal{Q}_0 \cap \mathcal{P}_\infty) \vee \mathcal{T}_\infty \vee \mathcal{Q}_\infty$, and dually $\mathcal{P}_\infty = \mathcal{P}_0 \vee \mathcal{T}_0 \vee (\mathcal{Q}_0 \cap \mathcal{P}_\infty)$.

In order to show that for M indecomposable at least one of the values $(\underline{\dim}\, M)\iota_0$, $(\underline{\dim}\, M)\iota_\infty$ is non-zero, consider the case $M \in \mathcal{P}_\infty$. Then either $(\underline{\dim}\, M)\iota_\infty < 0$ or else $M|A_\infty = 0$ and $(\underline{\dim}\, M)\iota_0 < 0$. Similarly, for $M \in \mathcal{Q}_0$, either $(\underline{\dim}\, M)\iota_0 > 0$ or else $M|A_0 = 0$ and $(\underline{\dim}\, M)\iota_\infty > 0$. Since A-mod $= \mathcal{Q}_0 \vee \mathcal{P}_\infty$, we know that always one of the values is non-zero. This finishes the proof.

Of course, one of the immediate consequences is that $h_\infty\iota_0 \neq 0$, since h_∞ is the dimension vector of an indecomposable module, and $h_\infty\iota_\infty = 0$. Also, we have shown in this way that for any indecomposable A-module M, its index is defined; let us recall that

$$\operatorname{index}(M) = -\frac{(\underline{\dim}\, M)\iota_0}{(\underline{\dim}\, M)\iota_\infty} .$$

The following table lists the sign of index(M) in various cases:

M in	$(\underline{\dim}\, M)\iota_0$	$(\underline{\dim}\, M)\iota_\infty$	index(M)
$\mathcal{P}_0$	< 0	≤ 0	negative or ∞
$\mathcal{T}_0$	$= 0$	< 0	0
$\mathcal{Q}_0 \cap \mathcal{P}_\infty$	> 0	< 0	positive
$\mathcal{T}_\infty$	> 0	$= 0$	∞
$\mathcal{Q}_\infty$	≥ 0	> 0	negative or 0

Thus, the module class which remains to be investigated, $\mathcal{Q}_0 \cap \mathcal{P}_\infty$, can be characterized as being given by the indecomposable A-modules with index in $\mathbb{Q}^+$. And of course we can divide $\mathcal{Q}_0 \cap \mathcal{P}_\infty$ further according to the index

(*)
$$\mathcal{Q}_0 \cap \mathcal{P}_\infty = \bigvee_{\gamma\in\mathbb{Q}^+} \mathcal{T}_\gamma$$

where, for $\gamma \in \mathbb{Q}^+$ we denote by $\mathcal{T}_\gamma$ the module class given by all indecomposable A-modules of index γ. According to 5.1.2, we also may describe $\mathcal{T}_\gamma$ as being given by all indecomposable A-modules M with $(\underline{\dim}\, M)\iota_\gamma = 0$, and, similarly we may define $\mathcal{P}_\gamma$ and $\mathcal{Q}_\gamma$ as being given by the indecomposable A-modules M with $(\underline{\dim}\, M)\iota_\gamma < 0$, or > 0, respectively. Note that for any $\gamma \in \mathbb{Q}^+$, we have

$$\mathcal{P}_\gamma = \mathcal{P}_0 \vee \bigvee_{0\leq\beta<\gamma} \mathcal{T}_\beta \qquad \mathcal{Q}_\gamma = \bigvee_{\gamma<\delta\leq\infty} \mathcal{T}_\delta \vee \mathcal{Q}_\infty .$$

[Namely, let $\gamma = \frac{\gamma_\infty}{\gamma_0}$, with $\gamma_\infty \in \mathbb{Z}$, $\gamma_0 \in \mathbb{N}_0$ and such that γ_0, γ_∞ have no common divisor $\neq 1$. Since $\gamma \in \mathbb{Q}^+$, actually both $\gamma_0, \gamma_\infty \in \mathbb{N}_1$. Using the table above, $\gamma_0(\underline{\dim}\, M)\iota_0 + \gamma_\infty(\underline{\dim}\, M)\iota_\infty$ is < 0 for $M \in \mathcal{P}_0 \vee \mathcal{T}_0$, and is > 0 for $M \in \mathcal{T}_\infty \vee \mathcal{Q}_\infty$. Also, if $M \in \mathcal{T}_\beta$ for some $0 < \beta < \gamma$, then $-\frac{(\underline{\dim}\, M)\iota_0}{(\underline{\dim}\, M)\iota_\infty} = \beta < \frac{\gamma_\infty}{\gamma_0}$ gives after multiplication with γ_0 (> 0) and $(\underline{\dim}\, M)\iota_\infty$ (< 0)

$$-\gamma_0(\underline{\dim}\, M)\iota_0 > \gamma_\infty(\underline{\dim}\, M)\iota_\infty ,$$

thus $(\underline{\dim}\, M)\iota_\gamma < 0$. Similarly, for $\gamma < \delta < \infty$, and $M \in \mathcal{T}_\delta$, we obtain $(\underline{\dim}\, M)\iota_\gamma > 0$. Altogether, we see that $\mathcal{P}_0 \vee \bigvee_{0\leq\beta<\delta} \mathcal{T}_\beta \subseteq \mathcal{P}_\gamma$, and $\bigvee_{\gamma<\delta\leq\infty} \mathcal{T}_\delta \vee \mathcal{Q}_\infty \subseteq \mathcal{Q}_\gamma$. On the other hand, given an indecomposable module M in $\mathcal{P}_\gamma$, then it does not belong to $\mathcal{T}_\gamma$ nor to $\mathcal{Q}_\gamma$, thus also not to any $\mathcal{T}_\delta$ with $\gamma < \delta \leq \infty$, nor to $\mathcal{Q}_\infty$. According to (1) and to (*), it follows that M lies in $\mathcal{P}_0$ or in one of $\mathcal{T}_\beta$, $0 \leq \beta < \delta$. Similarly, any indecomposable module from $\mathcal{Q}_\gamma$ has to lie in one of $\mathcal{T}_\delta$, $\gamma < \delta \leq \infty$ or in $\mathcal{Q}_\infty$. this finishes the proof.]

The crucial result of this chapter is the following:

(2) For any $\gamma \in \mathbb{Q}^+$, the module class $\mathcal{T}_\gamma$ is a (sincere, stable) tubular $\mathbb{P}_1k$-family of type $\mathbb{T}$, separating $\mathcal{P}_\gamma$ from $\mathcal{Q}_\gamma$, and it is controlled by the restriction of χ_A to the set of all elements $x \in K_0(A)$ of index γ.

We should add that of course all components contained in $\mathcal{Q}_0 \cap \mathcal{P}_\infty$ are stable, since the projective modules belong to $\mathcal{P}_0 \vee \mathcal{T}_0$, the injective modules to $\mathcal{T}_\infty \vee \mathcal{Q}_\infty$. Also, as soon as we know that some $\mathcal{T}_\gamma$, $\gamma \in \mathbb{Q}^+$, is a tubular family, separating $\mathcal{P}_\gamma$ from $\mathcal{Q}_\gamma$, then $\mathcal{T}_\gamma$ has to be sincere: [just factor maps from projective modules to injective modules through $\mathcal{T}_\gamma$].

The proof of this result will be given in section 5.7. Note that in this way, we determine completely the structure of all components of A-mod. Before we state this explicitly , let us mention a second result which will be shown at the same time in 5.7.

(3) Theorem. Any tubular algebra is also cotubular.

(4) Theorem. Let A be a tubular algebra of type $\mathbb{T}$. Then A-mod has the following components: first, a preprojective component $\mathcal{P}_0$, then, for any $\gamma \in \mathbb{Q}_0^\infty$, a separating tubular $\mathbb{P}_1k$-family $\mathcal{T}_\gamma$, all but $\mathcal{T}_0$ and $\mathcal{T}_\infty$ being stable of type $\mathbb{T}$, and, finally, a preinjective component $\mathcal{Q}_\infty$.

We may visualize the structure of A-mod as follows:

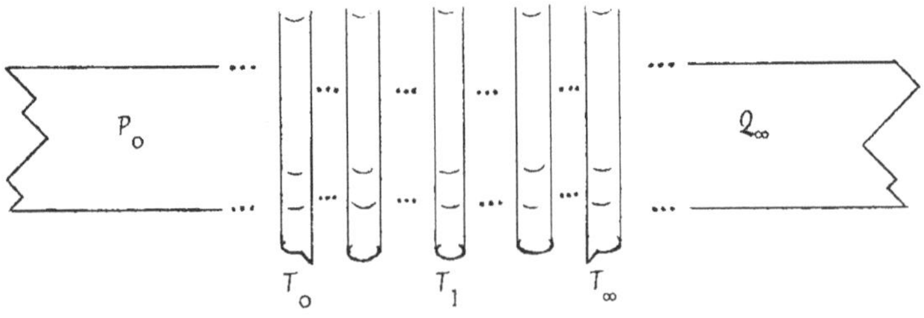

with non-zero maps going only from left to right: given X,Y indecomposable, with $\mathrm{Hom}(X,Y) \neq 0$, then either X,Y belong to the same component, or else $X \in \mathcal{P}_0$, $Y \notin \mathcal{P}_0$, or else $X \notin \mathcal{Q}_\infty$, $Y \in \mathcal{Q}_\infty$, or else $X \in \mathcal{T}_\gamma$, $Y \in \mathcal{T}_\delta$ and $\gamma < \delta$.

Another visualization, may be less personal, is obtained by just drawing in the plane $\mathbb{Z}^2 \subseteq \mathbb{R}^2$ the pairs $(x\iota_0, x\iota_\infty)$ which may occur for the dimension vectors $x = \underline{\dim}\, X$ for the indecomposable A-modules X. Of course, modules with the same index give rise to points lying on the same line through the origin.

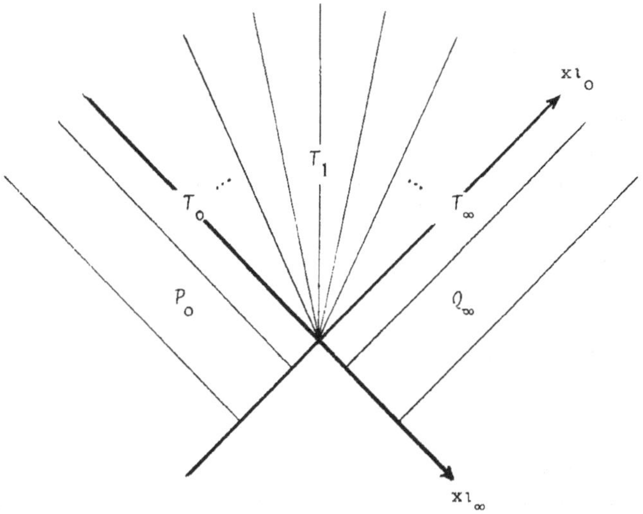

Note that for X indecomposable in P_0, we have $-d_0 \le (\underline{\dim}\, X)\iota_0 \le -1$, with d_0 always being bounded by 6, but usually, we can choose for d_0 even some smaller number, namely the largest component of h_0. [We only have to note that for $X = \tau^{-m}P(a)$, with $m \in \mathbb{N}_0$, and a a vertex of A_0,

$$(\underline{\dim}\, X)\iota_0 = \langle h_0, \underline{\dim}\, \tau^{-m}P(a)\rangle = \langle h_0, p(a)\rangle$$

and $\langle p(a), h_0\rangle$ is just the component with index a of h_0.] Similarly, for Y indecomposable in Q_∞, we have $1 \le (\underline{\dim}\, X)\iota_\infty \le d_\infty$ for some constant $d_\infty \le 6$, and we may choose as d_∞ the largest component of h_∞.

Along the way of proving (2) and (3), we will obtain the following result which in itself seems to be of interest:

(5) Theorem. If A is a tubular algebra, and ${}_AT$ is a tilting module in P_∞, then also $\operatorname{End}({}_AT)$ is a tubular algebra.

The sections 5.3 and 5.7 will be devoted to the proof of the results (2), (3) and (5). Let us give a short outline of the content of these sections. The proof of (2) and (3) will be given in 5.7. In 5.6, we consider some special algebras $C(4,\lambda)$, $C(6)$, $C(7)$, $C(8)$, which are both tubular and cotubular, and which will be shown to satisfy (2). In fact, we will use the main theorem 3.4, in order to construct the tubular family T_1, and then we use shrinking functors in order to shift T_1 to any T_γ, $\gamma \in \mathbb{Q}^+$. These shrinking functors are special tilting functors, they are defined and studied in 5.4. Two additional properties of stable tubular families will be needed there, they are derived in 5.3. The proof of (2) is a reduction to the special algebras considered in 5.6, the reduction being carried out in 5.7. In order to prove (3) for an algebra A, we realize A^{op} as the endomorphism ring of a tilting module $\mathrm{End}(_BT)$, where B is both tubular and cotubular, and $_BT$ is a tilting module in the module class P^B_∞ of B-mod, and use theorem (5), or better, the corresponding assertion 5.5.1. The investigation in section 5.5 of endomorphism rings of tilting modules is carried out under the additional assumption that we deal with an algebra which is known to satisfy assertion (2). Of course, after completion of the proof of (2) and (3), we have obtained, in this way, also a proof of (5).

We are going to derive some consequences of the structure theorem for the module category of a tubular algebra.

(6) Theorem. Let A be a tubular algebra. Then A-mod is controlled by χ_A.

Proof. The indecomposable modules in a preprojective or a preinjective component are directing (2.2.4), thus their dimension vectors are roots (2.4.8). If $_AM$ is an indecomposable module in T_0, then either $_AM$ is an A_0-module, and then $\underline{\dim}\,_AM$ is a root or a radical vector, according to 4.3.3 , or else $\underline{\dim}\,_AM$ is a root according to 4.7.3. By duality, if $_AM$ is an indecomposable module in T_∞, then $\underline{\dim}\,_AM$ is a root or a radical vector. Finally, if $_AM$ is indecomposable in $Q_0 \cap P_\infty$, then $\underline{\dim}\,_AM$ is a root or a radical vector, according to (2).

Conversely, if h is a positive radical vector, then, $h = \alpha_0 h_0 + \alpha_\infty h_\infty$ with $\alpha_0, \alpha_\infty \in \mathbb{Q}_0^+$, not both zero, according to 5.1.1, in particular, the index of h is $\gamma = \frac{\alpha_\infty}{\alpha_0} \in \mathbb{Q}_0^\infty$. If $\gamma = 0$, then h is a positive multiple of h_0, thus there are infinitely many indecomposable A_0-modules H with $\underline{\dim}\,H = h$, according to 4.3.2'; and similarly for $\gamma = \infty$. If $\gamma \in \mathbb{Q}^+$, then we use (2) in order to produce infinitely many indecomposable A-modules H with $\underline{\dim}\,H = h$.

Next, let x be a positive connected root. There can be at most one indecomposable module X with $\underline{\dim}\,X = x$. Namely, let $\underline{\dim}\,X = x$ for some indecomposable X. If $X \in P_0$, then the uniqueness is asserted in 2.4.8, if $X \in T_0$, then it is asserted in 4.7.3. Of course, for $X \in Q_0 \cap P_\infty$, we use (2) (note that for $\gamma \in \mathbb{Q}^+$, the module class T_γ is given by all indecomposable modules of index γ). Using duality, we obtain corresponding unicity results for $X \in T_\infty \vee Q_\infty$.

Now assume there is given a positive connected root x of χ_A. Choose an A-module X with $\underline{\dim}\,X = x$, and with dim End(X) being minimal. Let $X = \bigoplus_{i=1}^{t} X_i$, with all X_i indecomposable. According to 2.3.1, we have $\mathrm{Ext}^1(X_i,X_j) = 0$ for $i \neq j$. Let $x_i = \underline{\dim}\,X_i$. Then, for $i \neq j$, we have $\langle x_i,x_j\rangle \geq 0$, since gl.dim.A ≤ 2, and also $\langle x_i,x_i\rangle = 0$ or 1, according to the first part of the proof. Thus

$$1 = \chi_A(x) = \sum_{i,j} \langle x_i, x_j \rangle$$

shows that at most one x_i is a root. Thus, assume $t \geq 2$, then at least one x_i, say x_t, is a radical vector. Since $\langle x_t, x_j \rangle = -\langle x_j, x_t \rangle$, for any j, it follows that all $\langle x_t, x_j \rangle = 0$. Let γ be the index of x_t. Then $\gamma \in \mathbb{Q}_0^\infty$, and $x_j \iota_\gamma = 0$ for all j. In particular, we also have $x\iota_\gamma = 0$. In case $\gamma \in \mathbb{Q}^+$, it follows from (2) that there is an indecomposable module X' in $\mathcal{T}_\gamma$ satisfying $\underline{\dim}\, X' = x$. Thus, it remains to consider the cases $\gamma = 0$ and $\gamma = \infty$. Let $\gamma = 0$. We have seen before that at most one x_i is a root. However, since not all x_i can be multiples of h_0, we conclude that there also is some i, say $i = 1$, with x_1 being a root. Thus $x = x_1 + \alpha h_0$ for some $\alpha \in \mathbb{N}_1$, and $x_1 \iota_0 = 0$. First, assume the restriction $X_1' = X_1 | A_0$ of X_1 to A_0 is non-zero, let $x_1' = \underline{\dim}\, X_1'$. There is an indecomposable A_0-module X' with $\underline{\dim}\, X' = x_1' + \alpha h_0$ and such that X_1' is a submodule of X', and the pushout

$$\begin{array}{ccc} X_1' & \longrightarrow & X_1 \\ \downarrow & & \downarrow \\ X' & \longrightarrow & Y \end{array}$$

yields an indecomposable module Y with $\underline{\dim}\, Y = x$. Next, let $X_1 | A_0 = 0$. It follows that the support of X_1 is contained in one branch K_z of the extension $A = A_0[E_z, K_z]_z$. Let K_z be a branch of A in ω_z. Note that the component of X_1 in ω_z is zero, since otherwise $\operatorname{Ext}^1(X_1, X_t) \neq 0$. But this implies that $x_1 + \alpha h_0$ is not connected. Similarly, one deals with the case $\gamma = \infty$.

(7) Let us mention several consequences of the structure theorem (4) which show the wealth of information available concerning modules over a tubular algebra. Always, <u>we assume</u> A <u>to be tubular</u>. We denote by $k[T]$ the ring of polynomials in one indeterminate T.

(a) <u>If</u> M <u>is an indecomposable</u> A-<u>module, then</u> $\operatorname{End}(M) = k[T]/\langle T^n \rangle$ <u>for some</u> $n \in \mathbb{N}_1$, $\dim \operatorname{Ext}^1(M,M) \leq \dim \operatorname{End}(M)$, <u>and</u> $\operatorname{Ext}_A^2(M,M) = 0$.

<u>Proof</u>. If M belongs to $\mathcal{P}_0$ or $\mathcal{Q}_\infty$, then apply 2.4.8 or the dual. If M belongs to a standard stable tube $\mathcal{T}$, then $\operatorname{End}(M) = k[T]/\langle T^n \rangle$, since $\mathcal{T}$ is an abelian serial category. Also, if M belongs to a tube $\mathcal{T}$ which is obtained by ray insertion from a stable tube, then $\mathcal{T}$ may be considered as a module class in a larger stable tube $\mathcal{T}'$ (see 4.6), thus $\operatorname{End}(M)$, calculated in $\mathcal{T}'$, is of the indicated form. Dually, one deals with the case of $\mathcal{T}$ being obtained by coray insertions.

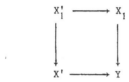

The modules M in $\mathcal{Q}_0$ satisfy inj.dim.$M \leq 1$, according to 2.4.1*, those in $\mathcal{P}_\infty$ satisfy proj.dim.$M \leq 1$, according to 2.4.1. Thus, always $\mathrm{Ext}_A^2(M,M) = 0$ for M indecomposable. Therefore

$$0 \leq \chi_A(\underline{\dim}\, M) = \dim \mathrm{End}(M) - \dim \mathrm{Ext}_A^1\,(M,M).$$

This finishes the proof.

(b) <u>If</u> $(M_1,\ldots,M_n)$ <u>is a cycle in</u> A-mod, <u>then all</u> M_i, $1 \leq i \leq n$, <u>belong to some tube</u> $\mathcal{T}_\gamma(\rho)$.

<u>Proof.</u> First of all, no M_i can belong to $\mathcal{P}_0$ or $\mathcal{Q}_\infty$, since the modules in $\mathcal{P}_0 \vee \mathcal{Q}_\infty$ are directing. Also, if $M_i \in \mathcal{T}_\gamma$, then $\mathrm{Hom}(M_i,M_{i+1})$ implies $M_{i+1} \in \mathcal{T}_\gamma \vee \mathcal{Q}_\gamma$ It follows that all M_i belong to some fixed $\mathcal{T}_\gamma$. Since for different tubes $\mathcal{T}_\gamma(\rho)$, $\mathcal{T}_\gamma(\rho')$ we have $\mathrm{Hom}(\mathcal{T}_\gamma(\rho), \mathcal{T}_\gamma(\rho')) = 0$, the modules M_i actually must belong to one tube.

(c) <u>If</u> M,N <u>are indecomposable</u> A-<u>modules and</u> $\mathrm{Hom}(M,N) \neq 0$, $\mathrm{Ext}^1(M,N) \neq 0$, <u>then</u> M,N <u>belong to some tube</u> $\mathcal{T}_\gamma(\rho)$.

<u>Proof.</u> $0 \neq \mathrm{Ext}^1(M,N) = D\,\overline{\mathrm{Hom}}(N,\tau M)$. Thus, we obtain the cycle $(M,N,\tau M,X)$, where X is indecomposable, with an irreducible map $X \longrightarrow M$.

5.3 Some further properties of stable tubular families

First, we are going to describe the quadratic forms which we encounter when considering standard stable tubular families.

Let us call a free abelian group of finite rank a lattice, and consider quadratic forms on lattices (with values in $\mathbb{Z}$). Given quadratic forms χ_1, χ_2, defined on the lattices U_1, U_2, respectively, we may consider the direct product $\chi_1 \amalg \chi_2$ defined on $U_1 \oplus U_2$, with values $(\chi_1 \amalg \chi_2)(u_1,u_2) = \chi_1(u_1) + \chi_2(u_2)$, where $u_1 \in U_1$, $u_2 \in U_2$. More generally, given quadratic forms χ_s, $1 \leq s \leq t$, with χ_s being defined on U_s, let $\coprod_s \chi_s$ be defined on $\oplus U_s$, with $(\coprod_s \chi_s)(u_1,\ldots,u_t) = \sum_s \chi_s(u_s)$. A sublattice W of the lattice V is called pure, provided V/W is torsionfree, thus again a lattice. We call a non-zero element $x \in U$ primitive, provided $\langle x\rangle$ is a pure subgroup of U. Thus, x is primitive if and only if $x = ny$ for some $n \in \mathbb{N}_1$, $y \in U$, implies $n = 1$. Given a quadratic form χ on the lattice V, and W a pure sublattice in the radical $\mathrm{rad}\chi$ of χ, then χ induces a quadratic form $\bar{\chi}$ on $\bar{V} = V/W$, namely $\bar{\chi}(\bar{v}) = \chi(v)$, where $\bar{v} = v + W$. Now assume χ_1,χ_2 are quadratic forms, defined on U_1,U_2, respectively, and that both are positive semi-definite with radical rank 1, and let h_1,h_2 be radical generators for χ_1,χ_2, respectively. We define the radical product $\chi_1 \amalg_r \chi_2$ to be the quadratic form defined on $U = U_1 \oplus U_2/\langle(h_1,-h_2)\rangle$ by $(\chi_1 \amalg_r \chi_2)\overline{(u_1,u_2)} = \chi_1(u_1) + \chi_2(u_2)$. [Thus $\chi_1 \amalg_r \chi_2$ is the form induced from the direct product $\chi_1 \amalg \chi_2$ of χ_1 and χ_2 on the factor group modulo $\langle(h_1,-h_2)\rangle$, this subgroup being a pure sublattice of $\mathrm{rad}(\chi_1 \amalg \chi_2)$.] Note that the form $\chi_1 \amalg_r \chi_2$ is, up to isometry, independent of the choice of the generators h_1,h_2. Since with χ_1,χ_2 also $\chi_1 \amalg_r \chi_2$ is positive semi-definite with radical rank 1, we may iterate this process of forming radical products. Equivalently, we may define directly the radical product $\coprod_{r,s} \chi_s$ of the positive semi-definite quadratic forms χ_s defined on U_s by forming $U = (\oplus U_s)/W$, where $W \subseteq \oplus_s \langle h_s\rangle$ is the kernel of the linear map $\oplus_s \langle h_s\rangle \longrightarrow \mathbb{Z}$ sending every h_s to 1,, with h_s being a radical generator for χ_s, for any s, and defining $(\coprod_{r,s} \chi_s)\overline{(u_1,\ldots,u_t)} = \Sigma\chi_s(u_s)$. [Again, this construction is, up to isometry, independent of the choice of the generators h_s. Actually, in the cases we will consider, there always will be a canonical choice for h_s. Namely, any U_s will have some fixed basis, with respect to which χ_s will be an integral quadratic form in the sense of 1.0, and the radical of χ_s will be generated by a positive vector. We will choose this positive radical generator as h_s.] We will denote the image of U_s in U again by U_s [note: the canonical map $U_s \longrightarrow U = (\oplus U_s)/W$ is an embedding and the restriction of $\coprod_{r,s} \chi_s$ to the image coincides with the form χ_s on U_s], thus U may be considered as a sum of subspaces U_s such that any two of these subspaces U_s, $U_{s'}$, with $s \neq s'$, intersect in the radical of $\chi = \coprod_{r,s} \chi_s$.

(1) Let χ_s, $1 \leq s \leq t$, be a positive semi-definite quadratic form on the lattice U_s, with radical rank 1. Let $\chi = \underset{s}{\underline{|r|}}\, \chi_s$ be the radical product, defined on U. Then χ is positive semi-definite with radical rank 1, any radical vector of χ belongs to all U_s, and any root of χ belongs to some U_s.

Proof. It remains to be shown that any root of χ belongs to some U_s. Let h_s be the chosen radical generator of χ_s; thus, in U, we have $h_s = h_{s'}$ for all s,s'. Now, let $\overline{(u_1,\dots,u_t)} \in (\oplus U_s)/W = U$ be a root for χ, thus $\Sigma\chi_s(u_s) = 1$. Since all χ_s are positive semi-definite, $\chi_s(u_s) \geq 0$ for all s, thus, there is precisely one s with $\chi_s(u_s) = 1$, and $\chi_i(u_i) = 0$ for $i \neq s$. It follows that for $i \neq s$, the element u_i belongs to $\mathrm{rad}\chi_i = \mathrm{rad}\chi \subseteq U_s$, thus $\overline{(u_1,\dots,u_t)}$ belongs to U_s.

(1') Corollary: Let χ_s, $1 \leq s \leq t$ be a positive semi-definite quadratic form on the lattice U_s, with radical rank 1. Let $\chi = \underset{s}{\underline{|r|}}\,\chi_s$ be the radical product, defined on U. Then the set of roots for χ in U is the disjoint union of the sets of roots for the various χ_s in U_s.

Proof. Any root for χ_s is a root for χ, and in this way, we obtain all roots for χ. Also, since for $s \neq s'$, the subspaces U_s and $U_{s'}$ intersect in $\mathrm{rad}\chi_s = \mathrm{rad}\chi_{s'}$, no root for χ_s coincides with a root for $\chi_{s'}$.

The radical products $\underset{s}{\underline{|r|}}\,\chi_s$ we are mainly interested in are those where all χ_s are quadratic forms of the form $\tilde{\mathbb{A}}_{n_s-1}$, for suitable n_s. Let us denote these forms explicitly. Let $e_i^{(s)}$, $1 \leq i \leq n_s$, be a canonical basis for a copy of $\tilde{\mathbb{A}}_{n_s-1}$, say given by the (not necessarily symmetric) bilinear form

$$\langle e_i^{(s)}, e_j^{(s)} \rangle = \begin{cases} 1 & i = j \\ -1 \quad \text{for} & i \equiv j-1 \pmod{n_s} \\ 0 & \text{otherwise} \end{cases}$$

and U_s the free abelian group with basis $e_i^{(s)}$, $1 \leq i \leq n_s$. Also, let $U^{(s)}$ be the sublattice generated by $e_i^{(s)}$, $2 \leq i \leq n_s$. Note that $h_s = \sum_{i=1}^{n_s} e_i^{(s)}$ belongs to the radical of the quadratic form $\tilde{\mathbb{A}}_{n_s-1}$, and actually is a generator of the radical. Also, $U_s = \langle h_s \rangle \oplus U^{(s)}$. Then, in the lattice U for the radical produced $\chi = \underset{s}{\underline{|r|}}\,\chi_s$, we have the following decomposition

$$U = (\mathrm{rad}\chi) \oplus \bigoplus_{s=1}^{t} U^{(s)} .$$

Let A be a finite-dimensional algebra, and we will assume that its Cartan matrix C_A is invertible. Given a stable tube $\mathcal{T}(\rho)$, say of rank n, and with the indecomposable modules $E_1,\dots,E_n$ lying on the mouth of $\mathcal{T}(\rho)$, then we denote $h(\mathcal{T}(\rho)) = \sum_{i=1}^{n} \underline{\dim}\, E_i$.

(2) Let $\mathcal{T}$ be the standard stable tubular family in A-mod, such that all indecomposable modules $M \in \mathcal{T}$ satisfy proj.dim.M = 1 = inj.dim.M. Assume that $\mathcal{T}$ is of type $\mathbb{T}_{n_1,\dots,n_t}$, say with non-homogeneous tubes $\mathcal{T}(\rho_s)$ of rank n_s, where $1 \le s \le t$, and let $E_i^{(s)}$, $1 \le i \le n_s$ be the indecomposable A-modules lying on the mouth of $\mathcal{T}(\rho_s)$. Let V be the sublattice of $K_0(A)$ generated by the dimension vectors $\underline{\dim}\, M$, with $M \in \mathcal{T}$, let $V^{(s)}$ be the sublattice of V generated by $\underline{\dim}\, E_i^{(s)}$, $2 \le i \le n_s$, and let $W = V \cap \operatorname{rad}\chi_A$. Then W is generated by the vectors $h(\mathcal{T}(\rho))$, with $\mathcal{T}(\rho)$ a tube in $\mathcal{T}$, $\dim V^{(s)} = n_s - 1$, and $V = W \oplus \bigoplus_{s=1}^{t} V^{(s)}$.

Proof. We may assume that $\tau E_i^{(s)} = E_{i-1}^{(s)}$, for all $1 \le i \le n_s$, with $E_0^{(s)} \approx E_{n_s}^{(s)}$, and let $e_i^{(s)} = \underline{\dim}\, E_i^{(s)}$. According to 3.1.3', the restriction of χ_A to $V^{(s)}$ is the quadratic form $\mathbb{A}_{n_s-1}$, thus $\dim V^{(s)} = n_s-1$. Since $\mathcal{T}$ is standard, we have $\operatorname{Hom}(\mathcal{T}(\rho_s),\mathcal{T}(\rho_{s'})) = 0 = \operatorname{Ext}_A^1(\mathcal{T}(\rho_s),\mathcal{T}(\rho_{s'}))$, for $s \neq s'$, thus, since proj.dim.$E_i^{(s)} = 1$, we have

$$\langle e_i^{(s)}, e_j^{(s')}\rangle = \dim \operatorname{Hom}(E_i^{(s)},E_j^{(s')}) - \dim \operatorname{Ext}^1(E_i^{(s)},E_j^{(s')}) = 0$$

for all i, j, provided $s \neq s'$, thus $\langle V^{(s)},V^{(s')}\rangle = 0$ for $s \neq s'$, and therefore, the sum of the sublattices $V^{(s)}$, $1 \le s \le t$, is direct. Also, the restriction of χ_A to any $V^{(s)}$, thus also to $\oplus V^{(s)}$, is positive definite, thus $W \cap (\bigoplus_{s=1}^{t} V^{(s)}) = 0$.

Let W' be the sublattice of V generated by all $h(\mathcal{T}(\rho))$, with $\mathcal{T}(\rho)$ a tube in $\mathcal{T}$. By construction, $h(\mathcal{T}(\rho))$ is the dimension vector of some module M_ρ in $\mathcal{T}$ with $\tau M_\rho \approx M_\rho$ [namely, $M_{\rho_s} = \bigoplus_{s=1}^{n_s} E_i^{(s)}$, in case $\rho = \rho_s$, and M_ρ the indecomposable module on the mouth of $\mathcal{T}(\rho)$, in case $\mathcal{T}(\rho)$ is of rank 1]. According to 2.4.4 and 2.4.d, we see that $h(\mathcal{T}(\rho))$ belongs to $\operatorname{rad}\chi_A$, thus $W' \subseteq W$. We claim that $V \subseteq W' + \sum_{s=1}^{t} V^{(s)}$, thus $W' = W$ and $V = W \oplus (\bigoplus_{s=1}^{t} V^{(s)})$. For the proof, consider any arbitrary element z of V, it is of the form $z = \underline{\dim}\, X - \underline{\dim}\, Y$, with X, Y in $\mathcal{T}$. Let $X = X' \oplus \left(\bigoplus_{s=1}^{t} X^{(s)}\right)$, and $Y = Y' \oplus \left(\bigoplus_{s=1}^{t} Y^{(s)}\right)$ with $X^{(s)}, Y^{(s)} \in \mathcal{T}(\rho_s)$, for $1 \leq s \leq t$, and X', Y' being the direct sum of modules belonging to tubes of rank 1. Note that $h' = \underline{\dim}\, X' - \underline{\dim}\, Y'$ belongs to W', and denote $h_s = h(\mathcal{T}(\rho_s))$. Now, $\underline{\dim}\, X^{(s)}$, $\underline{\dim}\, Y^{(s)}$ are integral linear combinations of the vectors $e_i^{(s)}$, therefore also $\underline{\dim}\, X^{(s)} - \underline{\dim}\, Y^{(s)}$, say

$$\underline{\dim}\, X^{(s)} - \underline{\dim}\, Y^{(s)} = \sum_{i=1}^{n_s} \beta_i^{(s)} e_i^{(s)}, \quad \text{with } \beta_i^{(s)} \in \mathbb{Z} .$$

Thus we have

$$\begin{aligned} z &= \underline{\dim}\, X - \underline{\dim}\, Y \\ &= \underline{\dim}\, X' - \underline{\dim}\, Y' + \sum_{s=1}^{t} \sum_{i=1}^{n_s} \beta_i^{(s)} e_i^{(s)} \\ &= h' + \sum_{s=1}^{t} \beta_1^{(s)} h_s + \sum_{s=1}^{t} \sum_{i=2}^{n_s} (\beta_i^{(s)} - \beta_1^{(s)}) e_i^{(s)} . \end{aligned}$$

Note that the summands h' and $\sum_{s=1}^{t} \beta_1^{(s)} h_s$ belong to W', and the remaining summands are in $\sum_{s=1}^{t} V^{(s)}$. This shows that $V \subseteq W' + \sum_{s=1}^{t} V^{(s)}$, and finishes the proof.

(2') Corollary. Let $\mathcal{T}$ be a (non-zero) standard stable tubular family in A-mod, such that all indecomposable modules $M \in \mathcal{T}$ satisfy proj.dim.M = 1 = inj.dim.M. Let V be the sublattice of $K_0(A)$ generated by the dimension vectors $\underline{\dim}\, M$, with $M \in \mathcal{T}$. If $\mathcal{T}$ is of type $\mathbb{T}_{n_1,\ldots,n_t}$, then the rank of V is $\geq 1 - t + \sum_{s=1}^{t} n_s$.

Proof. Since there is at least one tube $\mathcal{T}(\rho)$ in $\mathcal{T}$, we have $0 \neq h(\mathcal{T}(\rho)) \in V \cap \operatorname{rad}\chi_A = W$, thus W has rank at least 1. The assertion now follows from the decomposition $V = W \oplus \bigoplus_{s=1}^{t} V^{(s)}$ given in (2).

(2") Corollary. Let $\mathcal{T}$ be a standard stable tubular family in A-mod, such that all indecomposable modules $M \in \mathcal{T}$ satisfy proj.dim.M = 1 = inj.dim.M. Then almost all tubes in $\mathcal{T}$ are homogeneous.

Proof. If $\mathcal{T}(\rho_1), \ldots, \mathcal{T}(\rho_t)$ are pairwise different non-homogeneous tubes in $\mathcal{T}$, say with $\mathcal{T}(\rho_s)$ being of rank $n_s \geq 2$, then $t \leq \sum_{s=1}^{t} (n_s - 1) \leq -1 + \operatorname{rank} K_0(A)$, according to (2').

(3) Let $\mathcal{T}$ be a standard stable tubular family in A-mod of type $\mathbb{T}_{n_1,\dots,n_t}$, say with non-homogeneous tubes $\mathcal{T}(\rho_s)$ of rank n_s, where $1 \leq s \leq t$, and assume proj.dim.M = 1 = inj.dim.M, for any indecomposable module M in $\mathcal{T}$. Let U be a subspace of $K_0(A)$ of rank $1 + \sum_{s=1}^{t} (n_s-1)$, and assume $\underline{\dim}\, M \in U$ for any $M \in \mathcal{T}$. Then the following assertions are equivalent:

(i) If x is a connected positive root or a connected positive radical vector for χ_A, and x belongs to U, then $x = \underline{\dim}\, X$ for some indecomposable module $X \in \mathcal{T}$.

(i') If x is a root or a non-zero radical vector for χ_A, and x belongs to U, then $x = \underline{\dim}\, X$, or $x = -\underline{\dim}\, X$ for some indecomposable module $X \in \mathcal{T}$.

(ii) The vectors $h(\mathcal{T}(\rho_s))$, $1 \leq s \leq t$, are primitive in U, and the sublattice of U generated by the roots and the radical vectors for χ in U coincides with the sublattice of U generated by the dimension vectors $\underline{\dim}\, M$, with $M \in \mathcal{T}$.

Proof. Denote by V the sublattice of U generated by all dimension vectors $\underline{\dim}\, M$, with $M \in \mathcal{T}$. Also, let V' denote the sublattice of U generated by all roots and all radical vectors for χ in U. According to 3.1.3', $\chi_A(\underline{\dim}\, M) = 0$ or 1 for any indecomposable module in $\mathcal{T}$, thus $V \subseteq V' \subseteq U$.

For any $1 \leq s \leq t$, let $E_1^{(s)},\dots,E_{n_s}^{(s)}$ be the indecomposable A-modules lying on the mouth of $\mathcal{T}(\rho_s)$, with $\tau E_i^{(s)} = E_{i-1}^{(s)}$, for all $1 \leq i \leq n_s$, and $E_0^{(s)} \approx E_{n_s}^{(s)}$, and let $V^{(s)}$ be the sublattice of V generated by the vectors $e_i^{(s)} = \underline{\dim}\, E_i^{(s)}$, with $2 \leq i \leq n_s$, and $W = V \cap \operatorname{rad}\chi_A$. Note that $h(\mathcal{T}(\rho)) \in W$, for all tubes $\mathcal{T}(\rho)$ in $\mathcal{T}$, and let us denote $h(\mathcal{T}(\rho_s))$ just by h_s, $1 \leq s \leq t$. We have the direct decomposition

$$(*) \qquad V = W \oplus \bigoplus_{s=1}^{t} V^{(s)}, \text{ and } \dim V^{(s)} = n_s-1.$$

Since $0 \neq h_s \in W$, it follows that rank $W \geq 1$. However $V \subseteq U$, implies

$$1 + \Sigma(n_s-1) \leq \text{rank } W + \Sigma(n_s-1) = \text{rank } V \leq \text{rank } U = 1 + \Sigma(n_s-1),$$

thus rank W = 1, and V is of finite index in U. Let h be a generator for W. We can assume that h_1 is a positive multiple of h, say $h_1 = a_1 h$, for some $a_1 \in \mathbb{N}_1$. It follows that h is a positive connected vector in $K_0(A)$. Therefore, if we write $h_s = a_s h$, then also $a_s \in \mathbb{N}_1$, for $2 \leq s \leq t$. Also, the support of h contains the support of any $x \in V$. Namely, according to (*), we only have to consider $x = e_i^{(s)}$. However, in $K_0(A)$, we have $0 < e_i^{(s)} \leq h_s$, thus the support of $e_i^{(s)}$ is contained in the support of h_s, and h and h_s have the same support. Now, let us show the various implications:

(i) $\Rightarrow$ (ii): We have to show $V = V'$. Thus, let x be a non-zero radical vector of χ in U. Since V is of finite index in U, it follows that x is a rational multiple of h, thus x is connected and either positive or negative. Thus, according to (i), there is an indecomposable representation X in T with $\underline{\dim}\, X$ either $= x$ or $= -x$, but always it follows that $x \in V$. Similarly, let x be a root of χ in U. Adding some $n h_1$, for some large $n \in \mathbb{N}_1$, to x, we can achieve that $x + n h_1$ is positive. Also, since a non-zero multiple of x lies in V, the support of x is contained in the support of h_1, thus with h_1 also $x + n h_1$ is connected. Since $x + n h_1$ is a positive connected root in U, there is some indecomposable module X' in T with $\underline{\dim}\, X' = x + n h_1$. Since $\underline{\dim}\, X'$ and h_1 both lie in V, also x lies in V.

Next, we have to show that all h_s are primitive. Consider $e_1^{(s)} + h$. This is a positive connected root for χ in U, thus $e_1^{(s)} + h = \underline{\dim}\, Y$ for some indecomposable module Y in T. Now $e_1^{(s)} + h = (a_s+1)h - \sum_{i=2}^{n_s} e_i^{(s)}$ lies in $W \oplus V^{(s)}$, and neither in W nor in $V^{(s)}$. Since $\underline{\dim}\, Y \in (W \oplus V^{(s)}) \setminus W$, it follows that $Y \in T(\rho_s)$. Therefore Y is a linear combination of the $e_i^{(s)}$, $1 \le i \le n_s$, with non-negative integral coefficients, say $e_1^{(s)} + h = \sum_{i=1}^{n_s} \gamma_i e_i^{(s)}$ for some $\gamma_i \in \mathbb{N}_0$, and $\gamma_1 \geq 1$, since $e_1^{(s)} + h \notin V^{(s)}$. Therefore we have

$$\sum_{i=1}^{n_s} e_i^{(s)} = a_s h = a_s(\gamma_1 - 1)\, e_1^{(s)} + \sum_{i=2}^{n_s} a_s \gamma_i e_i^{(s)},$$

and since the vectors $e_i^{(s)}$, $1 \le i \le n_s$, are linearly independent, we obtain $a_s(\gamma_1 - 1) = 1$, and $a_s \gamma_i = 1$ for all $i \geq 2$, thus $a_s = 1$, and $h_s = h$.

However, h is primitive. Namely, if $h = n h'$ for some $h' \in U$, $n \in \mathbb{N}_1$, then with h also h' is a positive connected radical vector, thus the dimension vector of some indecomposable module in T, according to (i). Therefore $h' \in V \cap \operatorname{rad}\chi = W$, thus h' is a multiple of h, thus $n = 1$. This finishes the proof of one implication.

(ii) $\Rightarrow$ (i'). Since the vectors h_s are primitive, and positive multiples of h, it follows that $h_s = h$ for all s. As a consequence, the decomposition (*) shows that $\chi_A | V$ is the quadratic form $\underline{\lfloor r \rfloor}_{s}\, \tilde{\mathbb{A}}_{n_s - 1}$. Given any non-zero radical vector x in U, it is a multiple of h_1, thus either x or $-x$ is the dimension vector of some indecomposable module from $T(\rho_1)$. Similarly, if x is a root for χ in U then x belongs to V, and actually belongs to $W \oplus V^{(s)}$, for some s, according to (1). But a root for $\tilde{\mathbb{A}}_{n_s - 1}$ is either positive or negative, and a positive root for $\tilde{\mathbb{A}}_{n_s - 1}$ is of the form $\sum_{i=a}^{b} e_i^{(s)}$, where $a \le b$ are integers, and where $e_i^{(s)} = e_j^{(s)}$ for $i \equiv j \pmod{n_s}$. Of course, $\sum_{i=a}^{b} e_i^{(s)}$ is just the dimension vector of the indecomposable module $E_a^{(s)}[b-a+1]$ in $T(\rho_s)$, with $T(\rho_s)$-socle $E_a^{(s)}$ and $T(\rho_s)$-length $b-a+1$. This finishes the proof of the implication (ii) $\Rightarrow$ (i').

The implication (i') ⇒ (i) is trivial: if x is positive, then $-x$ cannot be the dimension vector of any module. This finishes the proof.

(4) Let $\mathcal{T}$ be a standard stable tube in A-mod, and let $\mathcal{M}$ be a module class in $\mathcal{T}$ closed under extensions and under $\mathcal{T}$-submodules. Assume that there exists an indecomposable module M in $\mathcal{M}$ with $\mathrm{Ext}_A^1(M,M) \neq 0$. Then, any module in $\mathcal{T}$ is generated by a module in $\mathcal{M}$, and given a module P with $\mathrm{Hom}(\mathcal{T},P) = 0$, then any map from P to a module in $\mathcal{T}$ factors through $\mathcal{M}$.

<u>Proof.</u> Let $\mathcal{T}$ be of rank n, say with indecomposable modules $E_1,\dots,E_n$ lying on the mouth of $\mathcal{T}$, with $\tau E_z = E_{z-1}$, $1 \le z \le n$, where $E_0 \approx E_n$. The indecomposable module in $\mathcal{T}$ with $\mathcal{T}$-socle E_z and $\mathcal{T}$-length ℓ will be denoted by $E_z[\ell]$. Also, let $E_z[0] = 0$. By assumption, $\mathcal{T}$ is standard, thus $\mathcal{T} \approx k(\Omega(n))$, and we may calculate dimensions of Hom-groups in $k(\Omega(n))$. In this way, we see that for $\ell < n$,

$$\mathrm{Ext}^1(E_z[\ell],E_z[\ell]) = D\,\overline{\mathrm{Hom}}(E_z[\ell],E_{z-1}[\ell]) = 0,$$

where we use 2.4.5. Thus, the $\mathcal{T}$-length of the given module M is $\ge n$.

First, we note that if $\mathcal{M}$ contains some indecomposable module $E_z[\ell]$, with $\ell \ge n$, then also $E_z[\ell+n]$ is in $\mathcal{M}$, since there is an exact sequence of the form

$$0 \to E_z[\ell] \to E_z[\ell+n] \oplus E_{z+n}[\ell-n] \to E_{z+n}[\ell] \to 0.$$

[Namely, the Auslander-Reiten sequences in $\mathcal{T}$ are of the form

$$0 \to E_z[\ell] \to E_z[\ell+1] \oplus E_{z+1}[\ell-1] \to E_{z+1}[\ell] \to 0$$

for all z, and all $\ell \ge 1$, thus we obtain a commutative square

$$\begin{array}{ccccc} & & E_z[\ell+1] & & \\ & \nearrow & & \searrow & \\ E_z[\ell] & & & & E_{z+1}[\ell] \\ & \searrow & & \nearrow & \\ & & E_{z+1}[\ell-1] & & \end{array}$$

which is both a pullback and a pushout. Fitting together the various squares starting at $E_{z'}[\ell']$, with $z \le z' < z+n$, and $z+\ell \le z'+\ell' < z+n-1$, we obtain a big square, both being pullback and pushout, thus giving the desired exact sequence]. Actually, since $\mathcal{M}$ is closed under submodules, it follows that $\mathcal{M}$ contains all modules in $\mathcal{T}$ with the same $\mathcal{T}$-socle as M.

Thus, let E_z be the $\mathcal{T}$-socle of M, thus all $E_z[\ell]$, $\ell \in \mathbb{N}_1$, are contained in $\mathcal{M}$. Given an arbitrary module T in $\mathcal{T}$, say with $\mathcal{T}$-top E_s, then N is an epimorphic image of any module $E_z[\ell]$ with $z+\ell-1 \equiv s(\mathrm{mod}\ n)$ and $\ell \ge \mathcal{T}$-length of T.

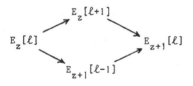

Finally, assume there is given a module P with $\mathrm{Hom}(\mathcal{T},P) = 0$, and a map $f : P \to T$, with T in $\mathcal{T}$. There is an epimorphism $\varepsilon : M' \to T$ with $M' \in \mathcal{M}$, and its kernel T' lies again in $\mathcal{T}$, since $\mathcal{T}$ is an abelian subcategory (T' even lies in $\mathcal{M}$). Now, $\mathrm{Ext}^1(P,T') = D\,\underline{\mathrm{Hom}}(\tau^- T',P)$, according to 2.4.5, but with T' also $\tau^- T'$ belongs to $\mathcal{T}$, thus, $\mathrm{Hom}(\tau^- T',P) = 0$, by assumption, and therefore the map f lifts along $\varepsilon : M' \to T$. This finishes the proof.

Let us add a related result which will be used in section 5.5.

(5) Let $\mathcal{T}$ be a standard, stable tubular family in A_0-mod, and $A = A_0[E_i,K_i]_{i=1}^t$ a tubular extension of A_0 using modules from $\mathcal{T}$. Let $\mathcal{T}(\rho)$ be a tube in $\mathcal{T}$, and $\mathcal{T}_0(\rho)$ the corresponding tube in A-mod (with $\mathcal{T}(\rho) = \mathcal{T}_0(\rho) \cap A_0$-mod). Let $\mathcal{M}$ be a modules class in $\mathcal{T}_0(\rho)$ closed under extensions and under factor modules inside $\mathcal{T}_0(\rho)$, and assume $\mathcal{M}$ contains a cycle. Then $\mathcal{M}$ contains an indecomposable module with dimension vector $h(\mathcal{T}(\rho))$.

<u>Proof.</u> Let $\mathcal{T}(\rho)$ be of rank n, thus $\Gamma\mathcal{T}(\rho) = \Omega(n)$. We know that $\mathcal{T}_0(\rho)$ is a standard component and $\Gamma\mathcal{T}_0(\rho) = \Omega(n)\,[e_z,B_z]_{z=1}^n$, where $e_1,\dots,e_n$ are the vertices on the mouth of $\Omega(n)$, and $B_1,\dots,B_n$ are (may be empty) branches (see 4.7.1, but note that for this assertion, it is sufficient to assume $\mathcal{T}$ to be standard, and not necessarily separating; or use directly 4.5.1). Let us denote $\Gamma = \Omega(n)[e_z,B_z]_{z=1}^n$, and let Γ' be the translation subquiver $\Gamma' = \Omega(n)[e_z,B_z^s]_{z=1}^n$ of Γ, as in 4.6.4, with B_z^s being the subspace branch of the same length as B_z. Note that any cyclic path in Γ actually runs inside Γ'. Let M_1,M_2 be two indecomposable modules in $\mathcal{T}_0(\rho)$, with $[M_1],[M_2] \in \Gamma'$, say $[M_1] = (\bar{z}_1,i_1,j_1)$, $[M_2] = (\bar{z}_2,i_2,j_2)$, in the notation of 4.6.4, where $z_1,z_2 \in \mathbb{Z}/n\mathbb{Z}$; $i_1,i_2 \in \mathbb{N}_1$, and $0 \le j_1 \le |B_{z_1}|$, $0 \le j_2 \le |B_{z_2}|$. Let us assume that in Γ', there exists a path from $[M_1]$ to some vertex $y \in \Gamma'$ along a coray, followed by a path of length ℓ, from y to $[M_2]$ along a ray. In particular, $y = (\bar{z}_2,i_2-\ell,j_2)$. Let M_1', M_2' be indecomposable modules in $\mathcal{T}_0(\rho)$ with $[M_2'] = (\bar{z}_2,i_2',j_2)$, and $[M_1'] = (\bar{z}_1,i_1+\ell,j_1)$. We claim that there exists an exact sequence in A-mod of the form

$$0 \to M_1 \to M_1' \oplus M_2' \to M_2 \to 0.$$

Observe that we obtain a commutative square

$$\begin{array}{ccccc} & & M_1' & & \\ & \nearrow & & \searrow & \\ M_1 & & & & M_2 \\ & \searrow & & \nearrow & \\ & & M_2' & & \end{array}$$

which is both a pullback and a pushout by fitting together various squares which stem from Auslander-Reiten sequences. [Namely, denote the indecomposable modules in $\mathcal{T}_0(\rho)$ just by the lables in Γ. Any arrow $\alpha : (\bar{z}_3,i_3,j_3) \to (\bar{z}_4,i_4,j_4)$ in Γ'

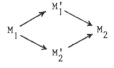

pointing to the mouth gives a pullback - pushout square

$$(*) \qquad \begin{array}{ccccc} & & (\bar{z}_3, i_3+\ell, j_3) & & \\ & \nearrow & & \searrow & \\ (\bar{z}_3, i_3, j_3) & & & & (\bar{z}_4, i_4+\ell, j_3) \\ & \searrow & & \nearrow & \\ & & (\bar{z}_4, i_4, j_4) & & \end{array}$$

by fitting together the squares arising from Auslander-Reiten sequences starting at the vertices $(\bar{z}_3, i_3+\ell', j_3)$, with $0 \leq \ell' < \ell$. Now, there is a path from $(\bar{z}_1, i_1, j_1)$ to (z_2, i_2', j_2) along a coray, thus using only arrows pointing to the mouth. Any arrow gives rise to a corresponding square (*), and they fit together in order to form the desired large pullback-pushout square.] Note that if M_1, M_2 belong to $\mathcal{M}$, then also M_1', M_2' belong to $\mathcal{M}$, since $\mathcal{M}$ is assumed to be closed under extensions. Now, let $(M_1, M_2, \ldots, M_n)$ be a cycle in $\mathcal{M}$, thus all M_i are indecomposable, $\operatorname{rad}(M_i, M_{i+1}) \neq 0$, for $1 \leq i \leq n$, with $M_{n+1} = M_1$. Since $\operatorname{rad}(M_i, M_{i+1}) \neq 0$, there is a path from $[M_i]$ to $[M_{i+1}]$, and at least one of these paths must contain an arrow pointing to infinity. Let i be the first index such that there is a path from M_i to M_{i+1} using at least one arrow pointing to infinity. Since $\operatorname{rad}(M_i, M_{i+1}) \neq 0$, there is a path from $[M_i]$ to some y along a coray, followed by a path of length $\ell \geq 1$ from y to $[M_{i+1}]$ along a ray. Thus, there is a path from M_1 to y along a coray, followed by a path of length ℓ from y to $[M_{i+1}]$ along a ray. The previous considerations show that with M_1 also an indecomposable module M_1' belongs to $\mathcal{M}$, where $[M_1] = (\bar{z}_1, i_1, j_1)$, and $[M_1'] = (\bar{z}_1, i_1+\ell, j_1)$, and with a cycle $(M_1', M_{i+1}, \ldots, M_n, M_1)$. By induction, we see that we obtain in $\mathcal{M}$ indecomposable modules M with $[M] = (\bar{z}_1, i, j_1)$, with i arbitrarily large. Now M has a submodule M' with $M' = (\bar{z}_1, 1, j_1)$, and the factor module M/M' again belongs to $\mathcal{T}_0(\rho)$, thus to $\mathcal{M}$. In case $i \geq 2$, we have $[M/M'] = (\overline{z_1+1}, i-1, 0)$, in particular, it is an A_0-module. Note that M/M' is an indecomposable module in the stable tube $\mathcal{T}(\rho)$ of $\mathcal{T}(\rho)$-length $i-1$. Thus, if $i \geq n+1$, then M/M' has as factor module in $\mathcal{T}(\rho)$ an indecomposable module M'' in $\mathcal{T}(\rho)$ of $\mathcal{T}(\rho)$-length n, thus $\underline{\dim}\, M'' = h(\mathcal{T}(\rho))$. Of course, also M'' belongs to $\mathcal{M}$. This finishes the proof.

Even the special case of $A = A_0$ will be of interest, let us note this explicitly:

(5') Let $\mathcal{T}(\rho)$ be a standard stable tube in A-mod, and $\mathcal{M}$ a module class in $\mathcal{T}(\rho)$ closed under extensions and under factor modules inside $\mathcal{T}(\rho)$, and assume $\mathcal{M}$ contains a cycle. Then $\mathcal{M}$ contains an indecomposable module with dimension vector $h(\mathcal{T}(\rho))$.

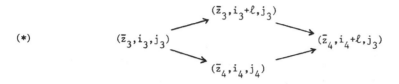

5.4 Shrinking functors

In the next sections, we will have to deal rather frequently with composition of functors, and here it seems convenient to denote the image of an object X under the functor F by XF, thus to denote the composition of the functors $F : K \longrightarrow K'$, $F' : K' \longrightarrow K''$ by FF'. We will use this convention for all functors, in particular also for the Auslander-Reiten translation τ, and the duality functor D.

In this section, A will always be a tubular extension of a tame concealed algebra A_o. A multiplicity-free tilting A-module $_AT = T_o \oplus T_p$, with T_o a preprojective tilting A_o-module, and T_p a projective A-module, will be called a shrinking module, and the corresponding functor $\Sigma = \Sigma_T = \mathrm{Hom}_A({}_AT_B,-) : A\text{-mod} \longrightarrow B\text{-mod}$, with $B = \mathrm{End}({}_AT)$ will be called a left shrinking functor. Note that T_p is uniquely determined by A_o and A, it is just the direct sum of all indecomposable projective A-modules P(a), with a outside A_o [Namely, no indecomposable direct summand T' of T_p can be an A_o-module, since T' would have to belong to $\langle T_o\rangle$. Since the number of indecomposable summands of T_o is rank $K_o(A_o)$, that of $T_o \oplus T_p$ is rank $K_o(A)$, it follows that any indecomposable A-module P(a) with a outside of A_o has to belong to $\langle T_p\rangle$]. Also, conversely, given any preprojective tilting A_o-module T_o, define T_p to be the direct sum of all indecomposable projective A-modules P(a) with a outside A_o, then $_AT = T_o \oplus T_p$ is a shrinking module [see 4.7.4]. We therefore may say that the corresponding left shrinking functor $\Sigma = \mathrm{Hom}_A({}_AT_B,-)$ is defined by T_o. Note that, if $\Sigma : A\text{-mod} \longrightarrow B\text{-mod}$ is a left shrinking functor defined by the A_o-module T_o, where A is a tubular extension of the tame concealed algebra A_o, then B is a tubular extension (of the same extension type as A) of the tame concealed algebra $B_o = \mathrm{End}(T_o)$. In this case, denote by h_o^A, h_o^B the minimal positive radical generators of A_o, and B_o, respectively, and let $\iota_o^A = \langle h_o^A,-\rangle : K_o(A) \longrightarrow \mathbf{Z}$, $\iota_o^B = \langle h_o^B,-\rangle : K_o(B) \longrightarrow \mathbf{Z}$. If we denote by $\sigma = \sigma_T : K_o(A) \longrightarrow K_o(B)$ the linear transformation corresponding to the tilting functor $\Sigma = \Sigma_T$ (see 4.1.7), then

$$h_o^A\sigma = h_o^B, \qquad \iota_o^A = \sigma\,\iota_o^B .$$

[Namely, the restriction of Σ to A_o-mod is just the tilting functor $\mathrm{Hom}_{A_o}({}_{A_o}T_{oB_o},-)$ from A_o-mod to $B_o\text{-mod} \subseteq B\text{-mod}$, and $h_o^A\sigma$ is positive again, thus $h_o^A\sigma = h_o^B$. Also, for any $x \in K_o(A)$, we have $x\iota_o^A = \langle h_o^A,x\rangle = \langle h_o^A\sigma,x\sigma\rangle = \langle h_o^B,x\sigma\rangle = x\sigma\iota_o^B$.]

Example. Let A be a tubular extension of a tame concealed algebra A_o. Then there is a left shrinking functor from A-mod to B-mod, where B is a tubular extension of a tame canonical algebra.

Proof. According to 4.3.2' and 4.3.5, there is a preprojective tilting

A_0-module T_0 with $\mathrm{End}(T_0)$ being canonical, thus T_0 defines the required left shrinking functor.

(1) Let ${}_AT$ be a shrinking module. Then $G({}_AT)$ contains all but at most finitely many indecomposable A-modules, and the indecomposable A-modules not contained in $G({}_AT)$ are preprojective A_0-modules.

Proof. Let M be an indecomposable A-module which does not belong to $G({}_AT)$. Then $0 \neq \mathrm{Ext}^1_A({}_AT,M) \approx D\,\mathrm{Hom}(M,\tau T) = D\,\mathrm{Hom}(M,\tau T_0)$, due to the fact that T_p is projective. Thus M is a predecessor of one of the indecomposable summands of the preprojective A_0-module τT_0. In particular, M itself is an A_0-module, and also preprojective.

Given an algebra B, and a linear form $\iota : K_0(B) \longrightarrow \mathbb{Z}$, let P^B_ι, T^B_ι, Q^B_ι be the module classes of all indecomposable B-modules M satisfying $(\underline{\dim}\, M)\iota < 0$, $= 0$, or > 0, respectively. For B being both tubular and cotubular, and $\gamma \in \mathbb{Q}^+$, we will denote $P^B_{\iota_\gamma}$, $T^B_{\iota_\gamma}$, $Q^B_{\iota_\gamma}$, just by P^B_γ, T^B_γ, Q^B_γ, as in 5.2 [however, we warn that, in this case, for $\gamma = 0$ and $\gamma = \infty$, we only have the inclusions $T^B_0 \subset T^B_{\iota_0}$, $T^B_\infty \subset T^B_{\iota_\infty}$, but not equality]. Let us call T^B_ι ι-separating provided T^B_ι separates P^B_ι from Q^B_ι. Also, we say that T^B_ι is ι-controlled provided it is controlled by the restriction of χ_B to $\mathrm{Ker}\,\iota$. Of course, in case we consider a tubular and cotubular algebra B and $\iota = \iota_\gamma$ for some $\gamma \in \mathbb{Q}^+$, then we just say γ-separating, and γ-controlled.

(2) Let A be a tubular algebra, ${}_AT$ a shrinking module, $B = \mathrm{End}({}_AT)$, and assume B is also cotubular. Let $\Sigma = \Sigma_T$, and $\sigma = \sigma_T$. Let $\gamma \in \mathbb{Q}^+$, and assume T^B_γ is a tubular, γ-separating family. Assume there exists an indecomposable A-module N in $T^A_{\sigma\iota_\gamma}$ with $\mathrm{Ext}^1_A(N,N) \neq 0$. Then Σ defines an equivalence from $P^A_{\sigma\iota_\gamma} \cap G({}_AT)$ onto P^B_γ. Also, any module in T^B_γ is generated by a module in $T^B_\gamma \cap Y({}_AT)$. Finally, assume for some $\delta \in \mathbb{Q}^+$, with $\delta < \gamma$, the module class $T^A_{\sigma\iota_\delta}$ is a sincere stable tubular $\sigma\iota_\delta$-separating family in A-mod, then T^B_δ is a (sincere, stable) tubular, δ-separating family in B-mod, and equivalent, under Σ, to $T^A_{\sigma\iota_\delta}$. If $T^A_{\sigma\iota_\delta}$ is, in addition, $\sigma\iota_\delta$-controlled and of type $\mathbb{T}_{n_1,\dots,n_t}$, with $2 - t + \Sigma\, n_s = \mathrm{rank}\, K_0(A)$, then T^B_δ is δ-controlled (and of the same type).

Proof. First, let us show $P^B_\gamma \subseteq Y({}_AT)$. The indecomposable A-modules M which do not belong to $G({}_AT)$, are preprojective A_0-modules, in particular, $\mathrm{Ext}^1_A(M,M) = 0$. As a consequence, the given module N belongs to $G({}_AT)$. It follows that $\underline{\dim}\, N\Sigma = (\underline{\dim}\, N)\sigma$, thus $N\Sigma \in T^B_\gamma$, say it belongs to the tube $T^B_\gamma(\rho_0)$. Since also $\mathrm{Ext}^1_B(N\Sigma,N\Sigma) \neq 0$, we can apply 5.3.4 to $M = T^B_\gamma(\rho_0) \cap Y({}_AT)$. Given $P \in P^B_\gamma$, let Q be its injective envelope. Then $Q \in Q^B_\gamma$, since the injective modules for B being cotubular belong to $T^B_\infty \vee Q^B_\infty$, and therefore to any Q^B_γ, where $\gamma \in \mathbb{Q}^+$. We can factor the inclusion map $P \longrightarrow Q$ through $T^B_\gamma(\rho_0) \cap Y({}_AT)$ and obtain in this way a monomorphism from P to a module in $Y({}_AT)$. Since $Y({}_AT)$ (as the set of torsionfree modules of a torsion pair) is closed under submodules, it follows that $P \in \mathcal{Y}({}_A T)$.

Given an indecomposable module M in $\mathcal{G}({}_A T)$, we have $\underline{\dim}\, M\Sigma = (\underline{\dim}\, M)\sigma$, thus M belongs to $\mathcal{P}^A_{\sigma\iota_\gamma}$ if and only if $M\Sigma$ belongs to $\mathcal{P}^B_\gamma$. This shows that Σ defines an equivalence from $\mathcal{P}^A_{\sigma\iota_\gamma} \cap \mathcal{G}({}_A T)$ onto $\mathcal{P}^B_\gamma \cap \mathcal{Y}({}_A T) = \mathcal{P}^B_\gamma$.

Since $\mathcal{P}^B_\gamma \subseteq \mathcal{Y}({}_A T)$, and $\mathcal{T}^B_\gamma$ is γ-separating, we have $\mathcal{X}({}_A T) \subseteq \mathcal{T}^B_\gamma \vee \mathcal{Q}^B_\gamma$. The modules in $\mathcal{X}({}_A T)$ are just those B-modules which are generated by $(A_A)D\Sigma\tau^-$. Given any tube $\mathcal{T}^B_\gamma(\rho)$ in $\mathcal{T}^B_\gamma$, let $X(\rho)$ be the largest direct summand of $(A_A)D\Sigma\tau^-$ belonging to $\mathcal{T}^B_\gamma(\rho)$, thus $\mathcal{T}^B_\gamma(\rho) \cap \mathcal{X}({}_A T)$ consists precisely of those modules in $\mathcal{T}^B_\gamma(\rho)$ which are generated by $X(\rho)$. Since $\mathcal{T}^B_\gamma(\rho)$ is a serial abelian category, the indecomposable modules in $\mathcal{T}^B_\gamma(\rho)$ generated by some fixed module $X(\rho)$ from $\mathcal{T}^B_\gamma(\rho)$ are of bounded length. We claim that $\mathcal{T}^B_\gamma(\rho) \cap \mathcal{Y}({}_A T)$ contains indecomposable modules of arbitrarily large length. Namely, every indecomposable module Z in $\mathcal{T}^B_\gamma(\rho)$ has a torsion submodule Z' with torsionfree factor module Z/Z', relative to the torsion pair $(\mathcal{Y}({}_A T), \mathcal{X}({}_A T))$. Now, Z' being a submodule of Z, belongs to $\mathcal{P}^B_\gamma \vee \mathcal{T}^B_\gamma(\rho)$, and as a module in $\mathcal{X}({}_A T)$ to $\mathcal{T}^B_\gamma \vee \mathcal{Q}^B_\gamma$, thus $Z' \in \mathcal{T}^B_\gamma(\rho)$, and therefore Z/Z' is a factor object of Z inside the serial abelian category $\mathcal{T}^B_\gamma(\rho)$, thus indecomposable. In this way, we obtain indecomposable modules Y in $\mathcal{T}^B_\gamma(\rho) \cap \mathcal{Y}({}_A T)$ of arbitrarily large length, in particular also such modules Y with $\mathrm{Ext}^1(Y,Y) \neq 0$ [we only need that the $\mathcal{T}^B_\gamma(\rho)$-length of Y is $\geq$ rank $\mathcal{T}^B_\gamma(\rho)$]. Applying now 5.3.4 to $\mathcal{M} = \mathcal{T}^B_\gamma(\rho) \cap \mathcal{Y}({}_A T)$, it follows that any module in $\mathcal{T}^B_\gamma(\rho)$ is generated by a module from $\mathcal{T}^B_\gamma(\rho) \cap \mathcal{Y}({}_A T)$.

Now assume there is given $\delta \in \Phi^+$, with $\delta < \gamma$, and such that the module class $\mathcal{T}^A_{\sigma\iota_\delta}$ is a tubular, $\sigma\iota_\delta$-separating family. Since the components of $\mathcal{T}^A_{\sigma\iota_\delta}$ are tubes, and the preprojective A_0-modules form a preprojective component, we see that $\mathcal{T}^A_{\sigma\iota_\delta}$ contains no preprojective A_0-module, thus $\mathcal{T}^A_{\sigma\iota_\delta} \subseteq \mathcal{G}({}_A T)$. Also, since $\delta < \gamma$, we have $\mathcal{T}^A_{\sigma\iota_\delta} \subseteq \mathcal{P}^A_{\sigma\iota_\gamma}$ [namely, for M indecomposable in $\mathcal{T}^A_{\sigma\iota_\delta}$, the module $M\Sigma$ belongs to $\mathcal{T}^B_\delta \subseteq \mathcal{P}_\gamma$]. It follows that Σ defines an equivalence from $\mathcal{T}^A_{\sigma\iota_\delta}$ onto $\mathcal{T}^B_\delta$.

Next, we claim that $\mathcal{Q}^B_\delta = (\mathcal{Q}^B_\delta \cap \mathcal{Y}({}_A T)) \rfloor \mathcal{X}({}_A T)$. Namely, $\mathcal{X}({}_A T) \subseteq \mathcal{T}^B_\gamma \vee \mathcal{Q}^B_\gamma \subseteq \mathcal{Q}^B_\delta$, thus $(\mathcal{Q}^B_\delta \cap \mathcal{Y}({}_A T)) \rfloor \mathcal{X}({}_A T) \subseteq \mathcal{Q}^B_\delta$. On the other hand, let Q be indecomposable in $\mathcal{Q}^B_\delta$. Then Q has a torsion submodule Q' with Q/Q' torsionfree, relative to the torsion pair $(\mathcal{Y}({}_A T), \mathcal{X}({}_A T))$. Now, either Q itself belongs to $\mathcal{Y}({}_A T)$, thus $Q \in \mathcal{Q}^B_\delta \cap \mathcal{Y}({}_A T)$. Or else $Q \notin \mathcal{Y}({}_A T)$, then, in particular, $Q \notin \mathcal{P}^B_\gamma$, therefore $Q \in \mathcal{T}^B_\gamma \vee \mathcal{Q}^B_\gamma$, thus also $Q/Q' \in \mathcal{T}^B_\gamma \vee \mathcal{Q}^B_\gamma \subseteq \mathcal{Q}^B_\delta$. It follows that $Q/Q' \in \mathcal{Q}^B_\delta \cap \mathcal{Y}({}_A T)$, and $Q' \in \mathcal{X}({}_A T)$.

In order to see that $\mathcal{T}^B_\delta$ is δ-separating, we need to know that

$$(*) \qquad \mathrm{Hom}(\mathcal{Q}^B_\delta, \mathcal{P}^B_\delta) = \mathrm{Hom}(\mathcal{Q}^B_\delta, \mathcal{T}^B_\delta) = \mathrm{Hom}(\mathcal{T}^B_\delta, \mathcal{P}^B_\delta) = 0.$$

Using that $\mathcal{T}^A_{\sigma\iota_\delta}$ is $\sigma\iota_\delta$-separating, the equivalence $\Sigma : \mathcal{G}({}_A T) \longrightarrow \mathcal{Y}({}_A T)$ shows that

$$\mathrm{Hom}(\mathcal{Q}^B_\delta \cap \mathcal{Y}(_A T),\ \mathcal{P}^B_\delta) = \mathrm{Hom}(\mathcal{Q}^B_\delta \cap \mathcal{Y}(_A T),\ \mathcal{T}^B_\delta) = \mathrm{Hom}(\mathcal{T}^B_\delta,\ \mathcal{P}^B_\delta) = 0,$$

using that both $\mathcal{P}^B_\delta$ and $\mathcal{T}^B_\delta$ are contained in $\mathcal{Y}(_A T)$. However, since $(\mathcal{Y}(_A T),\ \mathcal{X}(_A T))$ is a torsion pair, and using again that $\mathcal{P}^B_\delta$ and $\mathcal{T}^B_\delta$ are contained in $\mathcal{Y}(_A T)$, we also obtain

$$\mathrm{Hom}(\mathcal{X}(_A T),\ \mathcal{P}^B_\delta) = \mathrm{Hom}(\mathcal{X}(_A T),\ \mathcal{T}^B_\delta) = 0,$$

and therefore we obtain (*) from $\mathcal{Q}^B_\delta = (\mathcal{Q}^B_\delta \cap \mathcal{Y}(_A T)) \lceil \mathcal{X}(_A T)$.

Next, let us show that $\mathcal{T}^B_\delta$ actually is a union of components of B-mod. Let Y be an indecomposable module in $\mathcal{T}^B_\delta$. Let $Y' \longrightarrow Y$ be its sink map in B-mod. Using (*), we see that $Y' \in \mathcal{P}^B_\delta \vee \mathcal{T}^B_\delta$, thus Y' belongs to $\mathcal{Y}(_A T)$. Let M, M' be modules in $\mathcal{G}(_A T)$ with $Y = M\Sigma$, $Y' = M'\Sigma$, and $g : M' \longrightarrow M$ with $g\Sigma$ the given sink map. Now g is a relative sink map inside $\mathcal{G}(_A T)$. However, since $M \in \mathcal{T}^A_{\sigma\iota_\delta}$, and $\mathcal{T}^A_{\sigma\iota_\delta}$ is a union of components of A-mod, all being contained in $\mathcal{G}(_A T)$, it follows that $M' \in \mathcal{T}^A_{\sigma\iota_\delta}$, thus $Y' \in \mathcal{T}^B_\delta$. Also, the kernel $M\tau_A$ of g goes over, under Σ, to the kernel of Σg, thus $M\tau_A\Sigma \approx M\Sigma\tau_B$, and this module again belongs to $\mathcal{T}^B_\delta$. In particular, since the modules in $\mathcal{T}^A_{\sigma\iota_\delta}$ are τ-periodic, there is some d with $M\tau^d_A \approx M$, thus $Y\tau^d_B = M\Sigma\tau^d_B \approx M\tau^d_A\Sigma \approx M\Sigma = Y$. Now, given an irreducible map $Y \longrightarrow Y''$, there is an irreducible map $Y'' \longrightarrow Y\tau^{d-1}_B$, and $Y\tau^{d-1}_B$ belongs to $\mathcal{T}^B_\delta$, thus, by the previous discussion, also Y'' belongs to $\mathcal{T}^B_\delta$. This shows that $\mathcal{T}^B_\delta$ is a union of components of B-mod. Since Σ defines an equivalence from $\mathcal{T}^A_{\sigma\iota_\delta}$ onto $\mathcal{T}^B_\delta$, it follows that $\mathcal{T}^B_\delta$ is a stable tubular family.

To finish the proof that $\mathcal{T}^B_\delta$ is δ-separating, fix some component $\mathcal{T}^B_\delta(\rho)$ of $\mathcal{T}^B_\delta$, and let $f : P \longrightarrow Q$ be a map, with $P \in \mathcal{P}^B_\delta$, $Q \in \mathcal{Q}^B_\delta$. We have to factor f through $\mathcal{T}^B_\delta(\rho)$. Now, P belongs to $\mathcal{Y}(_A T)$, thus, if also Q belongs to $\mathcal{Y}(_A T)$, then we use the functor Σ and the corresponding factorization property of the tubes in $\mathcal{T}^A_{\sigma\iota_\delta}$, in order to get the derived factorization. Thus, assume $Q \notin \mathcal{Y}(_A T)$, therefore $Q \in \mathcal{T}^B_\gamma \vee \mathcal{Q}^B_\gamma$. If $Q \in \mathcal{Q}^B_\gamma$, we can factor f through $\mathcal{T}^B_\gamma$, since $\mathcal{T}^B_\gamma$ is γ-separating, thus we can assume $Q \in \mathcal{T}^B_\gamma$. However, since every module from $\mathcal{T}^B_\gamma$ is generated by a module from $\mathcal{T}^B_\gamma \cap \mathcal{Y}(_A T)$, we can factor the map $f : P \longrightarrow Q$ through $\mathcal{T}^B_\gamma \cap \mathcal{Y}(_A T)$, using 5.3.4. Thus we have to consider a map from $\mathcal{P}^B_\gamma$ to $\mathcal{T}^B_\gamma \cap \mathcal{Y}(_A T) \subseteq \mathcal{Q}^B \cap \mathcal{Y}(_A T)$, and, as we have seen before, such a map may always be factorized through $\mathcal{T}^B(\rho)$.

Finally, let us assume that $\mathcal{T}^A_{\sigma\iota_\delta}$ is $\sigma\iota_\delta$-controlled. Note that the indecomposable modules M in $\mathcal{T}^A_{\sigma\iota_\delta}$ satisfy proj. dim. $M = 1 =$ inj. dim M, according to 3.1.4. Since $_A T$ is a tilting module, and $\mathcal{T}^A_{\sigma\iota_\delta} \subseteq \mathcal{G}(_A T)$, we know that $\chi_B(\underline{\dim}\, M\Sigma) = \chi_A(\underline{\dim}\, M)$, for any module in $\mathcal{T}^A_{\sigma\iota_\delta}$, according to 4.1.c". Therefore, given any indecomposable B-module X in $\mathcal{T}^B_\delta$, its dimension vector $\underline{\dim}\, X$ is either root or a radical vector. Conversely, given a positive radical vector x in $\mathrm{Ker}\ \iota_\delta$, and let $y = x\sigma^{-1}$, then y belongs to $U^A = \mathrm{Ker}\ \sigma\iota_\delta$. Since σ is an isometry from $(K_0(A), \chi_A)$

onto $(K_0(B),\chi_B)$, it follows that y belongs to the radical of χ_A, and also that $U^A \cap \mathrm{rad}\,\chi_A$ (being mapped under σ onto $\mathrm{Ker}\,\iota_\delta \cap \mathrm{rad}\,\chi_B$) has rank 1. Now, consider some tube $\mathcal{T}^A_{\sigma\iota_\delta}(\rho)$. Then $h^A = h(\mathcal{T}^A_{\sigma\iota_\delta}(\rho))$ is non-zero and belongs to $U^A \cap \mathrm{rad}\,\chi_A$, according to 5.3.2, thus with h^A, also y is connected, and either positive or negative. If y is positive, there are infinitely many indecomposable A-modules Y_i in $\mathcal{T}^A_{\sigma\iota_\delta}$ with $\underline{\dim}\,Y_i = y$, and, under Σ we obtain infinitely many indecomposable B-modules $Y_i\Sigma$ in $\mathcal{T}^B_\delta$ with $\underline{\dim}\,Y_i\Sigma = y\sigma = x$. On the other hand, if y would be negative, then there would exist $Y \in \mathcal{T}^A_{\sigma\iota_\delta}$ with $\underline{\dim}\,Y = -y$, and therefore $\underline{\dim}\,Y\Sigma = -y\sigma = -x$ would be positive, a contradiction. Similarly, let x be a positive root in $\mathrm{Ker}\,\iota_\delta$, and let $y = x\sigma^{-1}$. Then y is a root for χ_A in U^A. Since $\mathrm{rank}\,U^A = \mathrm{rank}\,K_0(A) - 1 = 1 + \Sigma(n_s-1)$, we can apply 5.3.3. By assumption, the condition (i) is satisfied, thus even (i'), therefore there exists an indecomposable module Y in $\mathcal{T}^A_{\sigma\iota_\delta}$ with $\underline{\dim}\,Y$ either $= y$ or $= -y$. If $\underline{\dim}\,Y = y$, then $\underline{\dim}\,Y\Sigma = y\sigma = x$, and $Y\Sigma$ is the required module. The other case $\underline{\dim}\,Y = -y$ would imply that $\underline{\dim}\,Y\Sigma = -y\sigma = -x$, and therefore x would be both positive and negative, impossible. Also, since Y is the unique indecomposable A-module in $\mathcal{T}^A_{\sigma\iota_\delta}$ with $\underline{\dim}\,Y = y$, it follows that $Y\Sigma$ is the unique B-module X in $\mathcal{T}^B_\delta$ with $\underline{\dim}\,X = x$. This finishes the proof of (2).

Note that in (2), we did not assume A to be cotubular, since we want to use it in the proof of theorem 5.2.3. In case both A and B are known to be tubular and cotubular, and $\Sigma : A\text{-mod} \longrightarrow B\text{-mod}$ is a left shrinking functor, with corresponding linear transformation $\sigma : K_0(A) \longrightarrow K_0(B)$, we will define below a map $\bar{\sigma} : \mathbb{Q}_0^\infty \longrightarrow \mathbb{Q}_0^\infty$. We denote by h_0^A, h_∞^A, and by h_0^B, h_∞^B the canonical radical elements of A, and B, respectively. Then $h_\infty^A\sigma$ belongs to $\mathrm{rad}\,\chi_B$, thus we can write it in the form

(a) $$h_\infty^A\sigma = \eta_0 h_0^B + \eta_\infty h_\infty^B, \quad \text{with } \eta_0 \in \mathbb{Q}_0^+,\ \eta_\infty \in \mathbb{Q}^+.$$

Proof. Choose an indecomposable A-module M, with $\underline{\dim}\,M = h_\infty^A$. Since $\mathrm{Ext}^1_A(M,M) \neq 0$, we know that M belongs to $\mathcal{G}({}_A T)$, in particular, $\underline{\dim}\,M\Sigma = (\underline{\dim}\,M)\sigma = h_\infty^A\sigma$. Also, $\mathrm{Ext}^1_B(M\Sigma, M\Sigma) \neq 0$, thus $M\Sigma$ belongs to $\mathcal{T}_0^B \vee (\mathcal{Q}_0^B \cap \mathcal{P}_\infty^B) \vee \mathcal{T}_\infty^B$, but actually, $M\Sigma$ cannot be in $\mathcal{T}_0^B$, since this is the image under Σ of $\mathcal{T}_0^A$ (or just using that h_0^A, h_∞^A are linearly independent). Therefore, $(\underline{\dim}\,M\Sigma)\iota_0^B > 0$, $(\underline{\dim}\,M\Sigma)\iota_\infty^B \leq 0$. Let $c^B = h_\infty^B\iota_0^B$, then $c^B > 0$, since h_∞^B is the dimension vector of some indecomposable module in $\mathcal{T}_\infty^B$. Now $(\eta_0 h_0^B + \eta_\infty h_\infty^B)\iota_0^B = \eta_\infty c^B$, and $(\eta_0 h_0^B + \eta_\infty h_\infty^B)\iota_\infty^B = -\eta_0 c^B$, using that $h_0^B\iota_\infty^B = -c^B$. Thus $\eta_\infty \in \mathbb{Q}^+$, and $\eta_0 \in \mathbb{Q}_0^+$.

We define $\bar{\sigma} : \mathbb{Q}_0^\infty \longrightarrow \mathbb{Q}_0^\infty$ by

$$\gamma\bar{\sigma} = \frac{\eta_\infty\gamma}{\eta_0\gamma + 1} = \frac{\eta_\infty}{\eta_0 + \frac{1}{\gamma}} \qquad \text{for } \gamma \in \mathbb{Q}_0^\infty,$$

where $h_\infty^A\sigma = \eta_0 h_0^B + \eta_\infty h_\infty^B$. According to (a), we know that $\gamma\bar{\sigma} \in \mathbb{Q}_0^\infty$, for any $\gamma \in \mathbb{Q}_0^\infty$. Let us mention some properties of $\bar{\sigma}$.

(b) $0\bar{\sigma} = 0$, $\infty\bar{\sigma} = \frac{\eta_\infty}{\eta_0}$, and $\bar{\sigma}$ defines an order preserving bijection between $\mathbb{Q}_0^\infty$ and the set

$$\{\gamma \in \mathbb{Q}_0^\infty \mid 0\bar{\sigma} = 0 \leq \gamma \leq \frac{\eta_\infty}{\eta_0} = \infty\bar{\sigma}\} .$$

In case $\infty\bar{\sigma} \neq \infty$, for some $\sigma = \sigma_T$, the functor Σ_T is said to be a <u>proper</u> shrinking functor.

(c) $$\iota_\gamma^A = \sigma\iota_{\gamma\bar{\sigma}}^B$$

<u>Proof.</u> Let $\gamma = \frac{\gamma_\infty}{\gamma_0}$, thus, for $x \in K_0(A)$, we have

$$\begin{aligned} x\iota_\gamma^A &= \langle\gamma_0 h_0^A + \gamma_\infty h_\infty^A, x\rangle \\ &= \langle\gamma_0 h_0^A\sigma + \gamma_\infty h_\infty^A\sigma, x\sigma\rangle \\ &= \langle\gamma_0 h_0^B + \gamma_\infty\eta_0 h_0^B + \gamma_\infty\eta_\infty h_\infty^B, x\sigma\rangle \\ &= \langle(\gamma_0+\gamma_\infty\eta_0)h_0^B + (\gamma_\infty\eta_\infty)h_\infty^B, x\sigma\rangle, \\ &= x\sigma\iota_{\gamma\bar{\sigma}}^B , \end{aligned}$$

since $$\frac{\gamma_\infty\eta_\infty}{\gamma_0 + \gamma_\infty\eta_0} = \frac{\eta_\infty}{\frac{\gamma_0}{\gamma_\infty} + \eta_0} .$$

In particular,

(d) $$\operatorname{index}(h_\gamma^A\sigma) = \gamma\bar{\sigma} , \quad \text{for any } \gamma \in \mathbb{Q}_0^\infty .$$

<u>Proof.</u> Since $h_\gamma^A\sigma$ is a non-zero element of $\operatorname{rad}\chi_B$, its index is defined, and $h_\gamma^A\sigma\iota_{\gamma\bar{\sigma}}^B = h_\gamma^A\iota_\gamma^A = 0$, thus the index of $h_\gamma^A\sigma$ is $\gamma\bar{\sigma}$.

We note the following corollary to the first part of (2):

(2') Let A,B be both tubular, and cotubular, let ${}_AT$ be a shrinking module with $B = \operatorname{End}({}_AT)$, let $\Sigma = \Sigma_T$, $\sigma = \sigma_T$, and assume $\gamma = \infty\bar{\sigma} < \infty$. Also assume that $\mathcal{T}_\gamma^B$ is a tubular, γ-separating family. Then $\mathcal{P}_\gamma^B \subseteq \mathcal{Y}({}_AT)$, thus Σ defines an equivalence from $\mathcal{P}_\infty^A \cap \mathcal{G}({}_AT)$ onto $\mathcal{P}_\gamma^B$.

<u>Proof.</u> We only have to take any indecomposable A-module N with <u>dim</u> $N = h_\infty^A$, then $\operatorname{Ext}_A^1(N,N) \neq 0$, and note that $\sigma\iota_\gamma^B = \sigma\iota_{\infty\bar{\sigma}}^B = \iota_\infty^A$.

In the second part of assertion (2), we have shown that separating tubular families in A-mod may give rise to separating tubular families in B-mod. We also will need the converse assertion showing that certain separating tubular families in B-mod give rise to separating tubular families in A-mod.

(3) Let A,B be both tubular, and cotubular, let ${}_A T$ be a shrinking module with $B = \mathrm{End}({}_A T)$, let $\Sigma = \Sigma_T$, $\sigma = \sigma_T$, and assume $\gamma = \infty\bar{\sigma} < \infty$. Also let $\mathcal{T}^B_\gamma$ be a tubular, γ-separating family. Assume that for some $\beta \in \mathbb{Q}^+$, the module class $\mathcal{T}^B_{\beta\bar{\sigma}}$ is a $\beta\bar{\sigma}$-separating tubular family in B-mod, then $\mathcal{T}^A_\beta$ is a β-separating tubular family in A-mod, and equivalent, under Σ, to $\mathcal{T}^B_{\beta\bar{\sigma}}$. If $\mathcal{T}^B_{\beta\bar{\sigma}}$ is, in addition, $\beta\bar{\sigma}$-controlled and of type $\mathbb{T}_{n_1,\ldots,n_t}$, with $2 - t + \Sigma n_s = \mathrm{rank}\, K_0(A)$, then $\mathcal{T}^A_\beta$ is β-controlled (and of the same type).

Proof. Since $\beta \in \mathbb{Q}^+$, we know that $\mathcal{T}^A_\beta$ does not contain preprojective A_0-modules, thus $\mathcal{T}^A_\beta \subseteq \mathcal{G}({}_A T)$. Also $\beta < \infty$ implies $\beta\bar{\sigma} < \infty\bar{\sigma} = \gamma$, thus $\mathcal{T}^B_{\beta\bar{\sigma}} \subseteq \mathcal{P}^B_\gamma \subseteq \mathcal{Y}({}_A T)$, the last inclusion being due to (2'). Since Σ defines an equivalence from $\mathcal{G}({}_A T)$ onto $\mathcal{Y}({}_A T)$, and $\sigma\iota_{\beta\bar{\sigma}} = \iota_\beta$, it follows that $\mathcal{T}^A_\beta$ is equivalent, under Σ, to $\mathcal{T}^B_{\beta\bar{\sigma}}$.

Next, we want to show that

$$(**) \qquad \mathrm{Hom}(\mathcal{Q}^A_\beta, \mathcal{P}^A_\beta) = \mathrm{Hom}(\mathcal{Q}^A_\beta, \mathcal{T}^A_\beta) = \mathrm{Hom}(\mathcal{T}^A_\beta, \mathcal{P}^A_\beta) = 0.$$

First of all, we claim that $\mathcal{P}^A_\beta = \mathcal{F}({}_A T) \,\rfloor\, (\mathcal{P}^A_\beta \cap \mathcal{G}({}_A T))$. Namely, $\mathcal{F}({}_A T)$ contains only preprojective A_0-modules, thus $\mathcal{F}({}_A T) \subseteq \mathcal{P}^A_\beta$. Thus, $\mathcal{F}({}_A T) \,\rfloor\, (\mathcal{P}^A_\beta \cap \mathcal{G}({}_A T)) \subseteq \mathcal{P}^A_\beta$. On the other hand, let M be indedomposable in $\mathcal{P}^A_\beta$, and let M' be its torsion submodule, M/M' torsionfree, relative to the torsion pair $(\mathcal{F}({}_A T), \mathcal{G}({}_A T))$. Either $M/M' \neq 0$, then $\mathrm{Hom}(M, \mathcal{F}({}_A T)) \neq 0$ shows that M itself, and therefore also M', are preprojective A_0-modules, thus $M' \in \mathcal{P}^A_\beta$, and $M \in \mathcal{F}({}_A T) \,\rfloor\, (\mathcal{P}^A_\beta \cap \mathcal{G}({}_A T))$, or else $M = M'$, and therefore $M \in \mathcal{P}^A_\beta \cap \mathcal{G}({}_A T)$.
For the modules in $\mathcal{P}^A_\beta \cap \mathcal{G}({}_A T)$, $\mathcal{T}^A_\beta = \mathcal{T}^A_\beta \cap \mathcal{G}({}_A T)$ and $\mathcal{Q}^A_\beta = \mathcal{Q}^A_\beta \cap \mathcal{G}({}_A T)$, we have

$$\mathrm{Hom}(\mathcal{Q}^A_\beta, \mathcal{P}^A_\beta \cap \mathcal{G}({}_A T)) = \mathrm{Hom}(\mathcal{Q}^A_\beta, \mathcal{T}^A_\beta) = \mathrm{Hom}(\mathcal{T}^A_\beta, \mathcal{P}^A_\beta \cap \mathcal{G}({}_A T)) = 0,$$

using the functor Σ and that $\mathcal{T}^B_{\beta\bar{\sigma}}$ is $\beta\bar{\sigma}$-separating. On the other hand, we also have

$$\mathrm{Hom}(\mathcal{Q}^A_\beta, \mathcal{F}({}_A T)) = \mathrm{Hom}(\mathcal{T}^A_\beta, \mathcal{F}({}_A T)) = 0,$$

using again that $\mathcal{T}^A_\beta \vee \mathcal{Q}^A_\beta \subseteq \mathcal{G}({}_A T)$, and that $(\mathcal{F}({}_A T), \mathcal{G}({}_A T))$ is a torsion pair.

Next, let us show that $\mathcal{T}^A_\beta$ actually is a union of components of A-mod. Let M be indecomposable in $\mathcal{T}^A_\beta$, and let $f : M \longrightarrow M'$ be its source map in A-mod. Using (**), we see that $M' \in \mathcal{T}^A_\beta \vee \mathcal{Q}^A_\beta$, in particular, $M' \in \mathcal{G}({}_A T)$. Now f is a relative source map in $\mathcal{Y}({}_A T)$, but since the source map for $M\Sigma$ in B-mod lies in $\mathcal{T}^B_{\beta\bar{\sigma}} \subseteq \mathcal{Y}({}_A T)$, it follows that $f\Sigma : M\Sigma \longrightarrow M'\Sigma$ is the source map for $M\Sigma$ in B-mod, thus $M'\Sigma \in \mathcal{T}^B_{\beta\bar{\sigma}}$, and therefore $M' \in \mathcal{T}^A_\beta$. Also, the cokernel of $f\Sigma$ is $M\Sigma\tau^-_B$, a module which again belongs to $\mathcal{T}^B_{\beta\bar{\sigma}}$, and it is the image under Σ of the cokernel $M\tau^-_A$ of f. Note that $M\Sigma$ is τ-periodic [namely, it belongs to a tube, and all components inside $\mathcal{Q}^B_0 \cap \mathcal{P}^B_\infty$ are stable], say $M\Sigma\tau^{-d} \approx M\Sigma$, thus $M\tau^{-d}\Sigma \approx M\Sigma\tau^{-d} \approx M\Sigma$, therefore $M\tau^{-d} \approx M$. As a consequence, if $M'' \longrightarrow M$ is the sink map for M, then the source map for

$M\tau^{-d+1}$ is of the form $M\tau^{-d+1} \longrightarrow M''$, and by the previous observations, $M'' \in \mathcal{T}^A_\beta$. This shows that $\mathcal{T}^A_\beta$ is a union of components of A-mod, and since it is equivalent to $\mathcal{T}^B_{\beta\bar\sigma}$, it is a tubular family.

It remains to be shown that every map from $\mathcal{P}^A_\beta$ to $\mathcal{Q}^A_\beta$ factors through any tube $\mathcal{T}^A_\beta(\rho)$ in $\mathcal{T}^A_\beta$. Thus consider a map $h : P \longrightarrow Q$, with P indecomposable in $\mathcal{P}^A_\beta$, and $Q \in \mathcal{Q}^A_\beta$. If P belongs to $\mathcal{G}({}_AT)$, then we factor $h\Sigma$ through $\mathcal{T}^B_{\beta\bar\sigma}$, and obtain as inverse image a factorization of h through $\mathcal{T}^B_\beta$. In case P does not belong to $\mathcal{G}({}_AT)$, then P is a preprojective A_0-module, and, of course, $Q \in \mathcal{Q}^A_0$, thus we can factor h through $\mathcal{T}^A_0$, say $h = h'h''$, with h'' going from $\mathcal{T}^A_0$ to $\mathcal{Q}^A_\beta$. Since $\mathcal{T}^A_0 \subseteq \mathcal{P}^A_\beta \cap \mathcal{G}({}_AT)$, we know that we can factor h'' through $\mathcal{T}^A_\beta(\rho)$.

Finally, we assume that $\mathcal{T}^B_{\beta\bar\sigma}$ is $\beta\bar\sigma$-controlled. Since $\mathcal{T}^B_{\beta\bar\sigma}$ is a sincere, stable, separating tubular family, we know that proj.dim.M = 1 = inj.dim.M for any indecomposable module M in $\mathcal{T}^B_{\beta\bar\sigma}$, according to 3.1.4. Since ${}_AT$ is a tilting module, and $\mathcal{T}^A_\beta \subseteq \mathcal{G}({}_AT)$, we have $\chi_A(\underline{\dim}\, M) = \chi_B(\underline{\dim}\, M\Sigma)$ for any $M \in \mathcal{T}^A_\beta$, thus, given an indecomposable $M \in \mathcal{T}^A_\beta$, then $M\Sigma$ is indecomposable in $\mathcal{T}^B_{\beta\bar\sigma}$, and it follows that $\underline{\dim}\, M$ is a radical vector or a root for χ_A. Conversely, given a positive radical vector x in Ker ι_β, and let $y = x\sigma$, then y belongs to $U^B =$ Ker $\iota_{\beta\bar\sigma}$, and y is a radical vector for χ_B. Also, since $U^B \cap \operatorname{rad} \chi_B$ has rank 1, we know that y is a rational multiple of any $h(\mathcal{T}^B_{\beta\bar\sigma}(\rho))$, where $\mathcal{T}^B_{\beta\bar\sigma}(\rho)$ is a tube in $\mathcal{T}^B_{\beta\bar\sigma}$, and therefore connected, and either positive or negative. Thus, since $\mathcal{T}^B_{\beta\bar\sigma}$ is $\beta\bar\sigma$-controlled, there are infinitely many indecomposable modules Y_i in $\mathcal{T}^B_{\beta\bar\sigma}$ with $\underline{\dim}\, Y_i$ being $= y$ or $= -y$. Let $X_i \in \mathcal{T}^A_\beta$ with $X_i\Sigma = Y_i$. Then we obtain infinitely many modules X_i with $\underline{\dim}\, X_i$ being $= x$ or $= -x$, however, the latter is impossible, since x is assumed to be positive. Similarly, let x be a positive root in Ker ι_β, and let $y = x\sigma$. Then y is a root for χ_B in U^B. Since rank U^B = rank $K_0(B)-1$ = $1 + \Sigma(n_s-1)$, we again apply 5.3.3, and see that with condition (i) also condition (i') is satisfied, thus there exists an indecomposable module Y in $\mathcal{T}^B_{\beta\bar\sigma}$ with $\underline{\dim}\, Y$ either $= y$ or $= -y$. Let $Y = X\Sigma$ for some X in $\mathcal{T}^A_\beta$, then X is indecomposable, and $\underline{\dim}\, X$ either $= x$ or $= -x$, but the latter is impossible. Also, since Y is the unique indecomposable module in $\mathcal{T}^B_{\beta\bar\sigma}$ with $\underline{\dim}\, Y = y$, it follows that X is the only indecomposable module in $\mathcal{T}^A_\beta$ with $\underline{\dim}\, X = x$. This finishes the proof of (3).

Recall that a left shrinking functor $\Sigma = \Sigma_T$ with corresponding linear transformation $\sigma = \sigma_T$ is called <u>proper</u>, provided $\infty\bar\sigma < \infty$. An immediate consequence of (3) is the following result which will be used in order to reduce the proof of 5.2.2 to the consideration of very special algebras.

(3') Let A,B be both tubular and cotubular of type $\mathbf{T}$, and $\Sigma : A\text{-mod} \longrightarrow B\text{-mod}$ a proper left shrinking functor. Assume, for every $\gamma \in \mathbb{Q}^+$, the module class T^B_γ is a γ-controlled and γ-separating tubular $\mathbb{P}_1k$-family of type $\mathbb{T}$ in B-mod. Then, for every $\gamma \in \Phi^+$, the module class T^A_γ is a γ-controlled and γ-separating tubular $\mathbb{P}_1k$-family of type $\mathbb{T}$ in A-mod.

Proof. We should note that for the extension type $\mathbb{T} = \mathbb{T}_{n_1,\dots,n_t}$ of a tubular algebra A, always $2 - t + \Sigma n_s$ is equal to the rank of $K_o(A)$.

In order to define right shrinking functors, we start with an algebra A which is a tubular coextension of a tame concealed algebra A_∞. Let T be a multiplicity-free cotilting A-module, with $T = T_q \oplus T_\infty$, where T_∞ is a preinjective cotilting A_∞-module, and T_q is an injective A-module. As above, T_q is uniquely determined by A and A_∞, it is just the direct sum of all indecomposable injective A-modules Q(a), with a outside A_∞. Let $B = \mathrm{End}({}_AT)$, and $B_\infty = \mathrm{End}({}_AT_\infty)$. Then B_∞ is a tame concealed algebra, and B is a tubular coextension of B_∞. The functor $\Sigma_r = (\Sigma_r)_T : \text{A-mod} \longrightarrow \text{B-mod}$ given by

$$\Sigma_r = \mathrm{Hom}_A(-, {}_AT_B) \cdot D = {}_B(TD)_A \otimes -$$

is called the right shrinking functor defined by T_∞. Note that this is the dual concept of a left shrinking functor. [Namely, given the tubular coextension A of the tame concealed algebra A_∞, and the cotilting module $T = T_q \oplus T_\infty$, as above, with $B = \mathrm{End}({}_AT)$, then A^{op} is a tubular extension of the tame concealed algebra $(A_\infty)^{op}$, thus let $(A^{op})_0 = (A_\infty)^{op}$. Also $TD = T_qD \oplus T_\infty D$ is a tilting A^{op}-module, with $T_\infty D$ being a preprojective $(A^{op})_0$-module, and T_qD a projective A^{op}-module, and $\mathrm{End}({}_{A^{op}}(TD)) = B^{op}$. Thus $\mathrm{Hom}({}_{A^{op}}(TD),-) : A^{op}\text{-mod} \longrightarrow B^{op}\text{-mod}$ is a left shrinking module, and, actually $\mathrm{Hom}({}_{A^{op}}(TD),-) = \mathrm{Hom}_A(-D, {}_AT)$, both considered as functors from A^{op}-mod to B^{op}-mod.] As a consequence, for any result concerning left shrinking modules, there is a corresponding result concerning right shrinking modules. In particular, given a right shrinking functor $\Sigma_r : \text{A-mod} \longrightarrow \text{B-mod}$, there is a corresponding linear transformation $\sigma_r : K_o(A) \longrightarrow K_o(B)$. In case both A and B are also tubular, there is the induced map $\bar{\sigma}_r : \Phi^\infty_o \longrightarrow \Phi^\infty_o$, with

$$\gamma\bar{\sigma}_r = \frac{\delta_\infty + \gamma}{\delta_o}, \qquad \text{for } \gamma \in \Phi^\infty_o,$$

where $h^A_o\sigma_r = \delta_o h^B_o + \delta_\infty h^B_\infty$.

5.5 Tilting modules for tubular algebras

We are going to prove theorem 5.2.5 for algebras which already are known to satisfy 5.2.2 and 5.2.3. We will need this result for the proof of 5.2.3 in section 5.7. Of course, after having established 5.2.2 and 5.2.3 in section 5.7, the following result reduces to assertion 5.2.5. – Note that given a tubular and cotubular algebra A, and a tilting module ${}_AT$, with ${}_AT$ belonging to $\mathcal{P}_\infty$, then ${}_AT$ cannot belong to $\mathcal{P}_0$, thus there are some $\beta \in \mathbb{Q}_0^+$ with $\mathcal{T}_\beta$ containing an indecomposable summand of ${}_AT$. Since there can be only finitely many such β, there is a maximal one, say γ, thus ${}_AT$ is contained in $\mathcal{P}_\gamma \vee \mathcal{T}_\gamma$ and not in $\mathcal{P}_\gamma$, or, equivalently, ${}_AT = T_o \oplus T_\gamma$, with $T_0 \in \mathcal{P}_\gamma$, and $0 \neq T_\gamma \in \mathcal{T}_\gamma$.

(1) Let A be a tubular and cotubular algebra of type $\mathbb{T}$, and assume, for any $\delta \in \mathbb{Q}^+$, the module class $\mathcal{T}_\delta$ is a δ-separating tubular $\mathbb{P}_1k$-family of type $\mathbb{T}$. Let ${}_AT$ be a tilting module in $\mathcal{P}_\infty$. Then $B = \mathrm{End}({}_AT)$ is a tubular algebra, again. More precisely, let ${}_AT = T_o \oplus T_\gamma$ with $T_0 \in \mathcal{P}_\gamma$, and $0 \neq T_\gamma \in \mathcal{T}_\gamma$, for some $\gamma \in \mathbb{Q}_0^+$. Then $B_0 = \mathrm{End}(T_0)$ is a tame concealed algebra, and B is a tubular extension of B_0, of the extension type $\mathbb{T}$.

<u>Proof.</u> Let $\Sigma = \mathrm{Hom}_A({}_AT_B,-) : A\text{-mod} \longrightarrow B\text{-mod}$ and $\sigma = \sigma_T$. Also, choose some tube $\mathcal{T}_\gamma(\rho_0)$ in $\mathcal{T}_\gamma$ of rank 1. [This is possible according to 5.3.2".] First of all, we note that both $\mathcal{Q}_\gamma$ and $\mathcal{T}_\gamma(\rho_0)$ are contained in $\mathcal{G}({}_AT)$, [namely, $\mathrm{Ext}_A^1(T,M) = D\,\mathrm{Hom}(M,\tau T) = 0$, for $M \in \mathcal{T}_\gamma(\rho_0) \vee \mathcal{Q}_\gamma$, since the indecomposable direct summands of τT belong to $\mathcal{P}_\gamma$ and to some non-homogeneous tubes in $\mathcal{T}_\gamma$.] Since the modules in $\mathcal{T}_\gamma(\rho_0)$, being generated by ${}_AT$, have to be generated by T_0, it in particular follows that $T_0 \neq 0$. Let $\mathcal{P}' = (\mathcal{P}_\gamma \cap \mathcal{G}({}_AT))\Sigma$. Since T_0 belongs to $\mathcal{P}_\gamma \cap \mathcal{G}({}_AT)$, we see that $\mathcal{P}' \neq 0$. Also, the B-modules in $\mathcal{P}'$ are actually B_0-modules [namely, for any $M \in \mathcal{P}_\gamma$, we have $\mathrm{Hom}(T_\gamma,M) = 0$]. Similarly, also the B-modules in $\mathcal{T}_\gamma(\rho_0)\Sigma$ are B_0-modules.

Let us consider the radical rad χ_{B_0}. First of all, $h_\gamma\sigma \in K_0(B)$ belongs to $K_0(B_0)$ [namely, for any direct summand T(i) of T_γ, we have

$$\langle \underline{\dim}\, T(i), h_\gamma\rangle = -\langle h_\gamma, \underline{\dim}\, T(i)\rangle = -(\underline{\dim}\, T(i))\iota_\gamma = 0,$$

thus all components of $h_\gamma\sigma$ outside B_0 are zero]. Also, $h_\gamma\sigma$ is actually positive and sincere in $K_0(B_0)$. [Namely, any direct summand T(i) of T_0 belongs to $\mathcal{P}_\gamma$, thus

$$\langle \underline{\dim}\, T(i), h_\gamma\rangle = -\langle h_\gamma, \underline{\dim}\, T(i)\rangle = -(\underline{\dim}\, T(i))\iota_\gamma > 0,$$

thus all components of $h_\gamma\sigma$ at the vertices of B_0 are positive]. Thus, $h_\gamma\sigma$ is a sincere, positive radical vector in $K_0(B_0)$. On the other hand, the vector $h_\infty\sigma$ of $K_0(B)$ does not belong to $K_0(B_0)$ [namely, let T(i) be an indecomposable direct summand of T_γ. Then $T(i) \in \mathcal{T}_\gamma \subseteq \mathcal{P}_\infty$, thus

$$\langle \underline{\dim}\, T(i), h_\infty\rangle = -\langle h_\infty, \underline{\dim}\, T(i)\rangle = -(\underline{\dim}\, T(i))\iota_\infty > 0,$$

thus all the components of $h_\infty\sigma$ outside B_0 are positive]. This shows that the radical of χ_{B_0} has rank 1 and has a sincere, positive generator.

Let us show that there are no cycles in $\mathcal{P}'$. For, a cycle inside $\mathcal{P}_\gamma \cap \mathcal{G}({}_AT)$ must be inside some tube $\mathcal{T}_\beta(\rho) \subseteq \mathcal{T}_\beta$, with $0 \leq \beta < \gamma$. Let $\mathcal{M} = \mathcal{T}_\beta(\rho) \cap \mathcal{G}({}_AT)$, this is a module class contained in $\mathcal{T}_\beta(\rho)$, closed under extensions and under factor modules in $\mathcal{T}_\beta(\rho)$, and contains, by assumption, a cycle. Thus, we can apply 5.3.5' in case $\beta > 0$, and 5.3.5 in case $\beta = 0$ in order to conclude that $\mathcal{M}$ contains an indecomposable module M with $\underline{\dim}\, M = h(\mathcal{T}_\beta(\rho))$. Now, $h(\mathcal{T}_\beta(\rho))$ is a positive multiple of h_β, say $= a\, h_\beta$ for some $a \in \mathbb{N}_1$ (we even have $a = 1$, according to 5.3.3, but we do not need this here), thus

$$\langle \underline{\dim}\, M, \underline{\dim}\, \tau T_\gamma\rangle = a\langle h_\beta, \underline{\dim}\, \tau T_\gamma\rangle = a(\underline{\dim}\, \tau T_\gamma)\iota_\beta > 0,$$

since $0 \neq \tau T_\gamma \in \mathcal{T}_\gamma \subseteq \mathcal{Q}_\beta$. Since proj.dim.$M \leq 1$, it follows that $\operatorname{Hom}(M, \tau T_\gamma) \neq 0$, but this is impossible, since $\tau T_\gamma \in \mathcal{F}({}_AT)$, $M \in \mathcal{G}({}_AT)$, and $(\mathcal{F}({}_AT), \mathcal{G}({}_AT))$ is a torsion pair.

Next, we want to show that $\mathcal{P}'' = (\mathcal{P}_\infty \cap \mathcal{G}({}_AT))\Sigma$ is closed under predecessors in B-mod. Let M be in $\mathcal{P}''$, and assume there is a non-zero map $f: N \to M$ with N an indecomposable B-module N. Let N' be the torsion submodule of N, with N/N' torsionfree, relative to the torsion pair $(\mathcal{Y}({}_AT), \mathcal{X}({}_AT))$. Let $N/N' = \oplus N_i$, with all N_i indecomposable. Now f vanishes on N', since $N' \in \mathcal{X}({}_AT)$, $M \in \mathcal{Y}({}_AT)$, thus f induces a non-zero map $\overline{f} = (\overline{f}_i)_i : \oplus N_i \longrightarrow M$, with $\overline{f}_i : N_i \longrightarrow M$, thus $\overline{f}_i \neq 0$ for some i, say $i = 1$. There are indecomposable A-modules X_1, X in $\mathcal{G}({}_AT)$ with $X_1\Sigma = N_1$, $X\Sigma = M$, and a map $X_1 \to X$ with image under Σ being $\overline{f}_1$, therefore $\operatorname{Hom}(X_1, X) \neq 0$. Since $X \in \mathcal{P}_\infty$, it follows that also X_1 is in $\mathcal{P}_\infty$. In particular, we have proj.dim.$X_1 \leq 1$. However, then $\operatorname{Ext}^1(N_1, N') = 0$ according to 4.1.6.d, thus $N = N_1 = X_1\Sigma$ belongs to $\mathcal{P}''$.

It follows that also $\mathcal{P}'$ is closed under predecessors. Namely, given $\operatorname{Hom}(M', M) \neq 0$, with M' indecomposable, $M \in \mathcal{P}'$, then with M also M' belongs to $\mathcal{P}''$, say $M' = X'\Sigma$, $M = X\Sigma$, where X' is indecomposable in $\mathcal{P}_\infty \cap \mathcal{G}({}_AT)$, and $X \in \mathcal{P}_0$. Since $\operatorname{Hom}(X', X) \neq 0$, it follows that $X' \in \mathcal{P}_0$, thus $M' = X'\Sigma \in \mathcal{P}'$. In particular, since $\mathcal{P}'$ is closed under predecessors and does not contain a cycle, it follows that any indecomposable module in $\mathcal{P}'$ is directing.

Also, let us show that $\mathcal{P}''$ is closed under irreducible maps in B-mod. Since we already know that $\mathcal{P}''$ is closed under predecessors, we only have to show that for an irreducible map $f: M \to N$, with $M \in \mathcal{P}''$, N indecomposable, also $N \in \mathcal{P}''$. Assume $N \notin \mathcal{P}''$.

Again, let N' be the torsion submodule of N, with N/N' torsionfree relative to the torsion pair $(\mathcal{Y}(_AT), \mathcal{X}(_AT))$, and let $p : N \longrightarrow N/N'$ be the projection map. Note that N/N' belongs to $(\mathcal{T}_\infty \vee \mathcal{Q}_\infty)\Sigma$. [Namely, decompose $N/N' = \oplus N_i$, with all N_i indecomposable, and assume some N_i, say N_1, is not in $(\mathcal{T}_\infty \vee \mathcal{Q}_\infty)\Sigma$, thus $N_1 = X_1\Sigma$ with $X_1 \in \mathcal{P}_\infty \cap \mathcal{G}(_AT)$, therefore proj.dim.$X_1 \leq 1$, thus $\mathrm{Ext}^1(X_1\Sigma, N') = 0$, according to 4.1.6.d, thus $N = N_1 = X_1\Sigma \in \mathcal{P}''$, a contradiction]. Since $M \in (\mathcal{P}_\infty \cap \mathcal{G}(_AT))\Sigma$, we can choose some $\gamma < \delta < \infty$, with $M \in (\mathcal{P}_\delta \cap \mathcal{G}(_AT))\Sigma$, and we can factor the induced map $fp : M \longrightarrow N/N'$ through $\mathcal{T}_\delta\Sigma$, say $fp = f_1f_2'$, with $f_1 : M \longrightarrow Y\Sigma$, $f_2' : Y\Sigma \longrightarrow N/N'$ with $Y \in \mathcal{T}_\delta$. Note that f_1 is not split mono. Since proj.dim.$Y = 1$, we have $\mathrm{Ext}^1(Y\Sigma, N') = 0$, thus we can lift f_2' to N, say $f_2' = f_2 p$, where $f_2 : Y\Sigma \longrightarrow N$. Since $Y\Sigma \in \mathcal{P}''$, and N is not in $\mathcal{P}''$, we conclude that f_2 is not split epi. Now, consider the map $f - f_1f_2$, and note that $(f - f_1f_2)p = 0$, thus $g = f - f_1f_2$ maps into N'. Let $M = X\Sigma$ with $X \in \mathcal{P}_\delta \cap \mathcal{G}(_AT)$, and let H be an indecomposable module in $\mathcal{T}_\delta$ with $\underline{\dim}\, H$ being a multiple of h_δ, say $\underline{\dim}\, H = b h_\delta$ for some $b \in \mathbb{N}_1$. Then

$$\langle \underline{\dim}\, H, \underline{\dim}\, X\rangle = b(\underline{\dim}\, X)\iota_\delta < 0,$$

and since proj.dim.$H \leq 1$, this means that $\mathrm{Ext}^1_A(H,X) \neq 0$. Take a non-split exact sequence

$$0 \longrightarrow X \longrightarrow X' \longrightarrow H \longrightarrow 0,$$

and apply Σ. Since all modules belong to $\mathcal{G}(_AT)$, we obtain a non-split exact sequence

$$0 \longrightarrow X\Sigma \xrightarrow{g_1} X'\Sigma \longrightarrow H\Sigma \longrightarrow 0.$$

Note that $g : X\Sigma \longrightarrow N'$ can be extended along g_1, say $g = g_1g_2$, since $\mathrm{Ext}^1(H\Sigma, N') = 0$, again using that proj.dim.$H \leq 1$. Note that g_2 is not split epi, since $X'\Sigma \in \mathcal{Y}(_AT)$, and $N' \in \mathcal{X}(_AT)$. Thus we have factored $f = [f_1 g_1]\begin{bmatrix} f_2 \\ g_2 \end{bmatrix}$ through $Y\Sigma \oplus X'\Sigma$, and neither f_1 nor g_1 being split mono, and neither f_2 nor g_2 being split epi. This contradicts that f is irreducible.

Now, in order to show that $\mathcal{P}'$ is closed under irreducible maps, let $M \longrightarrow N$ be irreducible, $M \in \mathcal{P}'$, and N indecomposable. By the previous result, we know that $N = \Sigma Y$, with $Y \in \mathcal{P}_\infty \cap \mathcal{G}(_AT)$. Also, let $M = \Sigma X$, with $X \in \mathcal{P}_0 \cap \mathcal{G}(_AT)$, and $f : X \longrightarrow Y$ with $f\Sigma$ being irreducible. In particular, f is relative irreducible in $\mathcal{G}(_AT)$. Now, if Y belongs to $\mathcal{Q}_\gamma$, then we factor f through $\mathcal{T}_\gamma(\rho_0)$ and conclude that f cannot be relative irreducible in $\mathcal{G}(_AT)$. Also, if Y belongs to $\mathcal{T}_\gamma$, let $g : Y' \longrightarrow Y$ be the sink map for Y in A-mod, thus $f = f'g$ for some $f' : X \longrightarrow Y'$. Let Y'' be the torsion submodule of Y' with Y'/Y'' torsionfree, relative to the torsion pair $(\mathcal{F}(_AT), \mathcal{G}(_AT))$, and say with inclusion map $\mu : Y'' \longrightarrow Y'$. Since the image of f' is contained in Y'', we can write $f' = f''\mu$, thus $f = f'' \cdot (\mu g)$. In order to show that this is a proper factorization inside $\mathcal{G}(_AT)$, we show that Y'' belongs to $\mathcal{T}_\gamma$. Now Y'/Y'' belongs to $\mathcal{F}({}_AT)$, and, as a factor module of $Y' \in \mathcal{T}_\gamma$ to $\mathcal{T}_\gamma \vee \mathcal{Q}_\gamma$. Since $\mathcal{Q}_\gamma \subseteq \mathcal{G}({}_AT)$, we see that Y'/Y'' can have no non-zero summand in $\mathcal{Q}_\gamma$, thus $Y'/Y'' \in \mathcal{T}_\gamma$. Again, the existence of a proper factorization of f inside $\mathcal{G}({}_AT)$ contradicts that $f\Sigma$ is irreducible.

We also note that $\mathrm{Hom}(D(B_{0_{B_0}}), \mathcal{P}') = 0$. Namely, if Q is an indecomposable injective B_0-module, and H' is an indecomposable module in $\mathcal{T}_\gamma(\rho_0)$, then $\mathrm{Hom}(H'\Sigma, Q) \neq 0$, since H' is a sincere B_0-module. Now, if $\mathrm{Hom}(Q, \mathcal{P}') \neq 0$, then Q would belong to $\mathcal{P}'$, and also $H'\Sigma$ would belong to $\mathcal{P}'$, due to the fact that $\mathcal{P}'$ is closed under predecessors, impossible. Therefore $\mathrm{Hom}(Q, \mathcal{P}') = 0$, thus $\mathrm{Hom}(D(B_{0_{B_0}}), \mathcal{P}') = 0$.

Altogether, we conclude that we can apply 4.3.8 (to any component of B_0-mod contained in $\mathcal{P}'$) and conclude that B_0 is a tame concealed algebra and that $\mathcal{P}'$ is the set of all preprojective B_0-modules.

Next, we want to see that B is a tubular extension of B_0, of extension type **T**. The proof is the same as that given in 4.9.1, provided we show that for any tube $\mathcal{T}_\gamma(\rho)$ in $\mathcal{T}_\gamma$, the module class $(\mathcal{T}_\gamma(\rho) \cap \mathcal{G}({}_AT))\Sigma$ in B-mod is closed under irreducible maps, and the module class $(\mathcal{T}_\gamma(\rho) \cap \mathcal{G}({}_AT))\Sigma \cap B_0$-mod in B_0-mod is also closed under irreducible maps.

Thus, let $M = \Sigma X$ for some $X \in \mathcal{T}_\gamma(\rho) \cap \mathcal{G}({}_AT)$, and consider an irreducible map $N \longrightarrow M$ in B-mod with N indecomposable. Since $\mathcal{P}''$ is closed under predecessors, N belongs to $\mathcal{Y}({}_AT)$, say $N = \Sigma Y$ with $Y \in \mathcal{P}_\infty \cap \mathcal{G}({}_AT)$, thus actually in $\mathcal{P}_\gamma \vee \mathcal{T}_\gamma(\rho)$, since $\mathrm{Hom}(X,Y) \neq 0$. However, $Y \in \mathcal{P}_\gamma$ is impossible, since $\mathcal{P}'$ is closed under irreducible maps, therefore $Y \in \mathcal{T}_\gamma(\rho) \cap \mathcal{G}({}_AT)$. If there is an irreducible map $M \longrightarrow N$ in B-mod, with N indecomposable, then again $N \in \mathcal{Y}({}_AT)$, since $\mathcal{P}''$ is closed under irreducible maps, say $N = \Sigma Y$, with $Y \in \mathcal{T}_\gamma(\rho) \vee \mathcal{Q}_\gamma$, and let $f : X \longrightarrow Y$ be a map with $f\Sigma$ irreducible in B-mod. Assume $Y \in \mathcal{Q}_\gamma$. Consider the sink map $g : Y' \longrightarrow Y$, then $f = f'g$ for some $f' : X \longrightarrow Y'$, and since $Y' \in \mathcal{Q}_\gamma \subseteq \mathcal{G}({}_AT)$, we obtain a proper factorization $f\Sigma = (f'\Sigma)(g\Sigma)$ of $f\Sigma$, contradicting the fact that $f\Sigma$ was supposed to be irreducible. Therefore $Y \in \mathcal{T}_\gamma(\rho) \cap \mathcal{G}({}_AT)$.

Finally, consider $(\mathcal{T}_\gamma(\rho) \cap \mathcal{G}({}_AT))\Sigma \cap B_0$-mod, for some ρ, and denote this module class just by $\mathcal{T}'$. We claim that $\mathcal{T}'$ is a component in B_0-mod. Now, as in the proof of 4.9.1, one shows that $\mathcal{T}' \approx k(\Omega(n))$ for some $n = n(\rho)$, thus any indecomposable module in $\mathcal{T}'$ is contained in a cycle, thus it is a regular B_0-module; also, any two indecomposable modules in $\mathcal{T}'$ are connected by a path, thus $\mathcal{T}'$ is contained in a single regular component of B_0-mod, call it $\mathcal{T}''$. Now, $\mathcal{T}''$ is a stable tube, thus it is sufficient to show that given an irreducible map $M \longrightarrow N$ in B_0-mod with M,N indecomposable, and $N \in \mathcal{T}'$, also $M \in \mathcal{T}'$. Since $\mathcal{P}''$ is closed under predecessors, with N also M belongs to $\mathcal{P}''$, thus $M \in \mathcal{Y}({}_AT)$, and, since $\mathcal{P}'$ is closed under irreducible maps, $M = \Sigma X$ for some $X \in \mathcal{T}_\gamma(\rho) \cap \mathcal{G}({}_AT)$, thus $M \in \mathcal{T}'$. This finishes the proof.

5.6 Self-reproduction of tubular families

We are going to exhibit some tubular algebras $C = C(4,\lambda)$, $C(6)$, $C(7)$, and $C(8)$, with $C(4,\lambda)$ being defined for all $\lambda \in k \setminus \{0,1\}$, and for anyone of these algebras, we want to show, first that $\mathcal{T}_1$ is a 1-controlled and 1-separating tubular family, and then that one can use shrinking functors in order to shift $\mathcal{T}_1$ to any $\mathcal{T}_\gamma$, $\gamma \in \mathbb{Q}^+$, and that in this way any $\mathcal{T}_\gamma$, $\gamma \in \mathbb{Q}^+$ is seen to be a γ-controlled and γ-separating tubular family.

The algebras will be given by quivers and relations. In order to write down the relations, we will denote the path $\alpha_1\alpha_2\dots\alpha_t$ just by $\alpha_{12\dots t}$, and similarly, for β.

C	quiver	relations	h_o	h_∞	d	$\mathbb{T}$
$C(4,\lambda)$ $\lambda \neq 0,1$	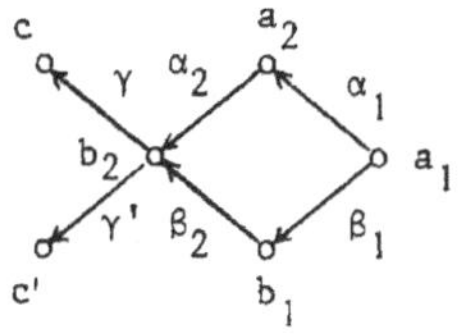	$(\alpha_{12}-\beta_{12})\gamma = 0$ $(\alpha_{12}-\lambda\beta_{12})\gamma' = 0$	${}^1_1 2^1_1 0$	${}^0_0 1^1_1 1$	2	(2,2,2,2)
$C(6)$	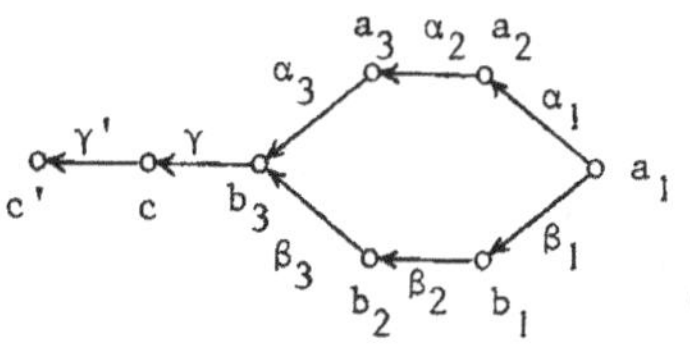	$(\alpha_{123}-\beta_{123})\gamma = 0$	$123^{21}_{21}0$	$001^{11}_{11}1$	3	(3,3,3)
$C(7)$	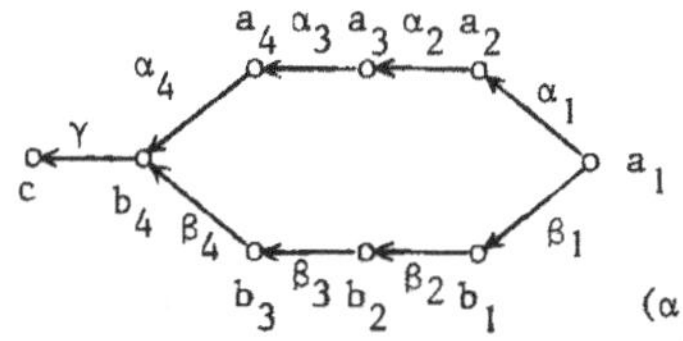	$(\alpha_{1234}-\beta_{1234})\gamma = 0$	$24^{321}_{321}0$	$01^{111}_{111}1$	4	(4,4,2)
$C(8)$	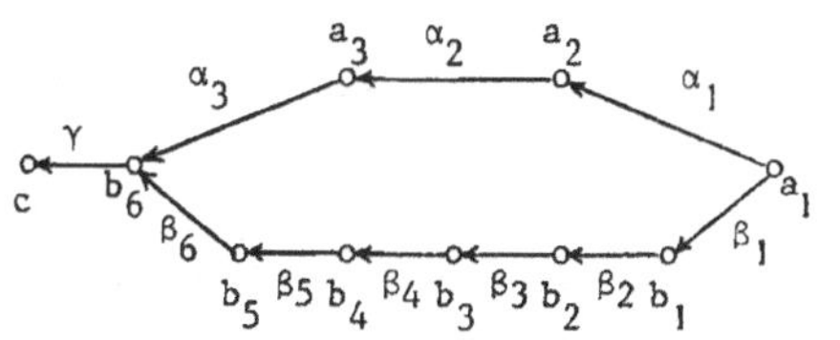	$(\alpha_{123}-\beta_{123456})\gamma = 0$	$36^{4\ 2}_{54321}0$	$01^{1\ 1}_{11111}1$	6	(6,3,2)

In the following, C will denote one of the algebras exhibited above. We will see below that C is both tubular and cotubular. We have listed the type $\mathbb{T}$ of C, the canonical radical generators h_0 and h_∞, and, in the column denoted by d, an important invariant, which will be used throughout. Of course, we denote by C_0, and C_∞ the restriction of C to the support of h_0, and h_∞, respectively; note that both C_0 and C_∞ are tame hereditary algebras, C_0 is given by a quiver of type $\widetilde{\mathbb{D}}_4$, $\widetilde{\mathbb{E}}_6$, $\widetilde{\mathbb{E}}_7$, or $\widetilde{\mathbb{E}}_8$, respectively, and C_∞ is given by a quiver of type $\widetilde{\mathbb{A}}_{22}$, $\widetilde{\mathbb{A}}_{33}$, $\widetilde{\mathbb{A}}_{44}$, or $\widetilde{\mathbb{A}}_{36}$, respectively. We denote by τ, τ_0, and τ_∞ the Auslander-Reiten translations of C, C_0, and C_∞, respectively.

(1) The C_0-module $R := \operatorname{rad} P(a_1)$ is simple regular of τ_0-period $d-1$, consequently C is a tubular algebra of extension type $\mathbb{T}$.

<u>Proof.</u> One easily checks that $\underline{\dim}\,R$ is equal to ${}^{1}_{1}2^{1}_{1}$, 112^{11}_{11}, 12^{111}_{111}, or 12^{11}_{11111}, respectively, and that R is indecomposable. In the cases $C(6)$, $C(7)$, $C(8)$, this immediately implies that R is simple regular of τ_0-period $d-1$. In case $C(4,\lambda)$, we have to check, in addition, that R has no submodule with dimension vector $\begin{smallmatrix}0&&0\\&1&\\0&&0\end{smallmatrix}$, $\begin{smallmatrix}1&&1\\&1&\\1&&1\end{smallmatrix}$, $\begin{smallmatrix}1&&1\\&1&\\0&&0\end{smallmatrix}$, $\begin{smallmatrix}0&&0\\&1&\\1&&1\end{smallmatrix}$, $\begin{smallmatrix}0&&1\\&1&\\1&&0\end{smallmatrix}$, or $\begin{smallmatrix}1&&0\\&1&\\0&&1\end{smallmatrix}$. In particular, we have to use that $\lambda \neq 0$ in order to exclude $\begin{smallmatrix}1&&1\\&1&\\0&&0\end{smallmatrix}$, and $\lambda \neq 1$ in order to exlude $\begin{smallmatrix}0&&0\\&1&\\0&&0\end{smallmatrix}$. Since C_0 is of tubular type $(2,2,2)$, $(3,3,2)$, $(4,3,2)$, or $(5,3,2)$, respectively, it follows that C is of tubular extension type $\mathbb{T}$.

(1*) The C_∞-module $Q(c)/E(c)$ is simple regular of τ_∞-period 1; in case $C(4,\lambda)$, the C_∞-module $Q(c')/E(c')$ is also simple regular of τ_∞-period 1, and not isomorphic to $Q(c')/E(c')$. As a consequence, C is a cotubular algebra of coextension type $\mathbb{T}$.

<u>Proof.</u> Always, $Q(c)/E(c)$ as C_∞-module is given by associating to any vertex the vector space k, and to any arrow the identity map 1_k. In case $C(4,\lambda)$, the C_∞-module $Q(c')/E(c')$ is given by

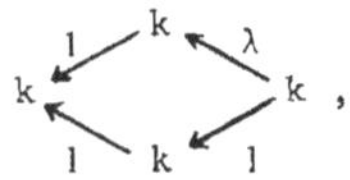

,

it is simple regular, since $\lambda \neq 0$, and non-isomorphic to $Q(c)/E(c)$, since $\lambda \neq 1$. Since C_∞ is of tubular type $(2,2)$, $(3,3)$, $(4,4)$, $(6,3)$, it follows that C is cotubular of coextension type $\mathbb{T}$.

(2) The module class $\mathcal{T}_1$ is a 1-controlled and 1-separating tubular $\mathbb{P}_1k$-family of type $\mathbb{T}$.

<u>Proof.</u> We want to apply the main theorem 3.4 to the algebra C_0. Let $Q_0(b_d)$ be the indecomposable injective C_0-module corresponding to the vertex b_d, and $W_0 = Q_0(b_d)\tau_0^{d-1}$. Let us indicate part of the preinjective component of C_0-mod, with the dimension vectors of all indecomposable injective modules, and of all modules of the form $Q_0(b_d)\tau_0^i$, $0 \leq i \leq d-1$.

C(4,λ)

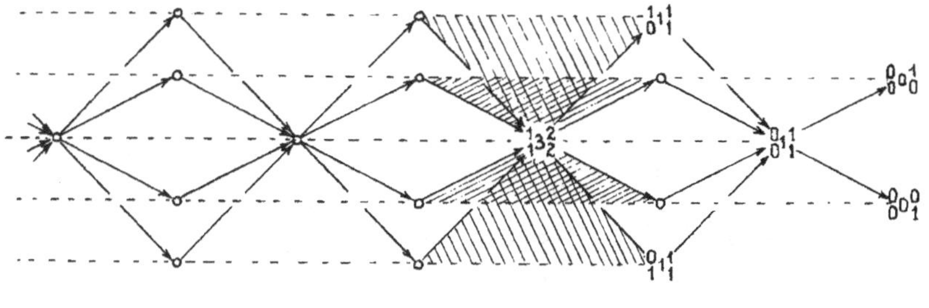

C(6)

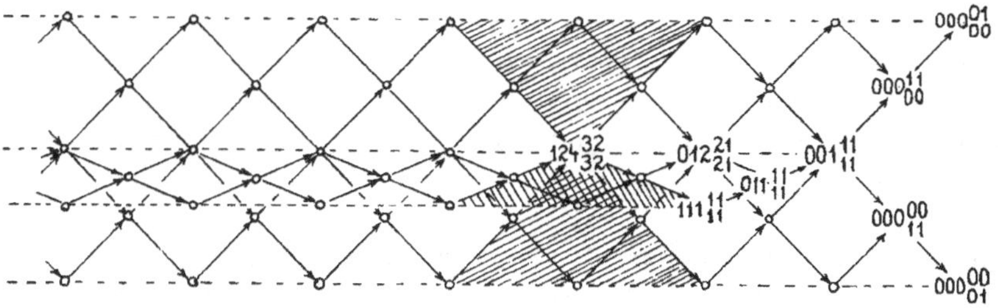

C(7)

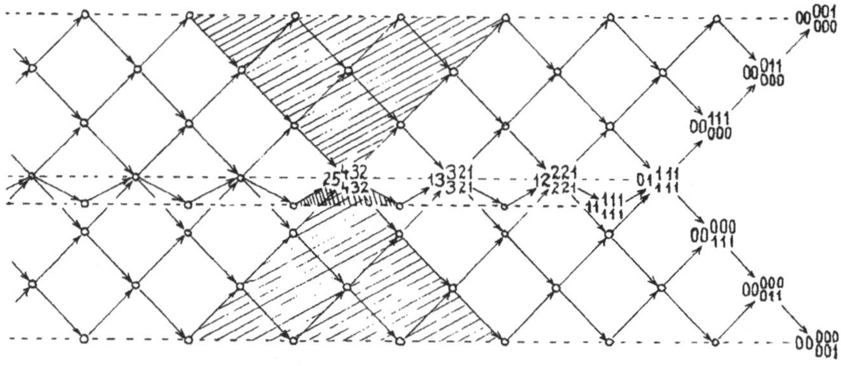

C(8)

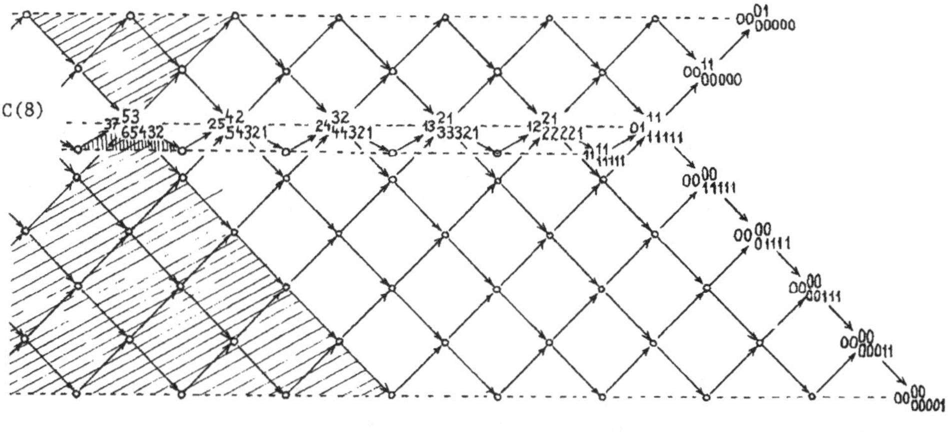

We see that W_0 is a sincere wing module of type $\mathbb{T}$, and it also is directing, since it belongs to the preinjective component of C_0-mod. We claim that W_0 is dominated by R. First of all,

$$(\underline{\dim}\, W_0)(I - \Phi_0^{-1}) = \underline{\dim}\, Q_0(b_d)\tau_0^{d-1} - \underline{\dim}\, Q_0(b_d)\tau_0^{d-2}$$

coincides with $\underline{\dim}\, R$. In order to see that proj.dim. $W_0(\rho) \leq 1$ for any $0 \neq \rho : R \to W_0$, we use (1*). Since C^{op} is a tubular algebra, we know that C has the sincere separating tubular family $\mathcal{T}_\infty$, separating $\mathcal{P}_\infty$ from $\mathcal{Q}_\infty$, and $\mathcal{P}_\infty$ contains all indecomposable modules M with $(\underline{\dim}\, M)\iota_\infty < 0$. Now, let $0 \neq \rho : R \to W_0$, so that the C-module $W_0(\rho)$ is indecomposable. Also, $\underline{\dim}\, W_0(\rho) = h_0 + h_\infty$, therefore $(\underline{\dim}\, W_0(\rho))\iota_\infty = (h_0 + h_\infty)\iota_\infty = h_0\iota_\infty = -d < 0$, thus $W_0(\rho)$ belongs to $\mathcal{P}_\infty$. It follows from 3.1.4 that proj.dim.$W_0(\rho) \leq 1$. This shows that W_0 is dominated by R, and we can apply the main theorem.

For anyone of these algebras C, we introduce functors Σ_ℓ, Σ_r : C-mod $\to$ C-mod with Σ_ℓ being a left shrinking functor, Σ_r a right shrinking functor.

Let Σ_ℓ be the left shrinking functor defined by $T_0 = ({}_{C_0}C_0)\tau_0^{-d+1}$. Note that ${}_{C_0}C_0$, and also $({}_{C_0}C_0)\tau_0^{-d+1}$ are slice modules in the preprojective component of C_0-mod. The shrinking module defined by T_0 is $T = T_0 \oplus P(a_1)$. In order to identify $\mathrm{End}({}_C T)$ with C, we first identify C_0 with $\mathrm{End}(({}_{C_0}C_0)\tau_0^{-d+1})$ using the usual right action of C_0 on ${}_{C_0}C_0$ and the functoriality of τ_0^{-d+1} (note that C_0 is hereditary). Using that τ_0^- is left adjoint to τ_0, we see that for any C_0-module M, we have

$$\mathrm{Hom}_{C_0}(T_0, M_0) = \mathrm{Hom}_{C_0}(({}_{C_0}C_0)\tau_0^{-d+1}, M_0) \approx \mathrm{Hom}_{C_0}({}_{C_0}C_0, M_0\tau_0^{d-1}) \approx M_0\tau_0^{d-1}$$

as left C_0-modules. Thus, the functors $\mathrm{Hom}_{C_0}(T_0, -)$, and τ_0^{d-1} from C_0-mod to C_0-mod are equivalent. Since $R = \mathrm{rad}\, P(a_1)$ is τ_0-periodic of period $d-1$,

$$\mathrm{Hom}_C(T_0, P(a_1)) = \mathrm{Hom}_{C_0}(T_0, R) \approx R\,\tau_0^{-d+1} \approx R,$$

as left C_0-modules, and, of course, also as right $\mathrm{End}(P(a_1))$-modules (note that $\mathrm{End}(P(a_1)) = k$). It follows that the $\mathrm{End}(T_0)$-$\mathrm{End}(P(a_1))$-bimodule $\mathrm{Hom}_C(T_0, P(a_1))$ is isomorphic to the bimodule ${}_{C_0}R_k$, thus we obtain an isomorphism of k-algebras $C = C_0[R] \approx \mathrm{End}({}_C T_0)[\mathrm{Hom}_C(T_0, P(a_1))] = \mathrm{End}({}_C T)$. Note that the restriction of Σ_ℓ to C_0-mod is given by τ_0^{d-1}, thus the linear transformation σ_ℓ corresponding to Σ_ℓ is given by $\sigma_\ell \mid K_0(C_0) = \Phi_0^{d-1}$, with Φ_0 being the Coxeter transformation of C_0, and $e(a_1)\sigma_\ell = e(a_1)$.

Let Σ_r be the right shrinking functor defined by $S_\infty = \bigoplus_{i=1}^{d'} Q(a_i)\tau_\infty \oplus \bigoplus_{i=1}^{d} Q(b_i)$, where $d' = 3$ in case $C = C(8)$ and $d' = d$ in all other cases. Note that τ_∞ denotes the Auslander-Reiten translation of C_∞, however, since $Q(a_i)\tau_\infty$ is also the τ-translate in C-mod, we may delete the index. Note that S_∞ is a slice module in the preinjective component of C_∞-mod. The right shrinking functor S defined by S_∞ is $Q(c) \oplus Q(c') \oplus S_\infty$ in the cases $C(4,\lambda)$ and $C(6)$, and $Q(c) \oplus S_\infty$ in the cases $C(7)$ and $C(8)$. In order to identify $\text{End}(_CS)$ with C, let us construct the modules $Q(a_i)\tau$ as pullbacks of β_1^* and $\alpha_i^* \ldots \alpha_1^*$ [recall that given an arrow $\delta : x \longrightarrow y$ in the quiver of C, we denote by $\delta^* : Q(y) \longrightarrow Q(x)$ the induced map]. For example, in case $C(6)$, we obtain the following diagram

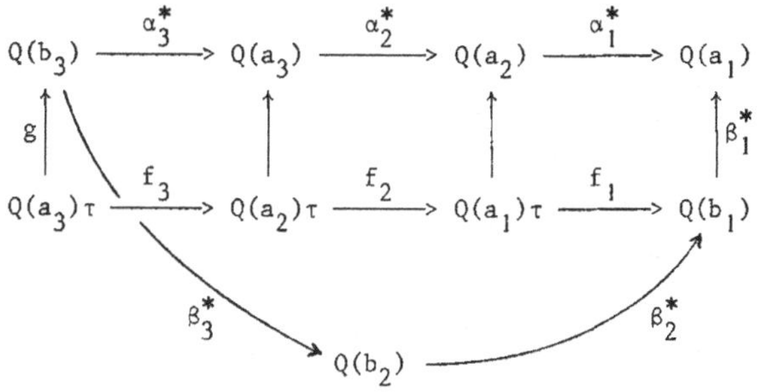

Here, the three squares are pullbacks. Also, consider the C-module $Q(c)$. The homomorphisms γ^* and $\gamma^*\beta_3^*\beta_2^*$ satisfy the relation

$$\gamma^*\alpha_3^*\alpha_2^*\alpha_1^* = (\gamma^*\beta_3^*\beta_2^*)\beta_1^*,$$

thus there exists a unique $h : Q(c) \longrightarrow Q(a_3)\tau$ such that $hg = \gamma^*$, $hf_3f_2f_1 = \gamma^*\beta_3^*\beta_2^*$. Thus we obtain the following set of indecomposable modules and maps

$$Q(c') \xrightarrow{(\gamma')^*} Q(c) \xrightarrow{h} Q(a_3)\tau \underset{g}{\overset{f_3}{\nearrow}} \begin{array}{c} Q(a_2)\tau \xrightarrow{f_2} Q(a_1)\tau \searrow^{f_1} \\ \\ Q(b_3) \xrightarrow{\beta_3^*} Q(b_2) \nearrow_{\beta_2^*} \end{array} Q(b_1)$$

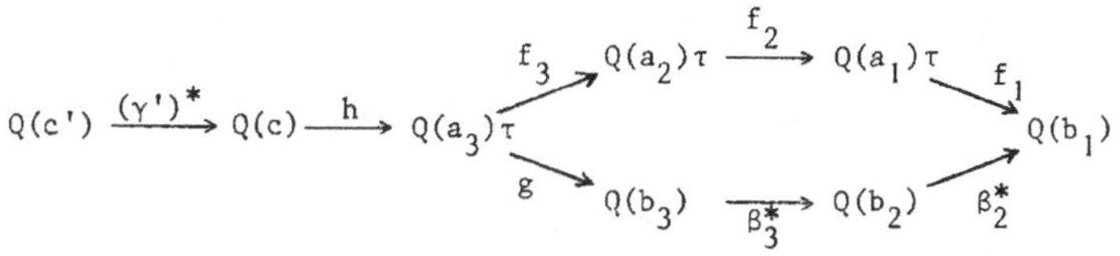

satisfying $hf_3f_2f_1 = \gamma^*\beta_3^*\beta_2^* = hg\beta_3^*\beta_2^*$, thus $h(f_3f_2f_1 - g\beta_3^*\beta_2^*) = 0$. Similarly, for $C(7)$ and $C(8)$ we obtain the following set of indecomposable modules and maps

$$Q(c) \xrightarrow{h} Q(a_4)\tau \underset{g}{\overset{f_4}{\nearrow}} \begin{array}{c} Q(a_3)\tau \xrightarrow{f_3} Q(a_2)\tau \xrightarrow{f_2} Q(a_1)\tau \searrow^{f_1} \\ \\ Q(b_4) \xrightarrow{\beta_4^*} Q(b_3) \xrightarrow{\beta_3^*} Q(b_2) \nearrow_{\beta_2^*} \end{array} Q(b_1)$$

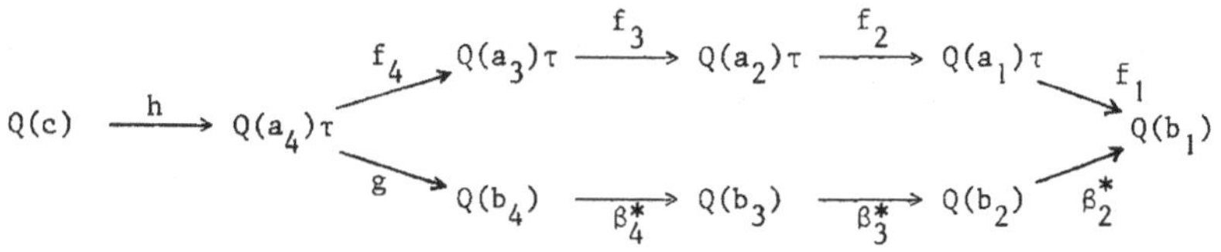

with $h(f_4f_3f_2f_1 - g\beta_4^*\beta_3^*\beta_2^*) = 0$, and

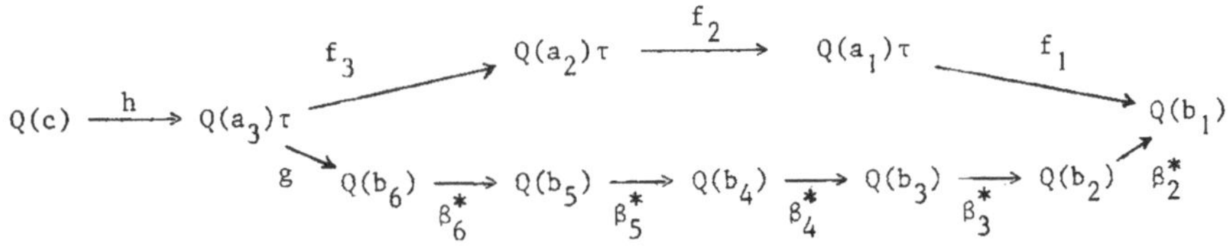

with $h(f_3f_2f_1 - g\beta_6^*\beta_5^*\beta_4^*\beta_3^*\beta_2^*) = 0$. Finally, for $C(4,\lambda)$, the construction of $Q(a_i)\tau$ gives the following diagram, both squares being pullbacks

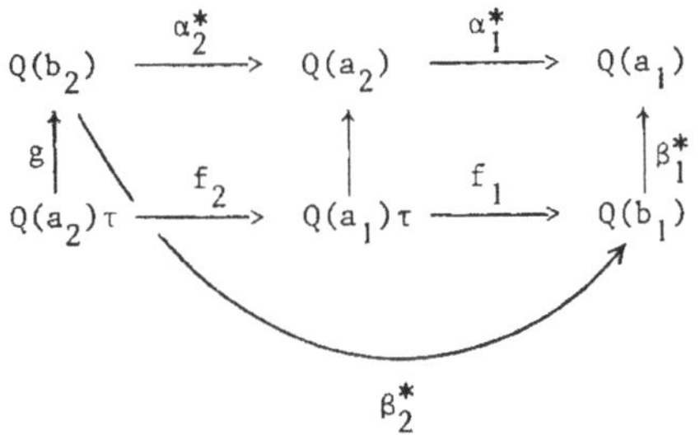

As before, starting with the homomorphism γ^*, we obtain $h : Q(c) \longrightarrow Q(a_2)\tau$ satisfying $hg = \gamma^*$, $hf_2f_1 = \gamma^*\beta_2^*$. In this case, there also is given $(\gamma')^* : Q(c') \longrightarrow Q(a_2)\tau$, satisfying the relation

$$(\gamma')^*\alpha_2^*\alpha_1^* = \lambda(\gamma')^*\beta_2^*\beta_1^* .$$

Using the pullback property, we obtain a unique $h' : Q(c) \longrightarrow Q(a_2)\tau$ satisfying both $h'g = (\gamma')^*$ and $h'f_2f_1 = \lambda(\gamma')^*\beta_2^*$. Thus we have $h'f_2f_1 = \lambda(\gamma')^*\beta_2^* = \lambda h'g\beta_2^*$. Thus we deal with the following set of indecomposable modules and maps

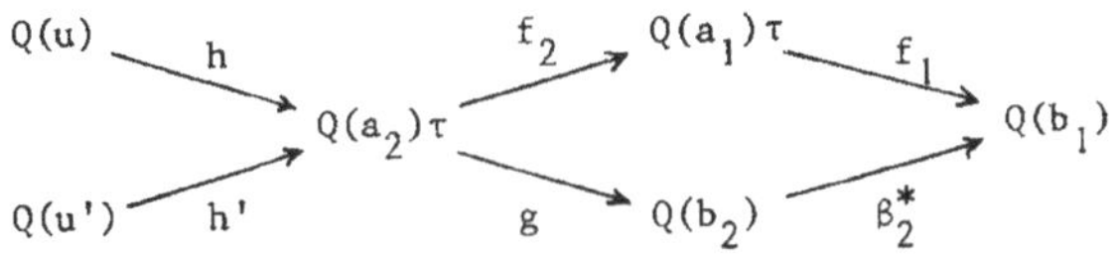

satisfying $h(f_2f_1 - g\beta_2^*) = 0$, and $h'(f_2f_1 - \lambda g\beta_2^*) = 0$. In all cases, we define a k-algebra morphism from C to $\mathrm{End}({}_CS)$ by sending the idempotents $(a_1|a_1)$; $(a_i|a_i)$, $2 \leq i \leq d'$; $(b_j|b_j)$, $1 \leq j < d$; $(b_d|b_d)$; $(c|c)$; and, if it exists, $(c'|c')$, to the canonical projections of ${}_CS$ onto $Q(b_1)$; $Q(a_{i-1})\tau$, $2 \leq i \leq d'$; $Q(b_{j+1})$, $1 \leq j < d$; $Q(a_{d'})\tau$; $Q(c)$; and, if it exists $Q(c')$, respectively, and by sending the radical generators α_i^*, $1 \leq i \leq d'$; β_j^*, $1 \leq j < d$; β_d^*; γ^*; and, if it exists, $(\gamma')^*$, to the endomorphisms of ${}_CS$ given by f_i; $1 \leq i \leq d'$; β_{j+1}^*, $1 \leq j < d$; g; h; and h' (in case $C(4,\lambda)$) or $(\gamma')^*$ (in case $C(6)$), respectively. One easily checks that these latter endomorphisms of ${}_CS$ generate the radical of $\mathrm{End}({}_CS)$, it follows that the constructed algebra morphism $C \longrightarrow \mathrm{End}({}_C S)$ is surjective, and a comparison of the dimensions of C and $\mathrm{End}({}_C S)$ shows that it is, in fact, bijective [namely, both C and $\mathrm{End}({}_C S)$ are obviously obtained from C_∞ by coextensions using modules with dimension vectors $1{}^{1\dots 1}_{1\dots 1}1$.] This shows that we can identify $\mathrm{End}({}_C S)$ with C, and, we always will use the identification constructed above. Let us write down the linear transformation σ_r corresponding to the shrinking functor $\Sigma_r = \mathrm{Hom}(-,{}_C S)D$. Its restriction to $K_0(C_\infty)$ is given by

$$\begin{pmatrix} & x_{d'} \ \dots \ x_2 & \\ y_d & & x_1 \\ & y_{d-1} \dots \ y_1 & \end{pmatrix} \longmapsto \begin{pmatrix} & z_{d'-1} \dots \ z_1 & \\ z_{d'} & & y_1 \\ & y_d \ \dots \ y_2 & \end{pmatrix},$$

where $z_i = x_{i+1} + y_1 - x_1$ for $1 \leq i < d'$, and $z_{d'} = y_d + y_1 - x_1$, whereas $e(c)\sigma_r = e(c)$, and, if it exists, $e(c')\sigma_r = e(c')$.

(3) We have $h_\infty \sigma_\ell = h_0 + h_\infty = h_0 \sigma_r$, thus $\beta \bar{\sigma}_\ell = \frac{\beta}{\beta+1}$, $\quad \beta \bar{\sigma}_r = \beta + 1$ for $\beta \in \mathbb{Q}_0^\infty$.

<u>Proof.</u> Again, denote by $Q_0(b_d)$ the indecomposable injective C_0-module corresponding to b_d, and $W_0 = \tau_0^{d-1} Q_0(b_d)$. Now $h_\infty \mid K_0(C_0) = \underline{\dim}\, Q_0(b_d)$, thus

$$\begin{aligned}(h_\infty \sigma_\ell) \mid K_0(C_0) = (h_\infty \mid K_0(C_0))\, \Phi_0^{-d+1} &= (\underline{\dim}\, Q_0(b_d))\, \Phi_0^{-d+1} \\ &= \underline{\dim}\, W_0 = (h_0 + h_\infty) \mid K_0(C_o).\end{aligned}$$

Since the coefficient of $e(a_1)$ is the same for both h_∞ and $h_0 + h_\infty$, and is not changed under σ_ℓ, it follows that $h_\infty \sigma_\ell = h_0 + h_\infty$.

The formula given above for the restriction of σ_r to $K_0(C_\infty)$ immediately shows that $h_0 \sigma_r = h_0 + h_\infty$.

The second assertion now follows from 5.4.

Consider now sequences $i_1, \dots, i_n \in \{\ell, r\}$, and let $\Sigma_{i_1 i_2 \dots i_n} = \Sigma_{i_1} \Sigma_{i_2} \cdots \Sigma_{i_n}$, $\sigma_{i_1 i_2 \dots i_n} = \sigma_{i_1} \sigma_{i_2} \cdots \sigma_{i_n}$, and $\bar{\sigma}_{i_1 i_2 \dots i_n} = \bar{\sigma}_{i_1} \bar{\sigma}_{i_2} \cdots \bar{\sigma}_{i_n}$. Note that $\Sigma_{i_1 \dots i_n}$ is an endofunctor of C-mod, $\sigma_{i_1 i_2 \dots i_n}$ is an endomorphism of $K_0(C)$, and $\bar{\sigma}_{i_1 i_2 \dots i_n}$ maps $\mathbb{Q}_0^\infty$ into itself. (Note that $\sigma_{i_1 i_2 \dots i_n}$ is invertible, however $\bar{\sigma}_{i_1 i_2 \dots i_n}$ is injective but for $n \geq 1$ not surjective). Of course, for $n = 0$, we deal with the identity functor of C-mod, the identity transformation of $K_0(C)$ and the identity map of $\mathbb{Q}_0^\infty$.

(4) Any element of $\mathbb{Q}^+$ can be written in a unique way as $1\, \bar{\sigma}_{i_1 \dots i_n}$ with $i_1 \dots i_n \in \{\ell, r\}$.

<u>Proof.</u> Given $\beta = \frac{\beta'}{\beta''} \in \mathbb{Q}^+$, with $\beta', \beta'' \in \mathbb{N}_1$ such that β', β'' have no proper common divisor, let $N(\beta) = \beta' + \beta''$. The proof will be by induction on $N(\beta)$.

Note that for $\beta = \frac{\beta'}{\beta''}$, we have $\beta\bar{\sigma}_{\ell} = \frac{\beta}{\beta+1} = \frac{\beta'}{\beta'+\beta''}$, and $\beta\bar{\sigma}_r = \beta+1 = \frac{\beta'+\beta''}{\beta''}$. In particular, $N(\beta\bar{\sigma}_{\ell}) = N(\beta\bar{\sigma}_r) > N(\beta)$, for any $\beta \in \mathbb{Q}^+$. The smallest possible value of $N(\beta)$ on $\mathbb{Q}^+$ is 2, namely for $\beta = 1$, and, in this case, we have to take the sequence with $n = 0$. Namely, for $n \geq 1$, we have $\mathbb{Q}^+\bar{\sigma}_{i_1\ldots i_n} \subseteq \mathbb{Q}^+\bar{\sigma}_{i_n}$, and $\mathbb{Q}^+\bar{\sigma}_{\ell} = \{\gamma \in \mathbb{Q} \mid 0 < \gamma < 1\}$, whereas $\mathbb{Q}^+\bar{\sigma}_r = \{\gamma \in \mathbb{Q} \mid 1 < \gamma\}$. Now assume the assertion is true for any γ with $N(\gamma) < m$, where $m > 2$. Let $\beta', \beta'' \in \mathbb{N}_1$ without proper common divisor, and $\beta = \frac{\beta'}{\beta''}$, $N(\beta) = \beta' + \beta'' = m$. First, consider the case $\beta' < \beta''$, then

$$\beta = \frac{\beta'}{\beta''} = \frac{\beta'}{\beta' + (\beta''-\beta')} = \frac{\beta'}{\beta''-\beta'}\bar{\sigma}_{\ell},$$

and $N(\frac{\beta'}{\beta''-\beta'}) = \beta'' < \beta' + \beta'' = N(\frac{\beta'}{\beta''}) = m$, thus by induction

$$\frac{\beta'}{\beta''-\beta'} = 1\bar{\sigma}_{i_1\ldots i_n}$$

for some unique sequence $i_1, \ldots, i_n \in \{\ell, r\}$, and

$$\beta = \frac{\beta'}{\beta''} = \frac{\beta'}{\beta''-\beta'}\bar{\sigma}_{\ell} = 1\bar{\sigma}_{i_1\ldots i_n\ell}.$$

Also, since $\beta' < \beta''$, we see that β lies not in the image of $\bar{\sigma}_r$, thus, if $\beta = 1\bar{\sigma}_{j_1\ldots j_t}$ for some sequence $j_1, \ldots, j_t \in \{\ell, r\}$, then $j_t = \ell$, and therefore $1\bar{\sigma}_{j_1\ldots j_{t-1}} = \frac{\beta'}{\beta''-\beta'} = 1\bar{\sigma}_{i_1\ldots i_{n-1}}$. By induction we have $t-1 = n-1$, and $i_s = j_s$ for all $1 \leq s \leq n-1$. This shows the unicity.

The second case of $\beta' > \beta''$ can be considered similarly. This finishes the proof.

(5) Assume C is one of the algebras $C(4,\lambda)$, $\lambda \in k \setminus \{0,1\}$, $C(6)$, $C(7)$, or $C(8)$, and of type $\mathbb{T}$. Then, for any $\beta \in \mathbb{Q}^+$, the module class $\mathcal{T}_\beta$ is a β-controlled and β-separating tubular $\mathbb{P}_1k$-family of type $\mathbb{T}$.

Proof. We write β in the form $1\bar{\sigma}_{i_1\ldots i_n}$, with $i_1, \ldots, i_n \in \{\ell, r\}$, and use induction on n. For $n = 0$, the assertion is given in (1), thus assume $n \geq 1$. Consider first the case $i_n = \ell$. We want to apply 5.4.2 for $A = C$, $B = C$, $\Sigma = \Sigma_{\ell}$, and $\gamma = 1$. According to (1), the assumptions on $\mathcal{T}^B_\gamma = \mathcal{T}^C_1$ are satisfied, also any indecomposable C-module N with $\underline{\dim}\, N = h_\infty$ satisfies $(\underline{\dim}\, N)\sigma_{\ell}\iota_\gamma = 0$, since $\gamma = 1 = \infty\bar{\sigma}_{\ell}$, and $\mathrm{Ext}^1_C(N,N) \neq 0$. Let $\delta = 1\bar{\sigma}_{i_1\ldots i_{n-1}}$, thus $\delta < 1 = \gamma$, and $\beta = \delta\bar{\sigma}_{\ell}$, $\sigma_{\ell}\iota_\beta = \sigma_{\ell}\iota_{\delta\sigma_{\ell}} = \iota_\delta$ (see 5.4.c). By induction, $\mathcal{T}_\delta$ is a δ-controlled, δ-separating tubular $\mathbb{P}_1k$-family of type $\mathbb{T}$, thus 5.4.2 asserts that $\mathcal{T}_\beta$ is a β-controlled, β-separating tubular family equivalent to $\mathcal{T}_\delta$, thus also a tubular $\mathbb{P}_1k$-family of type $\mathbb{T}$. In case $i_n = r$, we similarly apply the dual of 5.4.2. This finishes the proof.

5.7 The general case

Here, we are going to complete the proof of the assertions 5.2.2, 5.2.3, and 5.2.5. The proof of 5.2.2. will be a reduction to the case of one of the algebras considered in the last sections. This reduction is done in three steps. Note that a canonical tubular algebra is also cotubular.

(1) Let A be a canonical tubular algebra. Then there exists a proper left shrinking functor from A-mod to C-mod, where C is one of the algebras $C(4,\lambda)$, $\lambda \in k \setminus \{0,1\}$; $C(6)$; $C(7)$; or $C(8)$.

Proof. Let A be of type $\mathbb{T}$, thus $A = A_o[R]$, with A_o being given by the star $\mathbb{T}$ with subspace orientation, and R is a coordinate module for A_o, or, equivalently, R is simple regular of period $d-1$, with $d = 2,3,4$, or 6, in case $\mathbb{T} = \widetilde{\mathbb{D}}_4$, $\widetilde{\mathbb{E}}_6$, $\widetilde{\mathbb{E}}_7$, or $\widetilde{\mathbb{E}}_8$, respectively, see 4.8.4. Denote by 0 the centre of the star $\mathbb{T}$, it is the unique sink in the quiver of A.

For any orientation of $\mathbb{T}$, there are slices in the preprojective component of A_o of this form. In particular, there is a preprojective tilting module T_o with $\mathrm{End}({}_{A_o}T_o) = C_o$, where C_o is of the form

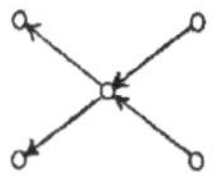 , 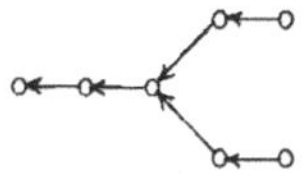 , 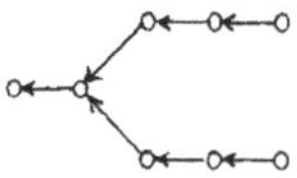 , or 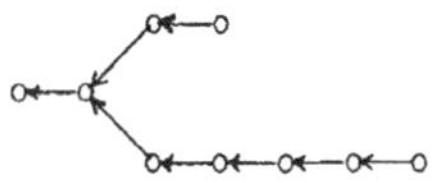 .

In addition, we can assume that T_o contains as a direct summand the module $\tau^{-d+1}P(0)$. As a consequence, $\mathrm{Hom}_{A_o}(T_o,R)$ being always simple regular of period $d-1$, has the central component of its dimension vector equal to 2, namely

$$\mathrm{Hom}_{A_o}(\tau^{-d+1}P(0),R) \approx \mathrm{Hom}_{A_o}(P(0),\tau^{d-1}R)$$

is 2-dimensional. Thus $\mathrm{Hom}_{A_o}(T_o,R)$ is simple regular with dimension vector

$${}^{1}_{1}2^{1}_{1}\ , \quad 112^{11}_{11}\ , \quad 12^{111}_{111}\ , \quad \text{or} \quad 12^{11}_{11111}\ ,$$

respectively. Thus, $C = \mathrm{End}(T_o)\,[\mathrm{Hom}(T_o,R)]$ is of the form $C(4,\lambda)$, $C(6)$, $C(7)$, or $C(8)$, respectively, and the left shrinking functor Σ_T defined by T_o goes from A-mod to C-mod.

In order to see that Σ_T is proper, let $\sigma = \sigma_T$ be the corresponding linear transformation. Let us denote by ω the extension vertex of A, and also of B. Then the component $(h_\infty^B\sigma)_\omega$ of $h_\infty^B\sigma$ at ω coincides with the corresponding component $(h_\infty^A)_\omega = d$ of h_∞^A. Note that this immediately implies that $h_\infty^A\sigma \neq h_\infty^C$

[since $(h_\infty^C)_\omega = 1$ in all cases], and, since both $h_\infty^A\sigma$, and h_∞^C are positive and primitive, we must have $h_\infty^A\sigma = \eta_0 h_0^C + dh_\infty^C$ with $\eta_0 \neq 0$, thus $\infty\bar{\sigma} = \frac{d}{\eta_0} < \infty$.

Remark. An actual calculation shows that

$$h_\infty^A\sigma = (d-1)h_0^C + dh_\infty^C ,$$

thus we have $\infty\bar{\sigma} = \frac{d}{d-1}$.

Proof. We have noted above that $(h_\infty^A\sigma)_\omega = d$, thus it remains to consider the restriction of $h_\infty^A\sigma$ to $K_0(C_0)$. This is most easily achieved as follows: Let Δ be the quiver of A_0, it is just $\mathbb{T}$ with subspace orientation. The preprojective component of A_0 may be identified with $\mathbb{N}_0\Delta^*$, and inside $\mathbb{N}_0\Delta^*$ we have the tilting set corresponding to the tilting module T_0 (unique up to a symmetry of $\mathbb{N}_0\Delta^*$). Of course, the full subquiver of $\mathbb{N}_0\Delta^*$ given by this tilting set may be identified with the dual of the quiver of C_0. We construct the additive function f on $\mathbb{Z}\mathbb{T} = \mathbb{Z}\Delta^*$ which has as restriction to $0\times\Delta^*$ just $h_\infty^A \mid A_0$, and $h_\infty^C \mid C_0$ will be just the restriction of f to the tilting set defining C_0. In the four cases, f is as follows (where we give only the values of f on the vertices lying between $0\times\Delta^*$ and the chosen tilting set):

$\mathbb{T}$		$h_\infty^A\sigma$
$\tilde{D}_4$	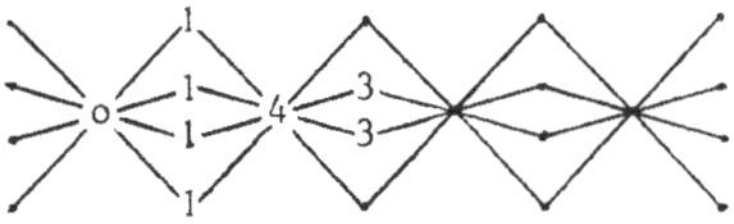	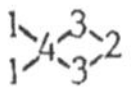
$\tilde{E}_6$	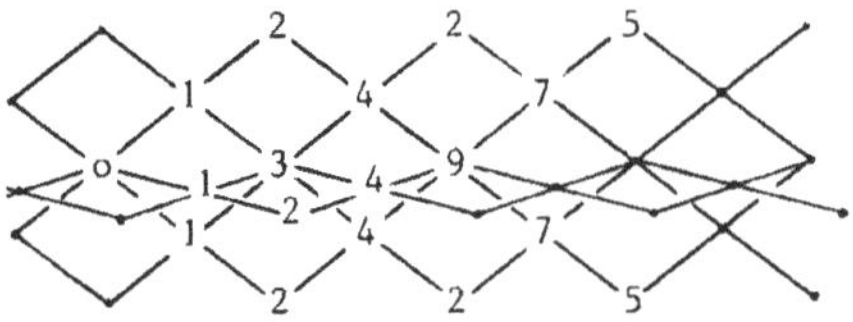	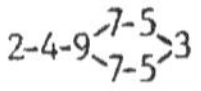
$\tilde{E}_7$	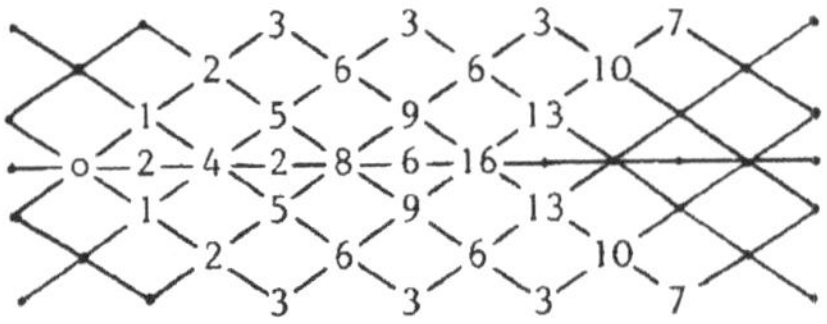	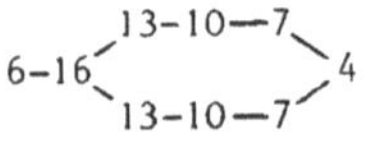
$\tilde{E}_8$	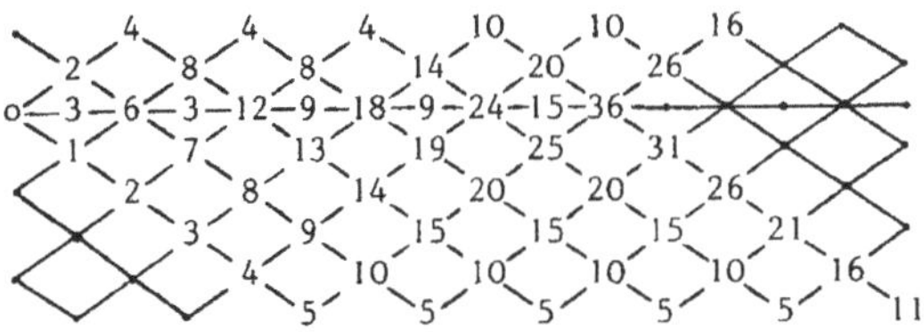	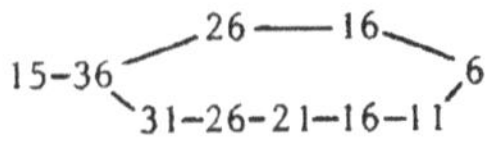

(2) Let B_0 be a tame concealed bush algebra, let M be a coordinate module for B_0, and $B = B_0[M]$. Then B is both tubular and cotubular. Assume that B itself is not a canonical algebra. Then there exists a proper left shrinking functor to a canonical tubular algebra C.

<u>Proof.</u> According to 4.8.1, the algebra B^{op} is a tubular extension of a canonical algebra $(B_\infty)^{op}$ of extension type $\mathbf{T}$. We assume that B itself is not canonical, or, equivalently, that B_0 is not given by $\mathbf{T}$ with subspace orientation. Note that this also is equivalent to the fact that B is a proper coextension of B_∞. Of course, this implies that the quadratic form χ_{B_∞} of B_∞ is semi-definite with radical rank 1, thus B_∞ is tame concealed. It follows that B is cotubular.

Let z be the centre of the quiver of B_0. There exists a preprojective tilting B_0-module T_0 with $C_0 = \mathrm{End}({}_{B_0}T_0)$ being given by the quiver $\mathbb{T}$ with subspace orientation, and such that $P(z)$ is a direct summand of T_0. Let ${}_BT$ be the shrinking B-module defined by T_0, it follows that $\mathrm{Hom}({}_{B_0}T_0, M)$ is a coordinate module for B_0, and thus $C = \mathrm{End}({}_BT)$ is a canonical tubular algebra.

It remains to be shown that $\Sigma = \Sigma_T$ is proper. We claim that $\sigma = \sigma_T$ satisfies $\infty\bar{\sigma} = 1$. Namely, the minimal positive radical generator h_∞^B in $K_0(B_\infty)$ has all its components (in B_∞) equal to 1, in particular $(h_\infty^B)_z = 1$, $(h_\infty^B)_\omega = 1$, with ω being the extension vertex of the extension $B_0[M]$ (or also of B_∞ considered as an extension of a Dynkin bush by a coordinate module). Now, the component $(h_\infty^B\sigma)_\omega$ is not changed, thus $(h_\infty^B\sigma)_\omega = 1$. Also, since we assume that $P(z)$ is a direct summand of T_0, we have

$$(h_\infty^B\sigma)_z = \langle p(z), h_\infty^B\rangle = (h_\infty^B)_z = 1.$$

Thus, $h_\infty^B\sigma$ has value 1 both at the source and at the sink of the quiver of C, thus

$$h_\infty^B\sigma = \frac{1}{d}(h_0^C + h_\infty^C),$$

(where $d = 2,3,4,$ or 6, as usual), therefore $\infty\bar{\sigma} = 1$.

Note that if B is a tubular algebra which is a tubular extension of some tame concealed, canonical algebra, then B^{op} is the one-point extension of a tame concealed bush algebra by a coordinate module (according to 4.8.1), thus by the previous consideration, B^{op} is both tubular and cotubular, therefore B is both tubular and cotubular.

(3) Let A be a tubular algebra. Then there exists a left shrinking functor $\Sigma : A\text{-mod} \longrightarrow B\text{-mod}$, where B is a tubular extension of a tame concealed, canonical algebra. In case A is known to be also cotubular, we can find such a left shrinking functor which is proper.

Proof. Let A_0 be of tubular type $\mathbb{T}$. Fix some vertex a belonging to A_0 and A_∞. Let $A_0(a)$ be obtained from A_0 by deleting a, thus $A_0(a)$ is representation finite, since A_0 is minimal representation infinite. The component P_0 of all preprojective A_0-modules contains a module class M with ΓM being a full translation subquiver of ΓP_0, and such that we can identify $\Gamma M = \mathbb{N}_0 \widetilde{\mathbb{T}}$. There are only finitely many indecomposable $A_0(a)$-modules, thus there is a subclass M' with $\Gamma M' = [m,\infty)] \widetilde{\mathbb{T}}$ for some $m \in \mathbb{N}_0$ such that M' contains no indecomposable $A_0(a)$-module. According to the remark after 4.3.1, we can choose a tilting set $\{t_1,\dots,t_n\}$ in $\Gamma = [m,\infty) \widetilde{\mathbb{T}}$ such that $\mathrm{End}_{k(\Gamma)} (\oplus t_i)$ is a (tame) canonical algebra. The corresponding module T_0 in $M' \subseteq P_0$ is a tilting A_0-module. Thus, T_0 is a preprojective tilting A_0-module, with $\mathrm{End}(T_0)$ being tame concealed, canonical, and such that no indecomposable summand of T_0 is an $A_0(a)$-module. Let ${}_AT = T_o \oplus T_p$ be the shrinking A-module defined by T_0, let $B = \mathrm{End}({}_AT)$, and $\Sigma = \Sigma_T$, $\sigma = \sigma_T$.

Now assume, A is also cotubular. We claim that $\infty\bar{\sigma} < \infty$. Namely, let $T^A(\rho)$ be some homogeneous tube in T^A_∞, thus, for all modules $M \in T^A_\infty(\rho)$, the dimension vector $\underline{\dim}\, M$ is a multiple of h^A_∞. Consider an indecomposable direct summand $T(i)$ of ${}_AT$. If $T(i)$ is a summand of T_0, then $\mathrm{Hom}_A(T(i),Q(a)) \neq 0$, and $T(i) \in P_0 \subseteq P_\infty$, $Q(a) \in Q_\infty$ (since a is a vertex of A_∞), thus we can factor any map $T(i) \longrightarrow Q(a)$ through $T^A_\infty(\rho)$. In particular, $\mathrm{Hom}(T(i),T^A_\infty(\rho)) \neq 0$. Also, for $T(i)$ a summand of T_p, there is a vertex b outside of A_0, thus in A_∞, with $T(i) = P(b)$, and again $Q(b) \in Q_\infty$. Since $T(i) = P(b) \in T_0 \subseteq P_\infty$, we can factor a non-trivial map $P(b) \longrightarrow Q(b)$ through $T^A_\infty(\rho)$ and obtain again $\mathrm{Hom}(T(i),T^A_\infty(\rho)) \neq 0$. It follows that $h^A_\infty\sigma$ is a sincere vector in $K_0(B)$. Therefore $h^A_\infty\sigma \neq h^B_\infty$. Since both $h^A_\infty\sigma$ and h^B_∞ are primitive positive vectors, we must have $h^A_\infty\sigma = \eta_0 h^B_0 + \eta_\infty h^B_\infty$ with $\eta_0 \neq 0$, thus $\infty\bar{\sigma} \neq \infty$, and therefore Σ is proper.

(4) Proof of 5.2.2. We use 5.4.3' in order to extend the validity of the assertion from the algebras $C(4,\lambda)$, $C(6)$, $C(7)$, and $C(8)$ first to the canonical tubular algebras, according to (1), then to the one-point extensions of tame concealed bush algebras by coordinate modules, according to (2). Since the assertion 5.2.2 is self-dual, and the opposite algebra of a tubular algebra B which is a tubular extension of some tame concealed canonical algebra, is the one-point extension of a tame concealed bush algebra by a coordinate module, we also have shown assertion 5.2.2 for such an algebra B. Now we use again 5.4.3' together with (3) to deal with the general case of a tubular and cotubular algebra.

(5) Proof of 5.2.3. Let A be a tubular algebra. In (3), we have constructed a shrinking module ${}_AT$ with $B = \mathrm{End}({}_AT)$ being a tubular extension of a tame concealed, canonical algebra. In particular, B is not only tubular, but also cotubular. Let $\Sigma = \Sigma_T$, and note that ${}_AD\Sigma \in Q^B_0$. However, ${}_AD\Sigma = ({}_AT_B)D$ is a cotilting B-module, and its endomorphism ring is A. The dual result of 5.5.1 asserts that A is a cotubular algebra, and this is what we wanted to show.

5.8 Tubular vectorspace categories

A tubular extension $(K, |\circ|)$ of a critical directed vectorspace category of extension type (2,2,2,2), (3,3,3), (4,4,2), or (6,3,2) will be called a tubular vectorspace category, its extension type will be called just the type of $(K, |\cdot|)$. Note that the rank of $G(K)$ is 5,7,8, or 9, respectively. A tubular coextension $(K, |\cdot|)$ of a critical directed vectorspace category of coextension type (2,2,2,2), (3,3,3), (4,4,2), or (6,3,2) will be called a cotubular vectorspace category.

(1) Classification of the tubular vectorspace categories: there are two infinite families of tubular vectorspace categories of type (2,2,2,2), and 5, 10, and 24 tubular vectorspace categories of type (3,3,3), (4,4,2), and (6,3,2), respectively, and any tubular vectorspace category is also cotubular.

In order to list the tubular vectorspace categories, we use the conventions of 4.10.6. [Note that the tubular vectorspace categories of types (3,3,3), (4,4,2) and (6,3,2) all are (up to isomorphism) uniquely determined by the corresponding diagram, whereas the two diagrams of type (2,2,2,2) represent one-parameter families of pairwise non-isomorphic vectorspace categories.] Always, the row indexed by $C(i)$ contains the vectorspace categories which are tubular extensions of $C(i)$, and the column indexed by $C(j)$ contains the vectorspace categories which are tubular coextensions of $C(j)$.

The tubular vectorspace categories of type (2,2,2,2):

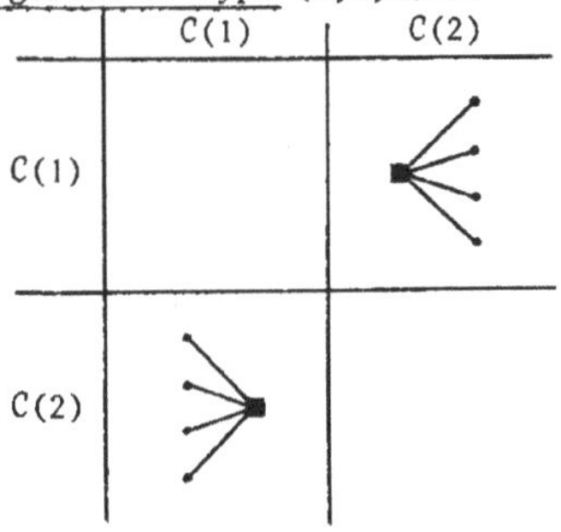

The tubular vectorspace categories of type (3,3,3):

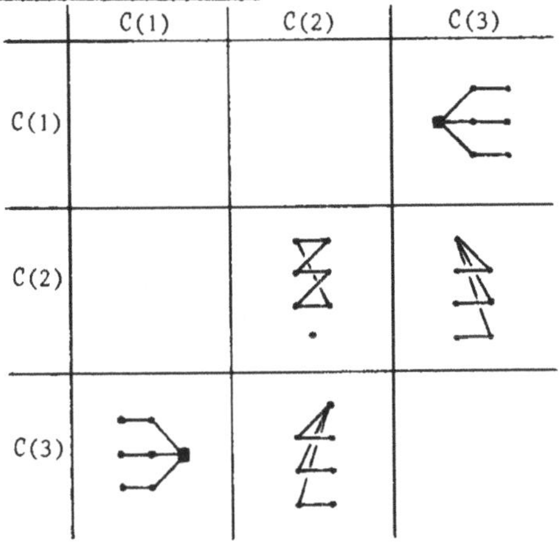

The tubular vectorspace categories of type (4,4,2):

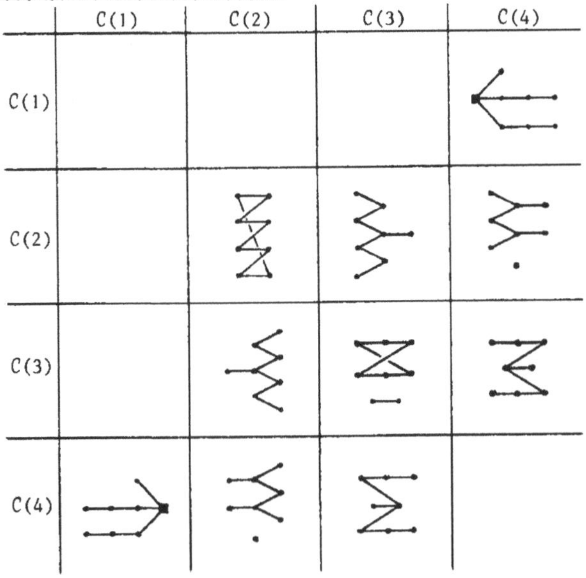

The tubular vectorspace categories of type (6,3,2)

	C(1)	C(2)	C(3)	C(4)	C(5)	C(6)
C(1)						
C(2)						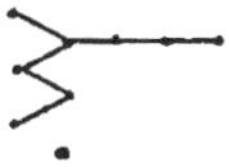
C(3)						
C(4)				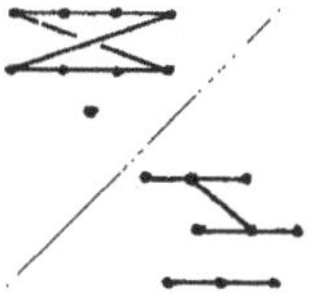		
C(5)						
C(6)		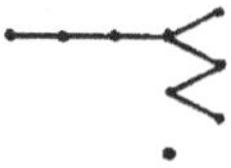				

Proof: This is an easy exercise, using the table of all simple regular objects in $\check{U}(K_o, |\cdot|_o)$, where $(K_o, |\cdot|_o)$ is a critical directed vectorspace category, as presented in 4.10.3. For example, consider the tubular extensions of $(K_o, |\cdot|_o) = C(3)$ of extension type (6,3,2), thus the third row of the table containing the tubular vectorspace categories of type (6,3,2). Let $V^{(1)}$, $V^{(2)}$, $V^{(3)}$ be one τ-orbit of simple regular objects in $\check{U}(K_o, |\cdot|_o)$. Since up to an isomorphism of $C(3)$, the three objects $V^{(i)}$ do not differ, we only have the following possibilities for $(K, |\cdot|)$: first of all, $(K, |\cdot|) = (K_o, |\cdot|_o)[V^{(1)},3]$; second there may be two of the $V^{(i)}$ involved, say $V^{(1)}$ and $V^{(2)}$, thus $(K,|\cdot|) = (K_o,|\cdot|_o)[V^{(1)},1][V^{(2)},2]$, or $= (K_o,|\cdot|_o)[V^{(1)},2],[V^{(2)},1]$. If all three $V^{(i)}$ are involved, then $(K,|\cdot|) = (K_o,|\cdot|_o)[V^{(1)}][V^{(2)}][V^{(3)}]$. This gives the four possibilities in the row indexed by $C(3)$ in the last table. Similarly, we consider the remaining cases.

Let $(K, |\cdot|)$ be a tubular vectorspace category, say a tubular extension of the critical directed vectorspace category $(K_0, |\cdot|_0)$, and a tubular coextension of the critical directed vectorspace category $(K_\infty, |\cdot|_\infty)$. Note that the radical $\mathrm{rad}\chi_K$ of χ_K has rank 2. [Namely, let $(K, |\cdot|) = (A\text{-inj}, \mathrm{Hom}(R,-))$ be an injective realization. Then $A[R]$ is a tubular algebra, and the quadratic forms χ_K and $\chi_{A[R]}$ are equivalent]. Let g_0 be the positive radical generator of χ_{K_0} in $G(K_0)\times\mathbb{Z} \subseteq G(K)\times\mathbb{Z}$, and g_∞ the positive radical generator of χ_{K_∞} in $G(K_\infty)\times\mathbb{Z} \subseteq G(K)\times\mathbb{Z}$. Then the elements g_0, g_∞ generate a subgroup of $\mathrm{rad}\chi_K$ of finite index, and $g_0 + g_\infty$ is sincere [since the radical of the quadratic form of a domestic tubular extension of a critical directed vectorspace category has rank 1]. Also, the non-sincere positive radical vectors are the positive multiples of g_0 and g_∞. The elements g_0, g_∞ are called the <u>canonical radical elements</u> in $G(K)\times\mathbb{Z}$. Given $\gamma \in \mathbb{Q} \cup \{\infty\}$, say $\gamma = \frac{\gamma_\infty}{\gamma_0}$ where $\gamma_\infty \in \mathbb{Z}$, $\gamma_0 \in \mathbb{N}_0$, and such that γ_0, γ_∞ have no common divisor different from 1, let $g_\gamma = \gamma_0 g_0 + \gamma_\infty g_\infty$.

(2) <u>Theorem: Let</u> $(K, |\cdot|)$ <u>be a tubular vectorspace category of type</u> $\mathbb{T}$. <u>For</u> $\gamma \in \mathbb{Q}_0^\infty$, <u>define object classes in</u> $\check{U}(K, |\cdot|)$ <u>as follows: an indecomposable object</u> X <u>belongs to</u> P_γ, T_γ, <u>or</u> Q_γ, <u>if and only if</u> $\langle g_\gamma, \underline{\dim}_K X\rangle < 0, =0,$ <u>or</u> >0, <u>respectively. Then</u> P_0 <u>is a preprojective component,</u> Q_∞ <u>is a preinjective component, and, for any</u> $\gamma \in \mathbb{Q}_0^\infty$,

$$P_\gamma = P_0 \vee \bigvee_{0\le\beta<\gamma} T_\beta, \quad Q_\gamma = \bigvee_{\gamma<\delta\le\infty} T_\delta \vee Q_\infty$$

<u>and</u> T_γ <u>is a separating tubular family, separating</u> P_γ <u>from</u> Q_γ. <u>All the tubular families</u> T_γ, <u>but</u> T_0 <u>and</u> T_∞, <u>are stable of type</u> $\mathbb{T}$. <u>Also,</u> $\check{U}(K, |\cdot|)$ <u>is controlled by</u> χ_K.

This means that the structure of the category $\check{U}(K, |\cdot|)$ for a tubular vectorspace category $(K, |\cdot|)$ is very similar to the structure of the module category of a tubular algebra. Actually, we should stress that in this case not only all T_γ with $\gamma \in \mathbb{Q}^+$, but also T_0 and T_∞ are characterized by the single condition $\langle g_\gamma, \underline{\dim}_K X\rangle = 0$ [whereas for a tubular algebra A, there always exist indecomposable A-modules M with $\langle h_0, \underline{\dim}\, M\rangle = 0$, but belonging to Q_∞, and also with $\langle h_\infty, \underline{\dim}\, M\rangle = 0$, but belonging to P_0].

<u>Proof:</u> Since $(K, |\cdot|)$ is cotubular, $(K, |\cdot|)$ is a tubular coextension of some critical directed vectorspace category $(K_\infty, |\cdot|_\infty)$. Let $(K, |\cdot|) = (A'\text{-inj}, \mathrm{Hom}(R,-))$ and $(K_\infty, |\cdot|_\infty) = (A'_\infty\text{-inj}, \mathrm{Hom}(R_\infty,-))$ be injective realizations, with both A', A'_∞ being basic. Let $A = A'[R]$, $A_\infty = A'_\infty[R_\infty]$. Then A_∞ is concealed, and A is a tubular coextension of A_∞, of coextension type $\mathbb{T}$. Note that for any non-zero object in $\check{U}(K, |\cdot|) \subseteq A\text{-mod}$, its restriction to $A_\infty\text{-mod}$ is non-zero. [Namely, consider $W = (W_0, W_\omega, \gamma_W) \in \check{U}(K, |\cdot|) = \check{U}(A'\text{-inj}, \mathrm{Hom}(R,-))$. Now, W_0 is an injective A'-module. An indecomposable injective A'-module is either an A'_∞-module

or its restriction to A'_∞ is of the form V_0, where (V_0,V_ω,γ_V) is a simple regular object in $\check{U}(K_\infty,|\cdot|_\infty)$, see 4.10.a. In particular, $V_0 \neq 0$. Thus, if $W_0 \neq 0$, also $W_0 \mid A_\infty = W_0 \mid A'_\infty \neq 0$, and therefore $W \mid A_\infty \neq 0$. On the other hand, the extension vertex of $A = A'[R]$ belongs to A_∞, thus, if $W_\omega \neq 0$, then again $W \mid A_\infty \neq 0$.]

Let h_0,h_∞ be the canonical radical elements for $K_0(A)$. Let ${}_AT$ be the canonical tilting module for $(K, |\cdot|)$. According to 4.10.3, we know that ${}_AT$ belongs to $\mathcal{P}^A_\infty$. There is $\alpha \in \mathbb{Q}^+_o$ such that ${}_AT = T_o \oplus T_\alpha$, where $T_0 \in \mathcal{P}^A_\alpha$, and $0 \neq T \in \mathcal{T}^A_\alpha$. Note that $\alpha > 0$. [Namely, A is tubular, thus a proper extension of some tame concealed algebra A_0. Since the extension vertex ω of $A = A'[R]$ is the unique source of the quiver of A, it follows that A' has as a direct factor a tubular extension of A_0. In particular, the restriction of an injective A'-module to A_0 is an injective A_0-module. Now, $T \mid A'$ is an injective A'-module, thus $T \mid A_0$ is an injective A_0-module. Since A^T is sincere, we also have $T \mid A_0 \neq 0$. Thus T does not belong to $\mathcal{P}^A_o \vee \mathcal{Q}^A_o$.] Let $C = \mathrm{End}({}_AT)$, and $C_0 = \mathrm{End}(T_0)$. Since ${}_AT \in \mathcal{P}^A_\alpha \vee \mathcal{T}^A_\alpha$, it follows that $\mathcal{Q}^A_\alpha \subseteq \mathcal{G}({}_AT)$. According to (the proof of) 5.5.1, $\mathcal{P}^A_\alpha \cap \mathcal{G}({}_AT)$ is a preprojective component of $\mathcal{G}({}_AT)$, namely the preprojective component of the tame concealed algebra C_0, and $\mathcal{T}^A_\alpha \cap \mathcal{G}({}_AT)$ is a tubular $\mathbb{P}_1k$-family of $\mathcal{G}({}_AT)$, being obtained from the tubular $\mathbb{P}_1k$-family of C_0-mod by ray insertions.

Let us denote by $\theta : \check{U}(K, |\cdot|) \longrightarrow \mathcal{G}({}_AT)$ the canonical equivalence, and by $\vartheta : G(A_0 - \mathrm{inj}) \times \mathbb{Z} \to K_0(A)$ the corresponding linear transformation. We recall from 2.5.a that ϑ is an isometry. Note that $g_\infty\vartheta = h_\infty$ [since g_∞ is the minimal positive radical element of $G_0(K_\infty) \times \mathbb{Z}$, and the restriction of ϑ to $G_0(K_\infty) \times \mathbb{Z}$ is the linear transformation $G_0(K_\infty) \times \mathbb{Z} \longrightarrow K_0(A_\infty)$ corresponding to the canonical embedding $\check{U}(K_\infty,|\cdot|_\infty) \longrightarrow A_\infty$-mod. Let $g_0\vartheta = \alpha_0 h_0 + \alpha_\infty h_\infty$, with $\alpha_o,\alpha_\infty \in \mathbb{Q}$; of course, $\alpha_0 \neq 0$, since g_0,g_∞ are linearly independent. Note that $\frac{\alpha_\infty}{\alpha_0} = \alpha$ [namely, since $(K, |\cdot|)$ is a tubular extension of some critical directed vectorspace category $(K_0,|\cdot|_0)$, there is some indecomposable object X in $K \setminus K_0$ such that X considered as an object in $\check{U}(K, |\cdot|)$, has a sink map of the form $V \longrightarrow X$ with V simple regular in $\check{U}(K_0,|\cdot|_0)$, thus $\langle g_0,\underline{\dim}\, V\rangle = 0$. Since $V\theta$ belongs to a component of $\mathcal{G}({}_AT)$ containing both projective vertices as well as cycles, we have $V\theta \in \mathcal{T}^A_\alpha \cap \mathcal{G}({}_AT)$. On the other hand,

$$0 = \langle g_0,\underline{\dim}\, V\rangle = \langle \alpha_0 h_0 + \alpha_\infty h_\infty, \underline{\dim}\, V\theta\rangle ,$$

thus the index of $V\theta$ is $\frac{\alpha_\infty}{\alpha_0}$, therefore $\alpha = \frac{\alpha_\infty}{\alpha_0}$.] As a consequence, we define $\bar{\vartheta} : \mathbb{Q}^\infty_o \longrightarrow \mathbb{Q}^\infty_o$ by

$$\gamma\bar{\vartheta} = \frac{\alpha_\infty + \gamma}{\alpha_0}, \quad \text{for } \gamma \in \mathbb{Q}^\infty_o ,$$

and then $g_\gamma\vartheta = g_{\gamma\bar{\vartheta}}$ for any $\gamma \in \mathbb{Q}^\infty_o$, in analogy to the dual assertion of 5.4.c. Of course, $\bar{\vartheta}$ defines an order preserving bijection between $\mathbb{Q}^\infty_o$ and the set $\{\gamma \in \mathbb{Q}^\infty_o \mid \alpha \leq \gamma\}$. It follows that θ defines equivalences $\mathcal{T}_\gamma \approx \mathcal{T}^A_{\gamma\bar{\vartheta}}$, for any $\gamma \in \mathbb{Q}^+ \cup \{\infty\}$. [For $\gamma = \infty$, we have to use in addition that for any indecomposable object W in $\check{\mathcal{U}}(K, |\cdot|)$, the restriction of $W\theta$ to A_∞ is non-zero, so that $\langle h_\infty, \underline{\dim}\, W\theta\rangle = 0$ actually implies $W\theta \in \mathcal{T}^A_\infty$.] Also, θ defines an equivalence $\mathcal{T}_0 \approx \mathcal{T}^A_\alpha \cap \mathcal{G}({}_A T)$. [Here, we use that $\alpha > 0$, so that for an indecomposable object W in $\mathcal{T}_0$, the condition $\langle h_0, \underline{\dim}\, W\theta\rangle = 0$ implies $W\theta \in \mathcal{T}^A_\infty$.] Of course, θ also defines equivalences $\mathcal{P}_0 \approx \mathcal{P}^A_\alpha \cap \mathcal{G}({}_A T)$, and $\mathcal{Q}_\infty \approx \mathcal{Q}^A_\infty \cap \mathcal{G}({}_A T)$. This proves all assertions but the last.

If $W \in \check{\mathcal{U}}(K, |\circ|)$ is indecomposable, then

$$\chi_K(\underline{\dim}_K W) = \chi_A(\underline{\dim}\, W\theta) = 0 \quad \text{or} \quad 1.$$

Conversely, given a positive radical vector g for χ_K, then $g = \gamma_0 g_0 + \gamma_\infty g_\infty$ for some $\gamma_0, \gamma_\infty \in \Phi_0^+$. Now, if $\gamma_\infty = 0$, then g is a multiple of g_0, thus there are infinitely many indecomposable objects W in $\check{\mathcal{U}}(K_0, |\circ|_0) \subseteq \check{\mathcal{U}}(K, |\cdot|)$ such that $\underline{\dim}_K W = g$. So assume $\gamma_\infty \neq 0$. Now, according to 2.5.b,c, we know that $g\vartheta$ is connected and positive. Since A-mod is controlled by χ_A, there are infinitely many indecomposable A-modules M such that $\underline{\dim}\, M = g\vartheta$. Since $\gamma_\infty \neq 0$, the index of $g\vartheta$ is $> \alpha$, thus all M belong to $\mathcal{Q}_\alpha \subseteq \mathcal{G}({}_A T)$, thus any $M = W\theta$ with $W \in \check{\mathcal{U}}(K, |\cdot|)$, and $\underline{\dim}_K W = g$. Next, let x be a positive root for χ_K. Again, $x\vartheta$ is connected and positive, thus there exists a unique indecomposable A-module N with $\underline{\dim}\, N = x\vartheta$. Assume in addition that $\langle g_0, x\rangle > 0$, thus $\langle h_\alpha, x\vartheta\rangle > 0$, therefore N belongs to $\mathcal{Q}_\alpha \subseteq \mathcal{G}({}_A T)$, thus there exists (again uniquely) $X \in \check{\mathcal{U}}(K, |\cdot|)$ with $X\theta = N$. It remains to consider the case that $\langle g_0, x\rangle \le 0$.

We now use instead of $(K, |\cdot|)$ the dual vectorspace category $(K^{op}, D|\cdot|)$. We may identify $G(K)$ and $G(K^{op})$, since the objects of K and K^{op} may be assumed to be the same. Let $X_1, \dots, X_n$ be the indecomposable objects of K, assumed to be pairwise indecomposable, thus the elements of $G(K) \times \mathbb{Z}$ are of the form $y = (y_1, \dots, y_n, y_\omega)$, with $y_1, \dots, y_n, y_\omega \in \mathbb{Z}$. Define a duality $* : G(K) \times \mathbb{Z} \longrightarrow G(K^{op}) \times \mathbb{Z}$ by mapping y to $y^* = (y_1, \dots, y_n, y^*_\omega)$, with

$$y^*_\omega = -y_\omega + \sum_{i=1}^{n} y_i \dim_k |X_i|.$$

Note that for $W \in \mathcal{U}(K^{op}, D|\cdot|)$, we have

$$\underline{\dim}_K(W^*) = (\underline{\dim}_{K^{op}} W)^*$$

(for the definition of W^*, see 2.5.1). Also, one easily checks that for $y, z \in G(K) \times \mathbb{Z}$,

$$\langle y, z\rangle = \langle z^*, y^*\rangle$$

(for example, one may use 2.5.3" and 2.5.1). Now, with $(K, |\cdot|)$ also $(K^{op}, D|\cdot|)$ is a tubular vectorspace category, and $(K^{op}, D|\cdot|)$ is a tubular coextension of $((K_0)^{op}, D|\cdot|_0)$. If we denote by g'_0, g'_∞ the canonical radical elements in $G(K^{op}) \times \mathbb{Z}$ for the quadratic form of $(K^{op}, D|\cdot|)$, then $g'_0 = (g_\infty)$, $g'_\infty = (g_0)$. Now, consider again the given positive root x for χ_K, satisfying $\langle g_0, x\rangle \le 0$.

Thus

$$\langle g'_\infty, x^*\rangle = -\langle x^*, g'_\infty\rangle = -\langle x^*, (g_o)^*\rangle = -\langle g_o, x\rangle \geq 0.$$

By the considerations above, applied to $(K^{op}, D|\cdot|)$, there exists a (unique) indecomposable object $N' \in \check{U}(K^{op}, D|\cdot|)$ satisfying $\underline{\dim}_{K^{op}} N' = x^*$. Of course, we even have $N' \in U(K^{op}, D|\cdot|)$, thus we can apply the duality functor $*$. Let $N = (N')^*$. Then

$$\underline{\dim}_K N = (\underline{\dim}_{K^{op}} N')^* = x^{**} = x.$$

This finishes the proof.

(3) Given a tubular vectorspace category $(K, |\cdot|)$, we may define its left reflection $\ell(K, |\cdot|)$ and its right reflection $r(K, |\cdot|)$ as follows: Assume

$$(K,|\cdot|) = (K_o,|\cdot|_o)[V^{(i)},m_i]_{i=1}^{t} = {}_{j=1}^{s}[n_j,W^{(j)}](K_\infty,|\cdot|_\infty),$$

with $(K_o,|\cdot|_o)$, and $(K_\infty,|\cdot|_\infty)$ both being critical directed vectorspace categories, and all $V^{(i)}$, $W^{(j)}$ simple regular objects. Then, by definition,

$$\ell(K,|\cdot|) = {}_{i=1}^{t}[m_i,V^{(i)}](K_o,|\cdot|_o),$$
$$r(K,|\cdot|) = (K_\infty,|\cdot|_\infty)[W^{(j)},n_j]_{j=1}^{s}.$$

Of course, both $\ell(K,|\cdot|)$ and $r(K,|\cdot|)$ are tubular vectorspace categories again, and have the same type as $(K,|\cdot|)$. Note that

$$r\ell(K,|\cdot|) = (K, |\cdot|) = \ell r(K, |\cdot|),$$

and, since there are only finitely many tubular vectorspace categories, they form finite orbits with respect to ℓ (and r), which we will call <u>reflection sequences.</u>

<u>The number of reflection sequences of tubular vectorspace categories of type (3,3,3), (4,4,2), and (6,3,2) is 2, 4, and 8, respectively.</u>

In order to write down the reflection sequences, we note the following: Given a tubular vectorspace category $(K,|\cdot|)$, it is the tubular extension of some uniquely defined critical directed vectorspace category $(K_o,|\cdot|_o) = C(i)$, and the tubular coextension of some, again uniquely defined critical directed vectorspace category $(K_\infty,|\cdot|_\infty) = C(j)$. If we fix the type of $(K,|\cdot|)$, then the pair (i,j) usually determines uniquely $(K,|\cdot|)$, the only exception being the case of $(K,|\cdot|)$ being of type (6,3,2), with $i = j = 4$, and then there are two possibilities for $(K,|\cdot|)$. Now, these two exceptional tubular vectorspace categories $(K,|\cdot|)$ both have the property that $r(K,|\cdot|) = (K,|\cdot|)$. In order to write down the remaining reflection sequences

$$(K^{(1)},|\cdot|^{(1)}),\ (K^{(2)},|\cdot|^{(2)}),\ldots,(K^{(n)},|\cdot|^{(n)}),$$

with $r(K^{(t)},|\cdot|^{(t)}) = (K^{(t+1)},|\cdot|^{(t+1)})$, and $r(K^{(n)},|\cdot|^{(n)}) = (K^{(1)},|\cdot|^{(1)})$, it is sufficient to give the sequence of numbers $i_1,\ldots,i_n$, where $(K_o^{(t)},|\cdot|_o^{(t)}) = C(i_t)$. These sequences are as follows:

<u>Tubular type</u> (3,3,3): 1,3,2,3.
2.

<u>Tubular type</u> (4,4,2): 1,4,3,2,3,4.
2,4.
2.
3.

<u>Tubular type</u> (6,3,2): 1,6,5,4,3,5,2,5,3,4,5,6.
2,6,4,6.
2,4.
3,6.
3.
5.

(keeping in mind the two additional reflection sequences which both would have to be labelled by the single number 4).

References and comments

Let us say that a partially ordered set S is tubular provided the vectorspace category $K(S)$ is tubular. The tubular partially ordered sets have been considered for the first time by Zavadskij and Nazarova [ZN]: they have shown that these partially ordered sets S are tame, and that the one-parameter families of indecomposable representations correspond bijectively to the positive radical vectors of the quadratic form χ_S. Later, Zavadskij characterized these partially ordered sets as the minimal non-domestic tame partially ordered sets of "finite growth". In [Ri4], also the remaining tubular vectorspace categories except those of type (2,2,2,2) have been shown to be tame. The complete classification of the indecomposable representations of a tubular partially ordered set has been given independently by Zavadskij, see [Z3], and we want to recommend his paper also for additional results concerning tame partially ordered sets, in particular, for his list of all sincere ones. It seems to be of interest to follow the reflection sequences of the tubular vectorspace categories, and the "patterns" produced in this way. These "patterns" were exhibited in [Ri4] (at that time, the "patterns" of type (2,2,2,2) were not yet known to be tame). The "patterns" we obtain in this way are those of "type" $(\tilde{\mathbb{D}}_4,1)$, with $\lambda \neq 0,1$ in k, we denote them just by $\textcircled{\mathbb{D}}_{4\lambda}$, and all those of "similarity type" $\textcircled{\mathbb{E}}_6$, $\textcircled{\mathbb{E}}_7$ and $\textcircled{\mathbb{E}}_8$. In Appendix A.3, we reproduce the list of these "patterns", always indicating in the right lower corner the corresponding reflection sequence. In drawing the various vectorspace categories in A.3, we again use the conventions of 4.10.6. Most of the vectorspace categories exhibited are of the form $K(S)$, with S a partially ordered set (S is infinite, but of width 3 or 4), and we recall that we just draw its Hasse diagram, but with edges from left to right instead of from below upwards (thus $a \bullet\!\!-\!\!\!-\!\!\bullet b$ indicates that $a < b$ and that $a \leq c < b$ implies $a = c$). In the remaining cases, we deal with vectorspace categories $(K,|\cdot|)$ such that for any indecomposable object X of K, we have $\dim_k|X| = 1$ or 2. Recall that the isomorphism classes of the indecomposable objects X of K are represented by dots $\bullet$ in case $\dim|X| = 1$, and by black squares $\blacksquare$ in case $\dim|X| = 2$. We will draw an edge from left to right, say $[X]$—$[Y]$, provided there is an irreducable map $X \longrightarrow Y$ in K, and we suppose that in this case $\dim K(X,Y) = 1$. In case we have the following part

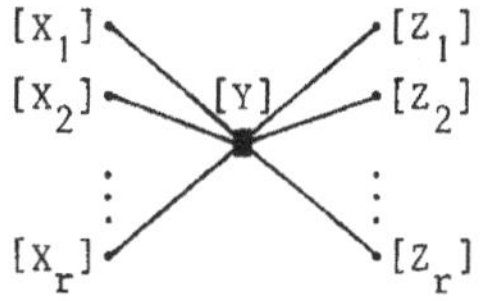

this shall include the informations that for non-zero maps $f_i:X_i \longrightarrow Y$, $g_i:Y \longrightarrow Z_i$, the image U_i of $|f_i|$ coincides with the kernel of $|g_i|$, and that these subspaces U_i of $|Y|$ are pairwise different, for $1 \leq i \leq r$. Always, one may choose a basis of $|Y|$ such that $|U_1| = k\times 0$, $|U_2| = 0\times k$, $|U_3| = \{(x,x) \mid x \in k\}$, and in case $\textcircled{\mathbb{D}}_{4\lambda}$ we assume, in addition that $|U_4| = \{(x,\lambda x) \mid x \in k\}$.

Examples of tubular algebras occurred in the work of Zavadskij [Z2] on tame quivers with a single zero relation, and of Shkabara [Shk] and Marmaridis [M1], dealing with tame quivers with commutativity relations. The notes [Ri4] are a report on algebras given by a quiver with one relation. In [BB], Brenner-Butler put quite a lot of emphasis on tame algebras with "corank 2 quadratic forms", in particular on the so-called "squids". As the examples show, their interest was directed towards the tubular algebras, and the tilting functors relating them. The use of shrinking functors in this Chapter 5 is influenced by their work.

6. Directed algebras

We are going to push forward the investigation of the directed algebras which was started in section 2.4. We will use several of the methods exhibited earlier. In particular, the proofs of the first results are based on tilting techniques and subspace categories.

6.1 The orbit quiver of a sincere directed algebra

(1) Proposition (Bautista-Larrion-Salmeron, Bongartz). The orbit graph $\overline{\mathcal{O}(B)}$ of a sincere directed algebra B is a tree with at most four endpoints.

Actually, we will show the following result which implies (1), according to 4.2.6" and 4.2.4"':

(1') Let B be a directed algebra, and P an indecomposable projective module which is not a proper predecessor of any indecomposable projective module, and such that $S = S(P\to)$ is a slice. Then, $\Delta(S)$ is a tree with at most three sinks (and just one source).

Proof: Note that P is of the form $P = P(\omega)$, where ω is a source of $\Delta(B)$, thus $B = B_o[R]$, where B_o is obtained from B by deleting ω, and $R = \text{rad}\, P(\omega)$. Of course, $R = \bigoplus_{i=1}^{d} R_i^{n_i}$, with $R_i = \tau X_i$, where $[X_i]$, $1 \le i \le d$, are the neighbors of $[P]$ in $\Delta(S)$, and n_i is the number of arrows $[P] \to [X_i]$. Since S is a slice in B-mod, it follows easily that τS is a slice in B_o-mod, and the $[R_i]$ are the sources of $\Delta(\tau S)$.

According to 2.5.8, we have

$$\text{B-mod} \approx \check{\mathcal{U}}(B_o\text{-mod},\text{Hom}(R,-)).$$

With B also B_o is a directed algebra, thus $(B_o\text{-mod},\text{Hom}(R,-))$ is a directed vectorspace category. According to 2.4.9', the algebra B is representation-finite, thus $(B_o\text{-mod},\text{Hom}(R,-))$ is subspace-finite. Therefore, we must have

(*) $\dim \text{Hom}(R,X) \le 1$ for any indecomposable X in B_o-mod,

since otherwise the full subcategory $\check{\mathcal{U}}(\langle X\rangle,\text{Hom}(R,-))$ will be subspace infinite. Consequently, all $n_i = 1$, since $\dim \text{Hom}(R,R_i) \ge \dim \text{Hom}(R_i^{n_i},R_i) = n_i$. It follows that we obtain a quiver Δ' isomorphic to $\Delta(S)$ by adding to $\Delta(\tau S)$ one additional vertex ω, and, for any $1 \le i \le d$, one arrow $\omega \to [R_i]$. Now assume $R_i \preceq Y$ for some indecomposable module Y in τS. Then $\dim \text{Hom}(R_i,Y)$ equals the number of paths in $\Delta(\tau S)$ from $[R_i]$ to $[Y]$, since the endomorphism ring of a slice module in τS is hereditary. Thus, there is precisely one path from R_i to Y in

$\Delta(\tau S)$. Also, there cannot exist an indecomposable module Y in τS with $R_i \preceq Y$, $R_j \preceq Y$ and $i \neq j$, since otherwise $\dim \operatorname{Hom}(R,Y) \geq 2$. It follows that the different R_i, $1 \leq i \leq d$, belong to different components Δ_i of $\Delta(\tau S)$. Of course, Δ_i has as vertices precisely the $[Y]$ with Y indecomposable in τS, and $R_i \preceq Y$. Since for any such Y there is precisely one path from $[R_i]$ to $[Y]$, we see that Δ_i is a tree. Since Δ' is obtained from the disjoint union of these Δ_i by adding ω and the arrows $\omega \to [R_i]$, $1 \leq i \leq d$, we see that Δ' is a tree. Also, the end-points of Δ' different from ω are given by the sinks of the various Δ_i, and, ω is an endpoint if and only if $d = 1$ (thus if and only if $R = \operatorname{rad} P$ is indecomposable). We claim that $\Delta(\tau S) = \bigsqcup \Delta_i$ has at most 3 sinks. Namely, assume there exist pairwise different indecomposable modules $Z_1,\ldots,Z_4$ in τS with all $[Z_i]$, $1 \leq i \leq 4$, being sinks. Then $\operatorname{Hom}(Z_i,Z_j) = 0$ for $i \neq j$. Thus, the vector-space category $(\langle Z_1,\ldots,Z_4\rangle, \operatorname{Hom}(R,-))$ is given by the partially ordered set $C(2)$, and therefore subspace infinite. This finishes the proof. [Note that the reference to 2.4.9' is not essential, it is sufficient to observe directly that vectorspace categories $C(1)$ and $C(2)$ obviously have cycles in their subspace categories.]

Remark. A directed algebra with orbit graph a tree is sometimes called a simply connected algebra. As we have seen above, any sincere directed algebra is simply connected. Note that a factor algebra of a simply connected algebra no longer has to be simply connected, an example is the algebra

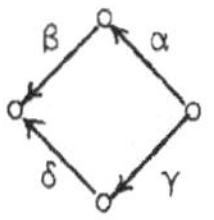

with $\alpha\beta = \gamma\delta$

which is sincere and directed, thus simply connected, with orbit graph $\mathbb{D}_4$ whereas

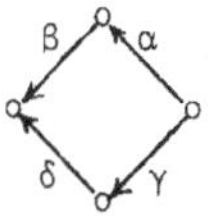

with $\alpha\beta = 0 = \gamma\delta$

has orbit graph $\tilde{\mathbb{A}}_3$, thus the orbit graph is not a tree. (Let us stress that a factor algebra of a directed algebra always is directed, again).

(2) Corollary. Let A be a directed algebra. Then A-mod has trivial modulation.

We recall that a Krull-Schmidt k-category K is said to have trivial modulation provided $\dim_k \operatorname{Irr}(X,Y) \leq 1$ for all indecomposable objects X,Y in K.

Proof. Let X,Y be indecomposable A-modules, and $\operatorname{Irr}_A(X,Y) \neq 0$. If $f : X \longrightarrow Y$ is irreducible, then f is either epi or mono. Let us suppose that f is mono. Let A' be the factor algebra of A modulo the ideal generated by the idempotents e of A with $eY = 0$. Then both X,Y are A'-modules, and Y is even sincere, as an A'-module. Also, $\mathrm{Irr}_{A'}(X,Y)$ maps surjectively onto $\mathrm{Irr}_A(X,Y)$. According to (1), we have $\dim \mathrm{Irr}_{A'}(X,Y) = 1$, thus also $\mathrm{Irr}_A(X,Y) = 1$.

(3) Using tilting functors, we are able to extend the results in 3.6 concerning maximal modules over directed algebras. We will see that given a maximal module M, then the number of neighbors of the τ-orbit of M in the orbit quiver furnishes important informations about M. In particular, it determines the number of exceptional vertices of $\underline{\dim}\, M$. [Given a directed algebra A, and a maximal module M, then according to 1.1 a vertex a in the support of M is called exceptional for M, provided $D_a\chi_A(\underline{\dim}\, M) \neq 0$. There are either one or two exceptional vertices, and in case of a unique exceptional vertex for M, there exist indecomposable modules N with the same support as M, and not isomorphic to M.]

Proposition. *Let B be a directed algebra, with orbit quiver $\mathcal{O}$. Let M be a sincere maximal B-module. Then, the τ-orbit of M has at most three neighbors in $\mathcal{O}$. Also*

(i) If the τ-orbit of M has precisely one neighbor in $\mathcal{O}$, then M is the only sincere indecomposable B-module, and M is projective-injective.

(ii) If the τ-orbit of M has precisely two neighbors in $\mathcal{O}$, then $\underline{\dim}\, M$ has two exceptional vertices b_1,b_2, and M is not projective-injective.

(iii) If the τ-orbit of M has three neighbors, then $\underline{\dim}\, M$ has a unique exceptional vertex, and M is not projective-injective.

Proof. The assertion (i) has been established in 3.6.2. Let us consider the other assertions. Let S be the slice constructed by 4.2.6 with slice module ${}_BS$, let $A = \mathrm{End}({}_BS)$, and ${}_AT_B = D({}_BS_A)$. Let $\Sigma = \mathrm{Hom}({}_AT,-)$, and σ the corresponding linear transformation. Then A has a unique sink a, and $\Sigma\, Q_A(a) = M$. Let $T' = \bigoplus_b T(b)^{d_b}$, with $d_b = D_b\chi(\underline{\dim}\, M)$. According to 3.6.1, $R = \Sigma T' = \bigoplus_b P_B(b)^{d_b}$ dominates M, thus $\underline{\dim}\, R = (\underline{\dim} M)(I-\Phi_B^{-1})$, with Φ_B the Coxeter transformation of B. Let $t' = \underline{\dim}\, T'$. Then

$$\begin{aligned} t' = \underline{\dim}\, T' &= (\underline{\dim}\, R)\sigma^{-1} = (\underline{\dim}\, M)(I-\Phi_B^{-1})\sigma^{-1} \\ &= (\underline{\dim}\, M)\sigma^{-1}(I-\Phi_A^{-1}) = (\underline{\dim}\, Q_A(a))(I-\Phi_A^{-1}) \\ &= q_A(a) + p_A(a), \end{aligned}$$

with $q_A(a) = \underline{\dim}\, Q_A(a)$, $p_A(a) = \underline{\dim}\, P_A(a)$, using 4.1.d, and 2.4.b.

We may assume that a is not an isolated vertex of A, and we denote by $a_1,\ldots,a_m$ the neighbors of a, say with arrows $\alpha_i : a_i \longrightarrow a$, where $1 \le i \le m$. Since a is a sink, $P_A(a)$ is a simple A-module, and thus the socle of $Q_A(a)$. Also, $Q_A(a)/P_A(a) = \bigoplus_{i=1}^m Q_A(a_i)$, with projection maps $Q_A(\alpha_i^*) : Q_A(a) \longrightarrow Q_A(a_i)$. The exact sequence

$$0 \longrightarrow P_A(a) \longrightarrow Q_A(a) \longrightarrow \bigoplus_{i=1}^m Q_A(a_i) \longrightarrow 0$$

gives an exact sequence

$$0 \longrightarrow \mathrm{Hom}(Q_A(a),P_A(a)) \longrightarrow \mathrm{Hom}(Q_A(a),Q_A(a)) \longrightarrow \mathrm{Hom}(Q_A(a), \bigoplus_{i=1}^m Q_A(a_i))$$

$$\longrightarrow \mathrm{Ext}_A^1(Q_A(a),P_A(a)) \longrightarrow 0,$$

and clearly $\mathrm{Hom}(Q_A(a),P_A(a)) = 0$. Also, $\mathrm{End}(Q_A(a)) = k$. and $\mathrm{Hom}(Q_A(a),Q_A(a_i))$ is one-dimensional, and generated by $Q_A(\alpha_i^*)$. It follows that $\dim \mathrm{Ext}_A^1(Q_A(a),P_A(a)) = m - 1$.

Since $T' \in \langle {}_AT\rangle$, and ${}_AT$ is a tilting module, we have $\mathrm{Ext}^1(T',T') = 0$. Thus

$$\begin{aligned}\dim \mathrm{End}(T') &= \chi_A(t') = \chi_A(p_A(a) + q_A(a)) \\ &= \chi_A(p_A(a)) + \chi_A(q_A(a)) + \langle p_A(a),q_A(a)\rangle + \langle q_A(a),p_A(a)\rangle \\ &= 1 + 1 + 1 - (m-1) = 4 - m,\end{aligned}$$

where we use that both $p_A(a)$ and $q_A(a)$ are roots, that $\mathrm{Hom}(P_A(a),Q_A(a)) = k$, that $\mathrm{Ext}^1(P_A(a),Q_A(a)) = 0 = \mathrm{Hom}(Q_A(a),P_A(a))$, and the above calculation of $\dim \mathrm{Ext}^1(Q_A(a),P_A(a))$.

Since $T' \neq 0$, we have $4-m > 0$, thus $m \le 3$, therefore, the τ-orbit of $[M]$ has at most 3 neighbors in $\mathcal{O}$. If $m = 3$, then $\mathrm{End}(T') = k$, thus there is precisely one vertex b of B with $d_b > 0$, thus there is precisely one exceptional vertex for $\underline{\dim}\, M$. Also, if $m = 1$ or $m = 2$, then $\dim \mathrm{End}(R) = \dim \mathrm{End}(T') = 3$ or $= 2$, respectively, and therefore R is the direct sum of two non-isomorphic indecomposable projective modules (note that $\mathrm{End}(P(b)) = k$ for any vertex b of B), thus there are two exceptional indices b_1, b_2 for $\underline{\dim}\, M$.

Finally, assume that M is projective-injective. Then $\mathrm{rad}\, M$ is indecomposable, since M is injective. Thus there is precisely one arrow ending in $[M]$, since M is projective. Similarly, there is precisely one arrow starting in M, namely $M \longrightarrow M/\mathrm{soc}\, M$. Also, $\tau(M/\mathrm{soc}\, M) = \mathrm{rad}\, M$. Since the τ-orbit of M consists just of M, we see that this τ-orbit is an endpoint of $\mathcal{O}$. This finishes the proof.

6.2 Sincere directing wing modules of Dynkin type

We are going to determine the algebras B having a sincere directing wing module of Dynkin type. We note that an algebra B with a sincere directing wing module of Dynkin type $\bar{\Delta}$ always is sincere and directing with orbit graph $\bar{\Delta}$ (this follows from 4.2.7, and 4.2.3) and any sincere maximal A-module again is a wing module (use 6.1.3). We will show that the algebras with a sincere directing wing module of Dynkin type can be read off from the list of all tame concealed algebras.

(1) Proposition. Let $\mathcal{B}$ be the set of all pairs (B,W), where B is a basic, sincere, directed algebra, with Dynkin orbit graph, and W a sincere maximal B-module which is a wing module. Let $\mathcal{C}$ be the set of all pairs (C,ω), where C is a basic, tame concealed algebra, and ω is a source of $\Delta(C)$ such that rad $P(\omega)$ is projective, and $h(C)_\omega = 1$, where $h(C)$ is the positive radical generator for χ_C. Let $(B,M) \in \mathcal{B}$. Then there exists a unique projective B-module R dominating M. Let $C = B[R]$, and ω the extension vertex. Then $(C,\omega) \in \mathcal{C}$, and, in this way, we obtain a bijection between $\mathcal{B}$ and $\mathcal{C}$.

Remark. A sincere directed algebra B with Dynkin orbit graph may have several sincere maximal B-modules, for example, the algebra B given by

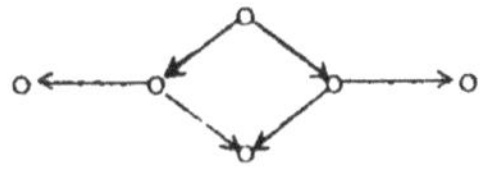

with the commutativity relation has orbit graph $\mathbf{E}_6$, and maximal B-modules with dimension vectors

$$1\ 2\ {}^{1}_{2}\ 2\ 1 \quad \text{and} \quad 1\ 2\ {}^{2}_{1}\ 2\ 1 .$$

On the other hand, a tame concealed algebra C may have several sources ω with rad $P(\omega)$ being projective and $h(C)_\omega = 1$, for example,

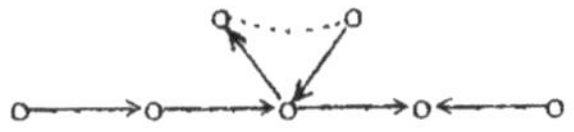

is of type $\tilde{\mathbf{E}}_6$ and the two endpoints of the horizontal branches are sources as required.

Proof. Let $(B,W) \in \mathcal{B}$, say, with orbit graph of Dynkin type $\bar{\Delta}$. The existence of R has been shown in 3.6.1. Of course, since R dominates W, the dimension vector $\underline{\dim}$ R is uniquely determined by W. Since the dimension vectors of the indecomposable projective modules are linearly independent, a projective module is uniquely determined by its dimension vector.

We want to show that B[R] is concealed. If $\bar{\Delta} = \mathbf{A}_n$, then B itself has to be hereditary, and $B[R]$ is tame hereditary, according to 3.6.4. Thus, assume $\bar{\Delta} \neq \mathbb{A}_n$. According to 4.2.7, we know that B is a tilted algebra: Let S be a slice module for the slice $S = S(W\to)$, and $A = \mathrm{End}({}_BS)$, ${}_AT = {}_ADS$. Then A is hereditary, of type $\bar{\Delta}$, and ${}_AT$ is a tilting module with $B = \mathrm{End}({}_AT)$. Also, $S = \{\Sigma_A Q \mid {}_AQ \text{ injective}\}$, where $\Sigma = \Sigma_T = \mathrm{Hom}({}_AT,-)$. Let a be the unique sink in the quiver of A, then $Q_A(a)$ is a wing module, and $\mathrm{Hom}({}_AT, Q_A(a)) = W$. Since we assume $\bar{\Delta} \neq \mathbb{A}_n$, the τ-orbit of $[W]$ has three neighbours, thus $\underline{\dim}\, W$ has a unique exceptional vertex (6.1.3), therefore R is indecomposable (3.6.1). Since R is also projective, there exists some indecomposable direct summand ${}_AM$ of ${}_AT$ such that $\Sigma_A M = {}_BR$. Since

$$\begin{aligned}\underline{\dim}_A M &= (\underline{\dim}\, R)\sigma_T^{-1} = (\underline{\dim}\, W')(I - \Phi_B^{-1})\sigma_T^{-1} \\ &= (\underline{\dim}\, W)\sigma_T^{-1}(I - \Phi_A^{-1}) \\ &= \underline{\dim}\, Q_A(a)\ (I - \Phi_A^{-1}),\end{aligned}$$

with σ_T the linear transformation corresponding to Σ_T, and where we use 4.1.d, it follows that

$$\underline{\dim}_A T' = 2\ \begin{matrix}1\cdots 1\\ 1\cdots 1\\ 1\cdots 1\end{matrix}\ .$$

Since ${}_AM$ is indecomposable, the algebra $A[M]$ is the canonical algebra of type $\bar{\Delta}$, thus $A[M]$ is concealed (4.3.5). Let ω be the extension vertex of $A[M]$. Then, clearly, ${}_AT \oplus P(\omega)$ is a tilting $A[M]$-module, since $M = \mathrm{rad}\, P(\omega)$ is a direct summand of $P(\omega)$, and ${}_AT \oplus P(\omega)$ belongs to the preprojective component of $A[M]$ (this is clear for $P(\omega)$, since $P(\omega)$ is projective; and the indecomposable summands T_i of ${}_AT$ belong to the preprojective component, since $\iota(\underline{\dim}\, T_i) < 0$, see the theorem in 3.7). It follows that $C = B[R] = \mathrm{End}({}_AT \oplus P(\omega))$ is a tame concealed algebra. We denote the extension vertex of $B[R]$ also by ω, and we see that $\mathrm{rad}\, P(\omega) = R$ is projective. The positive radical generator $h(A[M])$ of $\chi_{A[M]}$ has all its coefficients equal to 1. It follows that the coefficient of $h(A[M])\sigma_{T\oplus P(\omega)}$ at ω is again equal to 1. This shows that (C,ω) belongs to $\mathcal{C}$.

Conversely, assume there is given a tame concealed algebra C' and ω' a source of $\Delta(C')$ such that $\mathrm{rad}\, P(\omega')$ is projective and $h(C')_{\omega'} = 1$. Let B' be obtained from C' by deleting ω'. First of all, B' is directed. For, B'-mod is a full subcategory of B-mod closed under extensions, submodules and factor modules. Thus, if B'-mod would contain a cycle, this necessarily would be a cycle in some tube of B-mod, however then we obtain a contradiction, according to 5.3.5'. Let w' be the restriction of $h(C')$ to $K_0(B')$, thus $w' = h(C') - e(\omega)$, since $h(C')_\omega = 1$. We have $\chi_{B'}(w') = 1$, since $h(C')$ belongs to the radical of $\chi_{C'}$. Since gl.dim.$C' \leq 2$, also gl.dim.$B' \leq 2$. Since $\chi_{B'}$ is positive definite, there exists an indecomposable B'-module W' with $\underline{\dim}\, W' = w'$, according to 2.4.9. Therefore,

B' is sincere. Also, the orbit graph of B' gives a quadratic form $\mathbb{Z}$-equivalent to $\chi_{B'}$, thus since $\chi_{B'}$ is positive definite, also the quadratic form given by $\overline{\mathcal{O}(B')}$ is positive definite, therefore $\overline{\mathcal{O}(B')}$ is a Dynkin graph. Since rad $P(\omega')$ is projective, ω' is not connected to any other vertex by a dotted edge, thus $d_a = D_a\chi_B(w') \geq 0$ for all a in the support of w', therefore w' is a maximal positive root for the restriction $\chi_{B'}$ of χ_B to the support of w'. Thus, W' is a maximal module. It remains to be seen that W' is a wing module. For a in the support of w', we either have $d_a = 0$, or else a is an exceptional vertex for w'. Suppose that W' is not a wing module. Then clearly $\mathcal{O}(B') \neq \mathbb{A}_n$, and according to 3.6.2, there are two exceptional vertices $a \neq b$ for $\underline{\dim}\, W'$. It follows that rad $P(\omega') = P(a) \oplus P(b)$, and $C' = B'[P(a) \oplus P(b)]$. Of course, all indecomposable C'-modules with dimension vector $h(C') = w' + e(\omega')$ are regular, and therefore belong to tubes. Thus, we obtain a contradiction, according to the lemma which follows. This then finishes the proof.

(2) Let A_0 be a directed algebra, with orbit graph $\overline{\mathcal{O}(A_0)} \neq \mathbb{A}_n$. Let M_0 be a sincere maximal A_0-module, and assume its τ-orbit cuts off a branch from $\mathcal{O}(A_0)$. Let a,b be the exceptional vertices of $\underline{\dim}\, M_0$. Let $R = P(a) \oplus P(b)$, and $A = A_0[R]$, with extension vertex ω. Then there exists an indecomposable A-module M with $\underline{\dim}\, M = e(\omega) + \underline{\dim}\, M_0$ such that the component of A-mod containing M is not a tube.

Proof. According to 3.6.2 and interchanging, if necessary, a and b, either M_0 is projective - injective, and then $M_0 = P_0(a) = Q_0(b)$, or else M_0 has a wing θ, and the source of θ is $[P_0(a)]$, the sink of θ is $[Q_0(b)]$. According to 4.2.6, we know that $S(\to M_0)$ is a slice in A_0-mod. Since $\overline{\mathcal{O}(A_0)} \neq \mathbb{A}_n$, there are sectional paths

$$\begin{array}{lllllll} X_o \searrow & & & & & & \\ & X_1 \longrightarrow & X_2 \longrightarrow & \cdots & \longrightarrow & X_r = M_o & \\ X_{-1} \nearrow & & & & & & \end{array}$$

belonging both to $S(\to M_0)$ and to $S(\to Q_0(b))$, with $X_0 \not\approx X_{-1}$, and $r \geq 2$. Observe that $\dim \operatorname{Hom}(P_0(b),X_i) = \dim \operatorname{Hom}(X_i,Q_0(b)) = 1$ for all $-1 \leq i \leq r$, whereas $\operatorname{Hom}(P_0(a),X_i) = 0$ for $-1 \leq i < r$. We consider now A-modules, and use 2.5.8. In particular, we are interested in the A-modules $\overline{X}_i = (X_i, \operatorname{Hom}(R,X_i), 1)$, with $-1 \leq i < r$, and in the A-module $M = (M_0, \operatorname{Hom}(P_0(b),M_o), \iota)$, where ι is the canonical inclusion map. We obtain induced maps

$$\begin{array}{lllllll} \overline{X}_o \searrow & & & & & & \\ & \overline{X}_1 \longrightarrow & \overline{X}_2 \longrightarrow & \cdots & \longrightarrow & \overline{X}_{r-1} \longrightarrow & M, \\ \overline{X}_{-1} \nearrow & & & & & & \end{array}$$

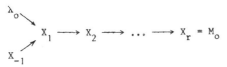

and all are irreducible in A-mod. For, we apply 2.5.5 to the A-modules $\overline{X}_i$, $-1 \leq i < r$, and we use the following observation: Let $f_i : X_i \longrightarrow Y_i$ be a source map for X_i in A_o-mod. Then the composition of any map $R \longrightarrow X_i$ with f_i factors through the inclusion $X_{i+1} \subseteq Y_i$, for $0 \leq i < r$, and through the inclusion $X_1 \subseteq Y_{-1}$, in case $i = -1$. Also, $\tau^-\overline{X}_o = \tau_o^- X_o$, and $\tau^-\overline{X}_{-1} = \tau_o^- X_{-1}$, according to 2.5.6. In particular, both $\tau_o^-\overline{X}_o$ and $\tau_o^-\overline{X}_{-1}$ belong to A_o-mod. Let $M' = M$, in case $r = 2$, and $M' = \overline{X}_2$, in case $r > 2$. Then M' is not projective (for, it is not an A_o-module, and also $M' \not\approx P(\omega)$, since the restriction of M' to A_o is indecomposable, whereas the restriction of $P(\omega)$ to A_o decomposes). We obtain from the irreducible map $\overline{X}_1 \longrightarrow M'$ an irreducible map $\tau M' \longrightarrow \overline{X}_1$, and $\tau M'$ is different both from $\overline{X}_o$ and from $\overline{X}_{-1}$. This shows that $\overline{X}_1$ has three direct predecessors in $\Gamma(A)$, thus the component containing $\overline{X}_1$ cannot be a tube. On the other hand, we have seen above that $\overline{X}_1$ and M belong to the same component of $\Gamma(A)$. This finishes the proof.

6.3 The large sincere directed algebras

Given a finite-dimensional algebra A, let us denote by n(A) the number of isomorphism classes of simple A-modules.

(1) Theorem (Bongartz). Let A be a sincere, directed algebra with $n(A) > 13$. Then A or A^{op} is an algebra presented in the table below.

In the table, we first give the quiver and its zero relations; in addition, one has to take all possible commutativity relations. The unoriented edges occurring in the list may be oriented arbitrarily [however, it is not allowed to replace a branch arbitrarily: the only possible changes of branches involving zero relations are already taken into account: the changes from (2) to (12), from (3) to (13) and from (4) to (14)]. We note that the algebras exhibited in the table have a unique sincere maximal module M, and we present in the second column its dimension vector. The third column gives the numbers of sincere positive roots, thus the number of isomorphism classes of sincere indecomposable modules. Here, m denotes the length of the largest branch. [Starting from the maximal root $z = \underline{\dim}\, M$, it is very easy to derive the remaining sincere positive roots using reflections σ_i]. The final column gives the orbit graph of the corresponding algebra. By a star, we indicate the possible positions of the τ-orbit of the sincere maximal module (it depends on the orientation of the unoriented edges). Observe that the orbit graph gives a quadratic form **Z**-equivalent to the quadratic form χ_A. (For, since A is sincere and directed, there is a slice $\mathcal{S}$ in A-mod, say with slice module S, and the underlying graph of the quiver of $\mathcal{S}$ coincides with $\overline{\mathcal{O}(A)}$. According to 4.1.7, the quadratic form of A is **Z**-equivalent to the quadratic form of End S.)

	A	$\underline{\dim}$ M	s	$\overline{\mathcal{O}(A)}$
(Bo1)	o—o- ··· -o—o	11···11	1	*—* ··· *—*
(Bo2)	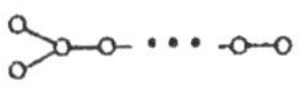	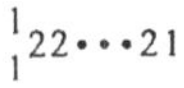	m	
(Bo3)	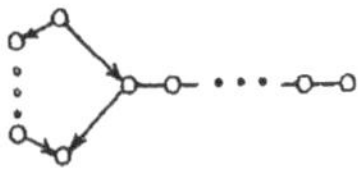	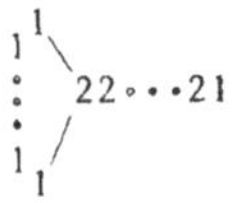	m	
(Bo4)	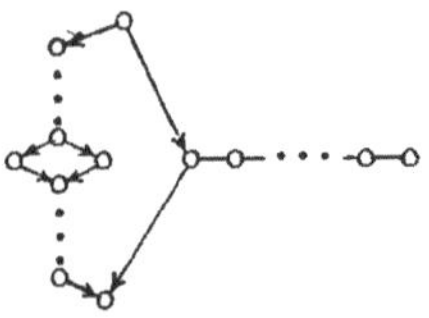	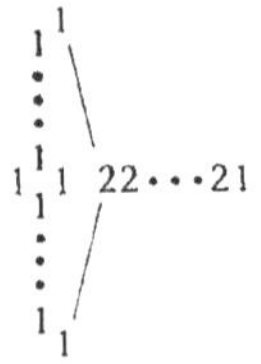	m	
(Bo5)	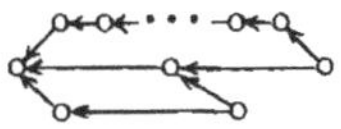	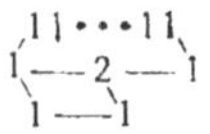	2	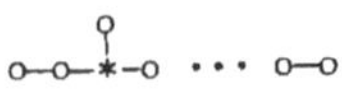

(Bo6)	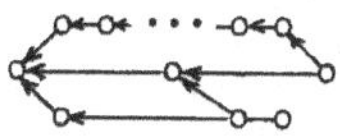	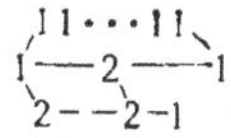	4	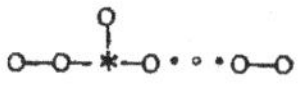
(Bo7)	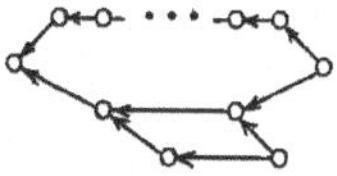	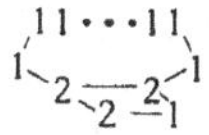	4	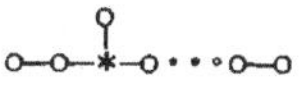
(Bo8)	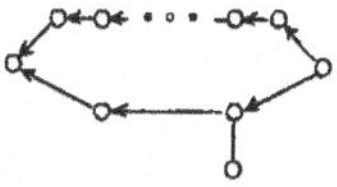	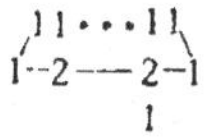	3	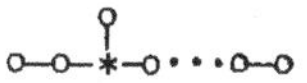
(Bo9)	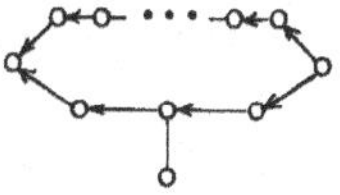	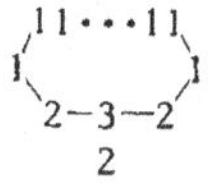	7	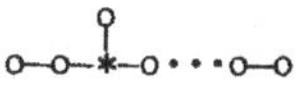
(Bo10)	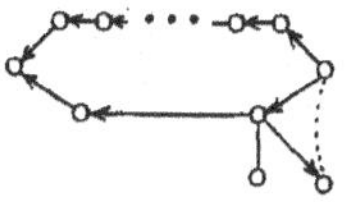	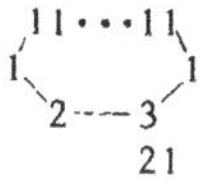	4	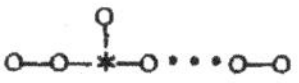
(Bo11)	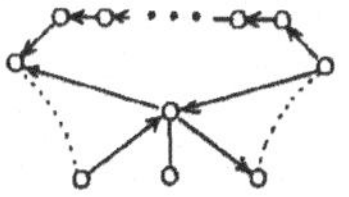	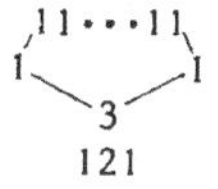	3	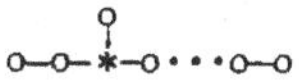
(Bo12)	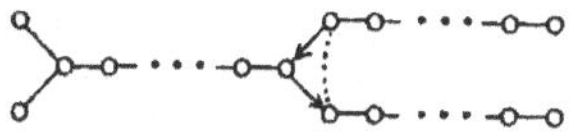	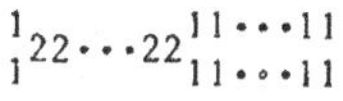	1	
(Bo13)	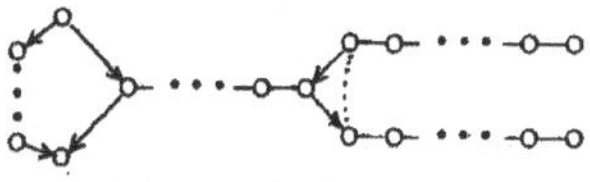	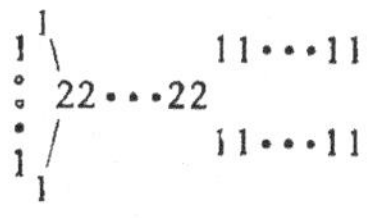	1	
(Bo14)	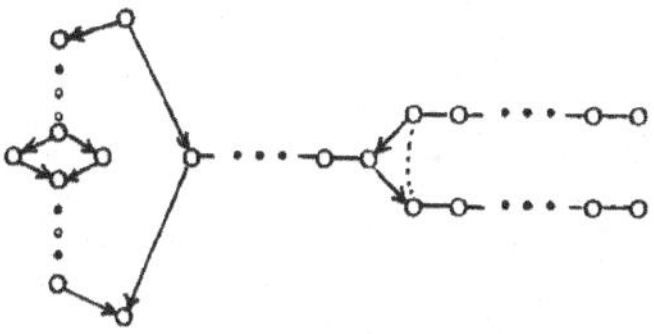	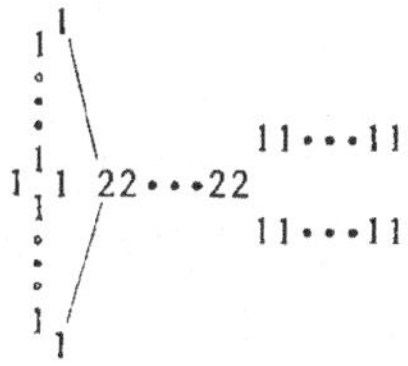	1	
(Bo15)	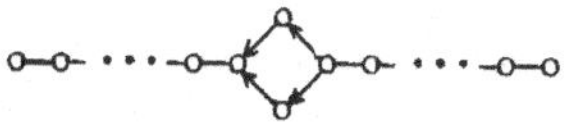	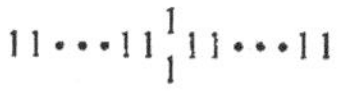	1	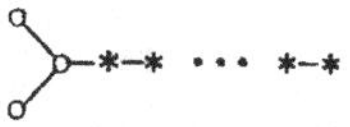

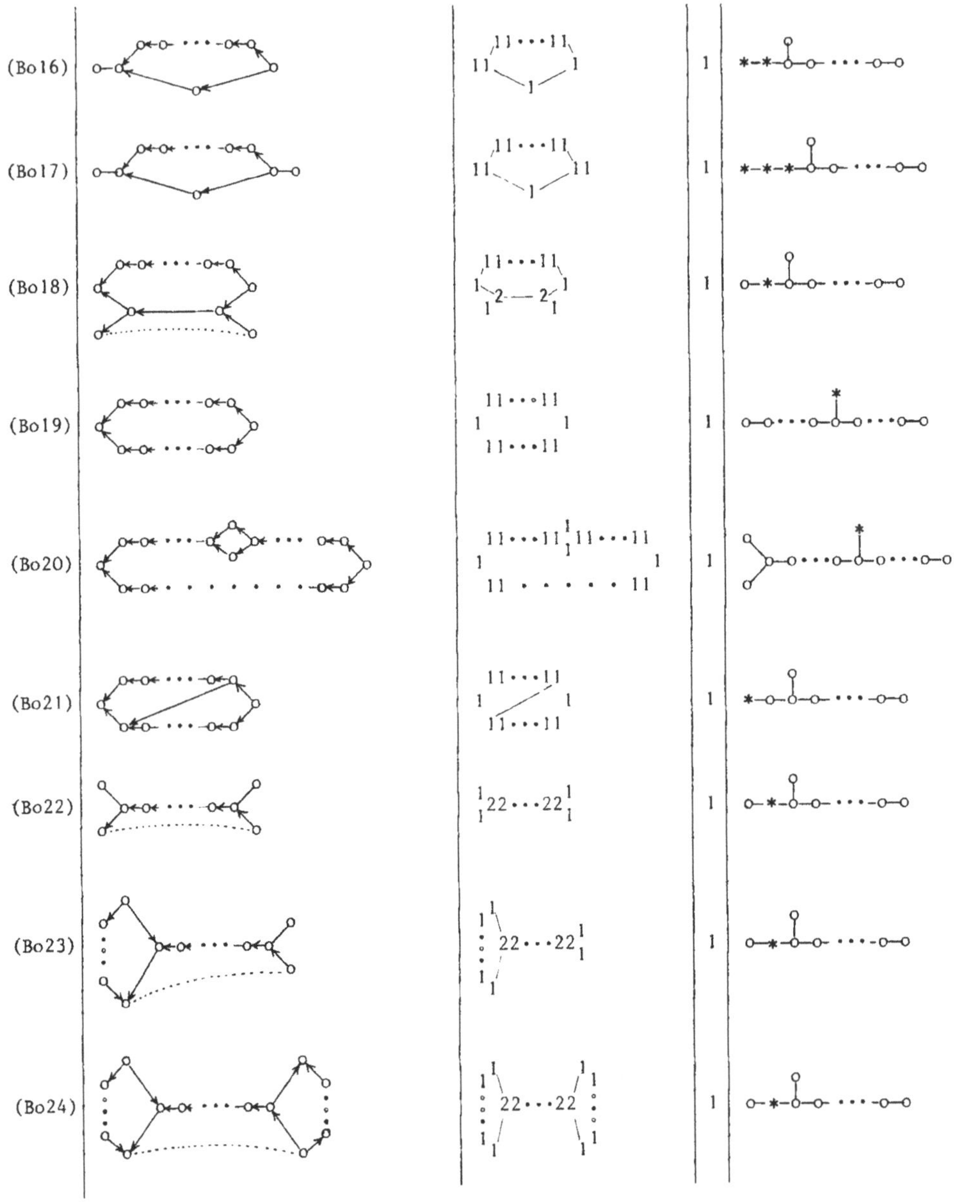

(2) <u>Corollary. Let</u> A <u>be a sincere directed algebra. Then there is a co-dimension 1 subspace</u> U <u>in</u> $K_o(A)$ <u>such that the restriction of</u> χ_A <u>to</u> U <u>is positive definite.</u>

<u>Proof.</u> We consider the quadratic form given by $\chi(\overline{\mathcal{O}(A)})$. If $n(A) > 13$, then we delete a suitable branching vertex of $\overline{\mathcal{O}(A)}$. On the other hand, given a tree with at most four endpoints (see 6.1.1) and at most 13 vertices, we always may delete

one vertex in order to obtain a disjoint union of Dynkin graphs.

A direct consequence of the classification and Ovsienko's theorem is also the following

(3) Corollary (Bongartz). Let A be a directed algebra, and M an indecomposable A-module. Then $|M| \leq 2 \cdot n(A) + 48$.

Proof of the corollary: Obviously, we may assume that A is sincere. We have $\underline{\dim}\, M \leq z$, where z is a maximal positive root for χ_A. If $n(A) \geq 14$, then A or A^{op} belongs to the list, and z is tabulated, thus we see that $|M| \leq 2n(A)-3$ (this bound is optimal for (Bo 2)). Now assume $n(A) \leq 13$. Let ω be an exceptional vertex for z, thus $z_\omega \leq 2$. According to theorem 1 of Ovsienko, $z_a \leq 6$ for all vertices a. Thus $|M| \leq \sum_a z_a \leq 2n(A) + \sum_{a \neq \omega} (z_a - 2) \leq 2n(A) + 12 \cdot 4$. Of course, the term 48 is not optimal, since never $z_a = 6$ for all $a \neq \omega$.)

(4) Outline of the proof of (1). It is easy to see that the only sincere directed algebras A with $\overline{\mathcal{O}(A)} = \mathbb{A}_n$ are the algebras (Bo1). It remains to consider the cases where $\mathcal{O}(A)$ has three or four endpoints. The possibilites for A in case $\mathcal{O}(A)$ has three endpoints are discussed in (6), (7), and (8). Assertion (10) gives a further restriction for the orbit quivers of sincere directed algebras having four endpoints. The possibilities for A in case $\mathcal{O}(A)$ has four endpoints, are discussed in (11). This then finishes the proof of (1). We need a preliminary result dealing with a maximal sincere module M whose τ-orbit cuts off a branch of $\mathcal{O}(A)$; this will be given in (5).

We introduce the following notation: Given an indecomposable A-module M, let $[[M]]$ denote its τ-orbit in $\Gamma(A)$. In case M belongs to a preprojective component, for example, if A is directed, then $[[M]]$ will be considered as an element of the orbit quiver $\mathcal{O}(A)$ of A. Also, we recall from 2.5, that given a one-point extension $A = A_0[R]$, we may consider the A-modules as triples $X = (X_0, X_\omega, \gamma_X)$, where X_0 is an A_0-module, X_ω a vectorspace, and $\gamma_X : X_\omega \to \mathrm{Hom}(R, X_0)$ a linear map. In particular, given an A_0-module Y, let $\bar{Y} = (Y, \mathrm{Hom}(R,Y), 1_{\mathrm{Hom}(R,Y)})$.

(5) Let A_0 be a directed algebra, let $R_1 \to R_2 \to \dots \to R_m$ be a sectional path, with R_i, $1 \leq i \leq m$, indecomposable, $[[R_1]]$ **endpoint**, and such that, for $1 \leq i < m$, $[[R_i]]$ cuts off a branch of $\mathcal{O}(A_0)$. Assume that $\tau^{-m+s}R_s$ is injective, for some $1 \leq s \leq m$. Let $A = A_0[R_1]$. Then there is at most one indecomposable sincere A-module, namely $\bar{R}_m$. Always, $[[\bar{R}_m]]$ cuts off a branch of $\mathcal{O}(A)$ of length m.

Proof. We may assume that m is minimal such that some $\tau^{-m+s}R_s$ with $1 \leq s \leq m$, is injective. Let $R_{ij} = \tau^{-i+1}R_{j-i+1}$, $1 \leq i \leq j \leq m$. Let $t = m-s+1$, thus R_{tm} is injective. Thus, in $\Gamma(A_0)$, there is the following part, which obviously is a wing for R_{1m}.

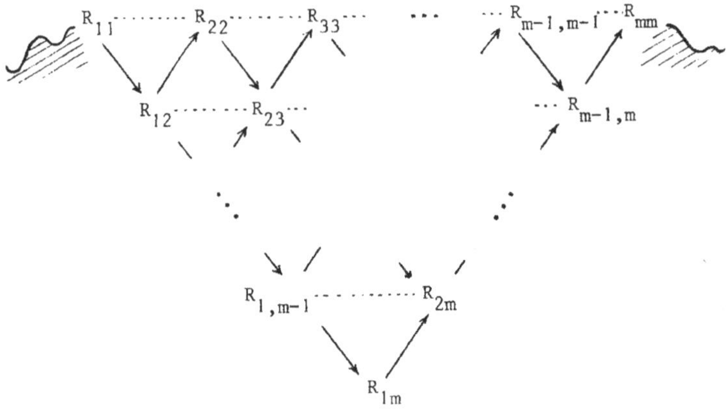

Note that $\mathrm{Hom}(R_1, R_{ij}) = k$ for $i = 1$, and $= 0$, otherwise. Thus, in $\Gamma(A)$, we obtain the following diagram

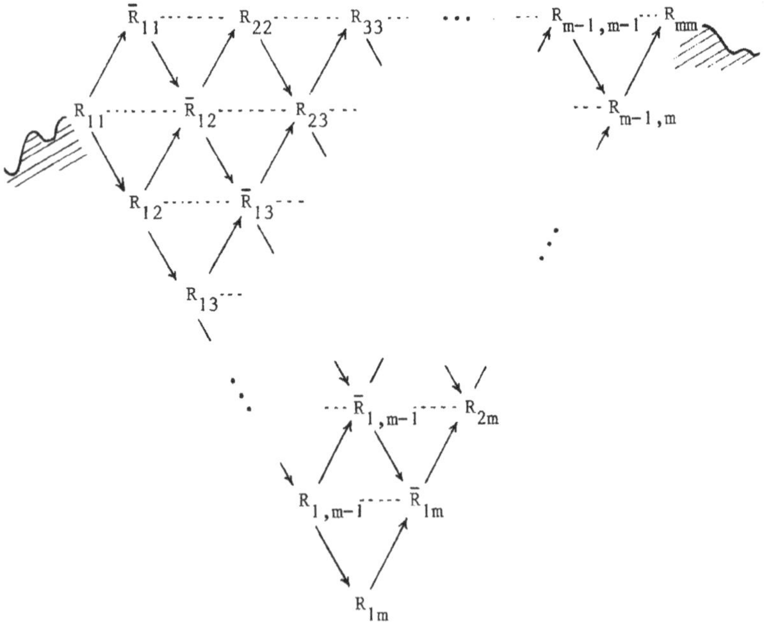

using 2.5.5. Note that $\bar{R}_{ij} = R_{ij}$ for $2 \le i \le j \le m$. Since R_{tm} is an injective A_o-module, $\bar{R}_{tm}$ is an injective A-module. Of course, with R_{tm} also R_{mm} is an injective A_o-module. Any indecomposable sincere A-module M must be a predecessor of the injective module $\bar{R}_{mm}$, and a successor of the projective module $\bar{R}_{11}$.

Thus M belongs to the modules exhibited above, and the only one having non-zero maps from $\bar{R}_{11}$ and with non-zero maps to $\bar{R}_{mm}$ is R_{1m}.

(6) Let A be a directed algebra with a sincere wing module. Then A or A^{op} is of the form (Bo*) with * = 1,2,3; 5,6,7,8,9,10, or 11, or else the type of the wing is given by $(m,3,2)$ with $m \leq 8$, or by $(m,4,2)$ with $m \leq 6$, or by $(m,3,3)$ with $m \leq 5$.

Proof. Given a quiver Δ, let $s(\Delta)$ be the sum of the numbers of sinks and of sources of Δ. We say that an algebra A has flat orientation provided $s(\Delta(A))$ does not decrease when changing the orientation of any branch of A. Note that if A has flat orientation, and if A is obtained from A' by adding a branch without zero relations in a, then this branch has either subspace, or factorspace orientation, and it has subspace orientation in case a is a source of A', whereas it has factorspace orientation in case a is a sink of A'.

An indecomposable projective module will be called final provided it is not a proper predecessor of any indecomposable projective module. According to 4.2.6", given a final indecomposable projective module P over a sincere directed algebra, then always $S(P\to)$ is a slice.

Now, let A be a directed algebra with a sincere wing module. We may assume that A has flat orientation. If the orbit graph $\overline{\mathcal{O}(A)}$ is of type $\mathbb{A}_n$, then clearly A is of the form (Bo1). Thus assume $\overline{\mathcal{O}(A)}$ is a proper star. We say that A is of level ℓ provided there exists a final indecomposable projective module P such that $[[P]]$ belongs to a branch of length ℓ, and given any final indecomposable projective module P', then $[[P']]$ belongs to a branch of length $\geq \ell$.

The colevel of A is, by definition, the level of A^{op}.

By assumption, the orbit graph $\mathcal{O}(A)$ of A is a star, and according to 6.1.3, it has precisely three branches. In case $\overline{\mathcal{O}(A)} = \mathbb{D}_n$, we know from 6.2.1 that A is of the form (Bo2), or (Bo3). Thus, we can assume that $\mathcal{O}(A) = \mathbb{T}_{n_1,n_2,n_3}$ with $n_1 \geq n_2 \geq n_3 \geq 2$, and $n_2 \geq 3$. We consider the various possibilities of (n_1,n_2,n_3), and use induction on the level of A.

Assume A is of level ℓ. We choose a final indecomposable projective A-module $P(\omega)$ such that $[[P(\omega)]]$ belongs to a branch of length ℓ. Since A has flat orientation, $[[P(\omega)]]$ actually is endpoint of $\mathcal{O}(A)$. Let A_o be obtained from A by deleting ω, and $R = \operatorname{rad} P(\omega)$. Let τ_o be the Auslander-Reiten translation in A_o-mod. We consider the slice $\tau S(P(\omega)\to)$ in A_o-mod. Let

$$X_{\ell-1} \to X_{\ell-2} \to \cdots \to X_1 \begin{matrix} \nearrow Y_2 \to \cdots \to Y_p \\ \searrow Z_2 \to \cdots \to Z_q \end{matrix}$$

be its quiver with $q \leq p$. In particular, $R = X_{\ell-1}$, and (ℓ,p,q) coincide, up to order with (n_1,n_2,n_3). Since A_o is directed and $A = A_o[R]$ is representation finite, $\dim \operatorname{Hom}(R,N) \leq 1$ for all indecomposable A_o-modules N. We consider the linear vectorspace category *) $K = \operatorname{Hom}(R,A_o\text{-mod})$, and the corresponding partially ordered set $S(K)$. In case $\tau_o^{-j}X_i \neq 0$, we write $X_{ij} = \tau_o^{-j}X_i$. In case $\operatorname{Hom}(R,X_{ij}) \neq 0$, we write $x_{ij} = [X_{ij}]$. If $\operatorname{Hom}(R,X_{ij}) \neq 0$, and X_{ij} is an injective A_o-module, we also will say that x_{ij} is injective. Also, let $x_i = x_{io}$. Similarly, we define Y_{ij}, Z_{ij}, y_{ij}, z_{ij}, and y_i, z_i.

According to (5), none of the modules X_{ij}, $1 \leq j < i$, is injective, thus for $1 \leq j \leq i$, X_{ij} is defined. Also, $\operatorname{Hom}(R,X_{ii}) \neq 0$ due to $p \geq 2$, $q \geq 2$, thus $S(K)$ contains the following convex subset:

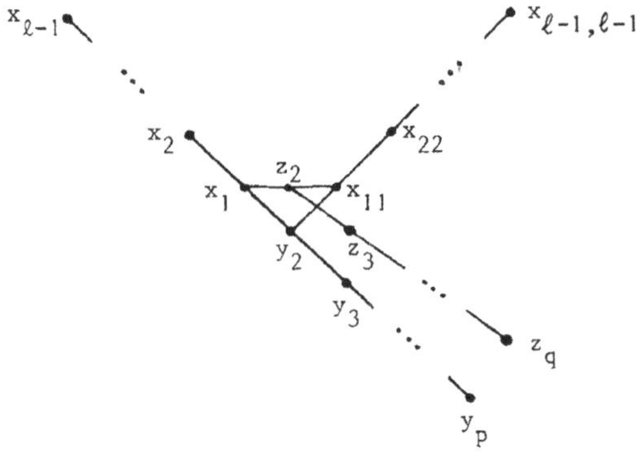

Note that the elements x_i $(1 \leq i \leq \ell-1)$ are comparable with all elements of $S(K)$, and the only elements of $S(K)$ incomparable with z_2 are the elements in the chain $\{y_2,y_3,\ldots,y_p\}$. As a consequence, we obtain in $\Gamma(A)$ the following part:

---

*) In case we deal with a vectorspace category $(K,|\cdot|)$ of the form $K = A_o\text{-mod}$, $|\cdot| = \operatorname{Hom}(R,-)$, we denote the corresponding vectorspace category $(K(|\cdot|),|\cdot|)$ as defined before 2.6.5 just by $\operatorname{Hom}(R,A_o\text{-mod})$. Thus, the objects of $\operatorname{Hom}(R,A_o\text{-mod})$ are just the objects of A_o-mod, but two maps $f,g : X \longrightarrow Y$ in A_o-mod are identified in $\operatorname{Hom}(R,A_o\text{-mod})$ provided $\operatorname{Hom}(R,f) = \operatorname{Hom}(R,g)$.

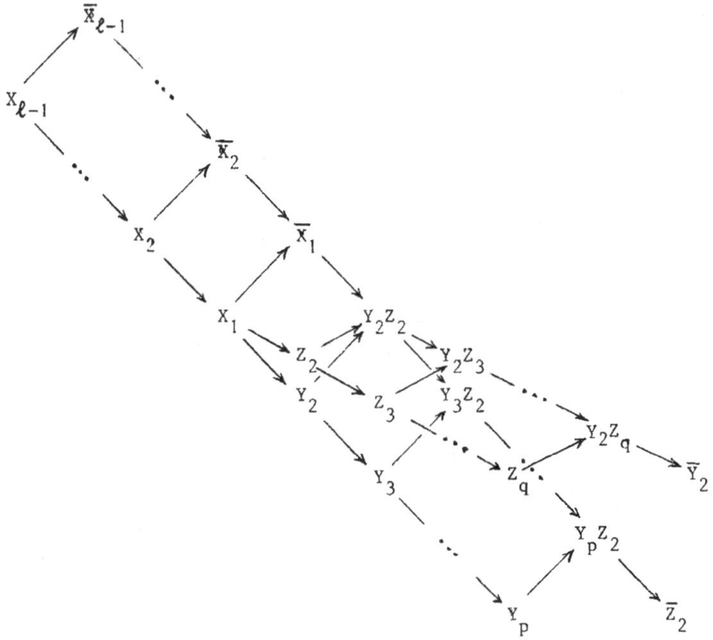

Here, given indecomposable A_o-modules Y, Z, with $\mathrm{Hom}(R,Y) = \mathrm{Hom}(R,Z) = k$, and $\mathrm{Hom}(Y,Z) = \mathrm{Hom}(Z,Y) = 0$, we denote by YZ the unique indecomposable A-module with restriction to A_o being $Y \oplus Z$. We can read from this diagram that $\tau^- Y_p = \overline{Z}_2$, and that no proper predecessor of $\overline{Z}_2$ can be injective.

<u>First case</u>: $n_3 \geq 3$. Consider the case $\ell = n_3$. If $n_2 \geq 4$, then $\{x_{11},x_{22},y_3,y_4,z_3,z_4\}$ is $C(3)$, impossible. Thus $q = n_2 = 3$. Assume $n_1 \geq 6$. Then z_2 has to be injective, since otherwise $\{x_{11},x_{22},z_{21},z_3,y_3,y_4,y_5,y_6\}$ gives $C(5)$. Also, y_2 has to be injective, since otherwise $\{x_{22},z_3,c_{21},c_4\}$ gives $C(2)$. But then also x_{11} has to be injective. (For, assume X_{11} is not injective. The source map for X_{11} is of the form $X_{11} \longrightarrow X_{22}$, since Y_2, Z_2 are injective, and this is also the sink map for X_{22}. The two sink maps $X_{11} \longrightarrow X_{22}$ and $X_{22} \longrightarrow X_{12}$ have zero composition, and this shows that X_{22} is the only indecomposable module with a non-zero map to X_{12}. Also, X_{12} is an injective A_o-module. If M is a sincere A-module, then the restriction M_o of M to A_o must have a direct summand of the form X_{12} or X_{22}. On the other hand, we also must have $\mathrm{Hom}(M_o,Y_2) \neq 0$, since Y_2 is an injective A_o-module. Now, the proper predecessors of y_2 are comparable with all elements of $S(K)$, and y_2 itself is incomparable only with z_2 and z_3. Thus, M cannot be indecomposable.) As a consequence, A_o is a bush algebra of type $(n_1,3,2)$, thus representation infinite, a contradiction. This shows that $n_1 \leq 5$.

Now assume $\ell > n_3$, in particular $\ell \geq 4$. If $q \geq 4$, then $\{x_{22}, x_{33}; y_3, y_4; z_3, z_4\}$ is of the form $C(3)$, impossible. Thus $q = 3$. If z_2 is not injective, and $p \geq 4$, then $\{x_{22}, y_3, z_{21}, z_4\}$ is of the form $C(2)$, impossible. If neither y_2 nor z_2 is injective, then $\{x_{22}, x_{33}; y_3, y_{21}; z_3, z_{21}\}$ gives $C(3)$. Thus (up to symmetry) we can assume that z_2 is injective. However, in this case $\tau^- Y_p = \bar{Z}_2$ is an injective A-module, and no injective A-module is a proper predecessor of $\bar{Z}_2$. Of course, since A is sincere, $S(\rightarrow \bar{Z}_2)$ is a slice. This shows that the colevel of A is 3.

<u>Second case:</u> $n_3 = 2$, $n_2 \geq 4$. First, assume we even have $n_2 \geq 5$. If $\ell = 2$, then $\{x_{11}; y_3, y_4, y_5; z_3, z_4, z_5\}$ is of type $C(4)$, impossible. Thus, consider the case $\ell > 2$. If y_2 is injective, then the colevel of A is 2. Thus, assume y_2 is not injective. Also, x_{11} cannot be injective, since otherwise A_o is a bush algebra of non-Dynkin type. But then $\{x_{33}, x_{44}; y_{21}, x_{12}; y_4, y_5\}$ is of type $C(3)$.

Thus $n_2 = 4$. Assume $n_1 \geq 7$. If level 2, then $\{x_{11}; y_3, y_4, y_5, y_6, y_7; z_3, z_4\}$ is of type $C(6)$. Assume the level is 4; we can assume that y_2 is not injective, since otherwise the colevel is 2. Also, x_{11} is not injective, since otherwise A_o is a bush algebra of non-Dynkin type. But then $\{x_{22}, x_{33}, x_{12}, y_{21}; y_4, y_5, y_6, y_7\}$ is of type $C(5)$. Finally, consider the case of level n_1. If y_2 is injective, then the colevel is 2, if z_2 is injective, then the colevel is 4. Thus, we can assume that Z_{21} is defined (however $\mathrm{Hom}(R, Z_{21}) = 0$). We claim that in case Z_{21} is an injective A_o-module (and therefore also an injective A-module), then the colevel of A is again 4. Namely, we obviously can construct in $\Gamma(A)$ the following part

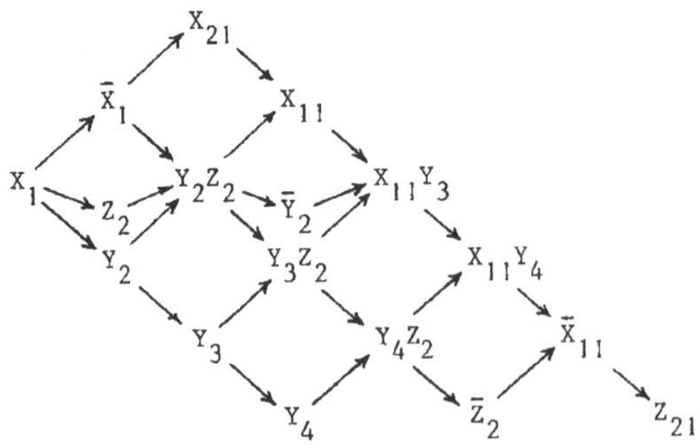

But if Z_{21} is not injective, then $\mathrm{Hom}(R, Z_{22}) = k$, thus z_{22} exists in $S(K)$, and $\{x_{33}, x_{44}, x_{55}; y_{21}, x_{12}, z_{22}; y_4\}$ is of type $C(4)$, impossible. Altogether we see that we must have $n_1 \leq 6$.

Third case: $n_3 = 2$, $n_2 = 3$. Assume $n_1 \geq 9$. First, consider the case $\ell = 2$. If y_2 is not injective, then $\{z_3; x_{12}, y_{21}; y_4, y_5, y_6, y_7, y_8\}$ is of type $C(6)$. Thus, assume y_2 is injective. In this case $S(K)$ is representation finite for all n_1. If z_2 also is injective, A_o is given by the quiver

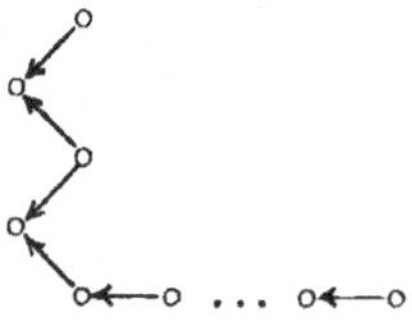

and $R = X_1$ is the unique sincere A_o-module, thus $A^{op} = $ (Bo5). If z_2 is not injective, then similarly we obtain as A the algebras (Bo8), with the two possible orientations of the free branch.

Next, let $\ell = 3$. We can assume that y_2 is not injective, since otherwise, the colevel of A is 2. If also y_3 is not injective, then $\{x_{22}; y_{21}, y_{31}; y_5, y_6, y_7, y_8, y_9\}$ is of type $C(6)$, impossible. Thus y_3 is injective. But then $S(K)$ is representation finite for all n_1. It remains to determine the possible cases for A_o and R such that $A = A_o[R]$ can have a sincere wing module. In $\Gamma(A)$ we calculate the following part (using the same conventions as before):

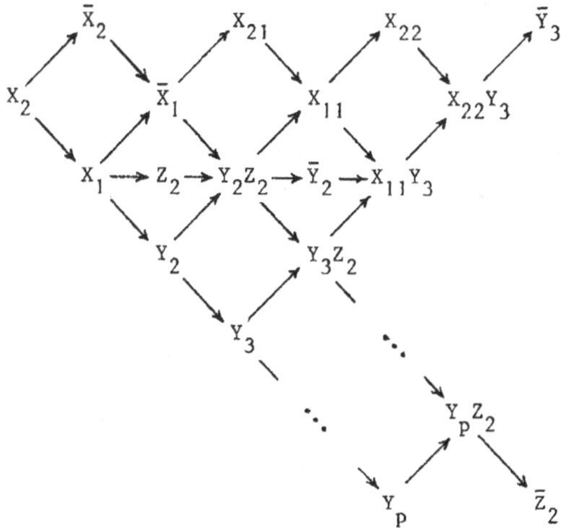

Since $\bar{X}_2$ is projective, and $\bar{Y}_3$ is injective, the only possible sincere wing modules are Y_2Z_2 and $X_{11}Y_3$. Actually, in case Z_2 is an injective A_o-module, only Y_2Z_2 remains as a candidate for a sincere wing module. If x_{11} is injective, then we obtain for A^{op} just the two algebras (Bo6), the orientation of the free branch being dependent on whether z_2 is injective or not. Thus, assume that x_{11} is not injective. If z_2 is injective, then x_{22}, x_{12}, y_{21} have to be injective (otherwise $Y_2 \oplus Z_2$ will not be a sincere A_o-module), thus A^{op} is of the form (Bo7). We now assume that z_2 is not injective, but that x_{12} is injective. Consider first the case of y_{21} being injective. If Z_{21} is injective, we obtain the two algebras (Bo10), if Z_{21} is not injective, then the two algebras (Bo9), always depending on whether x_{22} is injective or not. Second, assume that y_{21} is not injective. If $\tau^- Z_2$ is injective, we obtain the two algebras (Bo11), otherwise we obtain for A^{op} the two algebras (Bo10) (again depending on x_{22}). Now assume that x_{12} is not injective. Then neither X_{22} nor Z_{21} can be an injective A_o-module, since otherwise $\mathrm{Hom}(X_{11} \oplus Y_3, X_{13}) = 0$. If y_{21} is injective, then we obtain the case (Bo7). If y_{21} is not injective, then we obtain the two algebras (Bo6), depending on whether Y_{22} is injective or not. There are no further possibilities, due to the condition that $X_{11} \oplus Y_3$ (or $Y_2 \oplus Z_2$) has to be a sincere A_o-module.

Finally, let $\ell = n_1$. Successive calculation shows that none of the A_o-modules $Y_2, Z_2, X_{11}, Z_{21}, X_{22}$, or X_{33} can be injective, since otherwise the colevel of A would be 2 or 3. Namely, we obtain the following part of $\Gamma(A)$ (the calculation is done from left to right, at every step assuming that no proper predecessor is injective):

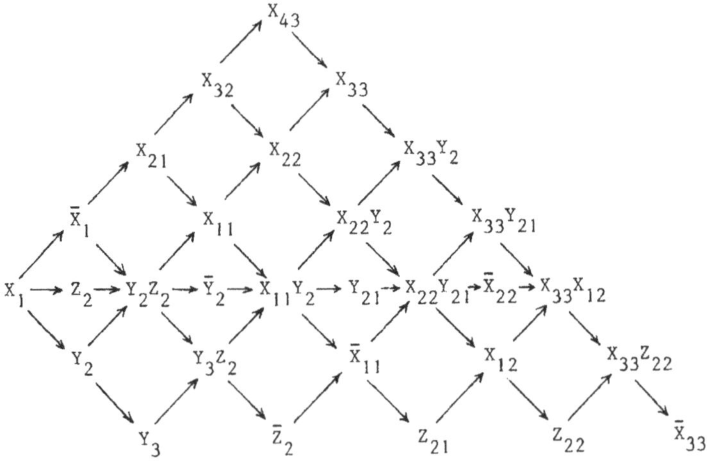

(note that $Z_{21} = \bar{Z}_{21}$, since $\mathrm{Hom}(R, Z_{21}) = 0$). Also, X_{12} cannot be injective, since otherwise A_o would be a bush algebra of non-Dynkin type. Thus, we obtain in $S(K)$ the subset $\{x_{55}, x_{66}, x_{77}, x_{88}; x_{23}, x_{34}, x_{13}, z_{22}\}$ of type $C(5)$, impossible. This finishes the proof.

Let A be a sincere, directed algebra. In case A has a unique **sincere** indecomposable module M, we say that M is a <u>characteristic</u> module. The dimension vector $\underline{\dim}\, M$ of a characteristic module M always has two exceptional vertices. For, a characteristic module is a maximal module, and if $\underline{\dim}\, M$ has a unique exceptional vertex, say a, then $\sigma_a \underline{\dim}\, M$ is again a sincere positive root and, according to 2.4.9, there is an indecomposable module M' with $\underline{\dim}\, M' = \sigma_a \underline{\dim}\, M$. As a consequence, it follows from 6.1.3 that the τ-orbit of a characteristic module M has at most two neighbors in $\mathcal{O}(A)$. On the other hand, if M is a sincere maximal module, and [[M]] cuts off a branch of $\mathcal{O}(A)$, then M is characteristic, according to 3.6.2. In particular, <u>if</u> $\mathcal{O}(A)$ <u>is a star, and</u> M <u>is a sincere maximal module, then either</u> M <u>is a wing module, or</u> M <u>is characteristic</u>.

Let M be a characteristic A-module. A branch of A of length ≥ 2 is said to be a <u>thin branch</u> provided $\dim M_i = 1$ for every vertex i of the branch. **Of course**, a thin branch never carries a zero relation. Note that the endpoint of a thin branch always is an exceptional vertex for $\underline{\dim}$ M. Let A be a directed algebra with a characteristic module M. According to 3.6.2, [M] has a wing θ in Γ(A), and the source of θ is of the form [P(a)], the sink of θ is of the form [Q(b)], where a,b are the exceptional vertices of $\underline{\dim}$ M. Let $P(a) \longrightarrow Y$ be irreducible, with Y indecomposable. If Y is projective, say $Y = P(a')$, then $a \longleftarrow a'$ is a thin branch, since obviously $\dim \mathrm{Hom}(P(a'),M) = 1$ (if Y is projective, then θ is of length ≥ 2). Also, if Y is not projective, and τY is projective, say $\tau Y = P(a'')$, then $a'' \longleftarrow a$ is a thin branch (for, $\dim \mathrm{Hom}(P(a''),M) = 1$, since [[M]] cuts off a branch). On the other hand, if neither Y, nor τY is projective, then $\mathrm{rad}\, P(a) = \tau Y$ is not projective, thus a does not cut off a branch of Δ(A). This shows that a belongs to a thin branch if and only if rad P(a) is projective. Also, if a does not belong to a thin branch, then P(a) is final. (For, if $P(a) \preceq P$ with P indecomposable projective, then $P \preceq M$ shows that P must belong to the wing θ, but then all predecessors of P in θ must be projective.)

(7) Let A be directed and with orbit graph $\overline{\mathcal{O}(A)}$ having three endpoints. Assume there exists a characteristic A-module and that A has no thin branches. If $\overline{\mathcal{O}(A)} = \mathbb{D}_n$, then A is of one of the following forms:

(Bo12') 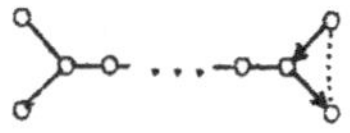

(Bo19') 

(Bo13') 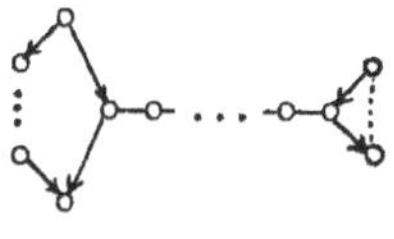

If $\overline{\mathcal{O}(A)} \neq \mathbb{D}_n$, and $n(A) \geq 13$, then A or A^{op} is of the form (Bo*) with $* = 18, 19, 21, 22, 23,$ or 24.

Proof. Let M be a characteristic A-module. We know that $[M]$ has a wing θ in $\Gamma(A)$, let $[P(a)]$ be the source, and $[Q(b)]$ the sink of θ. Since there are no thin branches, $P(a)$ is final, and $R = \operatorname{rad} P(a)$ is not projective. Similarly, $Z = Q(b)/\operatorname{soc} Q(b)$ is not injective. Let A_o be obtained from A by deleting the vertex a.

First case: Assume the branch of $\mathcal{O}(A)$ containing $[[M]]$ has length 2. Then $P(a) \approx M \approx I(b)$, since $[[M]]$ cuts off a branch, thus the A_o-module $\operatorname{rad} P(a)$ is sincere and injective. It follows that A is of the form (Bo19). Of course, $\overline{\mathcal{O}(A)} = \mathbb{D}_n$ if and only if A is of the form (Bo19').

Second case: Assume the branch of $\mathcal{O}(A)$ containing $[[M]]$ has length $m \geq 3$. Thus $\mathcal{O}(A_o)$ has three endpoints. We claim that A_o has a sincere wing module. Let

$$X_1 \xrightarrow{p_1} X_2 \xrightarrow{p_2} \cdots \xrightarrow{p_{m-1}} X_m = Q(b)$$

be the unique sectional path of length m ending in $Q(b)$, thus $[[X_1]]$ is the center of $\mathcal{O}(A)$. Assume the wing θ is of length ℓ and let $t = m-\ell$, thus $t \geq 1$ and $M \approx X_{t+1}$. Note that X_t is the restriction of M to A_o, since there is an exact sequence

$$0 \longrightarrow R \xrightarrow{[\iota *]} P \oplus X_t \xrightarrow{[p_t^*]} X_{t+1} \longrightarrow 0$$

with $\iota : R \longrightarrow P$ the inclusion map. It follows that the modules $\tau^i R$, with $1 \leq i \leq t-1$, cannot be projective, since $\operatorname{Hom}(\tau^i R, X_t) = 0$, for these i and X_t is sincere. As a consequence, all the maps p_i, with $1 \leq i \leq t-1$, are surjective. For, consider the Auslander-Reiten sequences

$$0 \longrightarrow \tau^i R \longrightarrow Y_{-i+1} \xrightarrow{e_{i-1}} \tau^{i-1} R \longrightarrow 0,$$

$1 \leq i \leq t-1$. Then, there is an exact sequence of the form

$$0 \longrightarrow Y_{-t+i+1} \xrightarrow{[e_{t-i-1} *]} \tau^{t-i-1} R \oplus X_i \xrightarrow{[p_i^*]} X_{i+1} \longrightarrow 0.$$

Note that there is the following part of $\Gamma(A)$:

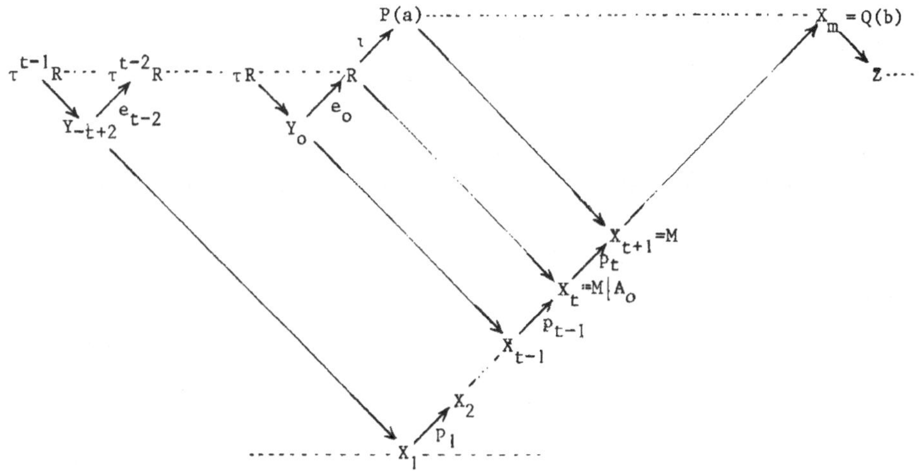

With X_t also X_1 is a sincere A_o-module. Thus, we have obtained an A_o-module X_1 which is sincere and with $[[X_1]]$ having three neighbors. It follows that X_1 is a wing module.

Consider the case $\overline{\mathcal{O}(A_o)} = \mathbb{D}_{n-1}$, where $n = n(A) = m+2$. According to 6.2.1, A_o is of the form (Bo2) or (Bo3). First, let $n \geq 6$ and assume that $[[R]]$ is the endpoint of the long branch. Since $\tau^{t-1}R$ is not projective, also Y_{-t+1} cannot be projective, thus $Y_{-t} := \tau_o Y_{-t+1}$ is non zero. For $i \geq 0$, define $Y_i = \tau_o^{-i} Y_o$. In $\Gamma(A_o)$, there is just one arrow starting in $[R]$, namely $[R] \longrightarrow [Y_1]$, and thus, in $\Gamma(A)$, there is the arrow $[P(a)] = [\bar{R}] \longrightarrow [\bar{Y}_1]$. We now use 2.5.6. It follows that $\tau^{-i}\bar{Y}_1 = Y_{i+1}$, for $1 \leq i \leq m-t-2$, and therefore $Y_{m-t-1} \approx X_{m-1}$. Also, Y_{m-t-1} cannot be injective as A_o-module. For, otherwise Y_{m-t-1} is injective also as A_o-module, since $\mathrm{Hom}(R, X_{m-1}) = 0$, but $\tau^- X_{m-1} = Z \neq 0$. Also, $\mathrm{Hom}(P(a), Z) = 0$ shows that Z is an A_o-module, thus $Y_{m-t-1} = \tau Z = \overline{(\tau_o Z)} = \tau_o Z$, or $Z = \tau_o^- Y_{m-t-1} = Y_{m-t}$. In $\Gamma(A)$, we have the following part

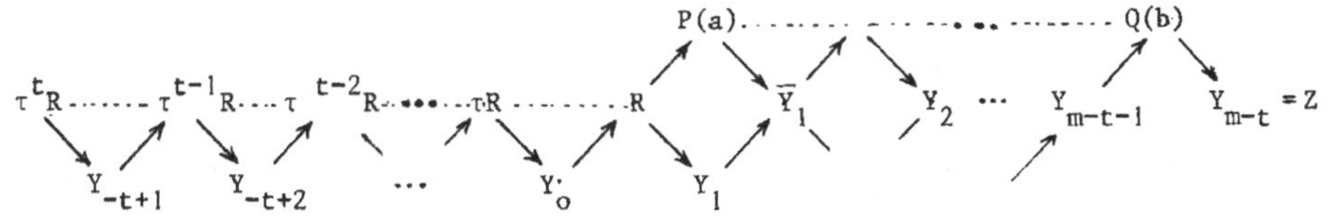

Now, in $\Gamma(A_o)$ we have m A_o-modules

$$Y_{-t+1},\ Y_{-t+2}, \dots,\ Y_{m-1}$$

belonging to one τ_o-orbit in the long branch of $\mathcal{O}(A_o)$. For $n(A_o) \geq 5$, and A_o of the form (Bo2) or (Bo3), one may check that the τ-orbits of the long branch of $\mathcal{O}(A_o)$ contain precisely m modules. Therefore, Y_{-t+1} is projective, and Y_{m-t} is an injective A_o-module. We claim that we must have $t = 1$. For, assume $t > 1$. Then $\mathrm{Hom}(R,Y_{m-t}) = 0$, thus $Z = Y_{m-t}$ is injective also as an A-module. A contradiction. Let $Y_o = P(u)$, $\tau R = P(u')$. We have an inclusion map $P(u') \to P(u)$, and its cokernel is R. It follows that $A = A_o[R]$ is obtained from A_o by adding the vertex a with an arrow $u \leftarrow a$, and a zero relation from a to a', thus A is of the form (Bo12) or (Bo13).

Next consider again the case $\overline{\mathcal{O}(A_o)} = \mathbb{D}_{n-1}$, but with $[[R]]$ the endpoint of a branch of length 2. We consider $\mathcal{S}(R\to)$ in A_o-mod, it is of the form

$$\begin{array}{ccccl} & & & \nearrow & Y_1 \longrightarrow \dots \longrightarrow Y_p \\ R & \longrightarrow & R_1 & & \\ & & & \searrow & Z \end{array}$$

Note that R_1 cannot be injective, since otherwise A would have a thin branch. Let $R_{11} = \tau^- R_1$. Consider the case of R being injective. In this case, R is sincere, thus there is some $1 \leq t < p$ such that the A_o-modules Y_i, $2 \leq i \leq t$ are not injective, whereas the A_o-modules Y_i, $t < i < p$ (and also the A_o-modules $\tau_o^- Y_i$, $2 \leq i \leq t$) are injective. Therefore A is of the form (Bo21). Now, we assume that R is not injective. Since A has a characteristic root, and R, R_1 are not injective, the module $R' = \tau^- R$ has to be injective, and R_1 has to be sincere. There exists $1 \leq t \leq p$ such that the A_o-modules Y_i, $2 \leq i \leq t$ are not injective, whereas the A_o-modules Y_i, $t < i \leq p$ are injective. If R_{11} is injective, then Z may or may not be injective, but all A_o-modules $\tau_o^{-1} Y_i$, $2 \leq i \leq t$, are injective, and we obtain as A^{op} the algebras (Bo23) in case $t < p$, and (Bo22) in case $t = p$. If R_{11} is not injective, then Z cannot be injective, and there is $1 \leq s < t$ such that the A_o-modules $\tau^- Y_i$, $2 \leq i \leq s$ are not injective, whereas the A_o-modules $\tau^- Y_i$, $s < i \leq t$ are injective. For $t = p$, we obtain the algebras (Bo23), for $t < p$ the algebras (Bo24). This finishes the proof of the case $\overline{\mathcal{O}(A_o)} = \mathbb{D}_{n-1}$. In particular, if even $\overline{\mathcal{O}(A)}$ is of the form $\mathbb{D}_n$, then the only possibilities are (Bo12') and (Bo13').

If we assume $\overline{\mathcal{O}(A_o)} \neq \mathbb{D}_n$, and $n = n(A) \geq 13$, then, according to (6), we know that A_o or A_o^{op} if of the form (Bo*) with $* = 5,6,7,8,9,10$, or 11.

Thus, let A_o or A_o^{op} be of the form (Bo*), with $* = 5,6,7,8,9,10$, or 11. Let N be a sincere A_o-module, with a wing θ' of length u' and assume that the sink of θ' is of the form $[Q(x)]$. Let $[R]$ be the source of θ', and assume the branch of $\mathcal{O}(A_o)$ containing $[[R]]$ is of length u. We claim that for $n(A_o) \geq 10$, the vectorspace category $\mathrm{Hom}(R, A_o\text{-mod})$ is subspace infinite, with the only exception

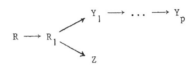

A_o = 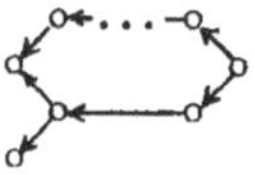 and $\underline{\dim}\, R = {}^{\,1}{}_{o}{}^{o \cdots o}_{1 \text{---} 1}{}^{o}$,

in which case $A_o[R]$ is of the form (Bo18). We list all possible cases for x and u', the vertex x is denoted by a star.

The cases with u = 2:

u' = 1,2 u' = 2

The cases with u = 3:

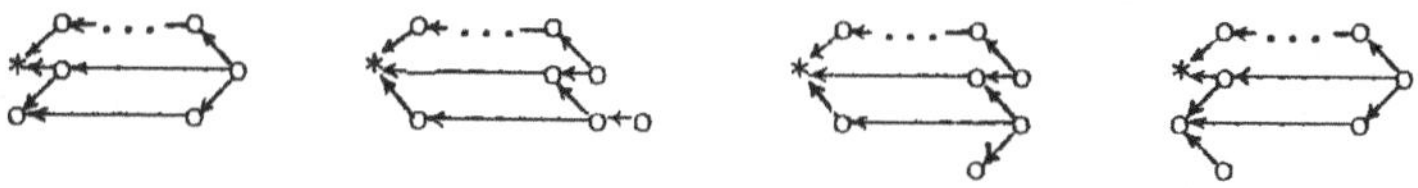

u' = 3 u' = 1,2,3 u' = 2,3 u' = 3

u' = 1,2,3 u' = 1,2,3 u' = 2,3 u' = 1,2,3

u' = 2,3 u' = 3 u' = 2,3

The cases with $u = n(A_o)-3$:

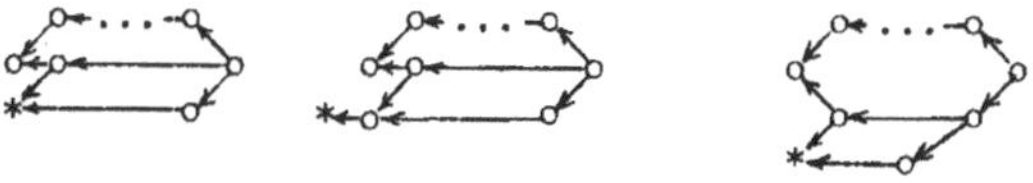

u' = u u' = u u' = u-1,u

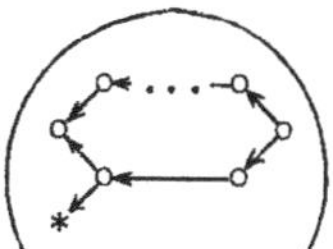

u' = u

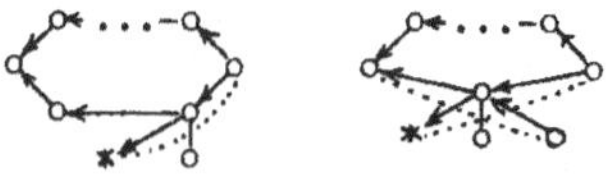

u' = u u' = u.

For $u = 2$, and $n(A_o) \geq 10$, the vectorspace category $Hom(R,A_o\text{-mod})$ is subspace infinite, since it contains the partially ordered set $C(3)$; namely, it starts as follows:

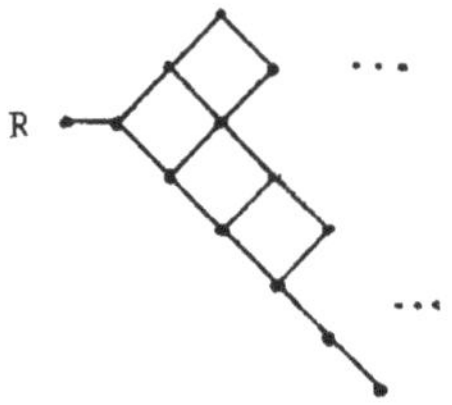

For $u = 3$, and $n(A_o) \geq 9$, the vectorspace category $Hom(R,A_o\text{-mod})$ is subspace infinite, since it contains $C(2)$. It starts as follows:

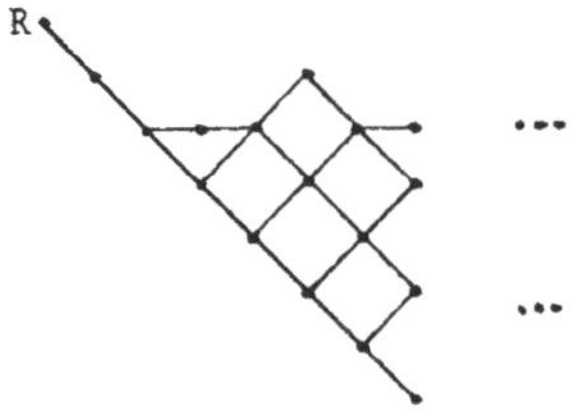

For $u = n(A_o)-3$, and $n(A_o) \geq 10$, the vectorspace category $Hom(R,A_o\text{-mod})$ is in all but the encircled case subspace infinite, since it starts with

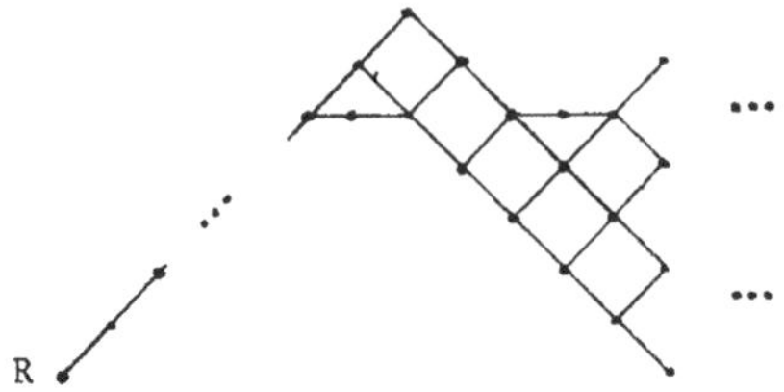

On the other hand, in the encircled case, $Hom(R,A_o\text{-mod})$ is given by the partially ordered set

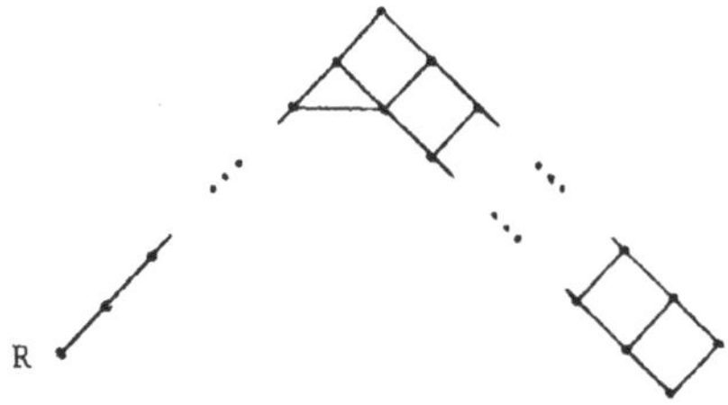

and therefore subspace finite. This finishes the proof.

(8) Let A be directed with orbit graph $\mathcal{O}(A)$ having three endpoints. Assume there exists a characteristic A-module and that A has at least one thin branch. If $n(A) \geq 13$, then A or A^{op} is of type (Bo*) with $* = 12, 13, 15, 16$, or 17, or of the form

Proof. The endpoint of a thin branch for M always is an exceptional vertex. Deleting all the vertices which cut off a thin branch, we obtain an algebra A', and the restriction M' of M to A' is a characteristic A'-module, but now without thin branches. Now, $[[M']]$ cuts off a branch of $\mathcal{O}(A)$, and $[M']$ has a wing in $\Gamma(A')$ say with source $[P_{A'}(a')]$ and sink $[Q_{A'}(b')]$. Note that $\mathrm{rad}\, P_{A'}(a')$ is not projective, since otherwise a' would cut off a thin branch for M'. The algebra A is obtained from A' by adding branches in a' and b', say of length s and t, respectively.

First, let us assume that $\overline{\mathcal{O}(A)} \neq \mathbb{D}_n$. Up to duality, we can assume $s \geq t$. By assumption, $s \geq 2$, and we can assume that the branch in a' has factorspace orientation. It follows that $P = P_A(a')$ is final, and we consider the slice $S(P\rightarrow)$, and its τ-translate $\tau S(P\rightarrow)$, which belongs to A_o-mod, where $A = A_o[R]$. The quiver of $\tau S(P\rightarrow)$ has two components, it is of the following form

$$V_2 \rightarrow V_3 \rightarrow \dots \rightarrow V_s$$

$$X_r \rightarrow X_{r-1} \rightarrow \dots \rightarrow X_1 \begin{array}{l} \nearrow Y_2 \rightarrow \dots \rightarrow Y_p \\ \searrow Z_2 \rightarrow \dots \rightarrow Y_q \end{array}$$

with $2 \leq q \leq p$, and $3 \leq p$. Let $A_o = A_1 \times A_2$, where X_r is an A_1-module, V_2 an A_2-module. We have $R = \mathrm{rad}\, P = X_r \oplus V_2$, and $n(A) = p+q+r+s-1$. As before, let $X_{ij} = \tau_o^{-j} X_i$, and let $x_{ij} = [X_{ij}]$ in case $\mathrm{Hom}(R, X_{ij}) \neq 0$, and $x_i = x_{io}$. Similarly, for V_i, Y_i, Z_i.

First, we note that $s \leq 5$, since otherwise $\{v_2,v_3,v_4,v_5,v_6;\ y_2,y_3;\ z\}$ is of type $C(6)$. If $s \geq 3$, then $q = 2$, since otherwise $\{v_2,v_3;\ y_2,y_3;\ z_2,z_3\}$ is of type $C(3)$. If $s \geq 4$, then $p = 3$, since otherwise $\{v_2,v_3,v_4;\ y_2,y_3,y_4;\ z_2\}$ is of type $C(4)$.

Consider the case of X_1 being injective. Then A_1 is a bush algebra of Dynkin-type. Assume $n(A_1) > 8$, thus A_1 is of type $\mathbb{A}_m$ od $\mathbb{D}_m$, thus either $r = 1$, or else $r = q = 2$ (since $\overline{\mathcal{O}(A)} \neq \mathbb{D}_n$). If $r = 1$, and $n(A) \geq 10$, then $q = 2$ and $s = 2$, (otherwise, we obtain a full subset of $\mathrm{Hom}(R, A_o\text{-mod})$ of the form $C(6)$

or $C(4)$), thus A^{op} is of the form (Bo16). If $r = q = 2$, then we obtain the algebras of the form (Bo17). On the other hand, if $n(A_1) \leq 7$, or both $n(A_1) = 8$ and $s \leq 5$ then $n(A) \leq 12$. It remains to consider the case $n(A_1) = 8$, and $s = 5$. It follows that $q = 2$, $p = 3$. No branch of A_1 can carry a zero relation, since otherwise we will not obtain an indecomposable sincere A-module. Thus, the only possible cases with $n(A) \geq 13$ are the algebras

Now assume, X_1 is not injective, thus $X_{11} \neq 0$. Note that $X_{rr} = 0$. For, assume $X_{r,r-1} \neq 0$, thus X_1 has in A_1-mod the following wing.

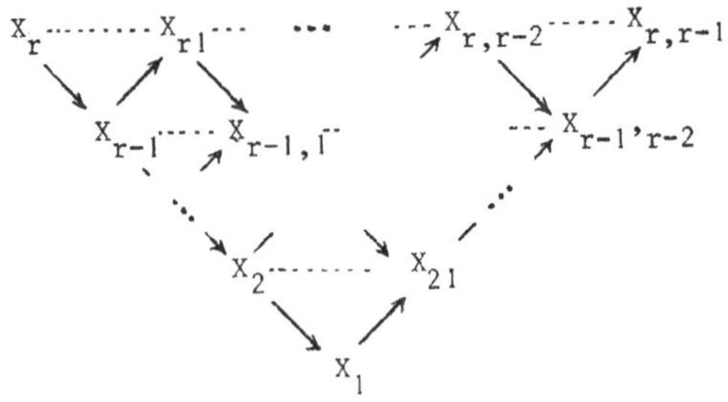

In $\Gamma(A)$, we obtain the following part

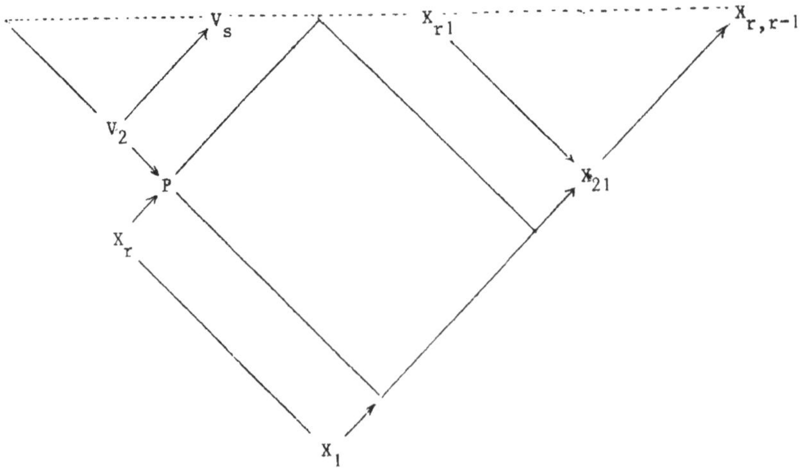

Let $P(a) = \tau^{s-2}V_s$, thus a is one of the exceptional vertices for M. Also according to 3.6.2, one of the modules $\tau^{-t}P(a)$, with $0 \leq t \leq r+s-2$ has to be an injective A-module. Now $\tau^{-(r+s-2)}P(a) = X_{r,r-1}$, thus $X_{r,r-1}$ is injective (as A-module, thus as A_1-module. Thus $X_{rr} = 0$. Let j be maximal with $X_{jj} \neq 0$, and $X_{j+1,j+1} = 0$. Then it is easy to see that A has a branch of length $r-j$ in b'. Of course, if $r-j \geq 2$, then this is a thin branch for M, thus $r-j = t \leq s$. It follows that always $r \leq s+j$, and therefore $n(A) \leq p+q+2s+j-2$. Since we assume that x_{11} exists, we must have $s \leq 4$. Otherwise, $\{v_2,v_3,v_4,v_5; x_{11},z_2,y_2,y_3\}$ would be of type $C(5)$. We consider the case $s = 4$. Then $X_{22} = 0$, since otherwise $\{v_2,v_3,v_4; x_{11},x_{22}; y_2\}$ is of type $C(4)$. Therefore $j = 1$, thus $n(A) \leq 12$. Next, let $s = 3$. Then $p = 3$, since otherwise $\{v_2,v_3; z_2,x_{11}; y_3,y_4\}$ is of type $C(3)$. We must have $j \leq 3$, since otherwise $\{v_2,v_3; z_2,x_{11},x_{22},x_{33},x_{44}; y_3\}$ is of type $C(6)$. Thus $n(A) \leq 12$.

Finally, let $s = 2$. We have $q = 2$, since otherwise $\{v_2,x_{11},y_3,z_3\}$ is of type $C(2)$. Also, $p \leq 6$, since otherwise $\{v_2; z_2,x_{11}; y_3,y_4,y_5,y_6,y_7\}$ is of type $C(6)$. If $p \geq 5$, then $j = 1$ since otherwise $\{v_2; z_2,x_{11},x_{22}; y_3,y_4,y_5\}$ is of type $C(4)$, therefore $n(A) \leq 11$. If $p = 4$, then $X_{44} = 0$, since otherwise $\{v_2; z_2,x_{11},x_{22},x_{33},x_{44}; y_3,y_4\}$ is of type $C(6)$. Thus $j \leq 3$, and $n(A) \leq 11$. It remains to consider the case $p = 3$. If Y_2 is not injective, then $j \leq 5$, since otherwise $\{v_2; x_{22},x_{33},x_{44},x_{55},x_{66}; y_3,y_{21}\}$ if of type $C(6)$, therefore $n(A) \leq 12$. Thus, let Y_2 be injective. If X_{11} is injective, then A_1 is a bush algebra, thus $r \leq 5$ and $n(A) \leq 10$. On the other hand, assume X_{11} is not injective, then $Hom(X_i,X_{1m}) = 0$ for all $m \geq 2$, and thus no X_i is sincere. However, using (5), we see that the restriction of M to A_1 has to be of the form X_i for some i. This contradiction finishes the proof.

It remains to consider the case $\overline{\mathcal{O}(A)} = \mathbb{D}_n$. In this case, the orbit quiver of A' is again of type $\mathbb{D}_m$, thus A' is known from the investigations in (7). Starting from A', with exceptional vertices a', b', we ask whether it is possible to add branches in a' and b' ("thin branches") but still obtaining a representation-finite algebra. Clearly, this works for algebras of the form (Bo12) and (Bo13), and we obtain algebras of the same form. Similarly, starting with the algebra

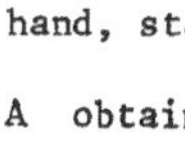 (a special case of (Bo19)), we obtain the algebras (Bo15). On the other hand, starting from an algebra 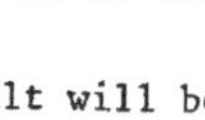 with at least five vertices, the algebras A obtained by adding thin branches no longer have orbit quivers of type $\mathbb{D}_n$.

The following technical result will be needed in our investigation of the sincere directed algebras with orbit quiver having four endpoints.

(9) Let A_o be a directed algebra. Let R be an indecomposable A_o-module such that $A_o[R]$ is sincere and representation finite. Let Q be an indecomposable

injective A_o-module with no proper predecessor in A_o-mod being injective. Assume $S_o(R\to) \cap S_o(Q\to)$ has at least two sinks. If $Q = X_o \to X_1 \to \dots \to X_r$ is a sectional path in $S_o(R\to)$, then $\tau_o^r X_r \neq 0$. If X is a proper successor of Q, and X belongs to $S_o(R\to)$, then $\mathrm{Hom}(Q,Q') \neq 0$ for any indecomposable injective A_o-module Q' with $X \preceq Q'$.

Proof. Let $Q = Q_o(x)$, with $x \in \Delta(A_o)$, and note that x is a sink of $\Delta(A_o)$. Let ω be the extension vertex of $A_o[R]$. If $R = Q$, then $P(\omega)$ is the only possible sincere $A_o[R]$-module, thus sincere. Therefore, $R = Q$ is a sincere A_o-module, thus $S_o(\to Q)$ is a slice in A_o-mod. Now assume $R \not\preceq Q$. According to (1'), $\mathcal{O}(A_o[R])$ is a tree, thus also $\mathcal{O}(A_o)$ is a tree. We can assume that Q belongs to $S_o(R\to)$, and since $\mathcal{O}(A_o)$ is a tree, and $R \not\preceq Q$, there is a unique sectional path $R = W_o \to W_1 \to \dots \to W_s = Q$. The source map for Q is of the form $Q \to \tau_o^- W_{s-1} \oplus Q'$, and the algebra B obtained from A_o by deleting x can be written in the form $B = B_1 \times B_2$, where $\tau_o^- W_{s-1}$ is a B_1-module, and Q' a B_2-module. By assumption, there exists an indecomposable $A_o[R]$-module $M = (M_o, M_\omega, \gamma_M)$ with M_o an A_o-module, M_ω a k-space, and $\gamma_M : M_\omega \longrightarrow \mathrm{Hom}(R,M_o)$ k-linear. Let $M_o = M_o' \oplus M_1 \oplus M_2$, where M_1 is a B_1-module, M_2 a B_2-module, and all indecomposable direct summands X of M_o' satisfy $\mathrm{Hom}(X,Q) \neq 0$, thus are of the form W_i, $0 \leq i \leq s$. Since $\mathrm{Hom}(M,\bar{Q}) \neq 0$, we have $M_o' \neq 0$. Now, let Q' be an indecomposable injective B_2-module, and assume $\mathrm{Hom}(Q,Q') = 0$. Observe that Q' is injective also as an A_o-module. Since M is sincere, we must have $\mathrm{Hom}(M_o,Q') \neq 0$. Before we proceed, let us stress that any map $W_i \longrightarrow U$, with U a B_2-module, factors through the modules W_j, with $i \leq j \leq s$. In particular, every map $W_i \longrightarrow Q'$ factors through Q, thus $\mathrm{Hom}(W_i,Q') = 0$. Since also $\mathrm{Hom}(M_1,Q') = 0$, we see that $M_2 \neq 0$. Fix some indecomposable summand W_i of M_o', and some indecomposable summand U of M_2. Since $w_i = [W_i]$ and $u = [U]$ are comparable in the partially ordered set $\mathcal{S}(K)$, where $K = \mathrm{Hom}(R,A_o\text{-mod})$, the support of M in $S(K)$ must contain two incomparable elements, say $v = [V]$ and $v' = [V']$ which are also incomparable with both w_i and u. Here, V, V' are indecomposable A_o-modules, successors of R, not predecessors of Q (since v, v' are incomparable with w_i), and not B_2-modules (again, since v, v' are incomparable with w_i). Thus, V,V' are B_1-modules. However, by assumption $S_o(P\to) \cap S_o(Q\to)$ has two sinks, say Y and Z, and these must be B_2-modules. It follows that K contains four pairwise incomparable elements, namely v, v', $[Y]$ and $[Z]$, thus $A_o[R]$ is not representation finite, a contradiction. This shows that $\mathrm{Hom}(Q,Q') \neq 0$ for any indecomposable injective B_2-module Q'. In particular, if X is a proper successor of Q, and X belongs to $S_o(P\to)$, then X is a B_2-module. Thus any indecomposable injective A_o-module Q'' with $X \preceq Q''$ also is a B_2-module, and therefore $\mathrm{Hom}(Q,Q'') \neq 0$. Also, let $Q = X_o \to X_1 \to \dots \to X_r$ be a sectional path in $S_o(R\to)$, and assume $\tau_o^r X_r = 0$. Then $\tau_o^{r'} X_r$ is indecomposable projective, for some $0 \leq r' < r$, say $\tau_o^{r'} X_r = P(a)$. On the other hand, a has to be a vertex of $\Delta(B_2)$, and we have $\mathrm{Hom}(P(a),Q) \approx D\,\mathrm{Hom}(Q,Q_o(a)) \neq 0$. However, $P(a)$ is not even predecessor of Q. This contradiction shows that $\tau_o^r X_r \neq 0$.

(10) Let A be a sincere directed algebra, and assume that $\mathcal{O}(A)$ has four endpoints. Let P be a final indecomposable projective module. Then $[[\mathrm{rad}\, P]]$ has at least three neighbors in $\mathcal{O}(A)$.

<u>Proof</u>. Let $P = P(\omega)$, and A_o obtained from A by deleting ω, say $A = A_o[R]$ and $R = \mathrm{rad}\, P(\omega)$. Note that R is an indecomposable A_o-module, since we assume that $\mathcal{O}(A)$ has four endpoints. We consider the slice $\tau S(P\to)$ in A_o-mod, let

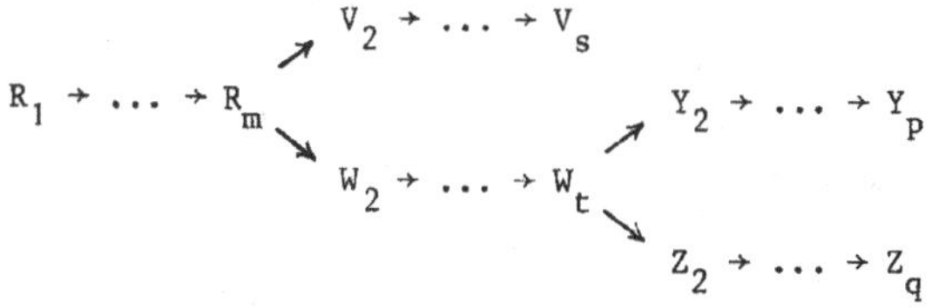

be the quiver of $\tau S(P\to)$, with $m,t \geq 1$, and $s,p,q \geq 2$. We have $R = R_1$, and the assertion claims that $m = 1$. Then, assume $m > 1$.

Let $W_1 = R_m$. One of the modules W_i, $1 \leq i \leq t$ has to be injective, since otherwise $\dim \mathrm{Hom}(R,\tau_o^- W_t) = 2$. Let i be minimal with W_i being injective.

First, assume that none of the modules R_i, $1 \leq i \leq m-1$, is injective. Then no proper predecessor of W_i is injective, thus we can apply (9) for $Q = W_i$, and conclude that both $\tau_o^{t-i+1} Y_2$ and $\tau_o^{t-i+1} Z_2$ are non-zero. Let $Q = Q_o(x)$ with $x \in \Delta(A_o)$, let C be obtained from A_o by deleting x, and $T = Q/\,\mathrm{soc}\, Q$, thus $A_o = [T]C$. Then $S(\mathrm{Hom}(C\text{-mod},T))$ contains four incomparable elements, namely the $[T]$ with $X = \tau_o^- R_{m-1}$, V_2, $\tau_o^{t-i} Y_2$, and $\tau_o^{t-i} Z_2$, and this contradicts that $A_o = [T]C$ is representation finite.

It follows that some R_i, $1 \leq i \leq m-1$, is injective. Choose i minimal. If $i > 1$, then again we obtain a contradiction, using (9). For, the modules $\tau_o^- R_{i-1}$, $\tau_o^{m-i} V_2$, $\tau_o^{m+t-i-1} Y_2$ and $\tau_o^{m+t-i-1} Z_2$ are pairwise incomparable and have non-zero maps to $R_i/\mathrm{soc}\, R_i$.

Thus $R = R_1$ is an injective A_o-module, and therefore P is an injective A-module. Note that this implies that P is the only sincere indecomposable A-module. As a consequence, R_1 is a sincere A_o-module. Thus A_o is a sincere directed algebra. Now assume $m \geq 3$. Then we can apply the dual considerations to the sincere directed algebra A_o, and the indecomposable, injective A_o-module R_1, and conclude that R_1 is a projective A_o-module. Let $R_1 = P(\omega')$. We see that

$\omega' \longleftarrow \omega$ is a branch at ω', and changing its orientation, we obtain an algebra A' which again is sincere and directed. The indecomposable projective A'-module $P_{A'}(\omega')$ is not a proper predecessor of any indecomposable projective A'-module and $S(P_{A'}(\omega')\rightarrow)$ has four sinks. This contradicts 6.1.1', therefore $m = 2$.

As we have seen, P is an injective A-module, say $P = Q(a)$, and we denote by B the algebra obtained from A by deleting both ω and a. Note that $\text{rad}\, P/\text{soc}\, P = R_2 = W_1$ is a B-module. Actually, all the indecomposable modules from $\tau S(P\rightarrow)$ with the only exception of R_1, are A_1-modules, and we obtain in this way a maximal slice in B-mod. Also observe that W_1 is a sincere indecomposable B-module, and that both $B[W_1]$ and $[W_1]B$ are representation-finite algebras (we have $[W_1]B = A_o$, and $B[W_1]$ is obtained from A by deleting a). As we have seen above, at least one of the modules W_i, $1 \leq i \leq t$, is an injective A_o-module, thus an injective B-module. Dually, one of the modules $\tau_B^{i-1}W_i$, $1 \leq i \leq t$, is a projective B-module. Choose i minimal, with W_i being injective, say $W_i = Q_B(b)$, and choose j minimal with $\tau_B^{j-1}W_j$ being projective, say $\tau_B^{j-1}W_j = P_B(b')$. We see that b is a sink of $\Delta(B)$, and the algebra obtained from B by deleting b is of the form $B_1 \times B_2$, where V_2 is a B_1-module, and Y_2 a B_2-module. Now, there is the sectional path

$$P_B(b') = \tau_B^{j-1}W_j \longrightarrow \tau_B^{j-2}W_{j-1} \longrightarrow \ldots \longrightarrow W_1 \longrightarrow V_2$$

thus $\text{Hom}(P_B(b'),V_2) \neq 0$, therefore b' belongs to $\Delta(B_1)$. However, one may verify without difficulty that also $\text{Hom}(P_B(b'),Y_2) \neq 0$.

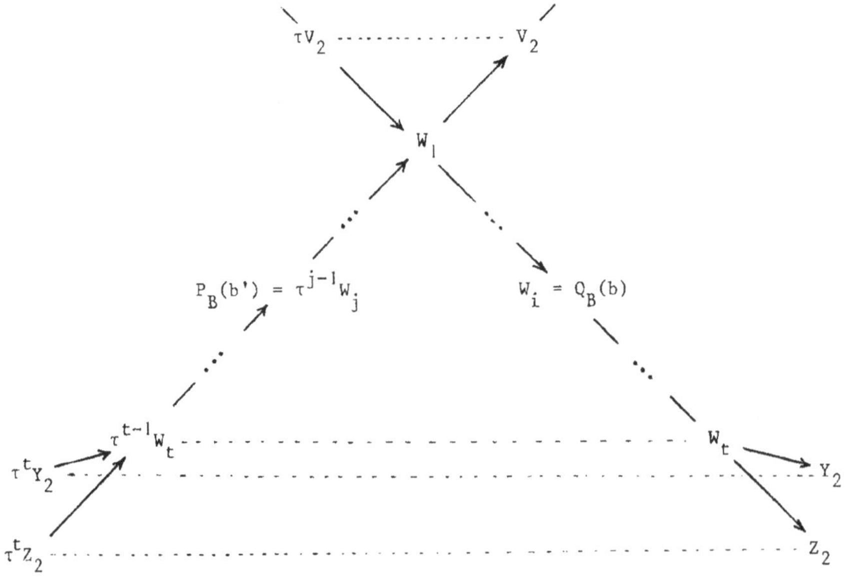

(Denote by H the subspace of $\mathrm{Hom}(P_B(b'),Y_2)$ given by all maps which factor through a module of the form $\tau_B^{2\ell+1}Y_2$, $\tau_B^{2\ell}Z_2$, $\tau_B^{\ell}Y_u$, or $\tau_B^{\ell}Z_u$, with $u \geq 3$. Then $\mathrm{Hom}(P_B(b'),Y_2)/H$ is one-dimensional). Thus b' also belongs to $\Delta(B_2)$. This contradiction establishes $m = 1$.

(11) Let A be a sincere directed algebra, and assume that $\mathcal{O}(A)$ has four endpoints. If $n(A) \geq 14$, then A is of the form $(Bo*)$ with $* = 4,14,20$.

Proof. Let $P(\omega)$ be a final indecomposable projective A-module, let A_o be obtained from A by deleting ω, and $R = \mathrm{rad}\, P(\omega)$. In A_o-mod, we have the slice $S_o(R\to)$, with quiver

$$R = W_1 \nearrow \begin{matrix} V_2 \to \cdots \to V_s \\ \\ W_2 \to \cdots \to W_t \end{matrix} \searrow \nearrow \begin{matrix} Y_2 \to \cdots \to Y_p \\ \\ Z_2 \to \cdots \to Z_q \end{matrix}$$

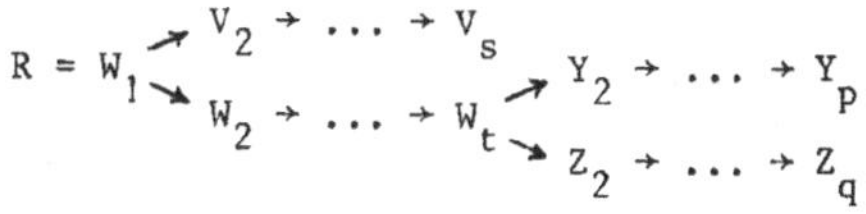

where $t \geq 1$; $s,p,q \geq 2$. One of the modules W_i, $1 \leq i \leq t$, has to be injective, since otherwise $\dim \mathrm{Hom}(R,\tau_o^- W_t) = 2$. We choose i minimal, thus $W_i = Q_o(a)$, for some a. Let B be the algebra obtained from A_o by deleting a, it decomposes $B = B_1 \times B_2$ with V_s a B_1-module, and Y_p,Z_q both B_2-modules. According to (9) we know that the restriction of $Q_o(a)$ to B_2 is sincere. On the other hand, $\Delta(A_o)$ has a branch at a of length $s+i-1$, such that all B_1-modules have support on this branch. Let us change this branch, if necessary, to a subspace branch; and we denote the algebra obtained in this way by A_o'. Let us collect some facts about A_o'. First of all, $\Delta(A_o')$ has a unique sink, the corresponding indecomposable injective A_o'-module is sincere, and A_o' has a branch of length ≥ 2 at the sink. Second, $\mathcal{O}(A_o')$ has precisely three endpoints. Since A_o' is a sincere directed algebra with a unique sink and with a branch of length ≥ 2 at the sink, we can use (7) and (8), and see that for $n(A_o') \geq 13$, one of A_o' or $(A_o')^{op}$ is of the form $(Bo*)$ with $* = 2,12,13,15,16$, or 17, or else the exceptional algebra listed in (8) with precisely 13 vertices. The cases $(Bo12)$, $(Bo13)$ cannot occur, since they have zero relations, thus not an indecomposable sincere injective module. Therefore, the only possibilities for A_o' are the following:

(Bo 2)

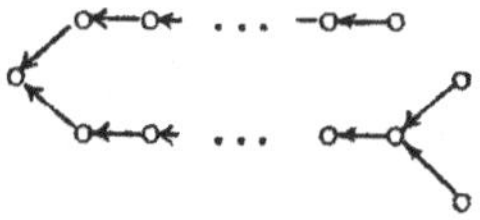

(Bo 15)

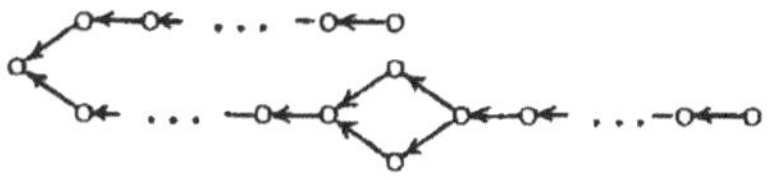

(Bo 16)

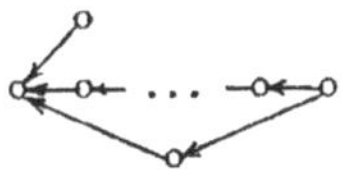

(Bo 17)

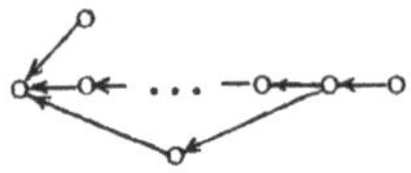

and the exceptional case with 13 vertices.

(E)

In the cases (Bo16) and (Bo17), the branch at the sink is of length 2, thus $s+i-1 = 2$, therefore $R = W_1$ is injective and $A_o = A'_o$. However, in both cases $A = A_o[R]$ is representation infinite for $n(A) \geq 10$, since the algebra obtained from A by deleting a is a bush algebra of type $\mathbb{T}_{m,3,2}$, or $\mathbb{T}_{m,4,2}$, respectively. Also, if A'_o is of the form (E), then $A_o[R]$ has a factor algebra which is a bush algebra of type $\mathbb{T}_{6,3,2}$. Thus, the only possibilities for A'_o are (Bo2) and (Bo15).

Consider first the case $i = 1$. Then $A_o = A'_o$, and $R = Q_o(a)$. We obtain as $A = A_o[R]$ the algebras of the form (Bo20).

Next, assume $i \geq 2$. In this case, we must have $s = 2$, since otherwise $[V_3]$, $[\tau_o^- W_1]$, $[Y_2]$ and $[Z_2]$ give a subset of type $C(2)$ in $\mathrm{Hom}(R, A_o\text{-mod})$. This shows that A_o is obtained from A'_o by replacing the branch (of length $i+1$) at the sink a by a suitable branch, and with a being a sink of the new branch. The vertex a has a unique neighbor a' in the new branch, and B_1 is a branch of A_o at a'. Note that $\underline{\dim}\, R$ is $= 1$ on a', on the vertices of $\Delta(B_2)$ and on the successors of a' in $\Delta(B_1)$, and $= 0$ outside. We consider the case of B_1 being a branch without zero relation. Then $A = A_o[R]$ is of the form (Bo4). In particular, the maximal root of χ_A has entries ≤ 2. It follows that replacing the branch B_1 by a branch with at least two zero relations, the new algebra will not have any sincere positive root. Thus, the only possibility for changing the branch B_1 by a branch with zero relations is (Bo14). This finishes the proof.

6.4 Auslander-Reiten sequences with four middle terms

The investigation in 6.3 of the sincere directed algebras A with $\mathcal{O}(A)$ having four endpoints can be used in order to show that for a directed algebra A, the middle term of an Auslander-Reiten sequence decomposes into at most four indecomposable summands. We need the following lemma which seems to be of interest in its own: it asserts that Auslander-Reiten sequences can always be calculated by replacing the given algebra by a sincere one.

(1) Let A be an algebra, and

$$0 \to X \to \bigoplus_{i=1}^{d} Y_i \to Z \to 0$$

an Auslander-Reiten sequence, with all Y_i indecomposable. Then the support of the dimension vector of at least one of the modules X, Y_i, Z contains the supports of all the others.

Proof. If all the maps $X \longrightarrow Y_i$ are epi, then choose the module X. The support of $\underline{\dim}\, X$ contains the support of any $\underline{\dim}\, Y_i$, and since Z is an epimorphic image of $\bigoplus Y_i$, the support of $\underline{\dim}\, Z$ is contained in the union of the supports of $\underline{\dim}\, Y_i$. Dually, if all the maps $Y_i \longrightarrow Z$, are mono, then choose the module Z. Now assume at least one map $X \longrightarrow Y_i$ is mono, say for $i = 1$, and at least one map $Y_j \longrightarrow Z$ is epi. Let $Y_1' = \bigoplus_{i=2}^{d} Y_i$. Since $X \longrightarrow Y_1$ is mono, the canonical map $Y_1' \longrightarrow Z$ also is mono, thus all the maps $Y_i \longrightarrow Z$, $i \geq 2$, are mono, thus $j = 1$. This shows that the supports of $\underline{\dim}\, X$ and $\underline{\dim}\, Z$ are contained in the support of $\underline{\dim}\, Y_1$, and that the supports of $\underline{\dim}\, Y_i$, $i \geq 2$, are contained in the support of $\underline{\dim}\, Z$. Thus, in this case, we will choose Y_1.

(2) Proposition. (Bautista-Brenner) Let A be a directed algebra, and $\bigoplus_{i=1}^{d} Y_i \longrightarrow Z$ a sink map. Then $d \leq 4$, and if $d = 4$, then Z is not projective, one of the Y_i is both projective and injective, and the remaining Y_i are neither projective nor injective, and pairwise non-isomorphic.

Proof. First, assume Z is projective. Since all the Y_i are factor modules of Z, we can assume that Z is sincere. This implies that Z is final, thus $\mathcal{S}(Z\to)$ is a tree with at most three sinks. As a consequence, Z has at most three neighbors in $\mathcal{O}(A)$, thus $d \leq 3$.

Now, consider the case of Z not being projective, and let X be the kernel of the map $\bigoplus Y_i \longrightarrow Z$, thus we deal with an Auslander-Reiten sequence. According to (1), we again may assume that A is sincere. Thus, $\mathcal{O}(A)$ is a tree with at most four endpoints. If $d \geq 4$, then $\mathcal{O}(A)$ has to be a star with four branches, and with [[Z]] being the center of the star. Consider this case, and let $P(\omega)$ be a final indecomposable projective A-module, and let A_o be obtained from A by deleting ω. According to 6.3.10, also [[R]] is the center of the star, where $R = \text{rad } P(\omega)$. Note that R has to be an injective A_o-module, since otherwise $\dim \text{Hom}(R, \tau_o^- R) = 2$. However, this implies that $P(\omega)$ is an injective A-module. Since $[[P(\omega)]]$ only consists of $[P(\omega)]$, we must have $Y_i = P(\omega)$ for some i, say $i = 1$. Since $\mathcal{S}(P(\omega)\rightarrow)$ is a slice, it contains the modules $\tau^- Y_i$, $2 \leq i \leq 4$; in particular, no Y_i, $2 \leq i \leq 4$, is injective. Since $\mathcal{S}(\rightarrow P(\omega))$ is a slice, it contains the modules τY_i, $2 \leq i \leq 4$; in particular, no Y_i, $2 \leq i \leq 4$, is projective

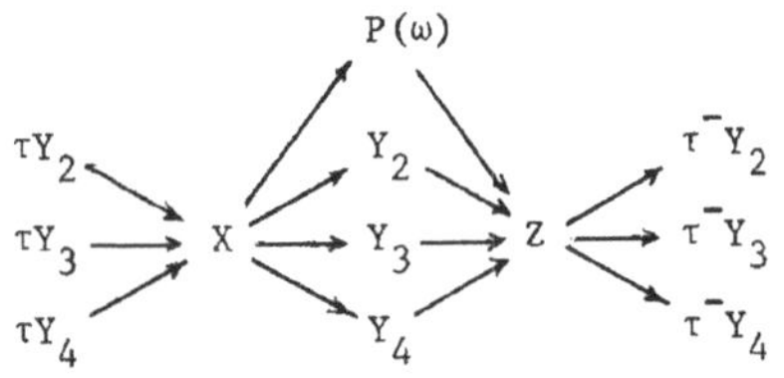

This finishes the proof.

Of course, there also is the dual assertion (2*) dealing with source maps.

Actually, one may give a complete classification of all sincere directed algebras having an Auslander-Reiten sequence with four middle terms.

(3) Let A be a sincere directed algebra, and assume $\mathcal{O}(A)$ is a star with four proper branches. Then A is one of the following fully commutative algebras:

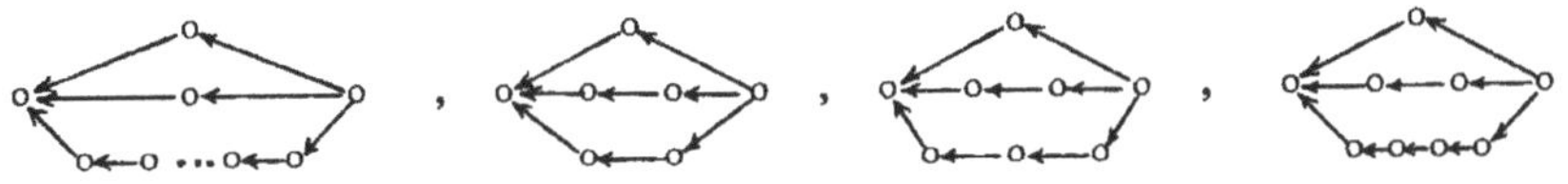

Proof. As above, let $P(\omega)$ be a final indecomposable projective A-module, $R = \text{rad } P(\omega)$. and let A_o be obtained from A by deleting ω. According to 6.3.10, $\tau\mathcal{S}(P(\omega)\rightarrow)$ is a star with three proper branches, and with center R. Also, as we have noted, R is an injective A_o-module. Thus A_o is a bush algebra with subspace branches, and the only possibilities for A_o to be representation-finite are the cases $\mathbb{D}_n$, $\mathbb{E}_6$, $\mathbb{E}_7$, $\mathbb{E}_8$. Also, R is the unique sincere injective A_o-module, thus A is as listed.

6.5 The inductive construction of sincere directed algebras

We are going to show that every sincere directed algebra A with $n(A) \geq 2$ can be obtained from a sincere directed algebra A_o by a one-point extension or a one-point coextension using an indecomposable A_o-module. This should be helpful for an inductive construction of the sincere directed algebras A with $n(A) \leq 13$.

(1) Let A be a directed algebra. Let Z be an indecomposable sincere A-module. Assume there exists a vertex a of $\Delta(A)$ such that $\sigma_a \underline{\dim} Z$ no longer is sincere. Then a is a sink or a source.

Proof. Let $z = \underline{\dim} Z$, thus z is a sincere positive root of χ_A. Suppose $\sigma_a z$ is no longer sincere. Now

$$\sigma_a z = z - (D_a\chi)(z)e(a),$$

and $|D_a\chi(z)| \leq 1$, according to 1.1.4, thus $z_a = 1$, and $2(e(a),z) = D_a\chi(z) = 1$. Assume a is neither a sink nor a source of $\Delta(A)$, thus there are vertices b,c and arrows $b \xrightarrow{\beta} a \xrightarrow{\gamma} c$. According to 2.4.7', the induced maps $\beta_Z : Z_b \longrightarrow Z_a$ and $\gamma_Z : Z_a \longrightarrow Z_c$ are non-zero. Since $\dim Z_a = z_a = 1$, it follows that β_Z is surjective, and γ_Z is injective. However, this implies that $\mathrm{Hom}(Z,E(a)) = 0$, $\mathrm{Hom}(E(a),Z) = 0$. Since Z is indecomposable and sincere, proj.dim $Z \leq 1$, and inj.dim $Z \leq 1$, see 2.4.7. Thus

$$\langle \underline{\dim}\, Z, \underline{\dim}\, E(a)\rangle = -\dim \mathrm{Ext}^1(Z,E(a)) \leq 0,$$

$$\langle \underline{\dim}\, E(a), \underline{\dim}\, Z\rangle = -\dim \mathrm{Ext}^1(E(a),Z) \leq 0,$$

therefore

$$2(e(a),z) = \langle \underline{\dim}\, Z, \underline{\dim}\, E(a)\rangle + \langle \underline{\dim}\, E(a), \underline{\dim}\, Z\rangle \leq 0,$$

a contradiction.

(1') Corollary. Let A be a sincere directed algebra, and M a sincere maximal module with two exceptional vertices. Then, any exceptional vertex is a sink or a source.

Proof. If a is an exceptional vertex, then $\sigma_a \underline{\dim} M$ no longer is sincere.

(2) Proposition. Let A be a sincere directed algebra, and $n(A) \geq 2$. Then A is a one-point extension or a one-point coextension of a sincere directed algebra A_o by an indecomposable A_o-module.

Proof. There exists an indecomposable sincere A-module Z such that $\sigma_a \underline{\dim} Z$ is no longer sincere. For, starting from a sincere maximal positive root y, we can apply reflections σ_i in order to obtain a sincere positive root z such that $\sigma_a z$ is no longer sincere, for some vertex a, see 1.1.5". Of course, there exists an indecomposable A-module Z with $\underline{\dim}\, Z = z$, according to 2.4.9.

Now assume, a is a source, let A_o be obtained from A by deleting a, and $R = \operatorname{rad} P(a)$. Let Z_o be indecomposable, with $\underline{\dim}\, Z_o = \sigma_a \underline{\dim}\, Z$. We claim that $\operatorname{Hom}(R, Z_o) \neq 0$. For, E(a) is injective, and $\operatorname{Hom}(R, E(a)) = 0$ gives $\langle \underline{\dim}\, R, \underline{\dim}\, E(a)\rangle = 0$, therefore

$$\begin{aligned}\langle \underline{\dim}\, R, \underline{\dim}\, Z_o\rangle &= \langle \underline{\dim}\, R, \underline{\dim}\, Z\rangle \\ &= \langle \underline{\dim}\, P(a), \underline{\dim}\, Z\rangle - \langle \underline{\dim}\, E(a), \underline{\dim}\, Z\rangle, \\ &\geq \langle \underline{\dim}\, P(a), \underline{\dim}\, Z\rangle = 1,\end{aligned}$$

where we use that $\operatorname{Hom}(E(a), Z) = 0$, since E(a) is injective, and $n(A) \geq 2$, and inj.dim $Z \leq 1$. Always, $\langle -, -\rangle$ denotes the bilinear form on $K_o(A)$ defined by C_A^{-T}. However, according to 2.5, the restriction of $\langle -, -\rangle$ to $K_o(A_o)$ is just the corresponding bilinear form defined by $C_{A_o}^{-T}$. Since Z_o is a sincere A_o-module, $\operatorname{inj.dim}_{A_o} Z_o \leq 1$, thus $\langle \underline{\dim}\, R, \underline{\dim}\, Z_o\rangle \geq 1$ shows that $\operatorname{Hom}(R, Z_o) \neq 0$. We may identify A-mod with $\check{U}(A_o\text{-mod}, \operatorname{Hom}(R,-))$, see 2.5.8. Note that we must have $\dim \operatorname{Hom}(R, Z_o) = 1$, since otherwise $(A_o\text{-mod}, \operatorname{Hom}(R,-))$ would not be subspace-finite. Of course, $(Z_o, \operatorname{Hom}(R, Z_o), 1_{\operatorname{Hom}(R, Z_o)})$ is just the A-module Z, since both are indecomposable and have the same dimension vector. Let $0 \neq \varphi : R \longrightarrow Z_o$. We claim that φ is mono. For, given an element r in the kernel of φ, the matrix $\begin{bmatrix} 0 & r \\ 0 & 0 \end{bmatrix} \in \begin{bmatrix} A_o & R \\ 0 & k \end{bmatrix}$ obviously annihilates the A-module $Z = \begin{bmatrix} Z_o \\ k \end{bmatrix}$, thus $r = 0$, according to 2.4.7". Now assume R is decomposable, say $R = R_1 \oplus R_2$, with $R_1 \neq 0 \neq R_2$. Since R is a submodule of Z_o, it follows that $\dim \operatorname{Hom}(R, Z_o) \geq 2$, a contradiction. This shows that $A = A_o[R]$ is a one-point extension of A_o by the indecomposable A_o-module R.

By duality, if a is a sink, then A is a coextension of some sincere directed algebra by an indecomposable module.

<u>Remark.</u> Given a sincere directed algebra A_o and an indecomposable A_o-module R, and $A = A_o[R]$, say with extension vertex ω, then $P(\omega)$ is always a final indecomposable projective module. We have seen above that for a sincere directed algebra A, the algebra A or its opposite always has a final indecomposable projective module $P(\omega)$ such that its radical is indecomposable, and the algebra obtained from A by deleting ω is again sincere. However, given an arbitrary final indecomposable projective module $P(\omega')$, say even with indecomposable $\operatorname{rad} P(\omega')$, then the algebra obtained from A by deleting ω' does not have to be sincere. A typical example is

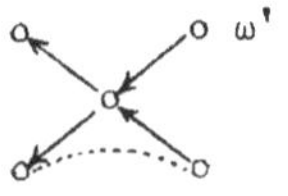

.

References

The first part of 6.1.1 asserting that the orbit graph of a sincere directed algebra is a tree, is due to Bautista-Larrion-Salmeron [BLS], the second part asserting that the orbit graph has at most four endpoints (or, stronger, the existence of a slice with one source and at most three sinks, see 6.1.1') is due to Bongartz [Bo2]. For the corollary 6.1.2, there are several proofs available (see [Ba1], [Rm], [Ri3]).

The classification of the large sincere directed algebras and the corollary 4.2.3 are due to Bongartz. In [Bo2], he showed that a sincere directed algebra A with n(A) > 336 belongs to one of the 24 families listed in 6.3.1. (We have reordered and renumbered the list. In particular, in [Bo2] the number (Bo 20) is split into two families depending on whether $\mathcal{O}(A)$ is a star or not, whereas the numbers (Bo 16) and (Bo 17) are considered as one family). The representation-finite algebras given by a fully commutative quiver (they always are sincere and directed) have been classified before by Loupias [L1], without any restriction on n(A). Of course, this includes all sincere directed algebras with an indecomposable projective-injective module. Also, given a sincere, directed algebra A with orbit graph having four endpoints, then the argument at the beginning of the proof of 6.3.11 can be used to show that A is obtained from a fully commutative quiver by changing at most one branch. In the general case, a classification of the remaining sincere directed algebras (those with $n(A) \leq 13$) has been announced by Bongartz, based on a computer search, but has not yet appeared.

The proof of theorem 6.3.1 given above relies on several results and techniques presented earlier in these notes. However, some of these references are rather marginal. In the course of the proof of 6.2.1, we refer to 5.3.5'; this is presented as a corollary of the rather technical 5.3.5, but may be proven directly without difficulty. Also, we use the classification of the tame concealed algebras of type $\tilde{\mathbb{D}}_n$, short proofs of this are published in [HV] and [Bo4].

The bound four for the number of middle terms in an Auslander-Reiten sequence (6.4.2) is due to Bautista-Brenner [BaB]. It is a special case of a general result which gives a bound on the "replication number" of any Euclidean subquiver inside the Auslander-Reiten quiver of a representation-finite algebra.

Let us note that some of the results presented in this chapter remain valid for arbitrary representation-finite (not necessarily directed) algebras, namely 6.1.2, and 6.4.2, using covering methods. We refer to [BG].

APPENDIX

A.1 The periodic additive functions on $\mathbb{Z}\Delta$, with Δ Euclidean

Given a stable translation quiver $\Gamma = (\Gamma_0, \Gamma_1, \tau)$, a function $f : \Gamma_0 \longrightarrow \mathbb{Z}$ is said to be an additive function on Γ, provided for every vertex c of Γ, one has

$$f(\tau c) + f(c) = \sum_{b \to c} f(b)$$

(the summation being formed over all arrows ending in c; thus, in case Γ has no multiple arrows, then we just have to sum over all $b \in c^-$). If f is an additive function on Γ, then we define its τ-translate $f \circ \tau$ by $(f \circ \tau)(c) = f(\tau c)$, for $c \in \Gamma_0$. Of course, $f \circ \tau$ is additive, again. Given $t \in \mathbb{N}_1$, an additive function f on Γ is said to be periodic with period t provided $f \circ \tau^t = f$. The set $A(\Gamma)$ of all additive functions on Γ is a subgroup in the group of all functions $\Gamma_0 \to \mathbb{Z}$ (with pointwise addition), and the periodic additive functions form a subgroup $A_p(\Gamma)$ of $A(\Gamma)$. Finally, an additive function f on Γ will be said to be positive provided $f \neq 0$, and $f(z) \in \mathbb{N}_0$ for all vertices z.

Now, consider the case $\Gamma = \mathbb{Z}\Delta$, with Δ a finite connected quiver without oriented cycles. Given an additive function f on $\mathbb{Z}\Delta$, let $f_z : \Delta_0 \longrightarrow \mathbb{Z}$ be given by the restriction of f to $z \times \Delta_0$, thus $f_z(a) = f(z,a)$ for $z \in \mathbb{Z}$, $a \in \Delta_0$. For a fixed z, we obtain in this way an isomorphism from $A(\mathbb{Z}\Delta)$ onto the group $\mathbb{Z}^{\Delta_0}$ of all integer valued functions on Δ_0, mapping f onto f_z. The following proposition is well-known (for a short proof, we refer to [HPR]).

Proposition. Let Δ be a finite connected quiver without oriented cycles. If there exists a positive, periodic, additive function on $\mathbb{Z}\Delta$, then Δ is Euclidean. Conversely, for Δ Euclidean, $A_p(\mathbb{Z}\Delta)$ is a pure sublattice of $A(\mathbb{Z}\Delta)$ of co-rank 1, generated by positive functions.

Let Δ be Euclidean, of type $\tilde{\mathbb{T}}_{n_1,\dots,n_t}$. We want to exhibit t positive additive functions $f_1,\dots,f_t$, of period $n_1,\dots,n_t$, respectively, such that these functions f_s, together with their τ-translates $f_s \circ \tau^j$ $(0 \le j \le n_s - 1,\ 1 \le s \le t)$ generate $A_p(\mathbb{Z}\Delta)$. Actually, we obtain a presentation of $A_p(\mathbb{Z}\Delta)$, using as relations that

$$\sum_{j=0}^{n_s - 1} f_s \circ \tau^j = \sum_{j=0}^{n_{s'} - 1} f_{s'} \circ \tau^j .$$

Actually, in the cases $\tilde{\mathbb{A}}_n$ and $\tilde{\mathbb{D}}_n$, we just deal with typical special cases:

$\tilde{\mathbb{A}}_{4,3} = \tilde{\mathbb{T}}_{4,3}$

f_1

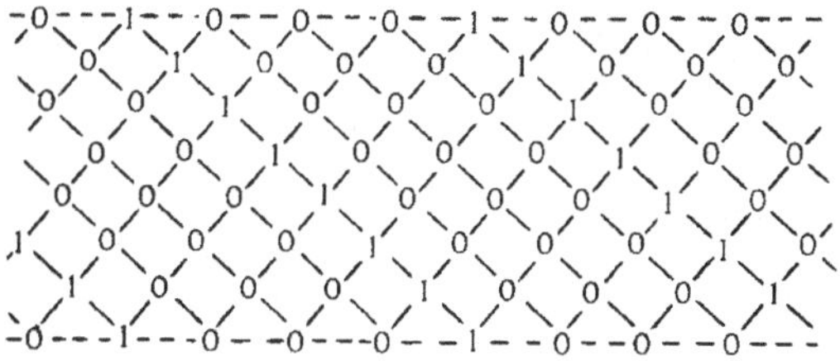

f_2

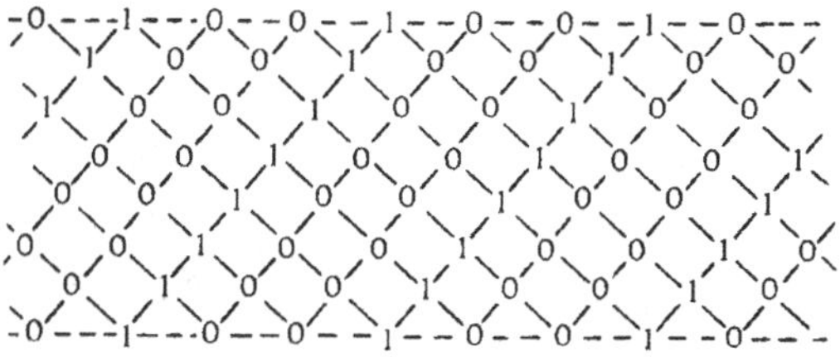

$\tilde{\mathbb{D}}_9 = \tilde{\mathbb{T}}_{7,2,2}$

f_1

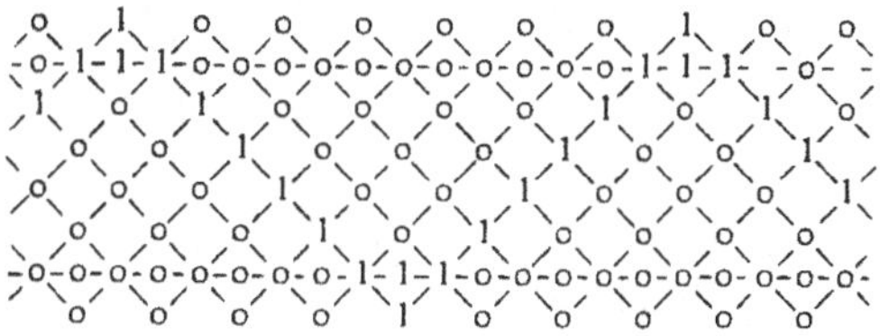

f_2

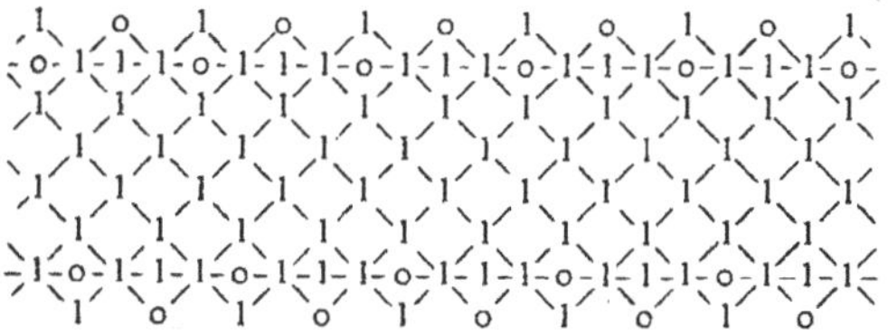

f_3

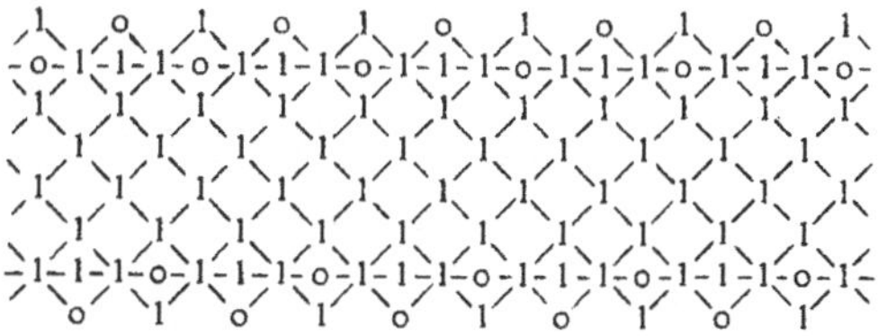

$\tilde{\mathbb{E}}_6 = \tilde{\mathbb{T}}_{3,3,2}$

f_1

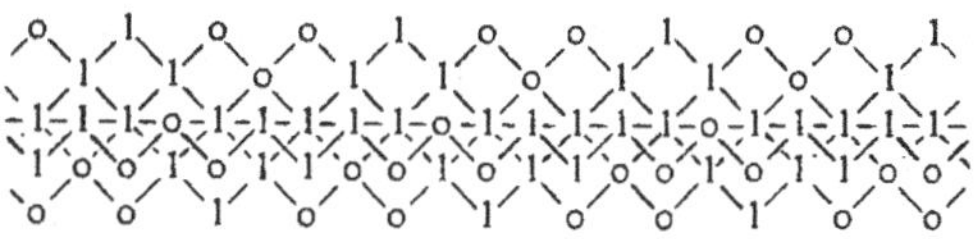

f_2

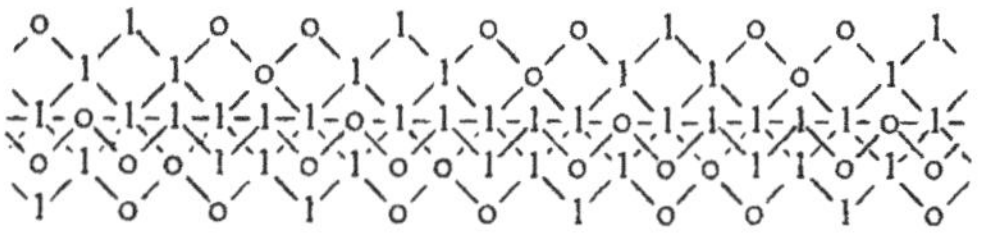

f_3

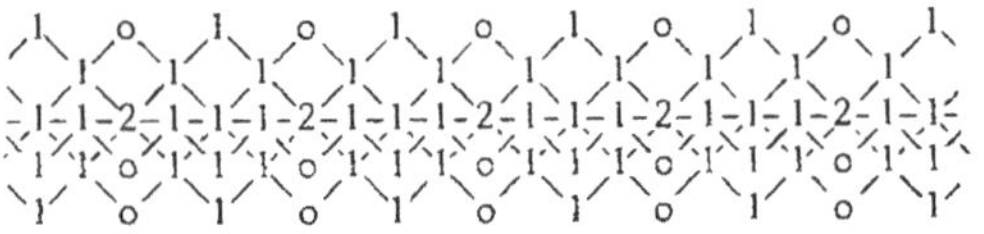

$\tilde{\mathbb{E}}_7 = \tilde{\mathbb{T}}_{4,3,2}$

f_1

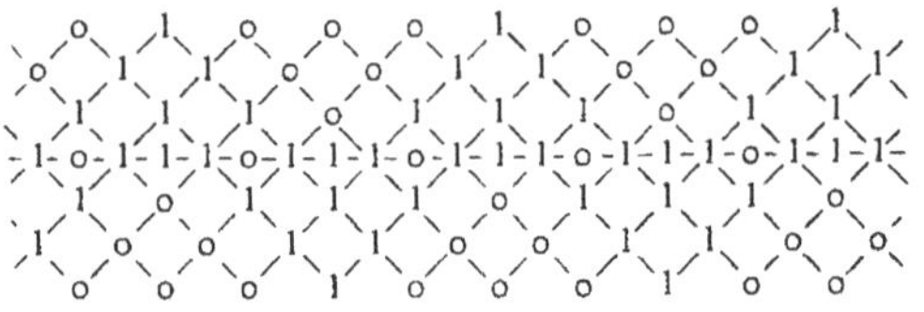

f_2

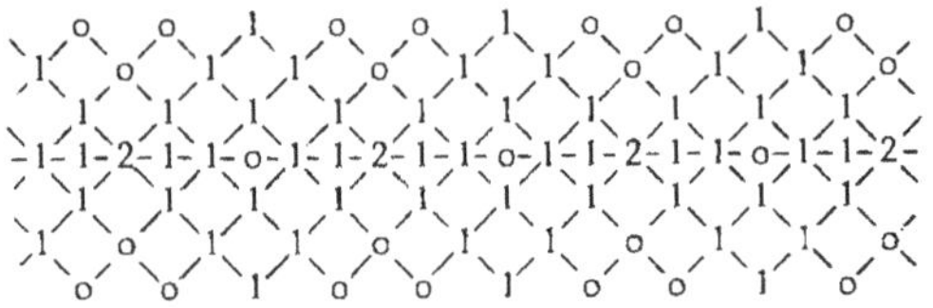

f_3

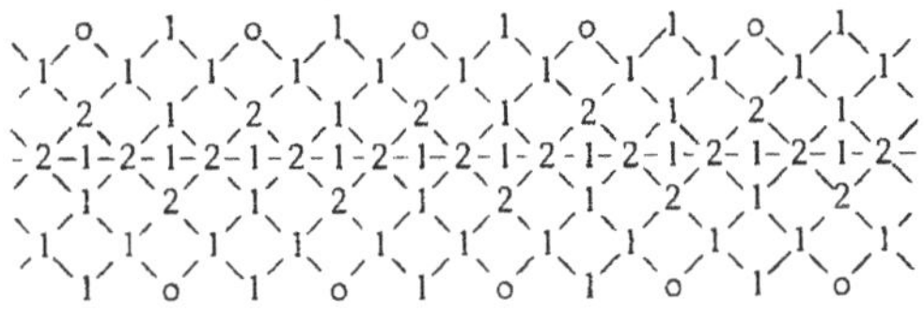

$\tilde{\mathbb{E}}_8 = \tilde{\mathbb{T}}_{5,3,2}$

f_1

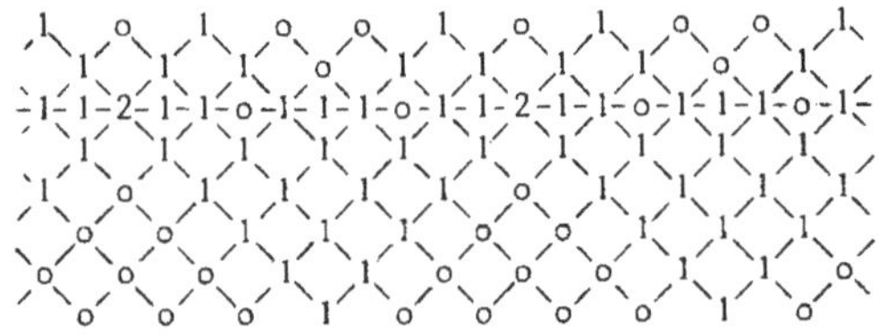

f_2

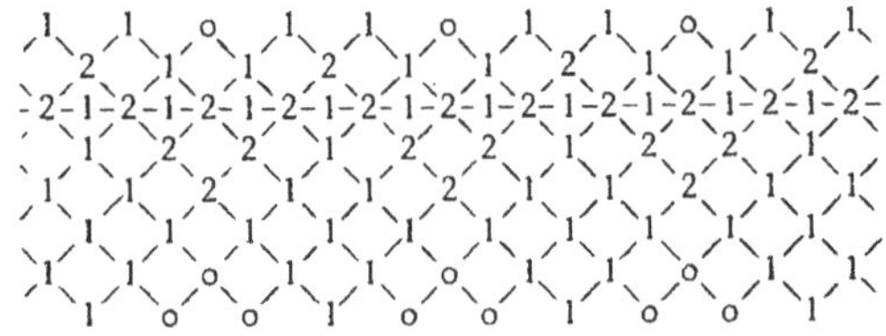

f_3

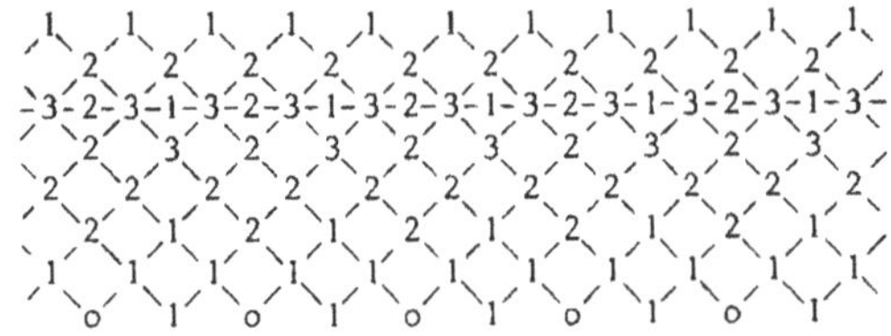

A.2 The frames of the tame concealed algebras (see 4.3.4)

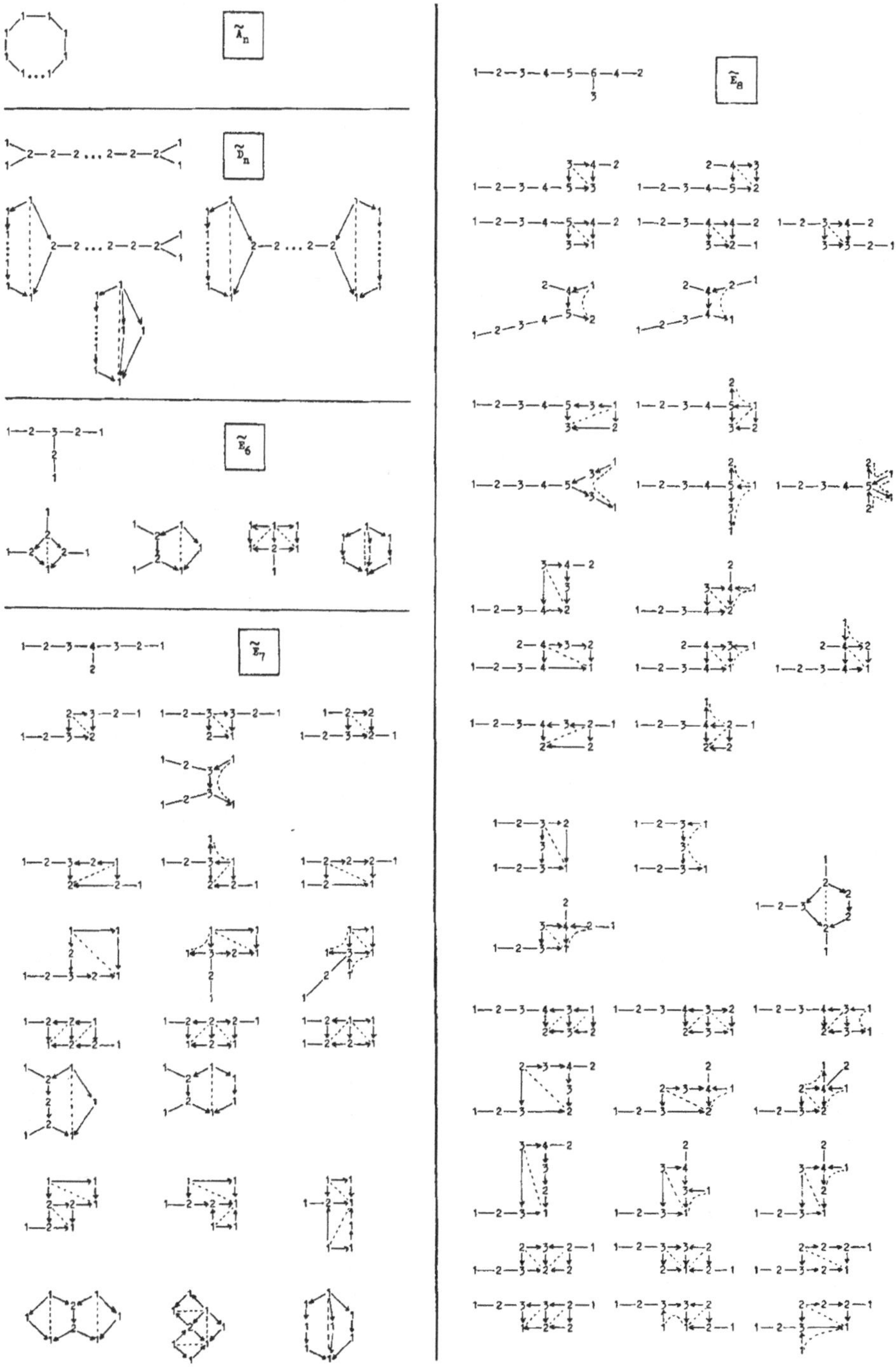

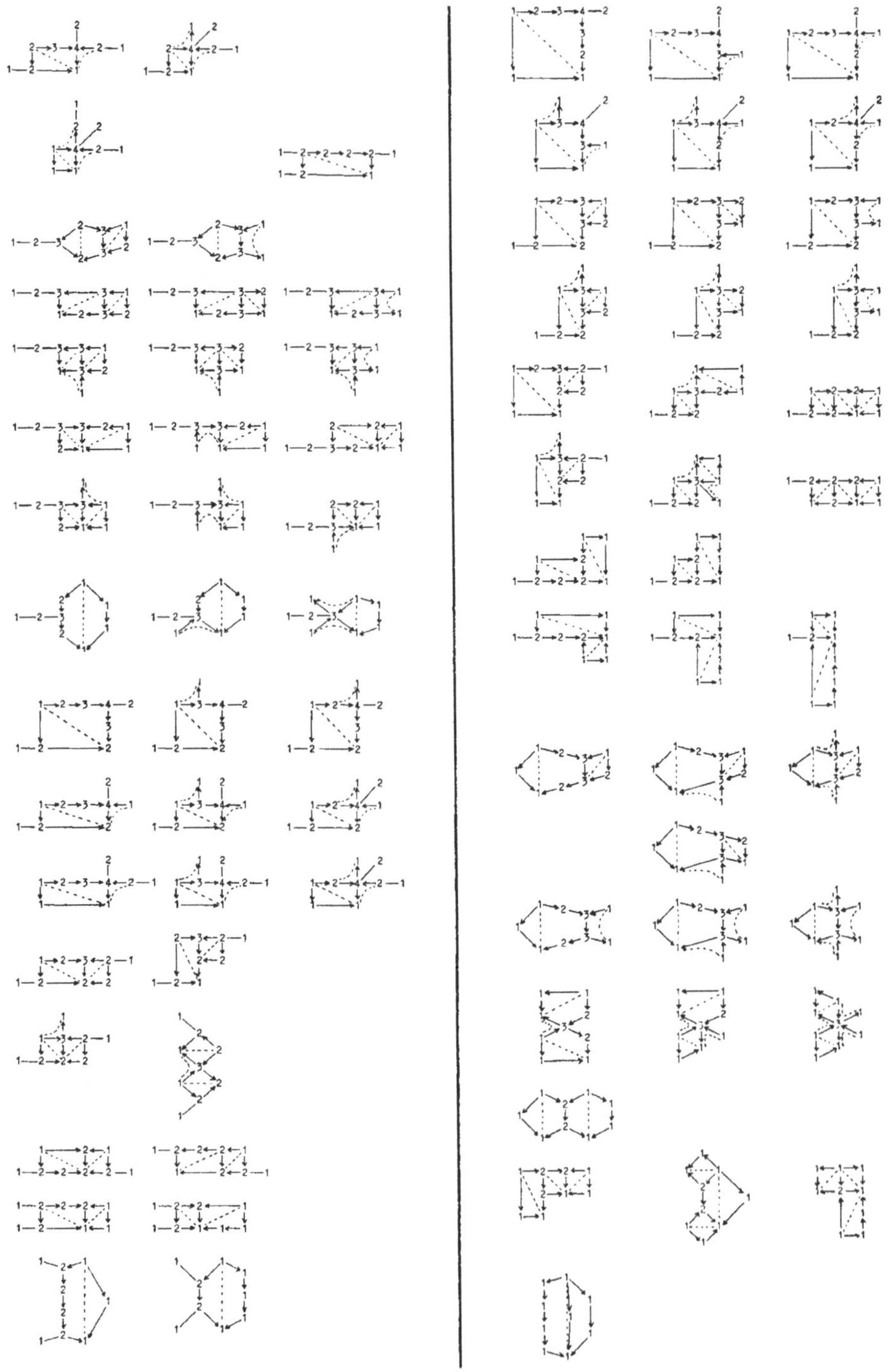

A.3 The tubular patterns (see p. 323)

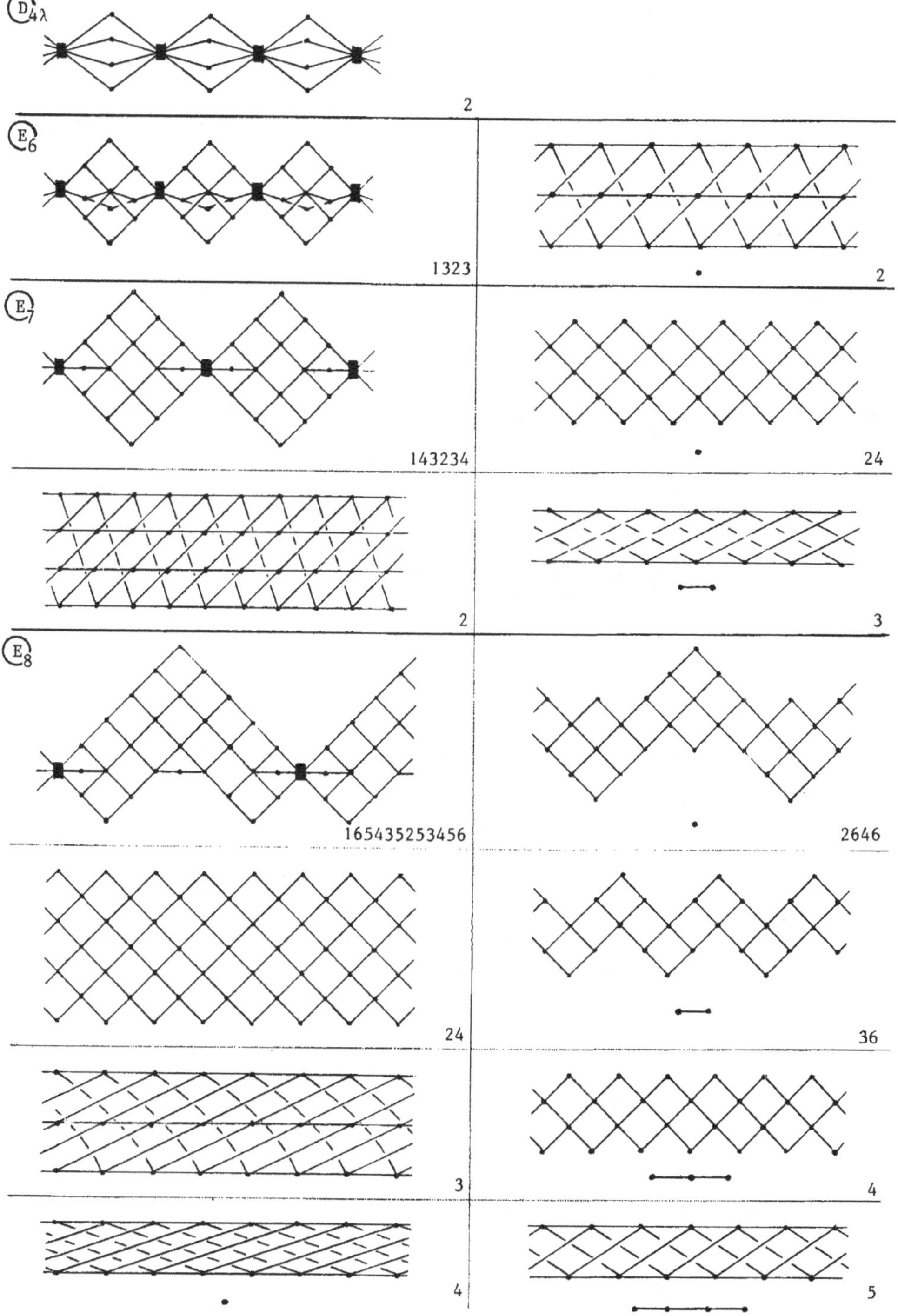

Bibliography

[AF] Anderson, F.W. and Fuller, K.R.: Rings and categories of modules. Springer (1974)

[A1] Auslander, M.: Representation dimension of artin algebras. Queen Mary College Mathematical Notes. London (1971)

[A2] Auslander, M.: Applications of morphisms determined by objects. In [Go], 245-327

[AL] Auslander, M., and Lluis, E. (editors): Representations of algebras. Springer Lecture Notes in Mathematics 903 (1981)

[AP] Auslander, M., and Platzeck, M.I.: Representation theory of hereditary artin algebras. In [Go], 389-424.

[APR] Auslander, M., Platzeck, M.I., and Reiten, I.: Coxeter functors without diagrams. Trans. Amer. Math. Soc. 250 (1979), 1-46

[AR1] Auslander, M. and Reiten, I.: Representation theory of artin algebras III. Comm. Algebra 3 (1975), 239-294

[AR2] Auslander, M. and Reiten, I.: Representation theory of artin algebras IV. Comm. Algebra 5 (1977), 443-518.

[AR3] Auslander, M. and Reiten, I.: Modules determined by their composition factors. Preprint

[AS] Auslander, M. and Smalø, S.O.: Preprojective modules over artin algebras. J. Algebra 66 (1980), 61-122.

[Ba1] Bautista, R.: Irreducible maps and the radical of a category. An. Inst. Mat. Univ. Nac. Autónoma México 22 (1982), 83-135

[Ba2] Bautista, R.: Sections in Auslander-Reiten quivers. In [DG2], 74-96

[BLS] Bautista, R., Larrión, F. and Salmerón, L.: On simply connected algebras. J. London Math. Soc.(2), 27 (1983), 212-220.

[BaB] Bautista, R., and Brenner, S.: Replication numbers for non-Dynkin sectional subgraphs in finite Auslander-Reiten quivers and some properties of Weyl roots. Proc. London Math. Soc.(3) 43 (1983), 429-462.

[BGP] Bernstein, I.N., Gelfand, I.M., and Ponomarev, V.A.: Coxeter functors and Gabriel's theorem. Uspechi Mat. Nauk 28 (1973), Russian Math. Surveys 28 (1973), 17-32.

[B] Bass, H.: Algebraic K-Theory. Benjamin. New York (1968)

[Bo1] Bongartz, K.: Tilted algebras. In [AL], 26-38

[Bo2] Bongartz, K.: Treue einfach zusammenhängende Algebren I. Comment. Math. Helv. 57 (1982), 282-330

[Bo3] Bongartz, K.: Algebras and quadratic forms. J. London Math. Soc. 28 (1983), 461-469

[Bo4] Bongartz, K.: Critical simply connected algebras. Manuscripta math. 46 (1984), 117-136.

[BG] Bongartz, K., and Gabriel, P.: Covering spaces in representation theory. Invent. math. 65 (1982), 331-378

[Bou] Bourbaki, N.: Groupes et algèbres de Lie. IV - VI. Hermann & Co, Paris 1960

[Br] Brenner, S.: Decomposition properties of some small diagrams of modules. Symposia Math. Ist. Naz. Alta Mat. 13 (1974), 127-141

[BB] Brenner, S., and Butler, M.C.R.: Generalizations of the Bernstein-Gelfand-Ponomarev reflection functors. In [DG3], 103-169

[Bü1] Bünermann, D.: Auslander-Reiten-Köcher von einparametrigen teilweise geordneten Mengen. Dissertation. Düsseldorf 1981

[Bü2] Bünermann, D.: Auslander-Reiten quivers of exact one-parameter partially ordered sets. In [AL], 55-61

[CE] Cartan, E., and Eilenberg, S.: Homological algebra. Princeton University Press. 1956

[CR] Curtis, C., and Reiner, I.: Representation theory of finite groups and associative algebras. New York (1962), and: Method of Representation Theory. Vol.I. New York (1981)

[DG1] Dlab, V., and Gabriel, P. (editors): Representations of Algebras. Springer Lecture Notes in Mathematics 488 (1975)

[DG2] Dlab, V., and Gabriel, P. (editors): Representation Theory I. Springer Lecture Notes in Mathematics 831 (1980)

[DG3] Dlab, V., and Gabriel, P. (editors): Representation Theory II. Springer Lecture Notes in Mathematics 832 (1980)

[DR] Dlab, V., and Ringel, C.M.: Indecomposable representations of graphs and algebras. Memoirs Amer. Math. Soc. 173 (1976)

[DF] Donovan, P., and Freislich, M.R.: The representation theory of finite graphs and associated algebras. Carleton Lecture Notes 5, Ottawa (1973)

[D1] Drozd, Ju.A.: Coxeter transformations and representations of partially ordered sets. Funkc. Anal. i Priložen 8 (1974), 34-42. Funct. Anal. and Appl. 8 (1974), 219-225

[D2] Drozd, Ju. A.: Tame and wild matrix problems. In [DG3], 242-258

[ER] d'Este, G., and Ringel, C.M.: Coherent tubes. J. Algebra 87 (1984), 150-201

[F] Fadeev, D.K. (editor): Investigations on Representation Theory. Zap. Naučn. Sem. LOMI 28 (1972), Engl. transl.: J. Soviet Math. 23 (1975), Number 5

[G1] Gabriel, P.: Unzerlegbare Darstellungen I. Manuscripta Math. 6 (1972), 71-103

[G2] Gabriel, P.: Indecomposable representations II. Symposia Math. Ist. Naz. Alta Mat. 11 (1973), 81-104

[G3] Gabriel, P.: Représentations indécomposables des ensembles ordonnés. Seminaire P. Dubreil. Paris (1972-73), 301-304

[G4] Gabriel, P.: Auslander-Reiten sequences and representation-finite algebras. In [DG2], 1-71

[Gm] Gantmacher, F.R.: Matrizenrechnung II. Berlin 1966

[GP] Gelfand, I.M., and Ponomarev, V.A.: Problems of linear algebra and classification of quadruples in a finite dimensional vector space. Coll. Math. Soc. Bolyai 5, Tihany (1970), 163-237

[Go] Gordon, R. (editor): Representation theory of algebras. Lecture notes in pure and applied mathematics 37 (1976)

[H] Heller, A.: Homological algebra in abelian categories. Ann. of Math. 68 (1958), 484-525

[Ha] Happel, D.: Tilting sets on cylinders. Proc. London Math. Soc. (to appear)

[HPR] Happel, D., Preiser, U., and Ringel, C.M.: Vinberg's characterization of Dynkin diagrams using subadditive functions with application to DTr-periodic modules. In [DG3], 280-294

[HR1] Happel, D., and Ringel, C.M.: Tilted algebras. Trans. Amer. Math. Soc. 274 (1982), 399-443

[HR2] Happel, D., and Ringel, C.M.: Construction of tilted algebras. In [AL], 125-144

[HV] Happel, D., and Vossieck, D.: Minimal algebras of infinite representation type with preprojective component. Manuscripta math. 42 (1983), 221-243

[Ka] Kac, V.: Root systems, representations of quivers and invariant theory. In: Invariant Theory. Springer Lecture Notes in Mathematics 996 (1983), 74-108

[Ke] Kerner, O.: Partially ordered sets of finite representation type. Communications Alg. 9 (1981), 783-809

[Kl1] Klejner, M.M.: Partially ordered sets of finite type. In [F], 32-41. Engl. transl. 607-615

[Kl2] Klejner, M.M.: On exact representations of finite type. In [F], 42-60. Engl. transl. 616-628

[K] Kronecker, L.: Algebraische Reduktion der Scharen bilinearer Formen. Sitzungsber. Akad. Berlin (1890), 1225-1237

[Kr] Krugliak, S.A.: Representations of the (p,p)-group over a field of characteristic p. Dokl. Acad. Nauk SSR 153 (1963), 1253-1256. Engl. Transl. Soviet Math. Dokl. 4 (1964), 1809-1813

[L1] Loupias, M.: Representations indécomposables des ensembles ordonnés finis. Thèse Tours (1975)

[L2] Loupias, M.: Indecomposable representations of finite ordered sets. In [DG1], 201-209

[M1] Marmaridis, N.: Darstellungen endlicher Ordnungen. Dissertation Zürich (1978). Summary: Représentations linéaires des ensembles ordonnés. C.R. Acad. Sc. Paris. t. 288 (1979)

[M2] Marmaridis, N.: Some remarks on reflection functors. In [AL], 211-221

[MR] Mitropolskij, Ju. A., and Rojter, A.V. (editors): Matrix problems. Akad. Nauk Ukr. SSR. Kiev (1977)

[N1] Nazarova, L.A.: Representations of quadruples. Izv. Akad. Nauk SSSR. Ser. Mat. 31 (1967), 1361-1377

[N2] Nazarova, L.A.: Representations of quivers of infinite type. Izv. Akad. Nauk SSSR. Ser. Mat. 37 (1973), 752-791

[N3] Nazarova, L.A.: The representations of partially ordered sets of infinite type. Izv. Akad. Nauk. SSSR. Ser. Mat. 39 (1975), 963-991

[NR1] Nazarova, L.A., and Rojter, A.V.: Representations of partially ordered sets. In [F], 5-31, Engl. transl. 585-607

[NR2] Nazarova, L.A., and Rojter, A.V.: Categorical matrix problems and the Brauer-Thrall conjecture. Inst. Math. Acad. Sci. Kiev (1973), German transl.: Mitt. Math. Sem. Gießen 115 (1975)

[Ot] Otraševskaja, V.V.: On the representations of the one-parameter partially ordered sets. In [MR], 144-149

[O] Ovsienko, S.A.: Integral weakly positive forms. In: Schur matrix problems, and quadratic forms. Kiev (1978), 3-17

[Qu] Quillen, D.: Higher algebraic K-theory I. In: Algebraic K-Theory I. Springer Lecture Notes in Mathematics 341 (1973), 85-147

[Rm] Riedtmann, Chr.: Algebren, Darstellungsköcher, Überlagerungen, und zurück. Comment. Math. Helv. 55 (1980), 199-224

[Ri1] Ringel, C.M.: Representations of K-species and bimodules. J. Algebra 41 (1976), 269-302

[Ri2] Ringel, C.M.: The rational invariants of tame quivers. Invent. math. 58 (1980), 217-239

[Ri3] Ringel, C.M.: Report on the Brauer-Thrall conjectures: Rojter's theorem and the theorem of Nazarova and Rojter. In [DG2], 104-136

[Ri4] Ringel, C.M.: Tame algebras. In [DG2], 137-287

[Ri5] Ringel, C.M.: Separating tubular series. In: Seminaire d'Algèbre P. Dubreil - M.P. Malliavin. Springer Lecture Notes in Mathematics 1029 (1983), 134-158

[Shk] Shkabara, A.S.: Commutative quivers of tame type I. Preprint, Kiev (1978)

[Si] Simson, D.: Representations of partially ordered sets, vector space categories, and socle projective modules. Preprint (1983)

[Sw] Swan, R.G.: Algebraic K-Theory. Springer Lecture Notes in Mathematics 76 (1968)

[V] Vossieck, D.: On indecomposables in preprojective components. To appear.

[Z1] Zavadskij, A.G.: Differentiation with respect to pairs of points. In [MR], 115-121

[Z2] Zavadskij, A.G.: Quivers with a distinguished path and no cycles which are tame. Preprint. Kiev (1977)

[Z3] Zavadskij, A.G.: Representations of partially ordered sets of finite growth. Kiev (1983)

[ZN] Zavadskij, A.G., and Nazarova, L.A.: Partially ordered sets of finite growth. Funkc. Anal. i Priložen 16 (1982), 72-73. Funct. Anal. and Appl. 16 (1982), 135-137

Index

Addendum to 4.2

The following general result is similar to 4.2.6, it deals with a modification of $S(\to M)$ and $S(M\to)$ where M is a sincere directing module, and no additional assumptions on the component of A-mod containing M is needed. It should be used as a replacement of both 4.2.6 and 4.2.7, and it seems to be of interest in its own. Given an indecomposable A-module M, let $S'(\to M)$ be the module class given by all indecomposable A-modules X with $X \preceq M$, and such that there is no indecomposable non-projective A-module Z satisfying both $X \preceq \tau Z$ and $Z \preceq M$. Dually, let $S'(M\to)$ be the module class given by all indecomposable A-modules Y with $M \preceq Y$ and such that there is no indecomposable non-projective A-module Z satisfying both $M \preceq \tau Z$ and $Z \preceq Y$.

Lemma. Let M be a sincere directing A-module. Then both $S'(\to M)$ and $S'(M\to)$ are slices. More generally, if X is indecomposable and in $S'(\to M)$, then $S'(X\to)$ is a slice; if Y is indecomposable and in $S'(M\to)$, then $S'(\to Y)$ is a slice.

Proof. Let X be indecomposable and in $S'(\to M)$. The conditions (α), (β), (γ) of a slice are obviously satisfied. In fact, there is the following stronger property:

(β') If N,S are indecomposable, $X \preceq N \preceq S$, and S is in $S'(X\to)$, then also N is in $S'(X\to)$.

In order to show (δ), let $S \in S'(X\to)$, and N indecomposable with an irreducible map $N \longrightarrow S$. Assume that N is not in $S'(X\to)$. According to (β'), we know that $X \not\preceq N$. In particular, N cannot be injective, since otherwise $X \preceq M \preceq N$, using the fact that M is sincere. Assume that also $\tau^- N$ does not belong to $S'(X\to)$. Since $X \preceq S \preceq \tau^- N$, there has to exist an indecomposable non-projective module Z with both $X \preceq \tau Z$ and $Z \preceq \tau^- N$. Fix a path $(Z_0, Z_1, \ldots, Z_t)$ with $Z = Z_0$, $Z_t = \tau^- N$. Note that $\mathrm{Hom}(Z_i, P) = 0$ for any $0 \le i \le t$, and any indecomposable projective module P, since otherwise we have both $X \preceq \tau Z$, and $Z \preceq Z_i \preceq P \preceq M$, using again the fact that M is sincere. This shows that $\tau Z_i \neq 0$ for all $0 \le i \le t$, and that $\mathrm{Hom}(Z_{i-1}, Z_i) = \underline{\mathrm{Hom}}(Z_{i-1}, Z_i)$ for $1 \le i \le t$. Therefore, $0 \neq \overline{\mathrm{Hom}}(\tau Z_{i-1}, \tau Z_i)$, for $1 \le i \le t$, according to 2.4.5. As a consequence, $(\tau Z_0, \tau Z_1, \ldots, \tau Z_t)$ is a path from $\tau Z = \tau Z_0$ to $\tau Z_t = N$, and therefore $X \preceq \tau Z \preceq N$. This contradiction finishes the proof of (δ). Similarly, $S'(\to Y)$ is a slice for any indecomposable module Y in $S'(M\to)$. In order to finish the proof of the lemma, we only have to note that a directing module M always belongs to both $S'(\to M)$ and $S'(M\to)$.

Remark. It follows easily that for M a sincere directing A-module, the module classes $S'(\to M)$ and $S(\to M)$ coincide, similarly, $S'(M\to)$ and $S(M\to)$ coincide. [For, according to 4.2.3, the modules in $S'(\to M)$ are images of the injective B-modules under a tilting functor, with B hereditary, thus the component of A-mod containing $S'(\to M)$ is standard, convex and directed. However, given an indecomposable module N in a standard, convex and directed component, then $S'(\to N) = S(\to N)$, and $S'(N\to) = S(N\to)$.]

Corollary. Let A be a finite-dimensional algebra with a sincere directing A-module. Then A is a tilted algebra.

Proof. Apply 4.2.3.